PROCESS MODELING OF FOREST GROWTH RESPONSES TO ENVIRONMENTAL STRESS

PROCESS MODELING OF FOREST GROWTH RESPONSES TO ENVIRONMENTAL STRESS

Edited by

Dr. Robert K. Dixon
Associate Professor, School of Forestry
Auburn University, Alabama, U.S.A.

Dr. Ralph S. Meldahl
Assistant Professor, School of Forestry
Auburn University, Alabama, U.S.A.

Dr. Gregory A. Ruark
Project Leader, USDA Forest Service
Research Triangle Park, North Carolina, U.S.A.

Dr. William G. Warren
Research Professor, Department of Forest Management
Oregon State University
Corvallis, Oregon, U.S.A.

TIMBER PRESS
Portland, Oregon

ISBN 0-88192-152-1
Printed in Hong Kong

TIMBER PRESS, INC.
9999 SW Wilshire
Portland, Oregon 97225

Library of Congress Cataloging-in-Publication Data

Process modeling of forest growth responses to environmental stress /
edited by Robert K. Dixon ... [et al.].
p. cm.
Includes bibliographical references.
ISBN 0-88192-152-1
1. Trees--Growth--Computer simulation. 2. Trees--Effect of stress on--Computer simulation. 3. Forests and forestry--Mensuration--Computer simulation. I. Dixon, Robert K.
SD396.P73 1990
634.9'285--dc20 89-20657
CIP

CONTENTS

PREFACE

The rapidly proliferating interest in modeling the impact of environmental stresses on forest growth revealed a need for a comprehensive treatise which included the disciplines of biometrics, pedology, plant ecology and physiology, and meteorology. The purpose of this compendium is to present a contemporary account of tree and forest process model development, structure and evaluation in relation to tree metabolism and structure. The primary concerns throughout are understanding the structure and metabolic processes of trees and the impact of atmospheric pollutants and other environmental stresses on the development of individual trees and forest stands. This book is purposely interdisciplinary and is intended to be useful as a reference for students and scientists and also as a guide for the development of environmental policy.

The chapters in this book were prepared by participants in the conference "Forest Growth: Process Modeling of Responses to Environmental Stress," held at Gulf Shores, Alabama, U.S.A., April 19–22, 1988. Thus, this book shares some characteristics with a conference proceedings. However, each contribution was reviewed by anonymous referees and an associate editor. A thorough technical review was requested for each manuscript, and all contributions were revised based upon review comments. The chapters were screened by the associate editors and assembled into the present format. To achieve uniform expression throughout, manuscripts were also reviewed and revised by the publisher. Ultimately, the content and views expressed in each chapter are those of the authors.

An effort was made to standardize terminology and nomenclature throughout the book. The many disciplines represented in this text made it difficult to standardize mathematic units and terms of expression. Authors were instructed to define terms used in their chapter to minimize communication difficulties. Special emphasis was given to the selection of reference material which is authoritative, well documented, and current. A glossary of abbreviations and acronyms and a table of conversion factors are included at the end of this volume to assist the reader.

We extend our sincere gratitude to each of the distinguished authors of this text. In addition, we offer thanks to Drs. Paul J. Hanson, Alan R. Ek, Dale S. Solomon, and Keith Paustian for assistance with preparation of the book. The contribution of the referees is acknowledged and appreciated. Mrs. Debbe Snowden, Ms. Julia C. McClelland, and Ms. Linda C. Kerr coordinated manuscript review and editing. The organizational and editing skills of Mrs. Karen Kirtley are gratefully acknowledged. Their multiple contributions are sincerely appreciated.

Robert K. Dixon
Ralph S. Meldahl
Gregory A. Ruark
William G. Warren
Auburn, Alabama, U.S.A.
December, 1988

ACKNOWLEDGMENTS

This book is an ancillary product of the conference "Forest Growth: Process Modeling of Responses to Environmental Stress," April 19–22, 1988, Gulf Shores, Alabama, U.S.A. The conference planning committee included Drs. J. Blake, T. Burk, T. Dell, R. Dixon, R. Goldstein, T. Hinckley, D. Johnson, R. Meldahl, G. Ruark, G. Somers, W. Warren, and C. Webb. Thanks are due to this group for their considerable efforts. Gratitude is also extended to Mr. R. Beadles, J. Jansen, G. Breece, and C. Geron for their help with implementing the event. Mrs. J. Ostmeyer, K. Somers, and M. Teeter, as well as Mr. R. Bolton, M. Miller, and B. Zutter, were responsible for conference implementation.

The financial support of the Forest Response Program of the National Acid Precipitation Assessment Program (jointly funded and managed by the USDA Forest Service and the U.S. Environmental Protection Agency), the Southern Company, and the Alabama Forestry Association is gratefully acknowledged. Appreciation is also extended to the School of Forestry and the Alabama Agricultural Experiment Station, Auburn University, Auburn, Alabama, U.S.A., for their assistance and cooperation.

GLOBAL CHANGE— A CHALLENGE TO MODELING

Michael A. Fosberg

Historically, forest managers and researchers have believed that forest productivity and health would be assured by prudent silviculture practices and protection from insects, disease, and fire. Much of our knowledge and many procedures for strategic planning and management are based on observations during the past 100 years. Because much of this knowledge, as collected in growth and yield tables, for instance, is empirically based, its validity is now being questioned under a changed climate and chemical environment. Our knowledge of the risks associated with management options is also based on past observations. These risks, too, will undoubtedly change and reduce our ability to choose the best option. Environmental changes will probably not occur uniformly over the globe, but instead will show regional variations in amount and direction of change. Our traditional knowledge base may not fit the events and effects of a future climate and chemical environment.

Atmospheric concentrations of greenhouse gases, carbon dioxide (CO_2), methane (CH_4), and nitrous oxides (NO_x), among others, are expected to double within the next century (Keeling, 1984; Freidli et al., 1986). It is anticipated that these greenhouse gases will result in a global warming of the atmosphere of 1–4°C. Warming will not be uniform over the globe. Minor temperature change is expected in the tropics, and a 7–10°C warming is expected in polar regions (Hanson et al., 1988). While there is consensus on the warming trend and the magnitude of the anticipated warming, there is far less agreement as to the effects on regional precipitation patterns (Dickenson, 1982). Also, temperature change predictions are only slightly greater than those changes which can be attributed to natural variability. Bernabo (1981) and Hepting (1963) have estimated natural variability to be about 1.5°C during the last three millennia.

Why, then, is the warming trend considered a potential problem? The significance lies in the rate of temperature change. Can vegetation migrate and soils evolve at a rate 10 times their historical rates? Consider loblolly pine (*Pinus taeda* L.), which has a historical migration rate of one to two km per year in response to climate change (Shugart et al., 1986). Is it reasonable to extrapolate from these observations and predict that loblolly pine will migrate 1,000 km in the next 100 years, assuming that climate change is the controlling factor? Or will loblolly pine migrate 100 km and be subjected to environmental stresses because of adverse climate? A number of similar scenarios can be constructed for disease, insects, and fire (Fosberg, in press).

Dr. Fosberg is with the USDA Forest Service, Washington, DC, 20090-6090 U.S.A.

In addressing the issue of global change, decision-makers will need answers to two underlying questions:

1. How will future physical and chemical changes alter the structure, function, and productivity of forest and range ecosystems?
2. What are the implications for forest and range management, and how must forest and range management activities be altered to sustain forest and range health and productivity?

These two questions imply that we anticipate benefits from research on global change and its effects. Again, I will use examples to illustrate the possible benefits of research and modeling efforts. In the field of forest fire protection, the fundamental laws of physics and chemistry are constant. Thus the thresholds of fire severity remain essentially unchanged. However, with the expected global warming, we would expect changes in fire frequency if those thresholds are exceeded. Using a temperature of 32°C or above as a surrogate, in the region of Washington, DC, U.S.A., we would expect the number of days at or above 32°C to change from the current climatological average of 34 days, to 42 days by the year 2000, and to an average of 67 days by the year 2030 (Hanson et al., 1988). Further research will help us to anticipate and understand this phenomenon.

Turning now to a second aspect of the management problem, let us consider prediction and predictability. Scientific investigation seeks to achieve order and understanding of chaotic natural phenomena. In particular, there is a strong desire to develop an understanding of cause-and-effect relations, which leads to deterministic predictions. Since in global change we are looking a century into the future, there will be numerous irreversible events which cannot be predicted. Also, it should be recognized that in a highly non-linear system, predictability, in a deterministic sense, may not be possible (Tennekes, 1988). The external driving force, the atmospheric general circulation model (GCM), is also highly non-linear.

Any given deterministic prediction will be associated with some level of uncertainty because of the numerous irreversible events which cannot be anticipated (e.g., fire, insect, or disease outbreaks). The timing and severity of these events will influence the deterministic prediction. These events can be described by a probability distribution of their frequency and severity. If a Monte Carlo technique is used to introduce these events into the prediction process, future scenarios will be expressed in terms of a probability distribution. Significant outputs from modeling efforts will be prediction of biotic and abiotic events in terms of their probability. Prediction of the probable distribution of events will enable resource managers to develop strategies to cope with changes in our atmosphere.

ECOSYSTEM MODELING

In order to synthesize and integrate the knowledge of individual plants and ecosystem processes into strategies for managing forests and related ecosystems, we will need to develop and utilize models of these basic processes. However, these detailed process models are not the end product.

The first step in modeling to reflect global atmospheric change is synthesis of models which address the biological and physical processes in individual organs and whole plants. These models describe our understanding of basic processes such as photosynthesis, respiration, and carbon allocation to various organs. Ultimately, models should be developed which characterize and predict whole-plant response to climate and atmospheric change.

The second step is concerned with modeling the ecosystem as a whole. It is unlikely that we will be able to assemble an ecosystem model by combining individual plant models. Our knowledge of plant processes is limited. Moreover, it is difficult to account for and assimilate the many interactions among plants in models of individual biological activity. Further development or refinement of empirical ecosystem models is constrained without consideration of plant processes.

A third step toward the desired end product in modeling, somewhat parallel to the previous step, is global change modeling. Current GCM's are very coarse, both spatially and temporally, representing thousands of square kilometers with single temperature and precipitation values. Temporally, at best they can represent seasonal change. A typical whole-plant or ecosystem model is based on spatial scales considerably smaller than those represented by GCM's. In order to successfully model future ecosystem response, this difference in scale and resolution must be resolved. Also, ecosystem models must be coupled to the global climate change models in order to provide for the exchange of energy and matter.

A fourth step is the realization that ecosystems are not necessarily deterministic. That is, there are numerous possible solutions, due to genetic differences in plants, the uncertainty of pathogen effects, and uncertainties as to how the ecosystem will respond to stress. The most useful output from the models of the physical and biological world are probability statements on the outcomes. These stochastic outputs not only more accurately reflect the real world, but they also provide the kinds of inputs needed for decision-making—for developing the economic and risk-analysis models that must be used to select management strategies.

Development of economic and risk-analysis models represents the fifth and final step. These models are necessary in order to develop strategies for managing natural ecosystems in a changing environment. Political decisions regarding resource management issues are ultimately based on analysis of socioeconomic and biophysical factors.

PRIORITY RESEARCH PROGRAM

The USDA Forest Service priority research program on forest-atmospheric interactions will use an ecosystem approach. Broadly speaking, the scientific questions to be addressed are

1. What processes are involved in atmospheric effects on forests?
2. How do atmospheric changes influence forest health and productivity?
3. How do forest management practices affect the atmosphere?

The physical and chemical climate determine, to a large extent, the structure, function, and productivity of forest, range, and related ecosystems. Our current understanding of the biogeophysical processes is based on historical data obtained under conditions of steady or slowly changing environments. These data tell us little about a future with significantly altered physical and chemical climates. Our research programs will therefore focus on four specific areas:

1. Processes of atmospheric effects on forests.
2. Effects of forest management on the atmosphere.
3. Long-term assessment of effects.
4. Development of management options and strategies.

These four areas have been broken down into 21 specific problem areas that will be addressed by USDA Forest Service scientists and their cooperators in other agencies and universities. These specific areas of concern are identified below.

Processes of Atmospheric Effects on Forests

The initial requirement is to determine what effects a changed atmospheric chemistry and increased variability of the physical and chemical climate will have on the ecosystem. To move us toward this end, we see the following tasks:

1. Determine the simultaneous responses of carbon, water, and nutrient cycles to altered physical and chemical atmospheric conditions. Refine understanding of the processes by which the atmosphere transports energy and materials within forest canopies, and how these materials are utilized in the photosynthesis process.
2. Develop methods to predict changes in
   a. Species composition and distribution resulting from global climate change.
   b. Water yields from forest and rangeland resulting from global climate change.
   c. Fire severity resulting from global climate change.
3. Determine the sensitivity of
   a. Aquatic ecosystems and fisheries to changes in the physical and chemical atmosphere.
   b. Wildlife species to changes in the physical environment and climate change.
   c. Wood, in terms of both quality and quantity, to altered climate and chemical atmosphere.

Effects of Forest Management on the Atmosphere

A second requirement is to determine what effects human activity in the forests will have on climate variability. The following are specific ancillary problems.

1. Refine estimates and mechanisms of emissions from prescribed fire and wildfire.
2. Develop an understanding of transport, dispersal, and removal of pollutants from prescribed fire and wildfire.
3. Determine effects of large-scale forest removal on carbon, oxygen, and water budgets and relative composition of the atmosphere.

Long-Term Assessment of Effects

A third need is to develop an appropriate assessment program which relates the improved understanding of the processes mentioned in the two previous sections to methods of quantitatively evaluating atmospheric variability in relation to management strategies.

1. Support long-term measurements of forest, range, and related ecosystems, particularly on experimental forest and rangeland and in research natural areas.
2. Identify critical variables necessary to assess ecosystem response to climate changes, and develop methodologies to monitor these variables on a wide basis, including monitoring within forest plans.
3. Identify ecological, physiological, and/or morphological characteristics that can serve as indicators of chemical and/or physical environmental changes.

Development of Management Options and Strategies

The final research requirement is to develop the methodologies needed to quantitatively use the knowledge of the processes and the data base derived from the monitoring activities in developing management strategies.

1. Develop and verify physiological models to determine the carbon, water, and other chemical budgets in forest stands.

2. Improve biophysical aspects of plant growth models, and incorporate physical and chemical environmental changes in these models.
3. Develop and/or improve
   a. Models of water quantity which reflect physical and chemical environmental changes.
   b. Physiologically based stand process models which are linked to growth and yield models.
   c. Models concerned with the impact of physical and chemical climate variations on wildlife and fish.
4. Develop techniques to improve interfaces between models at different spatial and temporal scales.
5. Develop economic and risk-analysis techniques to predict the impact of projected physical and chemical climate change on the ecosystems.

SUMMARY

The key points I want to emphasize are, first, that process modeling is an essential component of the USDA Forest Service priority research program. Second, we must develop an understanding of the basic processes in order to extrapolate beyond the range of historical data bases. Third, these process models are not the end product of the program. Such bottom-up models must complement top-down models, which are driven by the atmospheric models. These integrated models will be used to provide decision-makers with quantitative information. Lastly, it must be recognized that the end use will be input into socioeconomic and risk-analysis models.

LITERATURE CITED

Bernabo, J. C. 1981. Quantitative estimates of temperature changes over the last 2700 years in Michigan based on pollen date. *Quaternary Research* 15:143–159.

Dickenson, R. G. 1982. Modeling climate changes due to carbon dioxide increases. In: Clark, W. C., ed. *Carbon Dioxide Review: 1982.* Clarendon Press, New York, NY, U.S.A. Pp. 101–133.

Fosberg, M. A. In press. Forest productivity and health in a changing atmospheric environment. In: *Proceedings, Symposium on Climate and Geo-Sciences.* Reidel Publishing Company, Dordrecht, the Netherlands.

Friedli, H. L., H. Oeschaer, H. Siegenthaler, U. and B. Stauffer. 1986. Ice core record of the C^{13}/C^{12} ratio of atmospheric CO_2 in the past two centuries. *Nature* 324:237–238.

Hanson, J. F., I. Lagis, A. Lobedoff, D. Rind, D. Ruedy, R. and G. Russell. 1988. Prediction of near term climate evolution: what can we tell decision makers now? In: *Preparing for Climate Change.* Proceedings, Government Institute, Inc. Washington, DC, U.S.A., Pp. 35–47.

Hepting, G. H. 1963. Climate and forest diseases. *Annual Review of Phytopathology:* 1:31–50.

Keeling, C. 1984. *Atmospheric CO_2 Concentrations, Mauna Loa Observatory, Hawaii, 1958–1983.* U.S. Department of Energy Report NDP-011. Washington, DC, U.S.A.

Shugart, H. H., M. Y. Antonovsky, P. G. Jarvis, and A. P. Sangord. 1986. CO_2 climate change and forest ecosystems. In: Bolin, B., B. R. Doos, J. Jager, and R. A. Warrick, eds. *The Greenhouse Effect, Climate Change and Ecosystems.* John Wiley and Sons, New York, NY, U.S.A. Pp. 475–521.

Tennekes, H. 1988. The outlook: scattered showers. *Bulletin of the American Meteorological Society* 69:368–376.

PERSPECTIVES ON PROCESS MODELING OF FOREST GROWTH RESPONSES TO ENVIRONMENTAL STRESS

John I. Blake, Greg L. Somers, and Gregory A. Ruark

Abstract. Process modeling is a developing field of research that combines the knowledge and experience of scientists across a range of disciplines. The overall goal is to understand and predict the influence of biotic and abiotic factors on forest growth and development. A working definition of a process model, as well as a perspective on process modeling, is presented. Future challenges and opportunities for scientists and policy-makers in the natural resource arena are discussed.

> The objective of science is to describe nature and predict phenomena. It is impossible to do this from *a priori* theories. Using experiments alone, we would be lost in a bewildering array of disconnected facts. It is the zigzag between experiment and theory using mathematics as a language that has led to astounding recent achievements in physics (Segre, 1980).

Along with an increasing demand for renewable natural resources is a growing concern over the effects of a changing environment on the long-term productivity of the world's forest resources. Forest managers have always been concerned with predicting the effects of silvicultural treatments on forest yields. However, the acute effects of environmental stresses such as drought, pests, fire, pollutants, and other irregular occurrences have been given limited consideration (Grier, 1988). Studies have demonstrated the effects of atmospheric and climatic variations on individual tree characteristics over fairly small time intervals relative to the total cycle of forest development (Zahner, 1968). Unfortunately, the ultimate effect of these subtle variations on stand development is rarely evaluated at harvest or maturity. Management decisions depend upon our ability to predict the long-term impacts of these subtle abiotic and biotic perturbations utilizing information collected in limited observation intervals. One approach that can be used is to compartmentalize forest development into a set of fundamental functions representing the relationship between growth components and the environment. These functions or processes are quantitative expressions which form a model to predict responses to environmental stresses.

Drs. Blake and Somers are Assistant Professors at the School of Forestry, Auburn University, AL, 36849 U.S.A., and Dr. Ruark is a Project Leader for the USDA Forest Service at the Southeastern Forest Experiment Station, Research Triangle Park, NC, 27709 U.S.A. The authors would like to express their appreciation to Drs. Thomas Hinckley and Peter Farnum for ideas they expressed which contributed to the development of this paper.

Research utilizing predictive forest growth models is being conducted throughout the world. Results are being applied to many situations, but until recently, little attention had been given to predicting forest growth responses to air pollution as an environmental stress. To address this need, a conference entitled "Forest Growth: Process Modeling of Responses to Environmental Stress" was held at Gulf Shores, Alabama, U.S.A., April 19–22, 1988. The conference objective was to evaluate process modeling methodology as a means to understand and predict forest tree growth responses to environmental stresses, in particular, the potential impact of air pollutant stress on forest growth (Carey et al., 1986; Environmental Protection Agency, 1988). The conference brought together individuals from various scientific disciplines to represent a variety of prominent modeling approaches. Their goal was to provide information to policy-makers concerning the potential role of process modeling in assessing the impact of air pollution on forest growth. The effects of air pollutants cannot be evaluated independently of the other environmental factors which regulate forest development. Both historical and epidemiological studies indicate that certain environmental stresses can predispose trees to other growth-damaging agents (Wallace, 1978). The diversity of tree stress response phenomena and their interactions implies the need for detailed knowledge across many disciplines.

The purposes of this chapter are to provide a working definition of process modeling within the context of forest modeling and to present the advantages of a process modeling approach to forest growth predictions. In addition, three challenges to successful development of a process model of forest growth response to environmental stresses are discussed.

MODELS AND PROCESS MODELING ATTRIBUTES

Models can assume a variety of roles depending upon planned uses and objectives. In their simplest form, they may describe the dependency of one variable upon another, while more complex models are often used as calculation tools to predict behavior based on a convenient set of initial variables. They can be used to guide and prioritize experimental research by identifying critical underlying mechanisms which regulate behavior. The latter approach can be used to evaluate our understanding of the key relationships controlling response and their application to a broader range of conditions. Ultimately, the suitability of a model must be measured against the objective(s) for which it was developed.

Definitions of process modeling are often implied indirectly through association with a specific effort (e.g., Kimmins and Scoullar, 1984; Rennolls and Blackwell, 1986). More broadly, process modeling has been defined as a method by which the behavior of a system is represented by a set of component processes or compartments (Godfrey, 1983). Functional relationships and parameters are often assumed to be determinable under experimental conditions that are different from those in which the process behavior is simulated. An example is the prediction of moisture flow under field conditions from soil hydraulic parameters and plant water conductances determined under laboratory conditions (Molz, 1981). Most process models are also dynamic, such that behavior is regulated by feedback relationships among the components. The key assumption is that forest growth can be subdivided into several distinct components whose interdependencies are known and whose properties can be determined outside their functional environment.

In referring to process models, individuals have frequently focused on the difference between empirical and mechanistic approaches to characterize the relationships involved. In general, empirical procedures utilize observed correlations between variables to construct models. They are not intended to represent cause-and-effect relationships, but they do provide useful representations when cause-and-effect relationships are not well understood. Mechanistic approaches imply an underlying understanding of the cause-and-effect pathways within the system at the level at which the system is described. Although the quantitative expression of the model may be similar in both cases, mechanistic relationships are more easily generalized, at least conceptually. Most models incorporate empirical relationships with little loss in utility in order to provide valuable simplifications of complex processes or to improve precision (Lemeur et al., 1979). Consequently, most models represent practical compromises among alternative modeling approaches depending upon the modeler's objectives (Levins, 1986; Seligman, 1976).

WHY USE PROCESS MODELING?

Defining response in terms of a single growth variable in order to experimentally analyze the relationship between treatments and growth is often appropriate. Success, in terms of generality of the results, is dependent upon conducting experiments across the entire range of existing stand and environmental conditions. Further, it is generally assumed that soil, climatic, and management variables retain a constant relationship to growth over time. Aside from these considerations, there are practical limitations to a direct experimental approach. It can be cost-prohibitive to cover a sufficient range of conditions to achieve a meaningful response estimate, and control over the experimental variables may be difficult to achieve under field conditions. In some cases, it may be necessary to demonstrate that the mechanisms inferred from experiments are understood well enough for unambiguous cause-and-effect hypotheses to be generated. To achieve this, one must be capable of predicting observed behavior using the response mechanisms themselves. Subsequent testing of these predictions allows extension of the model to conditions outside the limits of the initial experiments. Currently, the quantitative description and simulation of the fundamental processes which characterize forest growth are increasing in importance in applied research. This approach is being used to extend results to new environments (Adams and Cady, 1988; Whistler et al., 1986) and to evaluate forest productivity constraints (Kimmins and Scoullar, 1984; McMurtrie et al., 1988).

CHALLENGES IN APPLYING PROCESS MODELING

A positive and constructive approach towards improving any methodology involves an evaluation of its apparent limitations. It is only through research efforts designed at overcoming the limitations that the method is improved and extended. As with any method, process modeling faces many challenges. The ability of a model to successfully predict forest growth behavior using environmental response processes depends upon several factors. The first is the modeler's perception of the basic structural-functional relationships or developmental strategy involved in forest growth (Givnish, 1983). The second relates to the available data bases and experimental methods which provide values for the parameters in the model. The third is a tractable quantitative expression which links processes together in a predictive and testable format. If any of these factors is sufficiently limiting, the effort to model growth can result in misleading predictions. Participants at the conference were asked to address these challenges within the context of their area of expertise. To clarify these factors, we will expand upon them briefly. A thorough discussion of similar problems from the systems engineering viewpoint is provided in Godfrey (1983).

Structural-Functional Relationships

Our understanding of essential relationships, which must be experimentally based and theoretically sound, is often incomplete. For example, the control of carbon partitioning among plant parts is so poorly understood that models of this process currently incorporate a variety of largely empirical functions (Cannell, 1985). Such limitations make it difficult to model forest growth and stress response using a process approach, particularly if the data base is limited. Problems of this nature can sometimes be overcome by altering the resolution in the temporal or spatial scale or by redefining state, rate, and driving variables in the model (Landsberg, 1986).

To give an example, a large number of processes have been described for tree nutrient uptake at the soil-root interface (Luxmoore and Stolzy, 1987). These processes can form a basis for modeling nutrient dynamics in forests. However, on the scale of hours, the number of parameters and their interrelationships, which must be estimated, can cause serious errors when simulating behavior over longer time periods. Also, the uncertainty about the quantitative importance of certain components, such as mycorrhizae, is large relative to their perceived function (Reid, 1985). Therefore the processes and probably the scales for expressing nutrient uptake must be defined differently. Representing nutrient uptake on a longer time scale by coupling the uptake of ions to annual fine-root activity might be a more tractable representation for forest stands. This method requires a knowledge of relevant time-dependent constraints such as elemental balances in the tissues (e.g., carbon to nitrogen), root biomass distribution, soil nutrient status, and water uptake. It would be reasonable to select this formulation for major cations provided that rooting densities and ion concentrations are sufficiently high to limit the importance of diffusion. The carbon cost for nutrient uptake must also be substantially greater for root growth-extension than it is for membrane transport of nutrients (Lambers et al., 1983).

Defining structural-functional relationships as a set of generic modular processes (e.g., photosynthesis, respiration, growth, allocation, and nutrient uptake) in modeling plant productivity is appealing from a practical viewpoint (Reynolds et al., 1987). However, in simulating the dimensional growth of trees, difficulties arise when a recursive structural expression compatible with function is sought for crown, stem, branch, or root growth. These expressions in themselves should result in a developmental strategy for controlling changes in the vertical and horizontal dimensions of a forest stand. One approach employing a physiological strategy to control tree development is the pipe concept (Shinozaki et al., 1964). The concept evolved from the close relationship observed between crown biomass and stem basal area (the codominant trees in Figure 1). This was interpreted as a constrained relationship between the cross-sectional area of stem tissue for water conduction and the transpirational area of foliage. When growth is modeled under this constraint, a recursive structure that is compatible with one function emerges (Valentine, 1988). Extreme deviations within the population can be interpreted as instability, implying shifts induced by the environment (the suppressed trees in Figure 1). There remains a need to test predictions and rectify the theory with other functional constraints, such as mechanical support.

Data Base Development and Methodology

Process modeling implies that model parameters, such as species-specific stomatal responses to vapor pressure, can be experimentally determined independently of the environmental conditions of the experiment. Few would accept this premise without serious reservations. For example, the observed relationships between photosynthetic rate and needle water potential for loblolly pine (*Pinus taeda* L.) seedlings imply a somewhat unique behavior under the conditions of each experiment (Figure 2) and under-

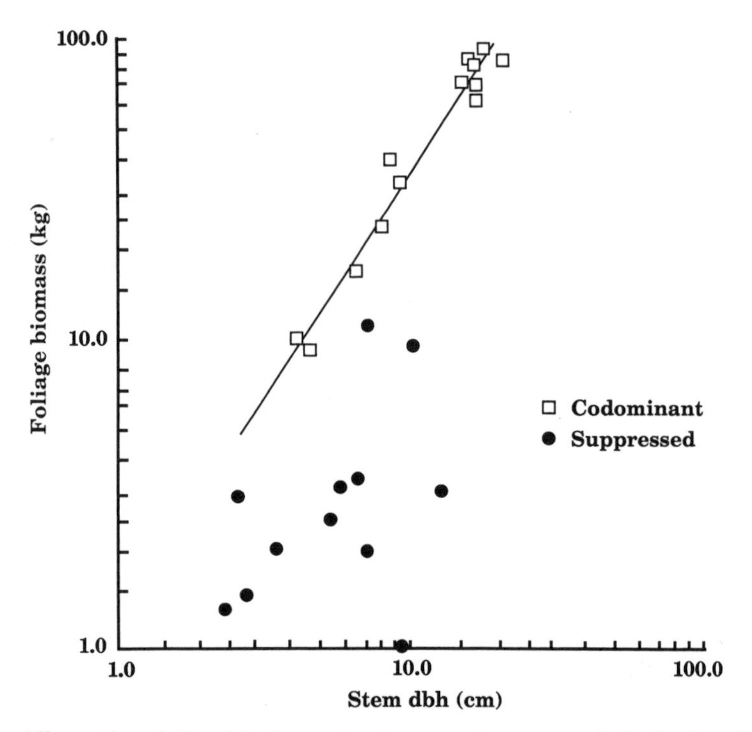

Figure 1. Allometric relationship for codominant and suppressed slash pine (*Pinus elliottii* Engelm.) from plantations ranging in age from 5 to 26 years (after Gholz and Fisher, 1982).

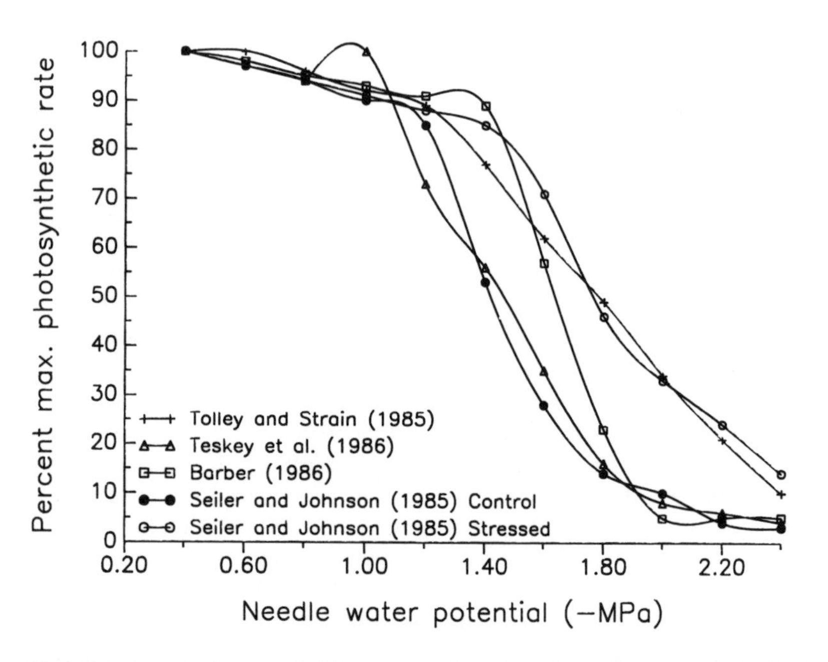

Figure 2. Variation in net photosynthetic rate as a function of needle water potential from five experiments with loblolly pine (*Pinus taeda* L.) seedlings. Published data fitted with cubic spline approximation.

score the importance of including measurements of the conditions under which the experiment was conducted. The problem is particularly apparent when dealing with adaptive stress responses in trees (Schulze and Hall, 1982). Adaptive stress responses are primarily responsible for the controversy over the application of results from seedling response studies to mature trees (Ford, 1987). For example, periodic drought stress is buffered more in larger trees because of stem tissue capacitance (Hinckley et al., 1978). Similarly, competition among trees within closed stands may alter the impact of pollutant stress at the stand level. There is no easy solution to this problem; but by iteratively adding quantitative expressions for each adaptive characteristic to the model and evaluating the impact, it may be possible to successively link limited experimental studies to a more comprehensive ecological perspective.

Environmental policy decisions require more than just estimates of the mean response. In order to estimate risk at a given level of pollutant exposure, information about the potential distribution of responses is needed to establish a reasonable (e.g., 90%) confidence interval. Given the distribution of model parameters, Monte Carlo (Gardner and Trabalka, 1985), first-order Taylor series approximations (Gertner, 1987), and the Fourier Amplitude Sensitivity Test (FAST) (McRae et al., 1982) can be used to establish the distribution of responses. Potential sources of variability in the model include natural population variation for each parameter value, measurement errors associated with determining parameters, and errors in model structure due to simplification and uncertainty about linkages (O'Neil and Gardner, 1979). Estimating the variance of model parameters from an experimental data set often results in underestimates of the true population variances, since experiments are generally designed to minimize variability. While estimates of the variance of each parameter can be determined by proper sampling of the population, estimates of covariance among parameters are more difficult to obtain. Commonly, processes are studied independently, providing no information about covariances that may exist among them. For example, respiration rate and photosynthetic rate functions are often derived independently on the assumption that they are related only through substrate availability, but research indicates that they are more closely linked (Graham and Chapman, 1979).

Quantitative Expression

The challenges of expressing a model in a quantitative format are being met by advances in computer technology and numerical methods. Increased memory, power, and speed are lowering the barriers to large, complex simulations. Software packages exist that contain numerical procedures to eliminate the restrictions of closed-form analytical solutions. Non-linear model estimation procedures have been refined, thereby reducing the need to rely exclusively on linear approximations. Advances in the mathematical modeling of physical systems are creating new methods for modeling non-linear dynamics in biological systems (Degn et al., 1987). The application of these techniques is frequently viewed as a means to increase the accuracy and precision of simulations. This is accomplished by increasing the degree to which realistic conditions are represented.

Unfortunately, errors in the quantitative expression of a model due to simplifications or resulting from uncertainty about the linkages among processes cannot be handled analytically. The consequences of simplification can sometimes be judged by comparing alternative formulations (Loehle, 1987). Estimating the impact of uncertainty in the linkages among processes can be established only through independent experiments on the model. When functions are derived independently, the impact of uncertainty in the linkages can be substantial. For example, Figure 3 shows the results of a linkage between site index, survival, and yield equations from different sources. Each of the equations is an accurate expression of the data used in its derivation; but when linked together, the

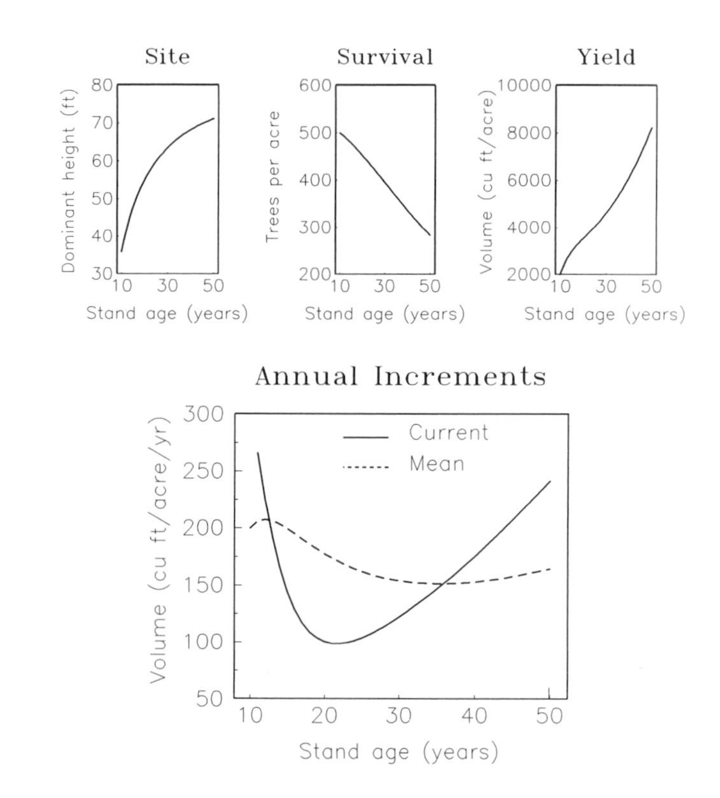

Figure 3. Results of linking a site index (Lenhart, 1971), survival (Cao et al., 1982), and cubic foot volume yield equation (Burkhart et al., 1972) to predict current and mean annual increments. All equations are for old-field loblolly pine (*Pinus taeda* L.) plantations.

resulting annual increments have increasing slopes in direct disagreement with observed data. One obvious challenge is to make the quantitative relationship among processes more transparent and thus minimize possibly serious misrepresentations of behavior.

CONCLUSIONS

Modeling is not merely a descriptive simplification of experimental observations, but an essential component of research for formalizing the structure of theory and hypothesis evaluation. We have described some of the challenges in applying process modeling as a means to understand and predict forest growth responses to environmental stress. The concepts presented here are intended as a framework for viewing the contributions of the various authors to this volume. A critical analysis of these contributions will reveal no panacea, only additional opportunities to realize the potential of a promising methodology.

LITERATURE CITED

Adams, N., and F. Cady. 1988. *Modeling Growth and Yield of Multipurpose Tree Species.* MPTS Network Technical Series, Vol. 1. Winrock Institute for International Development, Arlington, VA, U.S.A. 57 pp.

Barber, B. 1986. *Stomatal and Non-Stomatal Effects of Water Stress on Loblolly Pine* (*Pinus taeda* L.). Ph.D.

Thesis, Auburn University, AL, U.S.A.
Burkhart, H. E., R. C. Parker, M. R. Strub, and R. G. Oderwald. 1972. *Yields of Old-Field Loblolly Pine Plantations.* No. FWS-3-72, School of Forestry and Wildlife Resources, Virginia Polytechnic Institute and State University, Blacksburg, VA, U.S.A. 51 pp.
Cannell, M. G. R. 1985. Dry matter partitioning in tree crops. In: Cannell, M. G. R., and J. E. Jackson, eds. *Attributes of Trees As Crop Plants.* Institute of Terrestrial Ecology, Huntingdon, England, U.K. Pp. 160–188.
Cao, Q. V., H. E. Burkhart, and R. C. Lemin. 1982. *Diameter Distributions and Yields of Thinned Loblolly Pine Plantations.* No. FWS-1-82, School of Forestry and Wildlife Resources, Virginia Polytechnic Institute and State University, Blacksburg, VA, U.S.A. 62 pp.
Carey, A., A. C. Janetos, and R. Blair. 1986. *Responses of Forests to Atmospheric Deposition: National Research Plan for the Forest Response Program.* USDA-EPA, Washington, DC, U.S.A. 67 pp.
Degn, H., A. V. Holden, and L. F. Olsen, eds. 1987. *Chaos in Biological Systems.* Proceedings, NATO Advanced Science Institute Conference, Cardiff, South Glamorgan, Wales, U.K. Series A: Life Sciences, Vol. 138. Plenum Press, New York, NY, U.S.A. 323 pp.
Environmental Protection Agency. 1988. *Effects of Tropospheric Ozone on Forest Trees: A Research Plan.* United States Environmental Protection Agency, Corvallis Environmental Research Laboratory, Corvallis, OR, U.S.A. 81 pp.
Ford, E. D. 1987. *Mature Tree Response.* Proceedings of a workshop December 8–10, 1986, in St. Louis, MO, U.S.A. Sponsored by Southern Forest Commercial Research Cooperative, North Carolina State University, Raleigh, NC, U.S.A.
Gardner, R. H., and J. R. Trabalka. 1985. *Methods of Uncertainty Analysis for a Global Carbon Dioxide Model.* TR024, U.S. Department of Energy, Washington, DC, U.S.A. 41 pp.
Gertner, G. 1987. Approximating precision in simulation projections: an efficient alternative to Monte Carlo methods. *Forest Science* 33(1):230–239.
Gholz, H. L., and R. A. Fisher. 1982. Organic matter production and distribution in slash pine (*Pinus elliottii*) plantations. *Ecology* 63:1827–1839.
Givnish, T. J., ed. 1983. *On the Economy of Plant Form and Function.* Proceedings of the Sixth Maria Moors Cabot Symposium, "Evolutionary Constraints on Primary Productivity: Adaptive Patterns of Energy Capture in Plants," at Harvard Forest, Petersham, MA, U.S.A., August, 1983. Cambridge University Press, Cambridge, England, U.K. 717 pp.
Godfrey, K. 1983. *Compartmental Models and Their Applications.* Academic Press, New York, NY, U.S.A. 291 pp.
Graham, D., and E. A. Chapman. 1979. Interactions between photosynthesis and respiration in higher plants. In: Gibbs, M., and E. Latzko, eds. *Photosynthesis, II. Photosynthetic Carbon Metabolism and Related Processes. Encyclopedia of Plant Physiology,* New Series, Vol. 6. Springer-Verlag, Berlin, F.R.G. Pp. 150–161.
Grier, C. C. 1988. Foliage loss due to snow, wind, and winter drying damage: its effects on leaf biomass of some western conifer forests. *Canadian Journal of Forest Science* 18:1097–1102.
Hinckley, T., J. P. Lassoie, and S. W. Running. 1978. Temporal and spatial variations in the water status of forest trees. *Forest Science Monograph* No. 20.
Kimmins, J. P., and K. A. Scoullar. 1984. *FORCYTE-10: A User's Manual (Forest Nutrient Cycling and Yield Trend Evaluator).* Faculty of Forestry, University of British Columbia, Vancouver, B.C., Canada.
Lambers, H., R. K. Szaniawski, and R. De Visser. 1983. Respiration for growth, maintenance and ion uptake. An evaluation of concepts, methods, values and their significance. *Physiologia Plantarum* 58:556–563.
Landsberg, J. J. 1986. *Physiological Ecology of Forest Production.* Academic Press, London, England, U.K. 191 pp.
Lemeur, R., G. Gietl, and H. Hager. 1979. Comparison of radiation models. In: Halldin, S., ed. *Comparison of Forest Water and Energy Exchange Models.* Proceedings, IUFRO Workshop, Uppsala, Sweden, September 24–30, 1978. Developments in Agriculture and Managed Forest Ecology, No. 9. Elsevier Science Publications, Amsterdam, the Netherlands. Pp. 187–200.
Lenhart, J. D. 1971. Site index curves for old field loblolly pine plantations in the interior West Gulf Coastal Plain. Forestry Paper 8, Stephen F. Austin State University, Nagadoches, TX, U.S.A. 4 pp.
Levins, R. 1986. The strategy of model building in population biology. *American Scientist* 54:421–431.
Loehle, C. 1987. Errors in construction, evaluation and inference: a classification of sources of error in ecological models. *Ecological Modelling* 36:297–314.
Luxmoore, R. J., and L. H. Stolzy. 1987. Modeling belowground processes of roots, the rhizosphere, and soil communities. In: Wisiol, K., and J. Hesketh, eds. *Plant Modeling for Resource Management,*

Vol. II. CRC Press, Boca Raton, FL, U.S.A. Pp. 129–153.

McMurtrie, R. E., J. J. Landsberg, and S. Linder. 1988. Research priorities in field experiments on fast growing tree plantations: implications of a mathematical production model. In: Pereira, J. S., and J. J. Landsberg, eds. *Forest Biomass for Fibre and Energy.* Martinus Nijhoff Publications, the Netherlands.

McRae, G. J., J. W. Tilden, and J. H. Seinfeld. 1982. Global sensitivity analysis: a computational implementation of the Fourier Amplitude Sensitivity Test (FAST). *Computers and Chemical Engineering* 6:15–25.

Molz, F. J. 1981. Models of water transport in the soil plant system: a review. *Water Resources Research* 17:1245–1260.

O'Neill, R. V., and R. H. Gardner. 1979. Sources of uncertainty in ecological models. In: Zeigler, B. P., M. S. Elizas, G. J. Klir, and T. I. Oren, eds. *Methodology in Systems Modelling and Simulation.* North-Holland Publishing Company, the Netherlands. Pp. 447–463.

Reid, C. P. 1985. Mycorrhizae: a root-soil interface in plant nutrition. In: Todd, R. L., and J. E. Giddens, eds. *Microbial-Plant Interactions.* American Society of Agronomy Special Publication No. 47. Madison, WI, U.S.A. Pp. 29–50.

Rennolls, K., and P. Blackwell. 1986. *An Integrated Forest Process Model: Its Calibration and Predictive Performance.* Forestry Commission Research and Development Paper 148. Great Britain Forestry Commission, London, England, U.K. 31 pp.

Reynolds, J. F., J. W. Skiles, and D. L. Moorhead. 1987. SERECO: a model for the simulation of ecosystem response to elevated CO_2. Report 041, Series: Response of vegetation to carbon dioxide. U.S. Department of Energy, Office of Energy Research, Washington, DC, U.S.A.

Schulze, E.-D., and A. E. Hall. 1982. Stomatal responses, water loss and CO_2 assimilation rates of trees in contrasting environments. In: Large, O. L., P. S. Nobel, C. B. Osmond, and H. Ziegler, eds. *Physiological Plant Ecology,* II. *Water Relations and Carbon Assimilation. Encyclopedia of Plant Physiology,* New Series, Vol. 12b. Springer-Verlag, Berlin, F.R.G. Pp. 181–230.

Segre, E. 1980. *From X-rays to Quarks: Modern Physicists and Their Discoveries.* W. H. Freeman, San Francisco, CA, U.S.A. 332 pp.

Seiler, J. R., and J. Johnson. 1985. Photosynthesis and transpiration of loblolly pine seedlings as influenced by moisture stress conditioning. *Forest Science* 31:742–749.

Seligman, N. G. 1976. A critical analysis of some grassland models. In: Arnold, G. W., and C. T. de Wit, eds. *Critical Analysis of Systems Analysis in Ecosystems Research and Management.* Centre for Agriculture Publication and Documentation, Wageningen, the Netherlands. Pp. 84–97.

Shinozaki, K., K. Yoda, K. Hozumi, and T. Kira. 1964. A quantitative analysis of plant form: the pipe-model theory. I. Basic analysis. *Japanese Journal of Ecology* 14:97–105.

Teskey, R. O., J. A. Fites, L. J. Samuelson, and B. C. Bongarten. 1986. Stomatal and non-stomatal limitations to net photosynthesis in *Pinus taeda* L. under different environmental conditions. *Tree Physiology* 2:131–136.

Tolley, L. C., and B. R. Strain. 1985. Effects of CO_2 enrichment and water stress on gas exchange of *Liquidambar styraciflua* and *Pinus taeda* seedlings growing under different irradiance levels. *Oecologia* 65:166–172.

Valentine, H. T. 1988. A growth model of a self-thinning stand based on carbon balance and the pipe-model theory. In: *Forest Growth Modelling and Prediction,* Vol. 1. General Technical Report NC-120, USDA Forest Service North Central Forest Experiment Station, St. Paul, MN, U.S.A. Pp. 353–360.

Wallace, H. R. 1978. The diagnosis of plant diseases of complex etiology. *Annual Review of Phytopathology* 16:379–402.

Whistler, F. W., B. Acock, D. N. Baker, R. E. Fye, H. F. Hodges, J. R. Lambert, H. E. Lemon, J. M. McKinion, and V. R. Reddy. 1986. Crop simulation models in agronomic systems. *Advances in Agronomy* 40:141–207.

Zahner, R. 1968. Water deficits and growth of trees. In: T. T. Kozlowski, ed. *Water Deficits and Plant Growth,* Vol. II. Academic Press, New York, NY, U.S.A. Pp. 191–254.

SECTION I

Tree Metabolism and Growth

A myriad of metabolic processes influence the growth of trees and their response to environmental stresses. A comprehensive definition of these relationships (e.g., photosynthesis, respiration, resource allocation) has not been developed for woody plants. Process-based models offer a quantitative framework for linking the complex metabolic pathways within trees. Concomitantly, process modeling may be used to predict tree responses to a multitude of interacting biotic and abiotic stresses. The relative paucity of quantitative information on some aspects of tree metabolism (e.g., growth regulators and carbon allocation) has precluded widespread development and validation of complete whole-tree process models.

This section provides an overview of some of the principal processes involved in tree metabolism and growth. The authors describe concepts and discuss selected metabolic pathways which offer a basis for the development and validation of process models to predict tree growth. Considerable emphasis has been placed on the primary processes of carbon assimilation, allocation, and respiration. The role of nutrition and drought stress in tree metabolism and growth is also emphasized. General process models of selected metabolic pathways are presented. Specific quantitative approaches for characterizing and linking the metabolic pathways and growth patterns of trees via process models are partially addressed in this section.

CHAPTER 1

PHYSIOLOGICAL PROCESSES AND TREE GROWTH

Robert K. Dixon

Abstract. Vegetative and reproductive growth of trees is dependent on genetically controlled physiological processes influenced by environment. These conceptual processes are actually amalgams of complex biochemical and biophysical relationships which occur at several levels of organization (e.g., subcellular, cellular, tissue, organ, tree, and forest). Physiological activities which have received significant attention in modeling whole-tree growth include (1) photosynthesis, carbon partitioning, and respiration, (2) mineral metabolism, (3) growth regulation, and (4) intratree transport. Development of process-based models requires mathematical representation of limited physiological information from several levels of structural organization. Modeling of physiological processes can be constrained by the complexity of tree structure and our incomplete knowledge of interacting metabolic pathways. Thus, it is currently necessary to make assumptions or to simplify principal metabolic activities in order to develop whole-tree process models. Modeling of physiological processes and tree growth is a relevant analysis, simulation, and prediction tool if the underlying assumptions and limitations are realized.

INTRODUCTION

Tree growth is the result of myriad interacting physiological processes influenced by an inherited genetic constitution and the ambient environment. Some key physiological processes which may be useful in modeling aspects of whole-tree growth are presented in Table 1. Most of these physiological processes can be identified at different levels of organization: subcellular, cellular, tissue, organ, or whole-tree. Physiological activities such as photosynthesis, respiration, or transpiration are actually working concepts representing a complex mixture of biochemical and biophysical phenomena (Bonner and

Table 1. Principal physiological processes in tree growth and reproduction.

- **Photosynthesis:** Synthesis of carbon metabolites from carbon dioxide and water.
- **Mineral metabolism:** Incorporation of inorganic ions into organic compounds.
- **Respiration:** Oxidation of carbon metabolites for energy production.
- **Carbon allocation and assimilation:** Partitioning and conversion of carbon metabolites.
- **Absorption and accumulation:** Intake of water, minerals, gases.
- **Translocation:** Movement of inorganic and organic metabolites.
- **Growth regulation:** Interactive control of metabolite source and sink relationships by genes and growth-regulating substances.

Dr. Dixon is Associate Professor at the School of Forestry, Auburn University, Auburn, AL, 36849 U.S.A. This contribution is supported by the USDA-EPA Forest Response Program. The author thanks Dr. John I. Blake and Dr. Gregory A. Ruark for helpful discussions and critical review of the manuscript.

Varner, 1976; Jarvis and Leverenz, 1983). In order to analyze, simulate, and predict tree growth or behavior in response to environmental stress, it is necessary to acknowledge, understand, and link these metabolic pathways (Dale et al., 1985). The emerging utility of physiological process models in plant resource management was recently reviewed (Wisiol and Hesketh, 1987).

Metabolic pathways provide a useful map for studying physiological processes and for understanding and predicting the growth of whole trees (Johnson, 1987). Ideally, all metabolic pathways should be represented in models of tree growth response to environmental stress. However, this is neither practical nor desirable at present. The physiological processes and the underlying metabolic pathways by which trees detect environmental changes and modify the quality and/or quantity of growth are only partially understood (Ford and Kiester, this volume; Lavender, 1980). For example, basic concepts and hypotheses regarding plant growth and maintenance respiration continue to evolve (Lambers, 1984). While current physiological models are appealing and useful, they have proven difficult to validate experimentally (Bassow et al., this volume; Jarvis and Leverenz, 1983).

Two basic questions have emerged as scientists and engineers grapple with the challenge of modeling tree growth using physiological processes to evaluate their response to environmental stress:

1. What are the essential physiological processes in the vegetative and reproductive growth of a tree?
2. What biotic and abiotic factors modulate these processes?

Given our current knowledge of tree metabolism and growth response to environmental stress, many whole-tree modeling efforts have focused on five broad topics:

1. vegetative and reproductive growth;
2. photosynthesis, respiration, and carbon allocation;
3. mineral metabolism, especially nitrogen metabolism;
4. growth regulation; and
5. intra-tree transport of metabolites.

The objective of this chapter is to consider our current understanding of tree metabolism and growth with regard to relevant physiological processes which have been used to predict responses to environmental stress. A brief review of selected topics in tree physiology is offered, but the literature cited is not exhaustive. Topics and references were chosen to illustrate specific points and to provide sources of further information. An effort was made to examine past and current work in this emerging field and to identify areas of research which may be productive.

VEGETATIVE AND REPRODUCTIVE GROWTH

Trees have a complex annual vegetative growth cycle. Meristematic tissue, located in roots, shoots, leaves, and other tree parts, grows differentially in response to environmental stimuli. Both angiosperms and gymnosperms in temperate zones experience cyclic periods of growth and quiescence (or dormancy) due to environmental influences (Lavender, 1980). Specific biochemical mechanisms and anatomical features of root and shoot growth processes, including aspects specific to trees, are reviewed elsewhere (Cosgrove, 1988; Kramer and Kozlowski, 1979; Torrey and Clarkson, 1975).

Vegetative Growth

Shoot growth is the result of cell division, elongation, and maturation in apical and lateral meristems following bud opening (Cecich, 1980). Thus, vegetative buds (e.g., terminal, lateral, axillary, adventitious) are actually embryonic shoots. Shoot expansion is a result of internode elongation and is generally a function of cell turgor, metabolites, and growth regulators (Lavender, 1980). Shoot growth patterns have been classified on the basis of location, development, and type of bud. Dale (1988) reviewed the anatomical and biochemical aspects of leaf development.

Diameter growth of trees is a function of meristematic activity of the cambium (Worrall, 1980). The cambium is a cylindrical meristem located between the xylem and phloem of both roots and shoots. Cell types and tissues associated with cambial growth are described by Wilson (1970). Derivatives of the cambium, following an ordered sequence of events, differentiate into xylem and phloem cells. Following cell enlargement, secondary and tertiary walls are formed and lignified, and the protoplast ceases to function. Cambial activity and derivatives vary significantly between angiosperms and gymnosperms (Wodzicki, 1980).

Following establishment of the tap root, lateral root growth initiation and elongation begin. Lateral roots initiate from the pericycle, a zone of cells outside the stele. Initiation and elongation of lateral roots are a function of genetics, age, metabolic supply, and environment (Torrey and Clarkson, 1975). Radial growth of roots involves secondary growth of vascular tissues. Coutts (1987) recently reviewed developmental processes in tree root systems.

Most models of vegetative growth in whole trees or selected organs have been developed from the perspective of energy or mass balance. This approach is fairly well defined for some annual agronomic crops which grow indeterminately (Wisiol and Hesketh, 1987). Early attempts to represent tree growth using the mass-balance approach provide the basis for some current process models (West, 1987; McMurtrie and Wolf, 1983). Many whole-tree growth models based on mass balance only partially recognize or incorporate the full complement of interacting anatomical and physiological processes which result in cambium, leaf, root, or shoot growth. Moreover, some models assume physiologic processes are generic for both angiosperms and gymnosperms. A future challenge of process-based modeling is to represent and integrate the complex annual cycle of growth for a representative sample of tree species.

Our understanding of vegetative growth processes in relation to principal metabolic pathways (e.g., allocation pattern and respiration patterns) is limited (Isebrands et al., this volume; Cannell and Last, 1976). The relationship of physiological processes to tree structure is well defined in only a few species of selected genera (e.g., *Populus, Pinus*). Environmental factors directly influence interacting physiological processes but only indirectly influence vegetative growth of shoot and root systems (Borchert, 1975). Quantitative linkage of complex physiological processes to tree growth and yield patterns remains a formidable task. As future process-based models of tree growth are refined, the number of assumptions and simplifications regarding physiological processes and growth patterns can be reduced.

Reproductive Growth

Following a period of juvenile growth, temperate-zone trees achieve the capacity for sexual reproduction. Age of sexual maturity varies widely among genera. Moreover, sexual reproduction organs and patterns vary significantly between angiosperms and gymnosperms. Sequences common to reproductive growth include initiation of flower primordia, flowering, pollination, fertilization of gametes, embryo growth, and development of fruits or cones (Bernier, 1988). These complex and energy-intensive processes are the result of products from many metabolic pathways (Little, 1980).

The capacity of some tree genera to reproduce asexually is vital to their survival. The physiological processes which regulate sprouting ability are not fully known (Kramer and Kozlowski, 1979). Angiosperms sprout commonly but gymnosperms only rarely. Sprouting ability has been linked to tree age, metabolite supply, genetics, and environment. Disturbance or environmental stress stimulates sprouting in some angiosperms. Given the concerns regarding forest decline and mortality in relation to atmospheric pollutants, asexual reproduction patterns of trees subject to stress merit further consideration.

Although reproductive processes constitute a critical and environmentally sensitive period in the life cycle of angiosperms and gymnosperms, models of these processes at any level of organization (cellular, tissue, or whole-tree) are scarce (Bernier, 1988; Vincent and Pulliam, 1980).). Some recent models characterize or predict growth of established seedlings in response to environmental stress (e.g., Blake and Hoogenboom, 1988); but other processes such as flowering, pollination, seed development, or germination have received little attention. Reproductive growth requires substantial energy and metabolites and can significantly influence vegetative growth patterns (Johnson, 1987). Future refinements of mass-balance models in trees should consider variation in the allocation patterns and respiration rates of reproductive organs.

PHOTOSYNTHESIS, RESPIRATION, AND CARBON ALLOCATION

Photosynthesis

Because photosynthesis is the sole source of energy for trees, this process is of keen interest in the field of growth modeling. The working concept of photosynthesis is actually a series of biochemical and biophysical processes including:

1. capture of light energy by chloroplasts or other pigmented structures;
2. splitting of water and release of electrons;
3. production of energy (e.g., ATP) and reducing agents (e.g., NADPH) by electron transfer; and
4. expenditure of energy for carbon dioxide fixation and reduction.

The biochemical and biophysical dimensions of photosynthesis, including aspects specific to trees, have been reviewed (Govindjee, 1975; Schaedle, 1975). Tree and environmental factors significantly influence photosynthetic processes in angiosperms and gymnosperms (Dickson, 1986; Mäkelä, this volume).

Tree factors influencing photosynthetic processes include age, phenology, leaf morphology (e.g., sun versus shade leaves), crown architecture, and branch growth patterns. Some investigators assert that leaf size, shape, and orientation are the most significant factors in patterns of tree photosynthesis (Isebrands et al., this volume). Branch and leaf architecture form the structural basis for photosynthesis and related metabolic processes. Hanson et al. (1986) recently demonstrated the interdependence of crown architecture, phenology, and photosynthesis patterns in *Quercus* seedlings.

Photosynthesis is also influenced by a tree's environment. Edaphic and climatic factors can severely limit photosynthetic processes (Cannell and Last, 1976; Jarvis and Leverenz, 1983). Tree nutrition and water relations, the quality and quantity of light incident on tree foliage, and ambient temperature and ambient CO_2 levels all influence the rate and pattern of photosynthesis and gas exchange (Schulze, 1988; Spitters, 1986). Structural and functional compensatory factors in leaves may partially offset the effects of environmental stress on photosynthesis (Mann et al., 1980).

Within limits, our understanding of the biochemical and biophysical processes of photosynthesis is fairly complete. Conceptual models of photosynthetic events at various levels of organization (e.g., enzyme kinetics or the light reaction) have been developed (Govindjee, 1975). Future needs include the linking of leaf photosynthesis models to canopy, whole-tree, or stand models (McConathy and McLaughlin, 1987). Questions have emerged as efforts shift to develop process-based models of tree photosynthesis in response to environmental stress:

1. Is the environmental stress sufficient to temporarily or permanently impair the sequential biochemical and biophysical events of photosynthesis?
2. What is the resilience of a tree's photosynthetic apparatus in response to episodic stress (Boyer et al., 1986)?
3. How can gas exchange rates of foliage be integrated over time and space within a canopy (Gutschick, 1987)?

The paucity of information describing canopy photosynthesis under conditions of environmental stress is a major deficiency in the development of whole-tree growth models.

Respiration

Energy and metabolites are products of the photosynthesis process in leaves. However, these products are consumed by respiration in cells in all living organs of a tree. Respiration provides energy required for the maintenance of existing cells (maintenance respiration) and also for the production of new cells (growth respiration) (Gordon and Larson, 1968). A third component of respiration, associated with ion uptake, has been the subject of recent interest (Lambers, 1984). Environmental factors such as light intensity and photoperiod, temperature extremes, water stress, and mineral nutrition significantly influence the respiration rates of trees. Respiration rates in tree organs under environmental stress can be 3–5 times greater than rates in non-stressed organs.

Respiration can be classified into two primary steps, glycolysis and the Krebs cycle (Bonner and Varner, 1976). Glycolysis occurs in the cytoplasm, and the Krebs cycle in the mitochondria. The usual substrate for respiration is sugar, and the products are high-energy compounds such as ATP and NADPH (Penning de Vries, 1974). The biochemical pathways and products of respiration are reviewed by Lambers (1984). Models of energy flow have been developed for agronomic crops (Johnson, 1987; Thornley, 1972) and some woody plants (McMurtrie and Wolf, 1983). However, with respect to photosynthesis, detailed knowledge of respiration processes is limited. Many assumptions as well as simplifications regarding respiration have been incorporated into current process-based models of tree growth. Little emphasis has been placed on the respiration of tree organs, particularly in modeling response to environmental stress (McLaughlin et al., 1988). Our understanding of respiration-related processes remains a major void in the tree ecophysiology literature and hampers modeling efforts.

Carbon Allocation

Assimilate partitioning between sites of photosynthate production (source) and utilization (sink) is a major factor in tree growth. Loading, transport, and allocation of photosynthesis products from source to sink are a complex series of events. Giaquinta (1983) proposed a conceptual model of sucrose loading into the phloem, vascular transport, and unloading at the sink. Generally, photosynthetically active leaves produce carbon metabolites, and these products are transported to respiring organs and sites of assimilation. The remaining products of photosynthesis are stored in parenchyma cells as starch, fats, nitrogen compounds, hemicellulose, and other compounds (Gordon and Larson, 1968). At the peak of the growing season, the relative strength of sinks within trees can be ranked (in decreasing order of importance) as follows: reproductive growth, emerging shoots and leaves, mature leaves, cambial regions, root systems, and storage. This pattern is easily altered by environmental stress, and development of biologically meaningful carbon allocation coefficients remains a challenge for modelers.

A growing body of literature suggests that environmental stress significantly influences tree growth by altering carbon metabolite partitioning between organs (Reich and Amundsen, 1985; Little, 1980). The transport of assimilates from source to sinks within a tree represents an integrative step in the carbon cycle that can be characterized at various levels of structure (cell, tissue, organ, or whole-tree). At the organ level (e.g., the branch level), environmental stress can influence carbon partitioning by impacting phloem loading in the foliage or altering demand for assimilates at the sink (Giaquinta, 1983; Tingey et al., 1976). McLaughlin et al. (1988) presented a conceptual model of the impact of chronic environmental stress on tree foliage per branch, photosynthetic and respiration rates, carbon allocation, and vegetative growth. This approach is currently being expanded to model carbon allocation patterns for whole-trees (e.g., Boltz et al., 1986; Ford and Kiester, this volume).

MINERAL METABOLISM

Mineral nutrition is generally considered critical because an adequate supply of mineral elements is required for the vegetative and reproductive growth of trees. Over a dozen mineral elements, including nitrogen, phosphorus, potassium, magnesium, sulfur, calcium, iron, boron, manganese, zinc, copper, and molybdenum, are considered essential in metabolic pathways (Epstein, 1972). Mineral nutrients have many functions in plants, serving as constituents of cells and tissue, as biochemical catalysts, and in osmotic regulation (Clarkson and Hanson, 1980). Deficiencies and toxicities of minerals may significantly alter principal metabolic pathways linked to growth.

The uptake, transport, and metabolism of essential elements have been characterized for some tree species (Grigal, this volume; Kramer and Kozlowski, 1979). Considerable effort has focused on the macronutrients nitrogen, phosphorus, and potassium. Nitrogen is an essential component of amino acids, the building blocks of proteins. Examples of nitrogen-based compounds include purines, alkaloids, various proteins (enzymes), chlorophyll, and growth regulators. Phosphorus is also a component of proteins, together with cell membrane components and energy transport molecules. Both organic and inorganic phosphorus are transported in the vascular tissues of trees. Potassium is an essential element in enzyme activity, osmotic regulation, and carbon metabolism. Although little studied in tree metabolism, micronutrients are essential in the enzymatics of many metabolic pathways (Epstein, 1972).

The processes of mineral element uptake, assimilation, and cycling are of particular interest as robust models of trees, forests, and ecosystems are developed and hypotheses tested. The detail and scope of process-based models of tree mineral metabolism range from the ecosystem level (Goldstein et al., 1984) to the subcellular level (Oaks and Hirel, 1985). Model detail is dependent on the complexity of the process being simulated as well as the hypothesis examined.

Since nitrogen deficiency has a significant impact on many metabolic pathways in trees, models of nitrogen acquisition and assimilation have been developed (Gillespie and Chaney, 1989; Oaks and Hirel, 1985). Considerable emphasis has been placed on nitrate uptake by roots and nitrate reduction in leaves. The latter has been coupled to the photosynthetic light reaction and to mitochondrial respiration (Clarkson and Hanson, 1980; Penning de Vries, 1974). Although 40–75% of leaf nitrogen is located in the chloroplast, models linking nitrogen metabolism in tree leaves to the various steps of photosynthesis have proven difficult. Process-based models linking leaf nitrogen metabolism and energy balance require an understanding of compensatory and competitive interaction between the various steps in these two systems (Kenig, 1987).

Process-based models of whole-tree response to mineral toxicity have been developed in response to the need to characterize and predict the impact of acid deposition and various metals on forest ecosystems (Cronan, 1985). Considerable emphasis has been placed on aluminum toxicity in trees (Ohno et al., 1988). Aluminum, as Al^{+++}, can interfere with the physiological processes of cells, tissues, or organs. Apparently, Al^{+++} alters the architecture of cell membranes and thus influences the balance and flow of organic and inorganic substances of the metabolic pathways (Zhao et al., 1987). The presence of calcium, as Ca^{++}, in intercellular spaces can prevent this toxic effect of Al^{+++} on cell membranes. Subtle compensation mechanisms in mineral-nutrition metabolism, such as the interaction between Al^{+++} and Ca^{++}, complicate the development of valid and robust models of related physiological processes. As models of tree growth are refined, compensation and synergism of physiological processes merit careful consideration.

GROWTH REGULATION

The numerous physical and biochemical processes of tree development and growth are regulated by specific substances, generally termed plant growth regulators or hormones (Wareing, 1980). These substances include auxins, gibberellins, cytokinins, ethylene, and abscisic acid, which act singly or in concert to initiate and regulate the development and maintenance of well-proportioned and functional tree organs. Experimental evidence suggests that growth-regulating substances play a pivotal role in intratree communication, as well as in the regulation of physiological processes in response to environmental stress (Kossuth and Ross, 1987; Zeevart and Creelman, 1988). Growth regulator synthesis occurs in numerous locations in trees, and virtually every phase of tree development is sensitive to the five major growth substances. The biochemical synthesis of growth regulators, as well as their mode of action, is reviewed by Bonner and Varner (1976).

Growth is a very complex metabolic process, and control is a function of interrelated biochemical and biophysical events (Dale, 1988). Experimental evidence suggests competition for growth resources (water, minerals, light, CO_2) between various tree organs or growth phases (e.g., root or shoot; vegetative or reproductive growth). Growth substances apparently facilitate the allocation of resources (e.g., carbon metabolites) to various organs (Ho, 1988). Although sink competition may be amplified by growth-regulating substances, their role in prioritizing partitioning is generally unknown for specific developmental stages in trees.

Resource allocation among tree organs may be a function of differential sensitivity to growth-regulating substances. A preliminary model describing this complex system has been developed (Trewavas and Allan, 1987). Experimental evidence indicates that some cells, tissues, or organs are more sensitive to growth substances than others (Wareing, 1980). Tree response to environmental stress, in terms of internal resource allocation, is probably a function of growth-regulator action. This topic may merit greater emphasis in future process-based models of tree growth.

Ultimately, tree growth regulation and assimilate partitioning are a composite function of genetics and environment. Although the potential sink strength is genetically determined, the actual sink strength is determined by factors affecting rate-limiting processes within organs during development. There is no simple relationship between genetics, environmental influences, growth of a sink organ, and growth-regulating substances (Ho, 1988). Future modeling research should consider membrane transport processes, growth regulators, and enzymatic regulation of rate-limiting processes of developing organs. Models of these interacting biochemical pathways and physiological processes will help us to define and to stimulate the mechanisms which ultimately control growth.

INTRA-TREE TRANSPORT

Tree metabolism and growth are dependent on the efficient movement of organic compounds, minerals, and water. Long-distance, bi-directional, energy-efficient transport of metabolites is essential for the survival and growth of mature trees (Valentine, this volume). Acropetal transport from the root system to aerial portions of the tree occurs in the xylem. In contrast, the phloem carries metabolic products from leaves to other organs. Attributes of the vascular cambium and its components are reviewed by Little (1980).

Intra-tree transport of organic and inorganic substances, as well as a root-to-shoot feedback system, links many physiological processes (Borchert, 1975; Reynolds and Thornley, 1982). Minerals, carbon metabolites, nitrogen compounds, proteins, and growth-regulating substances are all subject to long-distance transport. The concentration of these substances varies diurnally and temporally. The solutes are carried along a water potential gradient from root to leaves (Hinckley et al., 1978; Newman, 1976; Tyree, 1988). Experimental evidence indicates that source-sink relationships significantly influence intra-tree transport. The relative distance from sink to source and sink strength (metabolic activity) influence transport patterns and partitioning of metabolites.

The direction and rate of carbon metabolite transport are influenced by principal physiological processes such as photosynthesis (Hanson et al., 1988). Developing leaves with significant metabolic activity import carbon from mature leaves until full expansion is achieved. Following full expansion, leaves become net bi-directional exporters of carbon metabolites. Metabolites exported acropetally are assimilated into developing tissues. As fully expanded leaves are overtopped by subsequent foliage, carbon metabolite export is primarily basipetal. The senescence of leaves is accompanied by a decline in photosynthesis and a decline in carbon metabolite export. These complex, experimentally derived transport patterns have yet to be fully incorporated into process-based models of tree growth.

A simulation model (ROOTSIMU) of loblolly pine (*Pinus taeda* L.) seedling root and shoot system development has been based on carbon metabolite and water transport (Blake and Hoogenboom, 1988). In this model, seedling carbon balance is a function of gross photosynthesis, maintenance respiration, growth efficiency, carbon allocation between root and shoot, and vegetative growth of the root and shoot systems. Seedling water potential is the primary variable controlling physiologic processes and growth rate in ROOTSIMU. Model components permit simulation of environmental stress factors on seedling carbon partitioning and vegetative growth (Hoogenboom and Huck, 1986). Further refinement of this model will depend on improved understanding of rate-limiting physiological processes.

INTEGRATING PHYSIOLOGICAL PROCESSES INTO WHOLE-TREE GROWTH MODELS

Forests are subject to increasing exposure to a wide array of environmental stresses, both biotic and abiotic. For example, atmospheric pollutants (e.g., O_3, NO_x, SO_x) interacting with other stress agents have many sites of action which alter the metabolic pathways of trees (McLaughlin et al., 1988). Thus, characterizing the principal physiological processes of trees requires the consideration of a myriad of biochemical pathways at many levels of structural organization. Mathematical or conceptual models provide a manageable framework for predicting and understanding key physiological processes in response to multiple stresses.

Tree structure and function can be considered hierarchical (e.g., cell, tissue, organ systems) and conceptually compartmentalized (e.g., discrete metabolic steps or pathways). Models provide a means to describe this system in quantitative terms if a process can be defined. Within any one level of hierarchy or structure, there may be several compartments representing physiological processes. Incorporating submodels, or modules, representing metabolic pathways into a hierarchy permits the representation of complex, interacting biochemical relationships. Several hierarchical and compartmentalized process-based models have been developed, including BRANCH (Ford and Kiester, this volume), CARBON (Bassow et al., this volume), ECOPHYS (Isebrands et al., this volume), and ROOTSIMU (Blake and Hoogenboom, 1988).

Models based on physiological processes provide a means to more fully analyze, simulate, and predict tree response to environmental stress. For example, mathematical models provide a framework for expressing current understanding, as well as knowledge gaps, regarding physiological processes in response to environmental stress. Preferably, models should contain sufficient detail to link observations at the whole-tree level with more detailed knowledge of underlying structural components and metabolic processes (Dale et al., 1985). Ultimately, the level of detail included is linked to the end use of the model (Wisiol and Hesketh, 1987).

In summary, modeling of physiological processes is a relevant, emerging tool in forest resource management. Observations which may merit consideration prior to the development of future models of physiological processes include the following:

1. Our current understanding of principal physiological processes is far from complete. Models can be used to summarize current understanding of concepts and to identify knowledge gaps with respect to physiological processes.
2. Physiological processes can be considered a series of interdependent biochemical or biophysical events. Mathematical models provide a hierarchical framework or conceptual compartments for organizing and analyzing metabolic pathways linked to tree growth.
3. Models usually incorporate many assumptions and are overly simplified. Future research should focus on reducing these limitations.
4. Most models of physiological processes are constructed from knowledge based on studies of juvenile trees. Validation of models with experimental data from mature trees is sorely needed.
5. Modeling objectives usually determine the model's mathematical complexity and ultimate utility for analysis, simulation, and prediction. Biological accuracy is sometimes sacrificed or lost in complex process models.
6. Biologically competent models of dynamic tree or forest processes will recognize that ontogeny, structure, and physiology are linked.

Teams of scientists representing a range of expertise (e.g., biological, computer, environmental sciences) will be required to address these topics and develop refined process-based models of tree growth which can answer complex environmental questions in a satisfactory manner.

LITERATURE CITED

Bassow, S. L., E. D. Ford, and A. R. Kiester. 1989. A critique of carbon-based tree growth models. In: Dixon, R. K., R. S. Meldahl, G. A. Ruark, and W. G. Warren, eds. *Process Modeling of Forest Growth Responses to Environmental Stress.* Timber Press, Portland, OR, U.S.A.

Bernier, G. 1988. The control of floral evocation and morphogensis. *Annual Review of Plant Physiology* 39:175–219.

Blake, J. I., and G. Hoogenboom. 1988. A dynamic simulation of loblolly pine (*Pinus taeda* L.) seedling establishment based upon carbon and water balances. *Canadian Journal of Forest Research* 18:833–850.

Boltz, B. A., B. C. Bongarten, and R. O. Teskey. 1986. Seasonal patterns of net photosynthesis of loblolly pine from diverse origins. *Canadian Journal of Forest Research* 16:1063–1068.

Bonner, J., and J. E. Varner, eds. 1976. *Plant Biochemistry.* Academic Press, New York, NY, U.S.A.

Borchert, R. 1975. Endogenous shoot growth rhythms and indeterminate shoot growth in oak. *Physiologia Plantaram* 35:152–157.

Boyer, J. N., D. B. Houston, and K. F. Jensen. 1986. Impacts of chronic SO_2, O_3, and SO_2 and O_3 exposures on photosynthesis of *Pinus strobus* clones. *European Journal of Forest Pathology* 16:293–299.

Cannell, M. G. R., and F. T. Last, eds. 1976. *Tree Physiology and Yield Improvement.* Academic Press, New York, NY, U.S.A. 567 pp.

Cecich, R. A. 1980. The apical meristem. In: Little, C. H. A., ed. *Control of Shoot Growth in Trees.* Canadian Forestry Service, Fredericton, New Brunswick, Canada. Pp. 1–11.

Clarkson, D. T., and J. B. Hanson. 1980. The mineral nutrition of higher plants. *Annual Review of Plant Physiology* 31:239–298.

Cosgrove, D. 1988. The biophysical control of plant cell growth. *Annual Review of Plant Physiology* 39:377–438.

Coutts, M. P. 1987. Developmental processes in tree root systems. *Canadian Journal of Forest Research* 17:761–767.

Cronan, C. S. 1985. Vegetation and soil chemistry of the ILWAS watersheds. *Water, Air and Soil Pollution* 26:355–371.

Dale, J. E. 1988. The control of leaf expansion. *Annual Review of Plant Physiology* 39:267–295.

Dale, V. H., T. W. Doyle, and H. H. Shugart. 1985. A comparison of tree growth models. *Ecological Modelling* 29:145–169.

Dickson, R. E. 1986. Carbon fixation and distribution in young *Populus* trees. In: Fujimori, T., and D. Whitehead, eds. *Proceedings IUFRO Conference on Crown and Canopy Structure in Relation to Productivity.* Forest and Forest Products Institute, Ibaraki, Japan. Pp. 409–426.

Epstein, E. 1972. *Mineral Nutrition of Plants: Principle and Perspectives.* Wiley, New York, NY, U.S.A.

Ford, E. D., and A. R. Kiester. 1989. Modeling the effects of pollutants on the processes of tree growth. In: Dixon, R. K., R. S. Meldahl, G. A. Ruark, and W. G. Warren, eds. *Process Modeling of Forest Growth Responses to Environmental Stress.* Timber Press, Portland, OR, U.S.A.

Giaquinta, R. T. 1983. Phloem loading of sucrose. *Annual Review of Plant Physiology* 34:347–387.

Gillespie, A. R., and W. R. Chaney. 1989. Process modeling of nitrogen effects on carbon assimilation and allocation: a review. *Tree Physiology* 5:99–112.

Goldstein, R. A., S. A. Gherini, C. W. Chen, L. Mok, and R. J. M. Hudson. 1984. Integrated acidification study (ILWAS): a mechanistic analysis. *Philosophical Transactions, Royal Society of London* 305:409–425.

Gordon, J. C., and P. R. Larson. 1968. Seasonal course of photosynthesis, respiration and distribution of ^{14}C in young *Pinus resinosa* trees as related to wood formation. *Plant Physiology* 43:1617–1621.

Govindjee. 1975. *Bioenergetics of Photosynthesis.* Academic Press, New York, NY, U.S.A.

Grigal, D. F. 1989. Mechanistic modeling of nutrient acquisition by trees. In: Dixon, R. K., R. S. Meldahl, G. A. Ruark, and W. G. Warren, eds. *Process Modeling of Forest Growth Responses to Environmental Stress.* Timber Press, Portland, OR, U.S.A.

Gutschick, V. P. 1987. Quantifying limits to photosynthesis. In: Wisiol, K., and J. D. Hesketh, eds. *Plant Growth Modeling for Resource Management.* CRC Press, Boca Raton, FL, U.S.A. Pp. 67–87.

Hanson, P. J., R. E. Dickson, J. G. Isebrands, T. R. Crow, and R. K. Dixon. 1986. A morphological index of *Quercus* seedling ontogeny for use in studies of physiology and growth. *Tree Physiology* 2:273–281.

Hanson, P. J., J. G. Isebrands, R. E. Dickson, and R. K. Dixon. 1988. Ontogenetic factors of CO_2 exchange of *Quercus rubra* L. leaves during three flushes of shoot growth. I. Median flush leaves. *Forest Science* 34:55–68.

Hinckley, T. M., J. P. Lassoie, and S. W. Running. 1978. Temporal and spatial variations in the water status of forest trees. *Forest Science Monograph* 24:1–72.

Ho, L. C. 1988. Metabolism and compartmentalization of imported sugars in sink organs in relation to sink strength. *Annual Review of Plant Physiology* 39:355–378.

Hoogenboom, G., and M. G. Huck. 1986. ROOTSIMU version 4.0: a dynamic simulation of root growth, water uptake, and biomass partitioning in a soil-plant-atmosphere continuum, update and documentation. Publication No. 109, Alabama Agricultural Experiment Station, Auburn University, Auburn, AL, U.S.A.

Isebrands, J. G., H. M. Rauscher, T. R. Crow, and D. I. Dickmann. 1989. Whole-tree growth process models based on structural-functional relationships. In: Dixon, R. K., R. S. Meldahl, G. A. Ruark, and W. G. Warren, eds. *Process Modeling of Forest Growth Responses to Environmental Stress.* Timber Press, Portland, OR, U.S.A.

Jarvis, P. G., and J. W. Leverenz. 1983. Productivity of temperate, deciduous and evergreen forests. In: Lange, O. L., P. S. Nobel, C. B. Osmond, and H. Zieger, eds. *Physiological Plant Ecology,* Vol. IV. *Encyclopedia of Plant Physiology,* New Series 12D. Springer-Verlag, New York, N.Y., U.S.A. Pp. 233–280.

Johnson, I. R. 1987. Models of respiration. In: Wisiol, K., and J. D. Hesketh, eds. *Plant Growth Modeling for Resource Management.* CRC Press, Boca Raton, FL, U.S.A. Pp. 89–108.

Jones, R. L. 1973. Gibberellins: their physiological role. *Annual Review of Plant Physiology* 24:571–598.

Kenig, A. 1987. The problem of quantifying limits to nitrogen use in relation to crop growth modeling. In: Wisiol, K., and J. D. Hesketh, eds. *Plant Growth Modeling for Resource Management.* CRC Press, Boca Raton, FL, U.S.A. Pp. 49–66.

Kossuth, S. V., and S. D. Ross, eds. 1987. *Hormonal Control of Tree Growth.* Martinus Nijhoff Publishers, Dordrecht, the Netherlands. P. 243.

Kramer, P. J., and T. T. Kozlowski. 1979. *Physiology of Woody Plants.* Academic Press, New York, NY, U.S.A. 811 pp.

Lambers, H. 1984. Respiration in intact plants and tissues: its regulation and dependence on environmental factors, metabolism and invaded organisms. In: Douce, R., and D. A. Day, eds. *Higher Plant Cell Respiration, Encyclopedia of Plant Physiology,* New Series, Vol. 19. Springer-Verlag, New York, NY, U.S.A.

Lavender, D. P. 1980. Effects of environment upon the shoot growth of woody plants. In: Little, C. H. A., ed. *Control of Shoot Growth in Trees.* Canadian Forestry Service, Fredericton, New Brunswick, Canada. Pp. 76–106.

Little, C. H. A., ed. 1980. *Control of Shoot Growth in Trees.* Canadian Forestry Service, Fredericton, New Brunswick, Canada.

Mäkelä, A. 1989. Modeling structural-functional relationships in whole-tree growth: resource allocation. In: Dixon, R. K., R. S. Meldahl, G. A. Ruark, and W. G. Warren, eds. *Process Modeling of Forest Growth Responses to Environmental Stress.* Timber Press, Portland, OR, U.S.A.

Mann, L. K., S. B. McLaughlin, and D. S. Shriner. 1980. Seasonal physiologic responses of white pine under chronic air pollution stress. *Environmental and Experimental Botany* 20:99–105.

McConathy, R. K., and S. B. McLaughlin. 1987. Measurement of gas exchange processes at the whole canopy level. In: Shriner, D. S., ed. *Acidic Deposition: Effects on Agricultural Crops.* Report EA-5149, Electric Power Research Institute, Palo Alto, CA, U.S.A.

McLaughlin, S. B., M. B. Adams, N. T. Edwards, P. J. Hanson, P. A. Layton, E. G. O'Neill, and W. K. Ray. 1988. Comparative sensitivity, mechanisms, and whole plant physiological implication of responses of loblolly pine genotypes to ozone and acid deposition. Publication 3105, Environmental Sciences Division, Oak Ridge National Laboratory, Oak Ridge, TN, U.S.A.

McMurtrie, R., and L. Wolf. 1983. Above- and below-ground growth of forest stands: a carbon budget model. *Annals of Botany* 52:437–448.

Newman, E. I. 1976. Water movement through root system. *Physiological Transactions, Royal Society of London* 273:463–478.

Oaks, A., and B. Hirel. 1985. Nitrogen metabolism in roots. *Annual Review of Plant Physiology* 36:345–365.

Ohno, T., E. I. Sucoff, M. S. Erich, P. R. Bloom, C. A. Buschena, and R. K. Dixon. 1988. Growth and nutrient content of red spruce seedlings in soil amended with aluminum. *Journal of Environ-*

mental Quality 17:666–672.
Penning de Vries, F. W. T. 1974. Substrate utilization and respiration in relation to growth and maintenance in higher plants. *Netherlands Journal of Agricultural Science* 22:40–44.
Reich, P. B., and R. G. Amundsen. 1985. Ambient levels of ozone reduce net photosynthesis in tree and crop species. *Science* 230:566–570.
Reynolds, J. F., and J. H. M. Thornley. 1982. A shoot:root partitioning model. *Annals of Botany* 49:585–597.
Schaedle, M. 1975. Tree photosynthesis. *Annual Review of Plant Physiology* 26:101–115.
Schulze, E.-D. 1988. Carbon dioxide and water vapor exchange in response to drought in the atmosphere and in the soil. *Annual Review of Plant Physiology* 39:247–274.
Spitters, C. J. T. 1986. Separating the diffuse and direct component of global radiation and its implication for modelling canopy photosynthesis. II. Canopy photosynthesis. *Agricultural and Forest Meteorology* 38:239–250.
Thornley, J. H. M. 1972. A balanced quantitative model for root:shoot ratios in vegetative plants. *Annals of Botany* 36:431–441.
Tingey, D. T., R. G. Wilhour, and C. Standley. 1976. The effect of chronic ozone exposure on the metabolite content of ponderosa pine. *Forest Science* 22:234–241.
Torrey, J. G., and D. T. Clarkson, eds. 1975. *The Development and Function of Roots.* Academic Press, New York, NY, U.S.A.
Trewavas, A., and E. Allan. 1987. An assessment of the contribution of growth substances to plant development. In: Wisiol, K., and J. D. Hesketh, eds. *Plant Growth Modeling for Resource Management.* CRC Press, Boca Raton, FL, U.S.A. Pp. 25–46.
Tyree, M. T. 1988. A dynamic model for water flow in a single tree: evidence that models must account for hydraulic architecture. *Tree Physiology* 4:195–217.
Valentine, H. T. 1989. A carbon-balance model of tree growth with a pipe-model framework. In: Dixon, R. K., R. S. Meldahl, G. A. Ruark, and W. G. Warren, eds. *Process Modeling of Forest Growth Responses to Environmental Stress.* Timber Press, Portland, OR, U.S.A.
Vincent, T. L., and H. R. Pulliam. 1980. Evolution of life history strategies for an asexual annual plant model. *Theoretical Population Biology* 17:215–231.
Wareing, P. F. 1980. Root hormones and shoot growth. In: Little, C. H. A., ed. *Control of Shoot Growth in Trees.* Canadian Forestry Service, Fredericton, New Brunswick, Canada. Pp. 173–183.
West, P. W. 1987. A model for biomass growth of individual trees in forest growth monoculture. *Annals of Botany* 60:571–577.
Wilson, B. F. 1970. *The Growing Tree.* University of Massachusetts Press, Amherst, MA, U.S.A.
Wisiol, K., and J. D. Hesketh, eds. 1987. *Plant Growth Modeling for Resource Management.* CRC Press, Boca Raton, FL, U.S.A. 348 pp.
Wodzicki, T. J. 1980. Control of cambial activity. In: Little, C. H. A., ed. *Control of Shoot Growth in Trees.* Canadian Forestry Service, Fredericton, New Brunswick, Canada. Pp. 173–183.
Worrall, J. G. 1980. The impact of environment on cambial growth. In: Little, C. H. A., ed. *Control of Shoot Growth in Trees.* Canadian Forestry Service, Fredericton, New Brunswick, Canada. Pp. 127–142.
Zeevart, J. A. D., and R. A. Creelman. 1988. Metabolism and physiology of abscisic acid. *Annual Review of Plant Physiology* 39:439–473.
Zhao, X., E. Sucoff, and E. J. Stadelman. 1987. Al+++ and Ca++ alteration of membrane permeability of *Quercus rubra* root cortex cells. *Plant Physiology* 83:159–162.

CHAPTER 2

A CARBON-BALANCE MODEL OF TREE GROWTH WITH A PIPE-MODEL FRAMEWORK

Harry T. Valentine

Abstract. The pipe-model theory is used to define a structural framework for a detailed derivation of a carbon-balance model of a tree. Growth of the model tree is measured in terms of average stem length (from leaves to feeder roots), basal area, total woody volume, and total carbon equivalents of dry matter. Within the pipe-model framework, the rate of consumption of carbon substrate for maintenance and renewal of live tissue increases with the average stem length of the model tree. Maximum average stem length occurs where the rate of production of carbon substrate equals the rate of maintenance respiration. An increase in the rate of feeder-root or leaf turnover or a decrease in the rate of substrate production may cause maximum average stem length to become shorter than the current actual length. This corresponds to situations in reality where stress-induced crown dieback or tree death should be expected.

INTRODUCTION

One of the important problems in connection with the development of carbon-balance models of plant growth has been the absence of a theoretical framework for allocating dry-matter production among the model plant's organs in order to conform the model plant to the shape and structural balance found in nature (Brugge and Thornley, 1985). A concise and elegant theory of plant form and structural balance was proposed by Shinozaki et al. (1964a, b) in terms of a pipe model. Recently, it was realized that this pipe-model theory furnishes both a structural framework and the dry-matter allocation rules needed for derivations of carbon-balance models of tree growth (Valentine, 1985; Mäkelä, 1986). Valentine (1988) also used the pipe-model theory to derive a carbon-balance model of the growth of an even-aged, self-thinning, mono-species stand. The pipe-model framework provided the derivation of growth models of average stem length, total basal area, and total volume of the stand.

West (1987) recently developed a tree-level model from the stand-level model of McMurtrie and Wolf (1983) and concluded the measure of tree size used in that model—dry matter—should be supplemented by other measures of tree size like height and diameter. I previously proposed growth models of height and basal area for individual trees (Valentine, 1985). The growth rate of basal area was specified as a function of the growth rate of average stem length, which was unknown. In the present paper, I simultaneously

Dr. Valentine is a Research Forester at the USDA Forest Service, Center for Biological Control of Northeastern Forest Insects and Diseases, 51 Mill Pond Road, Hamden, CT, 06514 U.S.A.

derive growth models of average stem length and basal area for individual trees by closely following the approach of my stand-level derivation (Valentine, 1988). A model of volume growth is also provided.

STRUCTURAL FRAMEWORK

To facilitate the derivation, I represent the quintessential elements of the form and structural balance of a tree by a pipe model (after Shinozaki et al., 1964a, b; Valentine, 1985, 1988). For this model, it is assumed that a tree is comprised of leaves, feeder roots, and two kinds of pipes: active and disused (Figure 1). The active and disused pipes, in aggregate, represent all the woody components of the tree: branches, bole, and transport roots. Active pipes extend from leaves to feeder roots, furnishing support and vascular connections. Disused pipes are vestiges of old active pipes that no longer connect leaves to feeder roots. An active pipe deactivates and becomes disused simultaneously with the death of its leaves and feeder roots. As a pipe converts from active to disused, its distal portion may be lost from the model tree, representing the shedding of a portion of dead branch. Analogous shedding is assumed to occur in the root sytem. The remaining portion of a disused pipe, which remains as part of the model tree and contributes to bole volume, is assumed to provide support.

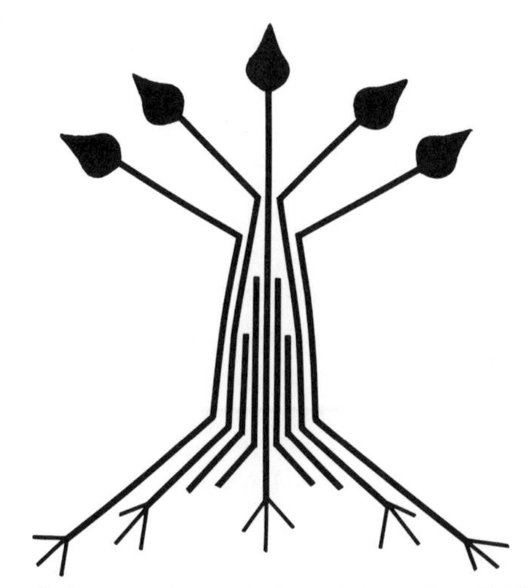

Figure 1. Active and disused pipes are theoretical modular units which represent the woody components of the model tree: branches, bole, and transport roots. The active pipes, in aggregate, represent conducting xylem or sapwood, connecting leaves to feeder roots. Disused pipes represent non-conducting xylem or heartwood and have no connections to leaves or feeder roots. All pipes are assumed to have constant cross-sectional area along their lengths.

In accordance with the pipe-model theory, I assume that the model tree maintains a constant amount of leaf dry matter per unit of cross-sectional area of active pipe. To maintain balance between leaves and feeder roots, I also assume a constant amount of feeder-root dry matter per unit of cross-sectional area of active pipe (after Valentine, 1985; Mäkelä, 1986). In some tree species, particularly conifers, the active pipes, in aggregate, may be assumed to be commensurate with the conducting xylem or sapwood (see, e.g., Waring et. al, 1982).

In order to derive equations describing the growth of the model tree, I assume that new woody volume is produced both by the extension of existing active pipes and the production of new active pipes. The possibility that all woody production is achieved solely through the production of new pipes is treated as a special case.

Structural Variables

Structural variables of the model are defined as follows:

A = total cross-sectional area of active pipes (active area)

X = total cross-sectional area of disused pipes at 1.3 m height (disused area)

B = total cross-sectional area of the bole at 1.3 m height (basal area)

L = average length of an active pipe

V = total woody (active plus disused) volume of the model tree

Each whole active pipe is assumed to have one unit of active area. The active volume of the model tree is AL.

As is evident, variables are denoted by uppercase Roman letters. Below, parameters are denoted by lowercase Roman letters. Greek letters denote quantities which are assumed to be constant but in reality may vary over time.

Growth Rates

Of prime interest are the growth rates of basal area (dB/dt), average active pipe length (dL/dt), and woody volume (dV/dt) of the model tree, where t is time (years). Equations for these rates are derived below so as to maintain the carbon balance of the model tree. However, the carbon-balance model furnishes the rates of production of new active pipes, and leaf and feeder-root dry matter, not growth rates. Therefore, it is necessary to establish the connection between the growth rates of B, L, and V and the production rate of new active pipes.

Area. The growth rate of the active area of the model tree is

$$dA/dt = dB/dt - dX/dt \tag{1}$$

Biologically, the growth rate of active area equals the rate of production of new active area minus the rate of conversion of old active area into disused area. The conversion of active area into disused area, which is permanent, has no effect on the basal area of the model tree. Basal area can only increase, and it does so at a rate equal to the rate of production of new active area. Therefore, dB/dt denotes both the growth rate of basal area and the rate of production of new active area, and dX/dt denotes both the growth rate of disused area and the rate of deactivation of old active area.

Length. New active volume is added to the model tree by the extension of existing pipes and by the production of new pipes. Active volume is lost through the deactivation of old active pipes. The production and deactivation of pipes complicates the formulation of the rate of change of average active pipe length because, on the average, deactivating pipes are apt to be shorter than L, and new pipes may be longer than L. To allow for this (or for the converse in a tree with crown dieback), I split dL/dt into two fractions (after Valentine, 1988), as follows:

$$dL/dt = dL_M/dt + dL_N/dt \tag{2}$$

where dL_M/dt is the rate of change of the average length of active pipes due to extension growth, and dL_N/dt adjusts L for production or deactivation of longer or shorter than average pipes.

An equation for dL_M/dt is obtained, along with an equation for dB/dt, in the carbon-balance formulation. An equation for dL_N/dt arises from definitions of d(AL)/dt, the growth rate of active volume. New active volume is added to the model tree by extension of existing active pipes at rate AdL_M/dt, and by the production of new active pipes at rate $\omega LdB/dt$, where ωL is the average length of a new pipe. Active volume is lost through the deactivation of pipes at rate $\theta LdX/dt$, where θL is the average length of a deactivating pipe. Therefore, the growth rate of the active volume of the model tree is

$$d(AL)/dt = AdL_M/dt+\omega LdB/dt-\theta LdX/dt \tag{3}$$

The time derivative of AL and Equations 1 and 2 furnish

$$d(AL)/dt = A(dL_M/dt+dL_N/dt)+L(dB/dt-dX/dt) \tag{4}$$

Subtracting Equation 4 from Equation 3 and solving for dL_N/dt obtains

$$dL_N/dt = L[(\omega-1)dB/dt+(1-\theta)dX/dt]/A \tag{5}$$

Volume. The growth rate of the woody volume of the model tree equals the rate of production of new active volume minus the rate of loss of portions of disused pipes, which represent dead branches and roots. Let $\nu LdX/dt$ denote the rate of loss of disused woody volume ($0 < \nu < \theta$), where νL is the average length of the shed fraction of a deactivating pipe. Therefore,

$$dV/dt = AdL_M/dt+\omega LdB/dt-\nu LdX/dt \tag{6}$$

CARBON BALANCE

Model Components

All the dry matter of the model tree is measured in carbon equivalents. The growth rate of the dry matter equals the rate of production of new dry matter minus the rate of loss of old dry matter from the shedding of portions of deactivating pipes and their leaves and feeder roots. The rate of production of dry matter (plus associated constructive respiration) by the model tree as a whole equals the rate of production of dry-matter substrate minus the rate of maintenance respiration. Symbolically,

$$\sum_{i=1}^{3} (G_i+C_i) = P - \sum_{i=1}^{3} R_i \tag{7}$$

where G_i = the rate of production of new dry matter; C_i = the rate of constructive respiration, which is proportional to G_i; and R_i = the rate of use of dry-matter substrate for maintenance and renewal of i (where i = 1, leaves; 2, feeder roots; and 3, active pipes, or wood). P = the rate of production of dry-matter substrate.

Production of new dry matter. Let z and f, respectively, denote the number of units of leaf and feeder-root dry matter per unit of cross-sectional area of active pipe maintained by the model tree, and let p denote the number of units of woody dry matter per unit volume of pipe. Therefore, the production rates of leaf, feeder-root, and active dry matter are

$$G_1 = zdB/dt; \quad G_2 = fdB/dt$$
$$G_3 = p(AdL_M/dt+\omega LdB/dt) \tag{8}$$

Constructive respiration. The number of units of dry-matter substrate consumed in the production of a unit of leaf, feeder-root, and active dry matter, respectively, are denoted r_z, r_f·, and r_p. Therefore, the rates of constructive respiration are

$$C_1 = r_z z dB/dt; \quad C_2 = r_f f dB/dt$$
$$C_3 = r_p p(AdL_M/dt + \omega L dB/dt) \tag{9}$$

Maintenance and renewal. Dry-matter substrate is consumed in respiration for the maintenance of leaf, feeder root, and active dry matter, and for renewing leaves or feeder roots of existing active pipes. The number of units of dry-matter substrate consumed per year for maintenance of each unit of leaf, feeder-root, and active dry matter is denoted, respectively, by b_z, b_f, and b_p.

If the number of annual rings of sapwood exceeds the number of age classes of leaves on a tree, then a dynamic pipe model of the tree should provide for the loss and replacement of the leaves attached to the active pipes. I assume that leaf dry matter is lost at rate zA/T_z, where T_z is the expected ultimate age of a leaf. The process of replacement (renewal) of this dry matter uses $1+r_z$ units of substrate for each unit of dry matter that is lost; one unit to replace the lost unit of dry matter, and r_z units which are consumed in constructive respiration during the renewal process. Therefore, the rate of use of substrate for renewal of leaf dry matter is $(1+r_z)zA/T_z$. Feeder roots may be renewed more frequently than leaves. I assume that substrate is used for renewal of feeder-root dry matter at rate $(1+r_f)fA/T_f$, where T_f is the expected ultimate age of a feeder root. Combining the rates of use of substrate for maintenance and renewal obtains

$$R_1 = zA[b_z+(1+r_z)/T_z]$$
$$R_2 = fA[b_f+(1+r_f)/T_f] \tag{10}$$
$$R_3 = b_p pAL$$

Production of dry-matter substrate. Production of dry-matter substrate is a function of the foliar dry matter, which is proportional to the active area of the model tree. Let a^* denote the maximum rate of substrate production per unit of active area, and let $I(t)$ denote a variable that scales the rate of substrate production, varying from 0 to 1, depending on factors such as insolation, water potential, nutrient availability, and the age structure of the foliage. Therefore,

$$P = a^*AI \tag{11}$$

Partitioning the Production

Substituting Equations 8, 9, and 10 with constant T_z and T_f, and Equation 11 into Equation 7 obtains

$$(z^*+\omega L)dB/dt + AdL_M/dt = A(aI - b^* - bL) \tag{12}$$

where

$$a = a^*/[p(1+r_p]$$
$$b^* = \{z[b_z+(1+r_z)/T_z]+f[b_f+(1+r_f)/T_f]\}/[p(1+r_p)]$$
$$b = pb_p/[p(1+r_p)]$$
$$z^* = [z(1+r_z)+f(1+r_f)]/[p(1+r_p)]$$

Using the partitioning method of Valentine (1988), the rate of production of new dry matter attributable to the extension of active pipes in Equation 12 is separated from that attributable to expansion caused by the production of new active pipes with partitioning coefficients (λ) and ($1-\lambda$):

$$AdL_M/dt = (1-\lambda)A(aI-b^*-bL) \tag{13}$$
$$(z^*+\omega L)dB/dt = \lambda A(aI-b^*-bL) \tag{14}$$

where $0 < \lambda < 1$. Solving for dB/dt and dL_M/dt obtains

$$dL_M/dt = (1-\lambda)(aI-b^*-bL) \quad (15)$$
$$dB/dt = \lambda A(aI-b^*-bL)/(z^*+\omega L) \quad (16)$$

Special case: $\lambda = 1$. Suppose that the number of annual rings of sapwood equals, rather than exceeds, the number of age classes of leaves on a tree. For a pipe model of such a tree, there is no need to allow for the extension growth of existing active pipes or the renewal of foliage attached to these pipes. Each newly produced active pipe is attached to z non-renewable units of leaf dry matter and f renewable units of feeder-root dry matter. The pipe deactivates when the leaf material dies. Production of all the new volume by the model tree is achieved through the production of new active pipes.

The absence of extension growth and leaf renewal simplifies many of the equations of the model. In Equation 10, the expression accounting for the cost of renewal of leaf dry matter, $zA(1+r_z)/T_z$, is dropped so that R_1 becomes

$$R_1 = zb_zA \quad (17)$$

and b*, which is defined under Equation 12, becomes

$$b^* = \{zb_z+f[b_f+(1+r_f)/T_f]\}/[p(1+r_p)] \quad (18)$$

The rate that average active pipe length changes due to extension growth, dL_M/dt, equals 0 in all equations in which it appears.

THE GROWTH MODEL

The growth rate equation for basal area of the model tree is Equation 16. Inserting Equations 5, 15, and 16 into Equation 2 gives an equation of the rate of change of average active pipe length of the model tree:

$$dL/dt = \{[z^*(1-\lambda)+(\omega-\lambda)L]dB/dt+\lambda(1-\theta)LdX/dt\}/\lambda A \quad (19)$$

Similarly, insertion of Equations 15 and 16 into Equation 6 to eliminate dL_M/dt gives an equation for the growth rate of the woody volume of the model tree:

$$dV/dt = \{[z^*(1-\lambda)+\omega L]/\lambda\}dB/dt-\nu LdX/dt \quad (20)$$

An equation of the growth rate of the total dry matter of the model tree (dW/dt) follows from the definitions of p, z, and f:

$$dW/dt = pdV/dt+(z+f)dA/dt$$
$$= \{z+f+p[z^*(1-\lambda)+\omega L]/\lambda\}dB/dt-(z+f+p\nu L)dX/dt \quad (21)$$

If $\lambda = 1$, then Equations 19, 20, and 21, respectively, reduce to

$$dL/dt = L[(\omega-1)dB/dt+(1-\theta)dX/dt]/A \quad (19a)$$
$$dV/dt = L(\omega dB/dt-\nu dX/dt) \quad (20a)$$
$$dW/dt = (z+f+pL)(\omega dB/dt-\nu dX/dt) \quad (21a)$$

DISCUSSION

Solution of the present model will require equations for both dX/dt, the rate of production of disused area, and I(t), the variable that scales the rate of production of carbon substrate. These equations remain to be formulated, and the former will probably depend on the latter. Mäkelä and Hari (1986) and West (1987) have suggested formulations of I(t) for tree-level carbon-balance models. Formulation of an equation for dX/dt will require some consideration of the interactions between a tree and its neighbors. Shinozaki et al. (1964b) indicated that the active area of a tree is approximated by the cross-sectional area of the stem at the base of the crown, inferring that disused pipes extend upward to that point. More accurately, disused pipes may extend upward past the base of the crown and into the basal portions of branches. In either event, however, it seems reasonable to assume that new disused area accrues as the crown recedes. The rate at which this happens depends on the degree to which the lowest branches are crowded or shaded by neighboring trees.

Maximum Size

The absence of equations for I(t) and dX/dt notwithstanding, it is still possible to analyze certain aspects of the model. Equation 7 indicates that when the rate of use of carbon substrate for maintenance and renewal of tissue by the model tree equals the rate of substrate production, then the production of new dry matter and associated constructive respiration are nil. This cessation of production and consequent attainment of maximum size corresponds to instances where Equations 12 through 16 equal 0, and

$$aI = b^* + bL \qquad (22)$$

Therefore, within the present pipe-model framework, the maximum size of the model tree is always attained when average active pipe length is

$$L = L_{max} = (aI - b^*)/b \qquad (23)$$

regardless of the values of A, B, and X. This essential result also was obtained by Valentine (1985) and Mäkelä (1986) with models derived from pipe-model theory under somewhat different assumptions.

In order to gauge the effect of variation in the rates of leaf turnover (T_z^{-1}) and feeder-root turnover T_f^{-1}) on maximum average pipe length, b* can be replaced in Equations 12 through 16, 22, and 23 by

$$b_1 + b_2/T_z + b_3/T_f \qquad (24)$$

where

$$b_1 = (zb_z + fb_f)/[p(1+r_p)]$$
$$b_2 = z(1+r_z)/[p(1+r_p)]$$
$$b_3 = f(1+r_f)/[p(1+r_p)]$$

With variable T_z or T_f or both, growth of the model tree ceases when

$$aI = b_1 + b_2/T_z + b_3/T_f + bL \qquad (25)$$

and maximum average pipe length is

$$L_{max} = (aI - b_1 - b_2/T_z - b_3/T_f)/b \qquad (26)$$

In reality, the values of I(t) and T_f may decrease in response to an acidifying environment or other factors. I(t) and T_z may be reduced by a defoliation caused by an insect or fungus, and in large angiosperms, this is often followed by crown dieback. Presumably, sudden reductions in the values of I(t), T_z, or T_f, singly or in combination, cause the level of substrate production to fall below the level required for maintenance of live tissue so that average active pipe length (average stem length from leaves to feeder roots) exceeds its maximum, i.e., $L > L_{max}$. When this happens, a shortening of average stem length (i.e., a dieback of the crown) to the degree that $L < L_{max}$ would not be unexpected. Reductions in $I(t)$, T_z, or T_f in any tree should reduce the rate of production of volume, and unless reversed, they should also cause a reduction of maximum average stem length.

LITERATURE CITED

Brugge, R., and J. H. M. Thornley. 1985. A growth model of root mass and vertical distribution, dependent on carbon substrate from photosynthesis and with non-limiting soil conditions. *Annals of Botany* 55:563–577.

Mäkelä, A. 1986. Implications of the pipe-model theory on dry matter partitioning and height growth in trees. *Journal of Theoretical Biology* 123:103–120.

Mäkelä, A., and P. Hari. 1986. Stand growth model based on carbon uptake and allocation in individual trees. *Ecological Modelling* 33:205–229.

McMurtrie, R., and L. Wolf. 1983. Above- and below-ground growth of forest stands: a carbon budget model. *Annals of Botany* 52:437–448.

Shinozaki, K., K. Yoda, K. Hozumi, and T. Kira. 1964a. A quantitative analysis of plant form: the pipe-model theory. I. Basic Analyses. *Japanese Journal of Ecology* 14(3):97–105.

Shinozaki, K., K. Yoda, K. Hozumi, and T. Kira. 1964b. A quantitative analysis of plant form: the pipe-model theory. II. Further evidence of the theory and its application in forest ecology. *Japanese Journal of Ecology* 14(4):133–139.

Valentine, H. T. 1985. Tree-growth models: derivations employing the pipe-model theory. *Journal of Theoretical Biology* 117:579–585.

Valentine, H. T. 1988. A carbon-balance model of stand growth: a derivation employing pipe-model theory and the self-thinning rule. *Annals of Botany* 62:389–396.

Waring, R. H., P. E. Schroeder, and R. Oren. 1982. Application of the pipe model theory to predict canopy leaf area. *Canadian Journal of Forest Research* 12:556–560.

West, P. W. 1987. A model for biomass growth of individual trees in forest monoculture. *Annals of Botany* 60:571–577.

CHAPTER 3

PHOTOSYNTHESIS, TRANSPIRATION, AND NUTRIENT UPTAKE IN RELATION TO TREE STRUCTURE

Pertti Hari, Eero Nikinmaa, and Maria Holmberg

Abstract. Photosynthesis and nutrient uptake are primary physiological processes in trees. Assuming that a tree is a well-balanced functional unit, it follows that there are interrelated patterns of photosynthesis, water uptake, water transport, and nutrient uptake. A highly simplified model of annual photosynthesis, nutrient uptake, and water transport in conifers is presented. An optimization problem is formulated as the maximization of photosynthesis. The problem is solved, and the result is shown as functions of the amount of transpired water and nutrient-uptake ability per unit of root mass.

INTRODUCTION

Environmental stress impacts tree photosynthesis, transpiration, and nutrient and water uptake (Luxmoore et al., 1986; Waring, 1987). Relatively few process models have been developed to characterize the primary processes of carbon fixation and allocation, as well as nutrient and water uptake and transport. Although physiological processes and tree structure are interrelated, their linkage has been difficult to model. The objective of this chapter is to analyze the implications of tree structure and photosynthate allocation using a simple model linking photosynthesis, transpiration, and nutrient uptake.

THE MODEL

Tree structure and function are significantly influenced by environmental conditions. For example, solar irradiance influences photosynthetic rate (Korpilahti, 1988), and initiation of metabolic growth processes depends on ambient temperature (Pietarinen et al., 1982). Regulation of physiological processes is also linked to tree structure (Luxmoore et al., 1986). We chose a one-year time scale to test our model linking tree structure and function. Separate simple models were constructed to describe tree photosynthesis, water uptake, water transport, and nutrient uptake. These models were linked by processes such as photosynthate allocation within the tree. The allocation of photosynthate was assumed to be a function of photosynthesis and the relative sink strength of various plant components.

Dr. Hari is at the Department of Mathematics, The Finnish Forest Research Institute, Unioninkatu 40 A, SF-00170 Helsinki, Finland. Mr. Nikinmaa and Ms. Holmberg are at the Department of Silviculture, Unioninkatu 40 B, SF-00170 Helsinki, Finland.

Photosynthesis

Photosynthesis and carbon allocation are primary metabolic processes in trees. We assume that photosynthetic processes and the amount of the annual photosynthate allocation are governed as follows:

Assumption 1: The amount of annual photosynthesis per unit mass of needles is dependent on the nutrient concentration of the needle and the amount of water transpired per unit mass of needle under a given set of environmental conditions.

The amount of photosynthetic production per unit mass of needles is P, the amount of transpired water per unit mass of needles is W, and the nutrient concentration of needles is N. N can be any nutrient; but in the present context, a general nutrient concentration is employed. The annual photosynthetic production is assumed to be dependent on the nutrient concentration of needles and on the amount of transpired water, as follows:

$$P = P_0 \times (1 - \frac{(N - N_0)^2}{N_0^2}) \times (1 - \frac{(W - W_0)^2}{W_0^2}) \qquad (1)$$

where P_0, N_0, and W_0 are model parameters: P_0 describes maximum photosynthesis under optimal environmental conditions, N_0 is the optimal concentration of nutrients in the leaves, and W_0 is optimum transpiration.

The model was constructed to correspond to observations reported in the literature. The approximation of transpired water is based on reports by Bengtson (1980) and Korpilahti (1988). The relationship of photosynthesis to nutrient concentration is based on measurements of photosynthetic rate (Sievänen et al., 1988). This characteristic is an essential feature of Equation 1. Behavior such as this can be introduced using several functions, but the qualitative features of the model analysis do not depend on the function applied in the approximation.

The values of the model parameters can be estimated from ecophysiological measurements of trees. The amount of annual photosynthesis under optimal conditions (P_0) can be obtained by integrating photosynthetic rate over a growing season (Korpilahti, 1988). The value of N_0 can be approximated using the mean nutrient concentration of leaves of individual trees grown under optimal conditions. The value of W_0 is determined by the partial pressure differences of ambient CO_2 and water vapor concentrations and leaf intercellular space (Hari et al., 1985). The present approximation of the annual value of W_0 was derived from field measurements (Korpilahti, 1988).

Water Transport

The pipe-model theory has received considerable attention in recent years (Shinozaki et al., 1964). This model links structural and functional processes in trees (Hari et al., 1985; Valentine, 1985; Mäkelä, 1986). The model assumes that the structure of the trees is linked to the requirements of water transport. However, the pipe-model theory is overly simplified. For example, the ratio of leaf mass to the water-transport capacity of the stem is also dependent on annual climate patterns and the microenvironment of the plant (Long and Smith, 1988). The pipe-model theory is introduced into the analysis by assuming that woody structures fulfill the functional requirements of water transport.

Assumption 2: The amount of photosynthate allocated to the water transport system is proportional to the amount of transpired water.

Some support for Assumption 2 comes from the observation that tree species growing in a dry environment require a larger water-conducting area per amount of leaves/needles than those growing in a relatively wet environment (Zimmermann, 1983; Kaufmann and

Troendle, 1981).

A tree water-transport system consists primarily of coarse roots, stem, and branches. The formation and maintenance of water-transport systems require photosynthate allocation to tissues. Let M denote the amount of photosynthate available for growth, a_1 the proportion of the photosynthate allocated to leaves, and A_T the proportion allocated to the water transport system. Assumption 2 can be expressed as follows:

$$A_T \times M = k_1 \times a_1 \times M \times W \tag{2}$$

where k_1 is a parameter describing the cost in photosynthate caused by transportation of a unit mass of water. The k_1 value is highly dependent on the size of the plant. This assumption follows from the pipe-model theory and can be formulated in different ways (Hari et al., 1985; Mäkelä, 1986; Valentine, 1985). All the formulations include the concept of functional relationships between the amount of needles and their transpiration rate, and the capacity of the wood to transport water.

Fine roots are important in water uptake, and the amount of transpired water is clearly connected to the amount of fine roots (Grigal, this volume; Kaufmann, this volume).

Assumption 3: The amount of water uptake by fine roots is proportional to the product of the mean water concentration in the soil and the amount of fine roots for water uptake.

Let A_W denote the amount of photosynthate allocated to fine roots for water uptake and v the mean water concentration in the soil. If the amount of transpired water equals the amount of water uptake by fine roots, then

$$a_1 \times M \times W = k_3 \times A_W \times M \times v \tag{3}$$

where k_3 is a parameter describing the specific water-uptake ability of fine roots. This conceptual model is a modification of that presented by Protopapas and Bras (1987).

The proportion of photosynthate allocated to the tree water uptake and transport system is obtained by adding the uptake and transport costs, i.e.,

$$a_2 = A_T + A_W \tag{4}$$

Nutrient Uptake

The mass flow of water from soil to roots provides a considerable supply of nutrients (Donahue et al., 1977).

Assumption 4: The amount of nutrients carried by the soil water to the roots can be considered a product of the amount of transpired water and nutrient concentration in soil solution.

Let N_w denote the amount of nutrients carried by the mass flow of water and C_W the concentration of nutrients in the soil solution. Then

$$N_w = a_1 \times M \times W \times C_W \tag{5}$$

Nutrients are not solely supplied by the flow of water into tree roots (Nye and Tinker, 1977; Donahue et al., 1977). Ions also move by diffusion or electrochemical gradients. In addition, root nutrient uptake changes the rhizosphere chemistry, which may enhance the uptake of additional nutrients from the surface of soil particles and organic material (Nye and Tinker, 1977).

Assumption 5: The amount of any additional nutrients taken up by roots is proportional to the product of the concentration of exchangeable or otherwise easily available ions and the amount of photosynthate allocated to root nutrient uptake.

Let N_a denote the amount of additional nutrients taken up and C_a the nutrient concentration of easily exchangeable ions in soil. Assumption 5 can be expressed as follows:

$$N_a = k_2 \times a_3 \times M \times C_a \tag{6}$$

where k_2 is a parameter describing the nutrient-uptake efficiency of fine roots. The allocation coefficient a_3 determines the proportion of photosynthate partitioned to fine roots which is required to guarantee a sufficient nutrient supply (in addition to the fine roots which are necessary for water uptake). A similar model for nutrient uptake is presented by Sievänen et al. (1988). The nutrients carried to the plant by mass flow and other processes are separated in the model to achieve clarity. In reality, it is difficult to make this distinction (Nye and Tinker, 1977).

The nutrient concentration of needles is assumed to be the amount of nutrients available divided by the amount of needles, assuming that the consumption of nutrients by other parts of the tree is non-significant. The uptake of nutrients is the sum of N_w and N_a. The nutrient concentration of needles is dependent on allocation as described by Equations 5 and 6.

$$N = \frac{N_a + N_w}{a_1 \times M} = \frac{k_2 \times a_3 \times C_a}{a_1} + W \times C_w \tag{7}$$

The Concept of Optimality

Use of the top-down modeling approach in biological systems assumes the presence of regulatory principles. These principles can be based on empirical observations. Another possibility is to use some general functional principles, which could be thought to have evolved during the evolution of species. For example, it could be assumed that certain, principal functions would be optimal or at least close to the optimum from the point of view of the whole plant. To evaluate the latter approach, we formulated the following optimality hypothesis.

Hypothesis: The carbohydrates are allocated in such a way that the amount of annual photosynthesis is maximized within the constraint of the amount of carbohydrates available for growth.

The amount of photosynthesis is the product of the amount of leaves and the amount of photosynthesis per unit of leaves. The amount of leaves is proportional to the allocation coefficient a_1. The hypothesis can be expressed as maximization of $a_1 \times M \times P$, i.e., the solution of

$$\frac{d(a_1 \times M \times P)}{da_1} = 0 \quad \text{and} \quad \frac{d^2P}{da_{12}} < 0 \tag{8}$$

The photosynthate available for growth (M) is allocated totally to growth of different tree parts, thus:

$$a_1 + a_2 + a_3 = 1 \tag{9}$$

Equations 2, 3 and 4 link the allocation coefficients of water uptake and transport, system and of leaves:

$$a_2 = a_1 \times W \times (k_1 + \frac{1}{k_3 \times v}) \tag{10}$$

Combining Equations 9 and 10 gives the allocation of photosynthate to roots, a_3, as a function of the allocation to leaves, a_1.

$$a_3 = 1 - a_1 \times (W \times (k_1 + \frac{1}{k_3 \times v}) + 1) \tag{11}$$

Using Equations 1, 7, and 11, the amount of annual photosynthesis per unit needle mass can be written as a function of a_1. Differentiating the total amount of photosynthesis, $a_1 \times M \times P$, with respect to the allocation of biomass to leaves, according to Equation 8, yields the solution

$$a_1^* = [(W \times (\frac{1}{k_3 \times v} + k_1 - \frac{C_W}{C_a \times k_2}) + 1) \times (\frac{W}{k_3 \times v} + \frac{2 \times N_0 - C_W \times W}{C_a \times k_2} + W \times k_1 + 1)]^{-0.5} \tag{12}$$

RESULTS AND DISCUSSION

Model parameter values are presented in Table 1. The soil parameter values correspond to Finnish podsol soils. The values for W were derived from field photosynthesis measurements of conifers at the University of Helsinki Field Station in central Finland (Korpilahti, 1988). The parameter value describing needle nutrient concentration corresponds to nutrient concentrations of trees on the most fertile sites in Finland.

Table 1. Model parameter values.

Parameter	Value	Dimension
N_0	0.03	kg nutrients/kg dw
W_0	250	kg H_2O/kg dw
P_0	3.1	kg C/kg dw
C_W	0.00001	kg nutrients/l H_2O
C_a	0.001	kg nutrients/l H_2O
v	0.3	kg H_2O/l soil
M	1.0	kg
k_1	0.004	kg dw/kg H_2O
k_2 Ca	0.001–0.1	kg nutrients/kg dw
k_3 v	1000	kg H_2O/kg dw
W	50–500	kg H_2O/kg dw

For parameters k_1, k_2, and k_3, no direct values from field measurements could be found in the literature. However, the values were derived indirectly from field experiments. The parameter value for k_1 depends strongly on the size of the tree. The trees were assumed to be 20 m tall. A value of 51.6 g needles/cm^2 for the foliage-biomass and sapwood-conducting-area relationship was assumed. The average density of wood, a value of 400 kg/m^3, was used (Kaipiainen and Hari, 1985). It was further assumed that sapwood remains functional for 20 years and needles for 3 years. The parameter values k_2 and k_3 were based on nutrient-uptake experiments done in solution culture (Cheeseman, 1985; Jensen and Pettersson, 1978; Nikinmaa, pers. comm.) and on measured transpiration rates (Nikinmaa, pers. comm.). In addition, it was assumed that the average functional period of feeder roots in podsol soils is approximately one month. It was assumed that roots can effectively exploit 1–10% of the soil volume of the top 30-cm layer. In the future, it will be necessary to thoroughly evaluate the validity of these parameter values.

Using the parameter values presented in Table 1, the value for photosynthate allocation to tree needles is 0.352. Since the second derivative of the amount of photosynthesis is −133, the value for a_1 represents the optimal solution of allocation under the assumptions stated.

By varying different parameter combinations, the model was used to examine the behavior of trees. The amount of water transpired by needles was selected since it is an essential process in a tree (e.g., photosynthesis, nutrient uptake, and water-transport structure). The parameter describing the nutrient-uptake ability of roots per unit mass of roots was another parameter tested. Proper values for the latter parameter have not been identified. Environmental changes due to acid deposition may strongly influence both values.

The allocation to leaves, to water uptake and transport, and to nutrient uptake according to Equations 10–12 is described in Figures 1, 2, and 3. When transpiration is increased, allocation to the water-transport structure is increased. The linear dependence of a_2 on the amount of transpiration (Equation 2) is the main cause for the changes in allocation in the model. Since allocation to the foliage decreases at the same time, the foliage/sapwood relation changes proportionally. However, there is evidence that the water-transport capacity of wood changes when the sapwood cross-section area increases (Tyree et al., 1985). Thus, the assumptions of linearity cannot always be assumed to be correct.

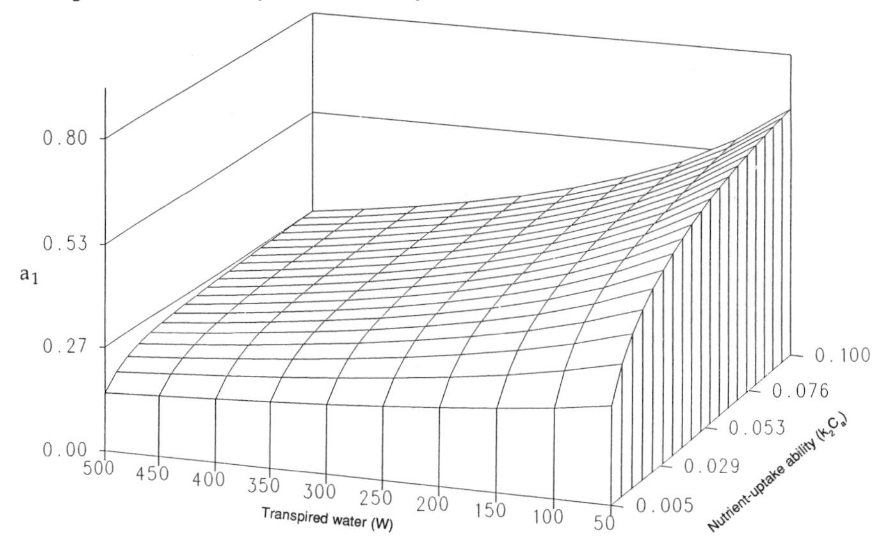

Figure 1. Allocation to leaves, a_1, as a function of transpired water, W, and nutrient-uptake ability of roots per unit root mass, k_2C_a.

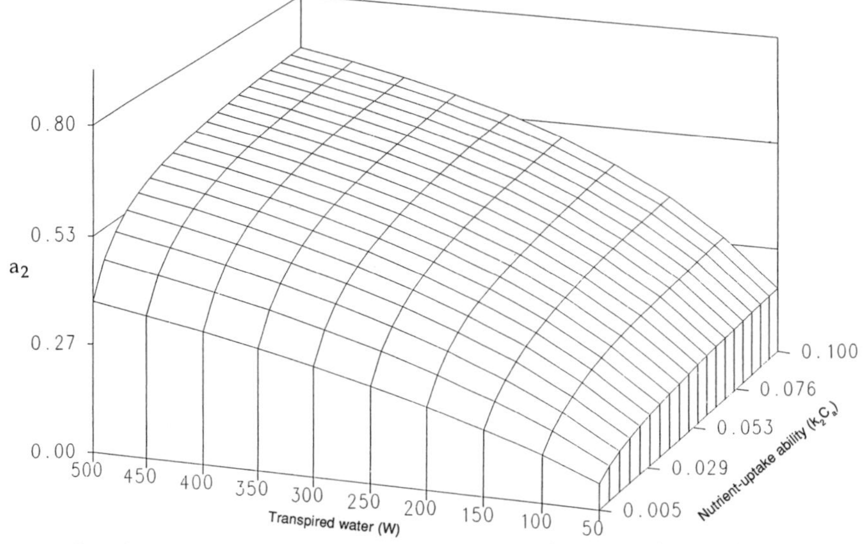

Figure 2. Allocation to water-transport structure, a_2, as a function of transpired water, W, and nutrient-uptake ability of roots per unit root mass, K_2C_a.

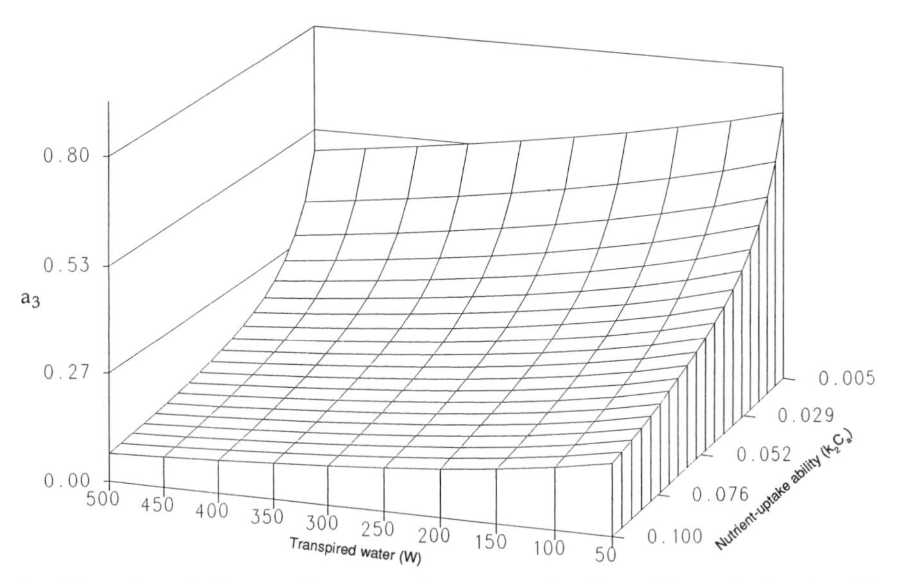

Figure 3. Allocation of photosynthate to roots, a_3, as a function of transpired water, W, and nutrient-uptake ability of roots per unit root mass, k_2C_a.

Improved soil nutrient availability increases allocation to the foliage and decreases allocation to the roots (Figures 1 and 3). Increasing allocation to foliage indirectly implies increasing allocation to the stem. Our results are similar to observations in fertilization experiments with nitrogen (Ingestad, 1979). Also, models which use Brouwer's (1962) functional-balance principle, as formulated by Davidson (1969), yield similar responses (Mäkelä, 1986). The combined effect of changing transpiration (W) and nutrient-uptake ability per unit root mass (k_2 C_a) can also be observed in Figures 1, 2, and 3. Allocation to needles is significantly affected by the nutrient-uptake ability per unit of root mass, when the transpiration is low rather than high (Figure 1). This result is expected, since with increasing transpiration, the nutrients carried by the mass flow of water may constitute a larger proportion of the nutrient requirement of the tree. Increases in transpiration normally reflect greater availability of water to the plant. Concomitantly, soil-water nutrient concentration C_W is changing, which is not considered in the simulations presented.

The model treats all nutrients collectively and equally, which is an oversimplification. Individual nutrients act in a different manner within a tree, and therefore their effect on growth and photosynthetic production differs. Similarly, their mobility in soil is different (Nye and Tinker, 1977). Therefore, key nutrients should be considered separately in a more complex model structure.

In contrast to needle allocation, allocation to the water uptake and transport system is more affected by the nutrient-uptake ability per unit of root mass when the transpiration rate is high (Figure 2). This is partially the result of the linear dependence of allocation to the water uptake and transport system on the allocation to foliage and the amount of water transpired. When allocation to the foliage is only slightly changed, the effect on allocation to the water uptake and transport system is large when transpiration is high. However, as mentioned earlier, the assumption of linear dependence between allocation to the water-transport system and amount of water transpired may not be justified.

Presently, only the structural formation of trees is considered. When a similar approach is used in more realistic models, respiration and possible photosynthate reserves must be considered. When annual patterns of growth are investigated, reserves play a decisive role in determining the amount of photosynthate available for growth. In the present model, it is assumed that the photosynthate available for growth, M, is approximately 50% of the total photosynthetic production (Linder, 1985).

Acid deposition is changing the properties of soil in large areas of Europe and North America. Soil models are available describing the changes in nutrient concentrations during acidification (Cosby et al., 1985; Holmberg et al., 1985; Reuss and Johnson, 1986). The stand development can also be described with dynamic models under the implicit assumption that the environmental factors are stable (Hari et al., 1985; Mäkelä and Hari, 1986; Mohren, 1987). One of the great problems in the quantitative analysis of acid deposition on forest growth is to link dynamic soil models and stand models. Equations 10, 11, and 12 provide a possible linkage, based in part on the main functional processes of a tree.

The model presented is relatively simple at the present stage and cannot accurately describe all functions of trees. However, it provides a general approach to coupling tree photosynthesis, transpiration, and nutrient uptake. The use of the optimality hypothesis offers a simple, holistic description of the mechanisms which regulate plant growth in a given environment. The problem with this approach is the proper choice of functions to be optimized and the constraints used. The applicability of this modeling approach can only be tested through rigorous experimental work.

LITERATURE CITED

Bengtson, C. 1980. Effects of water stress on Scots pine. In: Persson, T., ed. *Structure and Function of Northern Coniferous Forests: An Ecosystem Study. Ecological Bulletin* (Stockholm, Sweden) 32:205–215.

Brouwer, R. 1962. Distribution of dry matter in the plant. *Netherlands Journal of Agricultural Science* 10:361–376.

Cheeseman, J. M. 1985. An effect of NA^+ on K^+ uptake in intact seedlings of *Zea mays. Physiologia Plantarum* 64:243–246.

Cosby, B. J., G. Hornbhyer, J. Gelloway, and R. Wright. 1985. Modeling the effects of acid deposition: assessment of a lumped parameter model of soil water and stream water chemistry. *Water Resources* 21:51–63.

Davidson, R. L. 1969. Effect of root/leaf temperature differentials on root/shoot ratios in some pasture grasses and clover. *Annals of Botany* 33:561–569.

Donahue, J., H. Miller, and A. Schickluna. 1977. *Soils: An Introduction to Soils and Plant Growth.* Prentice Hall, New York, NY, U.S.A.

Grigal, D. F. 1989. Mechanistic modeling of nutrient acquisition by trees. In: Dixon, R. K., R. S. Meldahl, G. A. Ruark, and W. G. Warren, eds. *Process Modeling of Forest Growth Responses to Environmental Stress.* Timber Press, Portland, OR, U.S.A.

Hari, P., L. Kaipiainen, E. Korpilahti, A. Mäkelä, T. Nilsson, P. Oker-Blom, J. Ross, and R. Salminen. 1985. Structure, radiation and photosynthetic production in coniferous stands. Research Note 54, Helsinki, Finland. Department of Silviculture, University of Helsinki. 233 pp.

Hari, P., L. Kaipiainen, P. Heikinheimo, A. Mäkelä, E. Korpilahti, and J. Samela. 1986. Trees as a water transport system. *Silva Fennica* 20(3):205–210.

Holmberg, M., A. Mäkelä, and P. Hari. 1985. Simulation model of ion dynamics in forest soil. In: Troyanowsky, C., ed. *Air Pollution and Plants.* VCH, Weinheim, F.R.G. Pp. 236–239.

Ingestad, T. 1979. Nitrogen stress in birch seedlings, II. N, K, P, Ca and Mg Nutrition. *Physiologia Plantarum* 45:149–157.

Jensen, P., and S. Pettersson. 1978. Allosteric regulation of potassium uptake in plant roots. *Physiologia Plantarum* 42:207–213.

Kaipiainen, L., and P. Hari. 1985. Consistencies in the structure of Scots pine. In: Tigerstedt, P. M. A., P. Puttonen, and V. Koski, eds. *Crop Physiology of Forest Trees.* University Press, Helsinki, Finland. 336 pp.

Kaufmann, M. R. 1989. Ecophysiological processes affecting tree growth:water relationships. In: Dixon, R. K., R. S. Meldahl, G. A. Ruark, and W. G. Warren, eds. *Process Modeling of Forest Growth Responses to Environmental Stress.* Timber Press, Portland, OR, U.S.A.

Kaufmann, M., and C. Troendle. 1981. The relationship of leaf area and foliage biomass to sapwood conducting area in four subalpine forest tree species. *Forest Science* 27:477–482.

Korpilahti, E. 1988. Photosynthetic production of Scots pine in the natural environment. *Acta Forestalia Fennica* 202.

Linder, S. 1985. Potential and actual production in Australian forest stands. In: Landsberg, J. J., and W. Parsons, eds. *Research for Forest Management.* Division of Forest Research, CSIRO, Melbourne, Australia. 296 pp.

Long, J. N., and F. W. Smith. 1988. Leaf area—sapwood area relations of lodgepole pine as influenced by stand density and site index. *Canadian Journal of Forest Research* 18:247–250.

Luxmoore, R., J. Landsberg, and M. Kaufmann, eds. 1986. Coupling of carbon, water and nutrient interactions in woody plant soil system. *Tree Physiology* 2:1–467.

Mäkelä, A. 1986. Implications of the pipe-model theory on dry matter partitioning and height growth of trees. *Journal of Theoretical Biology* 123:103–120.

Mäkelä, A., and P. Hari. 1986. Stand growth model based on carbon uptake and allocation in individual trees. *Ecological Modelling* 33:205–229.

Mohren, G. M. I. 1987. Simulation of forest growth applied to Douglas fir stands in the Netherlands. Ph.D. Thesis, Agricultural University of Wageningen, Wageningen, the Netherlands.

Nye, P. H., and P. B. Tinker. 1977. *Solute Movement in the Soil-Root System.* Blackwell Scientific Publications, Oxford, England, U.K. 342 pp.

Pietarinen, I., M. Kanninen, P. Hari, and S. Kellomäki. 1982. A simulation model for daily growth of shoots, needles, and stem diameter in Scots pine trees. *Forest Science* 28:573–581.

Protopapas, A. L., and R. L. Bras. 1987. A model for water uptake and development of root systems. *Soil Science* 144:352–366.

Reuss, J., and D. Johnson. 1986. Acid deposition and the acidification of soils and waters. Springer-Verlag, New York, NY, U.S.A.

Shinozaki, K., K. Yoda, K. Hozumi, and T. Kira. 1964. A quantitative analysis of plant form: the pipe-model theory. I. Basic analyses. *Japanese Journal of Ecology* 14:97–105.

Sievänen, R., P. Hari, J. Orava, and P. Pelkonen. 1988. A model for the effect of photosynthate allocation and soil nitrogen on plant growth. *Ecological Modelling* 41:55–65.

Tyree, M. T., L. B. Flanagan, and N. Adamson. 1985. Response of trees to drought. In: Hutchinson, T. C., and K. M. Meena, eds. *Effects of Atmospheric Pollutants on Forests, Wetlands and Agricultural Ecosystems.* Springer-Verlag, Berlin, F.R.G.

Valentine, H. T. 1985. Tree-growth models: derivations employing the pipe-model theory. *Journal of Theoretical Biology* 117:579–585.

Waring, R. H. 1987. Characteristics of trees predisposed to die. *BioScience* 37:569–574.

Zimmermann, M. H. 1983. Xylem structure and the ascent of sap. Springer-Verlag, Berlin, F.R.G. 193 pp.

CHAPTER 4

A CRITIQUE OF CARBON-BASED TREE GROWTH MODELS

Susan L. Bassow, E. David Ford, and A. Ross Kiester

Abstract. Simulation models of the processes that control carbohydrate balance in coniferous trees are reviewed, and their suitability for assessing the effects of pollution is considered. Currently such models are at the forefront of attempts to simulate the growth process of trees, but they are not able to accurately predict the growth of forest trees under conditions of environmental stress such as an increased pollution load. Typically, model structures for growth based on carbon balance incorporate such features as constant allocation coefficients or regulation of foliage amount through a theoretical maximum at canopy closure. It is these features that render such models unable to predict the effects of pollution because the processes of compensation in relation to the pollution load are obscured. We present a model, CARBON, that prioritizes allocation among meristems according to functional requirements, but this is still insufficient to predict the effects of pollution. We discuss this problem and that of using a maximum foliage amount as a control parameter in canopy development and propose suggestions for an improved modeling framework for tree growth models. This should include the expression of translocation, storage, and utilization of carbohydrates in a system that is well defined spatially.

INTRODUCTION

Models that describe tree growth processes are increasingly needed for the prediction of forest growth and yield under varied environmental influences. Traditional growth and yield prediction techniques prove unable to answer many questions because they generally do not incorporate the effects of environmental variations, particularly variations in the pollution load, on timber yield. Ford and Bassow (in press) examine the traditional yield prediction models and their usefulness as predictive and interrogative tools. Only when growth models begin to incorporate the critical processes determining growth will they be able to answer more diverse questions as to the effects of chronic and acute pollution.

Our overall objective is to be able to simulate and assess the effects of pollution on tree growth. More specifically, we wish to address the question of how pollution effects may be

Ms. Bassow and Dr. Ford are Research Assistant and Director, respectively, at the Center for Quantitative Science, University of Washington, Seattle, WA, 98195 U.S.A. Dr. Kiester is Project Leader of the Synthesis and Integration Project, Forest Response Program, Environmental Protection Agency, Corvallis, OR, 97333 U.S.A. The research is sponsored by the Synthesis and Integration Project of the Joint U.S. Environmental Protection Agency USDA Forest Service Response Program, a part of the National Acid Precipitation Assessment Program. The research reported in this chapter has been funded (in part) by the U.S. Environmental Protection Agency under the cooperative agreement CR814640 to the University of Washington. It has been subjected to Agency review and approved for publication.

added into already existing carbon-based growth models, if this is possible at all. Much experimental evidence is being collected on the effects of pollutants on photosynthesis and foliage dynamics, particularly on the mortality rates of foliage in different age classes, and it is necessary to examine how this information may be used to assess the effects of pollutants on forest growth. In our model, CARBON, we attempt to express some of the interactions among plant parts comprehensively so that the expression of growth is versatile and useful for prediction under varying environmental conditions. In particular, we address the processes of allocation and foliage growth. We then describe the shortcomings of the model CARBON and explain why a radically new modeling perspective is necessary if we are to be able to predict growth under varying environmental conditions, including increased pollution loads.

PREVIOUS CARBON-BASED TREE OR FOREST GROWTH MODELS

Carbon-balance models described in the literature generally have a common goal, which is to explain the carbohydrate production and allocation of the tree or plant (e.g., Mäkelä, 1988; McMurtrie and Wolf, 1983; Reynolds and Thornley, 1982; Thornley, 1972). Because the models represent different levels of biological detail, however, they are useful in answering different questions. For a whole-tree carbon-balance model to be able to predict the effects of pollution, the model structure must represent the effects of environmental change on the growth process. Ford and Bassow (in press) present a more detailed discussion and comparison of existing dynamic growth models.

One of the early dynamic growth models, proposed by Thornley (1972), represents plant growth as the product of a functional balance between root mass and its role in nutrient uptake, and shoot mass and its role in photosynthate production. The concomitant growth of the root and the shoot in a potentially varying environment determines the structure of the plant. This carbon- and nitrogen-partitioning model was developed for the prediction and analysis of tomato plants, which grow indeterminately. For our purpose of predicting tree growth, though, it is important to consider the implications of determinate growth and also the accretion of the woody bole.

McMurtrie and Wolf (1983) present a model for tree growth that is structurally more complex and incorporates foliage, bole, and fine roots. Functionally, the model is less complex than that of Thornley (1972). The production, utilization, and allocation of carbohydrates are calculated in terms of carbon balance by a set of simple differential equations. Light interception affects the photosynthetic rate, and constant partitioning coefficients describe the growth of the various components. Respiration costs, litterfall, branch loss, and root death and turnover are also incorporated. Once canopy closure is reached, foliage and fine-root growth balance foliage and fine-root losses due to litterfall, consumption, and turnover. This is equivalent to assuming that a site index determines a limit to foliage amount which is *always* attained but never exceeded, and to assuming furthermore that roots are produced in proportion to foliage. Varying the partitioning coefficients in the model predicts slightly different patterns of growth, including the accumulation of woody tissue.

McMurtrie and Wolf's (1983) model has several problematic features. First, the use of constant partitioning coefficients is questionable. Cannell (1985) reviewed evidence that in trees, partitioning coefficients for the various plant parts vary with changes in internal and external conditions. Instead, it is likely that the photosynthate produced in the foliage is transported throughout the plant, as different growing meristems utilize carbohydrate for growth and respiration. Further, there is substantial evidence that plants show plasticity of form in response to environmental differences (Bradshaw, 1965). A model to simulate growth should reflect this dynamic process rather than assigning a summary statistic to partition the carbohydrate.

Second, McMurtrie and Wolf (1983) simulate the respiration of foliage, bole, and fine roots as functions of the biomass of the component parts. Their model predicts tree mortality as a consequence of respiration demands exceeding photosynthate production. Hence, accuracy in predicting respiration and its controls becomes very important. Kinerson (1975) suggests that bole respiration is better correlated with wood surface area than with wood biomass, because the most actively respiring tissue of the bole will be in the cambial layer and the outermost rings.

Finally, the model by McMurtrie and Wolf (1983) inadequately expresses foliage death and the concept of canopy closure. In the model, once canopy closure is reached, foliage birth is set equal to foliage death due to litterfall, thereby maintaining the total foliage amount at exactly the value at canopy closure.

By addressing these problematic issues, we attempt to advance the state of dynamic growth models. We find, though, that in spite of our efforts, this whole approach to modeling tree growth is structurally unsound for predicting the effects of pollution.

THE MODEL CARBON

The model CARBON runs on an annual time-step and calculates the tree's net production, respiration costs, and annual growth (Figure 1). Foliage amount is categorized by age. In the version shown in Figure 1, foliage is maintained on the branches for a maximum of 8 years. The sum of the products of age-specific photosynthetic rates and age-specific foliage amounts yields the annual primary production. The allocation of this photosynthate is prioritized: first to respiration of the whole tree, then in sequence to foliage growth, root growth, and wood growth. Respiration is determined for each component and then summed to evaluate the total respiration requirements. Respiration of foliage and respiration of fine roots are calculated as a function of biomass. Wood respiration is calculated as a function of bole surface area (Kinerson, 1975; Ågren et al., 1980).

The amount of foliage growth is determined by a series of limitations. When the canopy is small and expanding, the annual growth rate is limited by the total amount of photosynthates produced and the capacity of the branches to support new foliage. The latter is simulated by setting the maximum foliage increment to at most four times that of the previous year. In common with the model of McMurtrie and Wolf (1983), CARBON makes the assumption that the canopy is light-limited and that a maximum total foliage amount will be approached. As the total foliage amount approaches the ceiling level, the branching frequency is proportionately reduced. Presumably, canopy closure will reduce the relative rate of foliage growth.

Foliage death is calculated as a function of total live tree foliage. For foliage age 3 and older, a proportion of the foliage is subtracted as litterfall; foliage is maintained on the branch for at most 8 years. As the maximum total foliage amount is approached, the rate of foliage death increases proportionally. The combination of the foliage increment and death functions makes the total foliage amount approach a ceiling level. Simulations of the model show that the total foliage first overshoots the maximum then fluctuates around the given limit as a result of the capacitance effect of the age categories. In this example, the foliage amount limit is set at 50,000 g per tree; but if this value is halved or doubled, the resulting patterns of growth are very similar, with only slight shifts in time.

Total root mass is maintained in constant ratio to total foliage amount, in this example, 2:10. A set proportion of 0.4 of the fine roots are considered to die annually. Annual root growth must make up for that 0.4 loss and also increase to regain the 2:10 ratio with the growing proportion of the foliage mass.

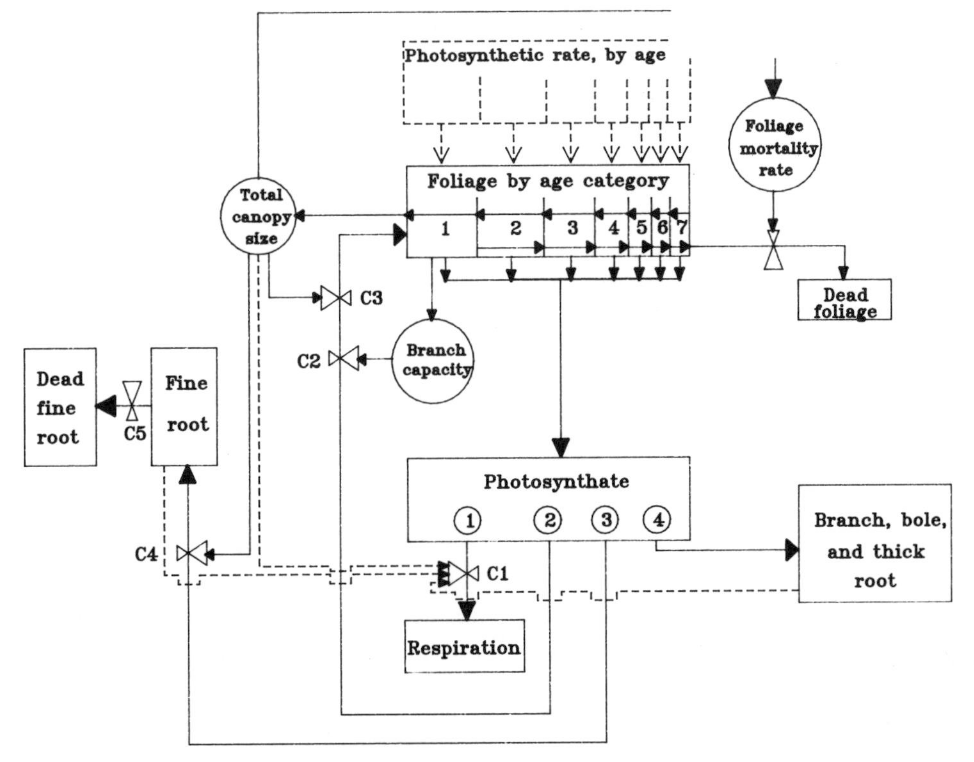

Figure 1. Schematic diagram of the model CARBON.

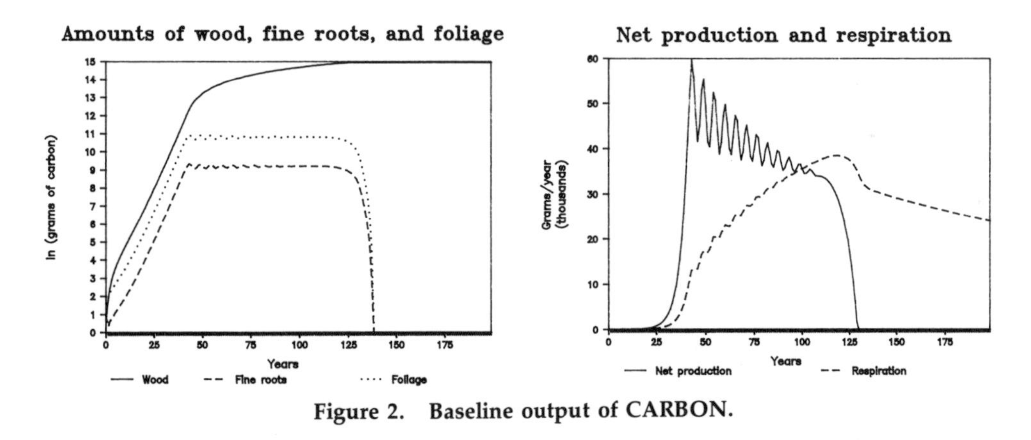

Figure 2. Baseline output of CARBON.

The remainder of the photosynthate is then allocated to wood growth. Wood growth, in terms of grams of carbon available, is determined by the simple conversion estimate of density, 0.35 g dw/cm^3 (Ågren et al., 1980). Radial increment can be determined from standard yield tables and a value of bole surface area estimated (e.g., Hamilton and Christie, 1971). All the wood is presently considered as bole.

Graphical display of the simulation output (Figure 2) shows the early growth of the tree then damped oscillations when the foliage amount reaches the given maximum foliage amount limit. This baseline output is simulated by CARBON with parameter values estimated from unpublished data: age-specific foliage respiration rates, [0.2, 0.1, 0.1, 0.1, 0.1, 0.1, 0.1] g C/g/year; wood respiration rate, 0.05 g C/cm^2/year; fine-roots respiration rate, 0.1 g C/g/year; age-specific photosynthetic rates, [2.0, 1.8, 1.6, 1.5, 1.3, 1.1,

1.0] g/g/year; annual root death, 40%; and maximum foliage limit, 50 kg. Notice that the respiration continues to increase, due to the accumulating wood, until the respiration requirement exceeds net production and the tree dies, i.e., when the net production falls to zero.

THE LIMITATIONS OF CARBON-BASED, WHOLE-TREE GROWTH MODELS

Limitations of the model CARBON fall into three main categories associated with the representation of (1) whole-tree respiration, (2) maximum foliage mass, and (3) allocation of photosynthate.

Respiration Rates

By calculating wood respiration as a function of bole surface area, we have made estimates of respiration costs incurred by the wood more realistic as compared to the model of McMurtrie and Wolf (1983), which utilizes wood biomass for calculating wood respiration. However, our method of estimating bole surface area, as a conversion from wood biomass to volume to surface area, could be improved by a more precise representation of tree morphology so that the surface area may be more accurately estimated.

A serious remaining problem is a lack of sufficient and precise estimates of the respiration rates of wood, foliage, and fine roots. Simulations of CARBON were made with different estimates of wood respiration rates, e.g., by Ågren et al. (1980) for 14-year-old Scots pine (*Pinus sylvestris* L.), 0.0195 g C/cm^2/year; and by Kinerson (1975) for loblolly pine (*Pinus taeda* L.), 0.000877 g C/cm^2/year. Using the former value (0.0195 g C/cm^2/year) in simulations of CARBON resulted in tree mortality at greater than 775 years (compiler limit to the run); and simulations using the latter value (0.000877 g C/cm^2/year) suggest an even longer time frame. Total respiration and net production asymptotically approach each other. These can be compared with the base run (Figure 1) in which the wood respiration rate was 0.05 g C/cm^2/year. In that case, the tree lives for approximately 100 years before respiration costs exceed net production. These estimates are orders of magnitude disparate, and the resultant growth patterns are considerably different. Clearly, this lack of data hinders progress toward versatile predictive growth models.

Maximum Foliage Biomass at Canopy Closure

As did McMurtrie and Wolf (1983), we assign an arbitrary value as the maximum canopy size, which, once attained, is roughly maintained as a value for a tree of mean size. This implicitly assumes that the canopy is limited by one or more environmental factors, such as radiation or water (e.g., Waring et al., 1978) and that the forest system constantly attempts to attain that value. Use of a maximum foliage amount in this way takes no account of the interaction among individual trees in the competition process, which may influence foliage amount.

The first challenge is to decide how a limit to total foliage amount should be expressed in a model. In an actual forest, maximum canopy amount may be approached and overshot (Jarvis and Leverenz, 1983). Canopy closure is the result of specific morphological and physiological processes; branching and foliage growth in a distinct and finite three-dimensional area. Representations of foliage growth should be comprised of a series of positive and negative feedbacks. Positive feedback would apply when there is unutilized area which receives sufficient incoming radiant energy. Conversely, negative feedback would apply when this limited area becomes increasingly exploited. The combination of these two feedbacks results in what appears to be an effective limit to total foliage amount.

In CARBON, we attempt to simulate these positive and negative feedbacks, as well as the more complex foliage interactions, rather than simply balancing foliage growth and foliage death as in McMurtrie and Wolf (1983). In CARBON, foliage growth rate is proportionately reduced and foliage death rate is proportionately increased as the total foliage amount approaches the arbitrarily set value of the maximum foliage amount, 50 kg per tree (Ford, 1982). The combination of the foliage-death and foliage-growth functions creates a capacitance, or buffer, which gives rise to the damped oscillations. But it should not be necessary to specify an arbitrary value for maximum foliage amount. A maximum foliage limit should be an outcome of the growth model, not a specification. Further, when a pollution load is received, there may be no maximum foliage amount, or the maximum may be different. A more precise description of the morphological construction of the growing tree and its competitive environment, and the implicit positive and negative feedbacks, is required. The problem remains that insufficient information is available regarding the positive and negative feedback of morphological constraints to canopy growth. Thus we are unable to model the processes accurately.

Allocation of Photosynthate

A third major problem with modeling the carbon balance of trees is the use of allocation coefficients. McMurtrie and Wolf (1983) use partitioning coefficients of constant value over the life of the tree to divide the net photosynthate among foliage, fine roots, and wood. The coefficients are merely estimates of the growth rates of the tree's component parts. However, it is likely that the coefficients change throughout the life of the tree—both on a short time scale due to seasonal fluctuations in growth rates, and on a long time scale due to slow changes in such conditions as nutrient availability or pollution load at a particular site. Allocation coefficients cannot be simple estimated parameters for a given tree species but instead should represent the growth process. Using constant partitioning coefficients is using only the end result proportions to infer the actual growth processes. For example, improved nutrition may result in increased aboveground production due to decreased partitioning to the roots. Additionally, water stress and shading may affect partitioning ratios. McMurtrie and Wolf (1983) demonstrate the resultant pattern of growth if different allocation coefficients are utilized, i.e., to simulate the effect of nutrient-poor soils. The structure of the model only allows for one set of partitioning coefficients to be assigned for a single model execution.

CARBON attempts to avoid the use of constant partitioning coefficients by prioritizing allocation of the photosynthate, first to respiration, then to foliage, then to fine roots, and finally to wood. The growth rules discussed above are used to determine the amounts of photosynthate utilized by each tree component. Although this approach avoids the problem of constant allocation coefficients, it has inadequacies. The order of priority of allocation may be inaccurate since priority is likely to change with the season (and is the basis of phenology) and with other ecological factors. Further, the whole concept of sequential allocation may misrepresent the growth process. A more complete approach would be to simulate the allocation process as a series of spatially located competing sinks, or growing meristems. This would call for an entire restructuring of the dynamic growth prediction model, which in turn would make the model more flexible for prediction purposes.

DISCUSSION

Several carbon-budget models exist that simulate patterns of production, allocation, and growth of trees with varying levels of morphological and biological detail (Mäkelä, 1988; McMurtrie and Wolf, 1983; Reynolds and Thornley, 1982; Thornley, 1972;

Valentine, 1985). These models are structured in such a way that they cannot readily be used to simulate the possible effects of pollution. CARBON was devised with the objective of creating a process-based model sensitive enough to predict the cumulative effects of pollution on tree growth. This may not be possible, however. Pollution effects cannot simply be overlaid onto the basic structure of a carbon-based growth model. The mechanisms determining the effects of the pollutants are hidden, and the compensatory actions of the tree obscure the direct pollution effects. We suggest that it is imperative to consider the morphology, as well as the physiology, of the tree in order to be able to detect pollution effects. If the morphology of a tree is considered, then the previously simple allocation coefficients or prioritization scheme becomes a complex network involving spatially defined photosynthate production, translocation, storage, retranslocation, and utilization by the growing and respiring meristems.

In general, if a model is to be used for growth prediction under varied environmental conditions, the following factors should be considered:

1. *Periodicities.* Seasonality of the component growth may stagger the photosynthate requirements of the tree's parts. For example, in the early spring the roots flush first, partially utilizing stored starches, then budbreak begins the period of rapid foliage growth (e.g., Little, 1970; Deans and Ford, 1986).

2. *Transport.* Transport requirements and limitations arise as the tree's morphology increases in size and complexity. When modeling the growth of a seedling into a large tree, these morphological changes should be considered. Intuitively, one knows that mechanically the transport of photosynthates becomes increasingly limited as the tree canopy grows further away from the roots.

3. *Interactive effects with the environment.* For example, an increased nitrogen level of the soil may act as a stimulus to root growth (e.g., Keyes and Grier, 1981). As the environment changes during the life of the tree, perhaps due to pollution deposition, the tree responds and alters its growth patterns.

4. *Interrelationships between storage and current photosynthesis.* Stored starches may play an important role during the time of rapid growth in the spring. Little and Loach (1973) found that the fluctuating amounts of starch in the 1-year-old needles of balsam fir correlates positively with the photosynthetic rate. They conclude that photosynthate accumulation in the form of starch does not influence photosynthetic rate. So the amount of stored starch does not act to decrease the required photosynthetic rates but acts as a supplemental carbohydrate source in times of rapid utilization.

LITERATURE CITED

Ågren, G. I., B. Axelsson, J. G. K. Flower-Ellis, S. Linder, H. Persson, H. Staaf, and E. Troeng. 1980. Annual carbon budget for a young Scots pine. In: Persson, T., ed. Structure and function of northern coniferous forests: an ecosystem study. *Ecological Bulletin* (Stockholm, Sweden) 32:307–313.

Bradshaw, A. D. 1965. Evolutionary significance of phenotypic plasticity in plants. *Advances in Genetics* 13:115–155.

Cannell, M. G. R. 1985. Dry matter partitioning in tree crops. In: Cannell, M. G. R., and J. E. Jackson, eds. *Attributes of Trees As Crop Plants.* Institute of Terrestrial Ecology, Huntingdon, England, U.K. Pp. 160–193.

Deans, J. D., and E. D. Ford. 1986. Seasonal patterns of radial root growth and starch dynamics in plantation-grown Sitka spruce trees of different ages. *Tree Physiology* 1:241–251.

Ford, E. D. 1982. High productivity in a polestage Sitka spruce stand and its relation to canopy structure. *Forestry* 55:1–17.

Ford, E. D., and S. L. Bassow. In press. Modeling the dependence of forest growth on environmental influences. In: Pereira, J. S., ed. NATO ARW: *Biomass Production by Fast-Growing Trees.*
Hamilton, G. J., and J. M. Christie. 1971. *Forest Management Tables (Metric).* Her Majesty's Stationery Office, England, U.K. 201 pp.
Jarvis, P. G., and J. W. Leverenz. 1983. Productivity of temperate, deciduous and evergreen forests. In: Lange, O. L., P. S. Nobel, C. B. Osmond, and H. Ziegler, eds. *Physiological Plant Ecology,* Vol. IV. *Ecosystem Processes: Mineral Cycling, Productivity and Man's Influence.* Springer-Verlag, Berlin, F.R.G. Pp. 233–280.
Keyes, M. R., and C. C. Grier. 1981. Above- and below-ground new production in 40-year-old Douglas-fir stands on low and high productivity sites. *Canadian Journal of Forest Research* 11:599–605.
Kinerson, R. S. 1975. Relationships between plant surface area and respiration in loblolly pine. *Journal of Applied Ecology* 12:956–971.
Little, C. H. A. 1970. Derivation of the springtime starch increase in Balsam fir (*Abies balsamea*). *Journal of Botany* 48:1995–1999.
Little, C. H. A., and K. Loach. 1973. Effects of changes in carbohydrate concentration on the rate of net photosynthesis in mature leaves of *Abies balsamea. Canadian Journal of Botany* 51:751–758.
Mäkelä, A. 1988. Models of pine stand development: an ecophysiological systems analysis. Research Note 62, Department of Silviculture, University of Helsinki, Helsinki, Finland. Pp. 1–54.
McMurtrie, R., and L. Wolf. 1983. Above- and below-ground growth of forest stands: a carbon budget model. *Annals of Botany* 52:437–448.
Reynolds, J. F., and J. H. M. Thornley. 1982. A shoot:root partitioning model. *Annals of Botany* 49:585–597.
Thornley, J. H. M. 1972. A balanced quantitative model for root:shoot ratios in vegetative plants. *Annals of Botany* 36:431–441.
Valentine, H. T. 1985. Tree-growth models: derivations employing the pipe-model theory. *Journal of Theoretical Biology* 117:579–585.
Waring, R. H., W. H. Emmingham, H. L. Gholz, and C. C. Grier. 1978. Variation in maximum leaf area of coniferous forests in Oregon, and its ecological significance. *Forest Science* 24:131–140.

CHAPTER 5

FLOW THROUGH CONIFER XYLEM: MODELING IN THE GAP BETWEEN SPATIAL SCALES

Jacqueline L. Haskins and E. David Ford

Abstract. Construction of process-based models of tree water relations requires a synthesis of whole-tree and cellular-level phenomena. Simulations indicate that much of the detail required to model movement between two cells must be retained in models of a small block of xylem tissue. Intertracheid connections must be an integral part of a tissue-level model, not a factor added on top of a pipe-flow model.

INTRODUCTION

At present, more is known about factors controlling transpiration and water uptake than about movement of water through the tree itself, in large part because water movement inside the tree has been relatively difficult to measure. Patterns of water movement in the tree result from patterns of water potential in the tree; these regulate many important spatial processes, including translocation.

Modeling movement through the tree involves structural considerations, e.g., tree morphology on a macroscopic and a microscopic scale, and functional considerations, especially embolism (the filling of a tracheid with vapor). By macroscopic morphology, we mean characteristics such as trunk height and diameter, crown shape, and rooting depth. By microscopic morphology, we mean characteristics such as diameter and length of tracheids and type, number, size, and arrangement of intertracheid pits. The microscopic morphology of a tree determines tissue conductance, and its macroscopic morphology determines the structure of the transport path and the spatial distribution of uptake and transpiration. Embolism decreases xylem conductivity and may be an important source of water capacitance from inextensible tissues within the tree (Zimmermann, 1983).

Tree water relations have been modeled at the level of the whole tree (Edwards et al.,

Ms. Haskins and Dr. Ford are Research Assistant and Director, respectively, at the Center for Quantitative Science, University of Washington, Seattle, WA, 98195 USA. The research is sponsored by the Synthesis and Integration Project of the Joint U.S. Environmental Protection Agency USDA Forest Service Response Program, a part of the National Acid Precipitation Assessment Program. Although the research reported in this chapter has been funded in part by the U.S. Environmental Protection Agency under the cooperative agreement CR814640 to the University of Washington, it has not been subjected to Agency review and therefore does not necessarily reflect the views of the Agency, and no official endorsement should be inferred. The authors wish to thank Susan Bassow, Rupert Ford, and J. Renée Brooks for helpful discussion, and Anne Avery for the computer support that made this work possible.

1986; Federer, 1979; Hatheway and Winter, 1981; Kowalik and Eckersten, 1984; Waring and Running, 1976) and at a cellular level (Bolton and Petty, 1978; Calkin et al., 1986; Chapman et al., 1977; Schulte and Gibson, 1988). The effectiveness of the whole-tree models is often hampered by a lack of understanding of flow at a finer spatial scale (Edwards et al., 1986), while the cellular-level models often fail to discuss implications for water flow within the whole tree.

Integration of these two spatial scales requires the development of a tissue-level model. A model of a small block of xylem would provide context to the cellular models and detail to the whole-tree models. Such a model could be used to explore the role of microscopic morphology and the role of embolism as well as the interaction between the two. As a preliminary to the creation of a process-based, tissue-level model, we report the results of simulations to investigate the level of detail required in such a model.

Xylem anatomy and morphology differ so greatly for conifers and hardwoods (Jane, 1970) that a tissue-level model cannot apply to both. We have chosen to model conifers.

Flow through the xylem has sometimes been modeled as the flow of pure water through a collection of impermeable, uniform, circular pipes of infinite, uninterrupted length, using the Hagen-Poiseuille equation:

$$\frac{\Delta P}{Q} = \frac{8\eta L}{\pi r^4} \tag{1}$$

where Q = the volumetric flow rate, ΔP = the pressure drop across length L, r = the capillary radius, and η = the viscosity of the liquid.

Such models greatly overpredict conductance (Pickard, 1981) because the passage between pits imposes a significant additional resistance (Petty, 1970; Smith and Banks, 1971). This pit resistance increases relative to lumen resistance as tracheid diameter increases (Bolton and Petty, 1978; Calkin et al., 1986).

Modified versions of the Hagen-Poiseuille equation are generally accepted as appropriate representations of flow through the tracheid lumen. However, we do not yet have satisfactory representations of flow through the pits. Bolton and Petty (1978) attempted a theoretical derivation of pit resistance. Their analysis required many approximations, and some regions within the pit were so difficult to model that they were simply excluded from the analysis. Hence Bolton and Petty's pit-resistance component was an acknowledged underestimate.

Schulte and Gibson (1988) attempted to derive pit resistance empirically. They used an enzyme digestion technique to remove primary wall material and estimated pit-membrane resistance as the difference in resistance before and after the digestion. However, the results obtained varied widely from species to species; for some species, there were no publishable results. As yet, there has been no reported application of this technique to species with pit tori. Thus this technique cannot yet provide reliable experimental measurements of pit resistance.

Given the difficulties involved in obtaining measures of pit resistance, there has been some tendency to leave this factor out of large-scale models, or to approximate it with a fixed parameter. Our purpose in the research described here is to determine whether a tissue-level model of flow would require inclusion of the pits as an integral part of the model, or whether flow could be described with a Hagen-Poiseuille model and some additional factor to reduce conductance.

Lumen diameter is a very important predictor of flow (Equation 1), and it correlates with other important parameters such as tracheid length and pitting characteristics (Calkin et al., 1986). Using a biophysical model which considers the pits in detail, Schulte and Gibson (1988) simulated the change in tracheid conductance with increasing tracheid diameter for six woody and herbaceous species and compared this to Hagen-Poiseuille predictions. Many of their simulated species curves first rise with an accelerating slope, then pass a point of inflection and level off. Thus at small diameters where lumen effects predominate, the general shape of the Hagen-Poiseuille curve is retained, although flows are lower. As diameter increases, the simulated curves depart completely from the Hagen-Poiseuille model, leveling off as pit effects assume predominance.

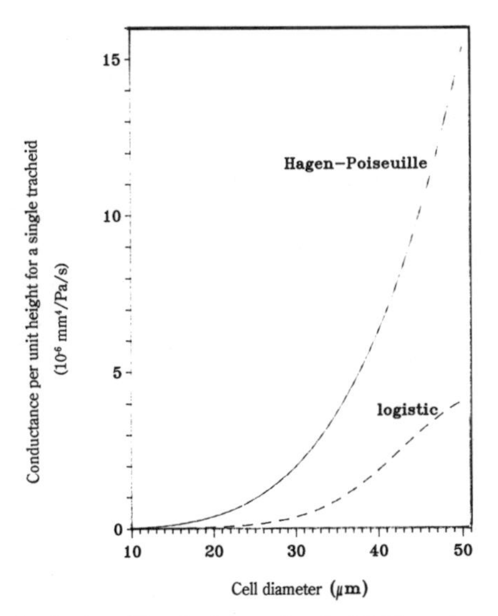

Figure 1. Hypothetical relationships between tracheid diameter and conductance. A logistic relationship was compared to the Hagen-Poiseuille model.

MODEL DESCRIPTION

Figure 1 illustrates two hypothetical relationships between tracheid diameter and tracheid conductance: the Hagen-Poiseuille model and a logistic model which produces a curve similar to those derived by Schulte and Gibson (1988). The Hagen-Poiseuille model considers only lumen effects; the logistic model emulates models which consider lumen effects, pit effects, and the interaction of the two as a function of trackeid diameter. Our purpose is to determine whether these two models have qualitatively different behaviors under various conditions of xylem structure and function, or whether it would be possible to construct a Hagen-Poiseuille–based model with behavior similar to the model in which pitting effects play an integral role.

The logistic curve in Figure 1 was produced by the following relationship:

$$C = \frac{.0005}{1 + 5000e^{-.2D}} \tag{2}$$

where C = the conductance per unit height in mm^4/Pa/s, and D = the tracheid diameter in μm.

Tracheid diameter varies considerably with tissue age and location in the tree, and also within a single tissue. We compare the tissue conductance that would be obtained under a variety of distributions of tracheid diameters for each of the two models of cell conductance. Diameter distributions of fern tracheids have been observed to be right-skewed (Calkin et al., 1986). We consider gaussian (normal) and right-skewed distributions of a variety of means, spreads, and skewnesses (Figure 2).

Embolism is simulated by removing the conductance of the largest-diameter cells,

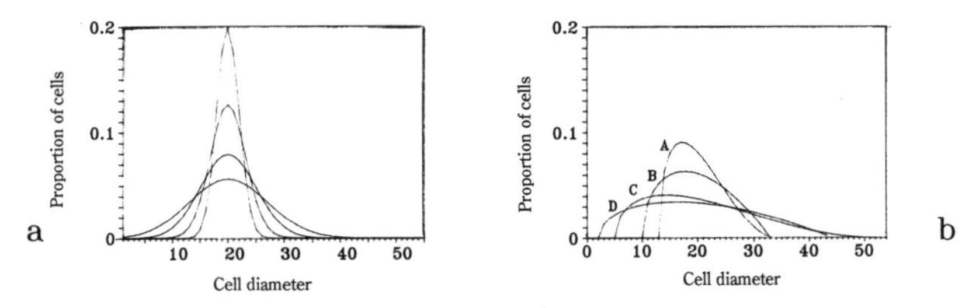

Figure 2. Simulated tracheid distributions of mean 20. Equivalent distributions with means of 30, 40, and 50 were also used in the simulations. (*a*) Gaussian distributions of mean 20. Variances are 4, 10, 25, and 50. (*b*) Right-skewed distributions of mean 20. All curves were of the general form $(D - DMIN + 1)^{a-1} * (DMAX + 1 - D)^{b-1}$ where D = tracheid diameter, DMIN = the smallest diameter tracheid, DMAX = the largest diameter tracheid, and a and b are shape parameters.

since embolism is thought to occur preferentially in larger conduits (see Pickard, 1981, Equation 15). Results shown are for 10% embolism, i.e., 10% of the cells have zero conductance.

To calculate tissue conductance, we sum the predicted conductance of each cell and convert to a per-area basis. This requires some assumption about the relationship between lumen diameter and tissue area. Based on cell wall thicknesses published in Ford et al. (1978), we use the following relationship:

$$\text{For cells of diameter } 20\mu\text{m or less, } A = (D + 14)^2 \quad (3)$$

$$\text{For remaining cells, } A = (D + 8)^2 \quad (4)$$

where A = the area of tissue in μm, and D = the tracheid diameter in μm.

RESULTS AND DISCUSSION

As expected, predicted conductance is considerably lower under the logistic model (Figures 3 and 4; note order of magnitude difference in scales). Of greater interest is our finding that the behavior of the two models is qualitatively different. Below its point of inflection, the logistic model mirrors the behavior of the Hagen-Poiseuille model; above its point of inflection, it behaves very differently.

Under the Hagen-Poiseuille model, conductance increases strongly with increasing mean, and generally increases with increasing maximum (Figure 3). Right-skewed distributions have higher conductance than gaussian distributions. Truncation of the distribution to simulate embolism, and the degree of embolism simulated, strongly affect the pattern of conductances. None of these generalizations holds true for the logistic model.

The behavior of the logistic model (Figure 4) is highly dependent on mean diameter. Few generalizations apply to all mean diameters; there are even exceptions to the generalization that increasing mean diameter increases conductance (Figure 4d). Prediction of the optimal diameter distribution, i.e., the distribution of greatest conductance, is little affected by simulated embolism.

We conclude that even when we move to a larger spatial scale than that of flow between two tracheids, we cannot move to a model in which pit effects are ignored or are simply added onto a Hagen-Poiseuille model. The complexities imposed by intertracheid connections have a multitude of ramifications at the tissue level, and these must be explored.

Whole-tree water relations models have been hampered in the past by a lack of detail

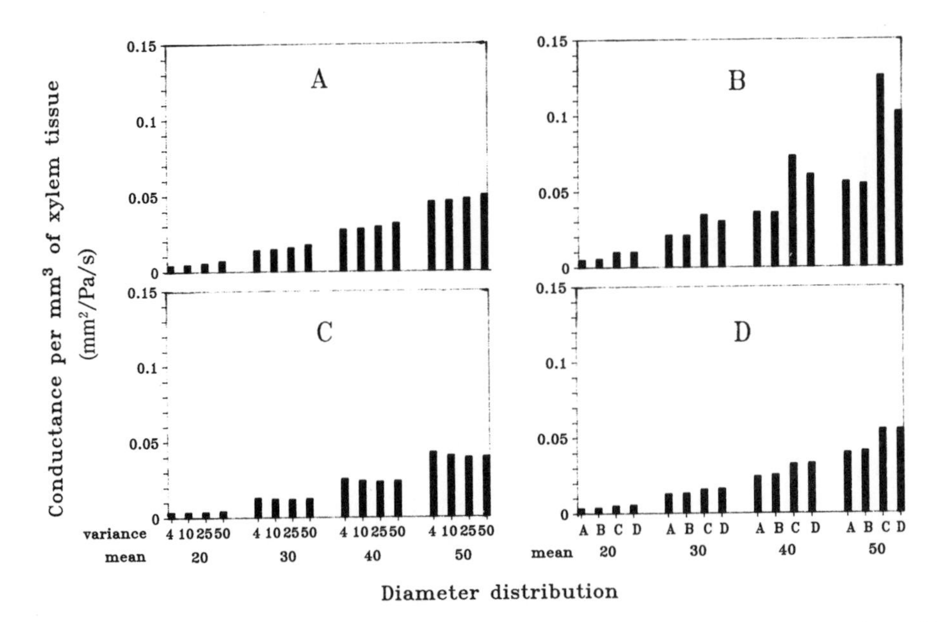

Figure 3. Tissue conductance predicted by the Hagen-Poiseuille model under a variety of diameter distributions. (*a*) Gaussian distributions, no embolism. (*b*) Right-skewed distributions, no embolism. (*c*) Gaussian distributions, 10% embolism. (*d*) Right-skewed distributions, 10% embolism.

regarding temporal and spatial variation in tissue conductance (Edwards et al., 1986). Tissue conductance is determined in large part by microscopic morphology, by embolism, and by the interaction between the two. To investigate how morphological characteristics of the tissue, such as the distribution of tracheid diameters, interact with embolism to determine tissue conductance, we need to model at a spatial scale intermediate to extant models. Such a tissue-level model must retain much of the anatomical detail of single-cell models.

Embolism affects tissue conductance not merely through the removal of conducting tissue, as in these simulations, but also through alteration of the pathway of flow. Pathway effects are likely to increase in importance as the number of embolisms in a tissue increases. Future work in this field should focus on process-based, tissue-level models, on the role of tracheid interconnections, and on the pathway of water flow in three-dimensional tracheid networks.

LITERATURE CITED

Bolton, A. J., and J. A. Petty. 1978. A model describing axial flow of liquids through conifer wood. *Wood Science and Technology* 12:37–48.

Calkin, H. W., A. C. Gibson, and P. S. Nobel. 1986. Biophysical model of xylem conductance in tracheids of the fern *Pteris vittata*. *Journal of Experimental Botany* 37:1054–1064.

Chapman, D. C., R. H. Rand, and J. R. Cooke. 1977. A hydrodynamical model of bordered pits in conifer tracheids. *Journal of Theoretical Biology* 67:11–24.

Edwards, W. R. N., P. G. Jarvis, J. J. Landsberg, and H. Talbot. 1986. A dynamic model for studying flow of water in single trees. *Tree Physiology* 1:309–324.

Federer, C. A. 1979. A soil-plant-atmosphere model for transpiration and availability of soil water. *Water Resources Research* 15(3):555–562.

Ford, E. D., A. W. Robards, and M. D. Piney. 1978. Influence of environmental factors on cell production and differentiation in the early wood of *Picea sitchensis*. *Annals of Botany* 42:683–892.

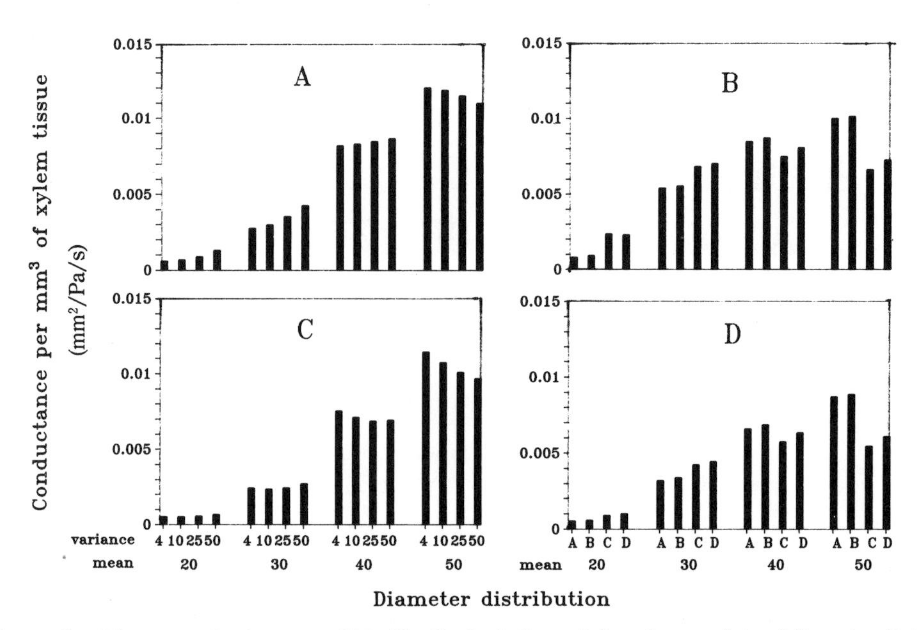

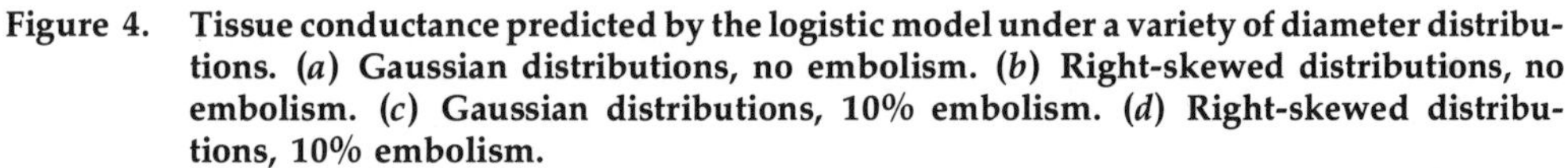

Figure 4. Tissue conductance predicted by the logistic model under a variety of diameter distributions. (*a*) Gaussian distributions, no embolism. (*b*) Right-skewed distributions, no embolism. (*c*) Gaussian distributions, 10% embolism. (*d*) Right-skewed distributions, 10% embolism.

Hatheway, W. H., and D. F. Winter. 1981. Water transport and storage in Douglas-fir: a mathematical model. *Mitteilungen Vienna Foretliche Bundes-Versuchsanstalt Maria brunn* 42:193–222.

Jane, F. W. 1970. *The Structure of Wood.* A. and C. Black, London, England, U.K.

Kowalik, P. J., and H. Eckersten. 1984. Water transfer from soil through plants to the atmosphere in willow energy forest. *Ecological Modelling* 26:251–284.

Petty, J. A. 1970. Permeability and structure of the wood of Sitka spruce. *Proceedings, Royal Society of London,* Series B 175:149–166.

Pickard, W. F. 1981. The ascent of sap in plants. *Progress in Biophysics and Molecular Biology* 37:181–229.

Schulte, P. J., and A. C. Gibson. 1988. Hydraulic conductance and tracheid anatomy in six species of extant seed plants. *Canadian Journal of Botany* 66:1073–1079.

Smith, D. N. R., and W. B. Banks. 1971. The mechanism of flow of gases through coniferous wood. *Proceedings, Royal Society of London,* Series B 177:197–223.

Waring, R. H., and S. W. Running. 1976. Water uptake, storage and transpiration by conifers: a physiological model. In: Lange, O. L., L. Kappen, and E.-D. Schulze, eds. *Water and Plant Life: Problems and Modern Approaches.* Springer-Verlag, New York, NY, U.S.A.

Zimmermann, M. H. 1983. *Xylem Structure and the Ascent of Sap.* Springer-Verlag, Berlin/Heidelberg, F.R.G.

CHAPTER 6

ECOPHYSIOLOGICAL PROCESSES AFFECTING TREE GROWTH: WATER RELATIONSHIPS

Merrill R. Kaufmann

Abstract. The effects of water stress on tree growth may be considered in the context of feedback mechanisms regulating the relative size of tree parts. The primary feedback mechanisms involve water stress and carbohydrate availability. Environmental effects, which include both prevailing conditions and episodic or catastrophic events, affect growth chiefly through internal effects on water stress or carbon balance and allocation. For predicting long-term growth with process-based models, relatively simple water-transport models may be useful for understanding how tree water stress is related to environmental conditions. Whatever the method used for evaluating the development of water stress, a growth model should address how the severity of stress affects internal processes and growth. These effects range from reduction of stomatal opening and loss of turgor in growing tissues to dieback of foliage and fine roots.

INTRODUCTION

Tree growth is the consequence of many physiological processes occurring under the constraints of environmental conditions and genetic characteristics of the species. It is the nature of plants to grow and take on a form that inadvertently subjects them to water stress. This is because plants lack a strategy—the capability to make intelligent decisions (Kramer, 1980). Through feedback mechanisms, water stress is one of the primary factors by which plant form and growth are regulated. It follows from this concept that there is no stress-free environment for trees, that at times water stress is an integral and limiting part of the growth process for each tree.

The ubiquitous nature of water stress makes this term difficult to define. Clearly there are developmental conditions under which the water status of the tree has no negative or limiting effects on physiology and growth. For example, the water status of a tree during mild weather following a significant rain may be highly favorable for growth. In contrast, other conditions exist in which water stress is clearly severe and limiting to growth, for example, during a prolonged drought characterized by little precipitation and hot, dry atmospheric conditions. Landsberg (1986, p. 156) writes,

> Defining "stress" is not simple. There has been much discussion about this at various times, some of it concerned with the definition of stress by analogy with the engineering concept. However, it is clear that there is no cut-off point, in terms

Dr. Kaufmann is Principal Plant Physiologist, USDA Forest Service Rocky Mountain Forest and Range Experiment Station, Fort Collins, CO, 80526 U.S.A.

of some measure of tissue water status, on one side of which plants are stressed, and on the other unstressed. Rather than attempting to define stress we should, perhaps, be more concerned to explain the effects of tissue water status on growth.

The complexity of defining water stress quantitatively stems partly from the fact that large differences exist in the sensitivity of plant tissues to water stress. These differences result both from phenological effects and from acclimation processes that alter tissue vulnerability to water deficits. Thus a foliage water potential of −2.0 MPa may have more serious consequences in the early part of the growing season when tissues are expanding than later when they are not, or when stress first occurs in tissue rather than after the tissue has been acclimated by several previous periods of water stress. Furthermore, other factors such as nutrient availability may cause trees to experience different levels of water stress under similar environmental conditions because of effects on the size of shoot and root systems, stomatal function, and so on.

Nonetheless, the researcher interested in modeling growth can judge which of the physiological responses are likely to be significant in the overall growth process for the level of complexity required in the model being considered. It is beyond the scope of this paper to discuss how cellular processes influenced by water stress affect specific growth processes during the course of a day (Dixon, this volume). Rather, attention will focus on integrated plant water relations and related physiological processes that are likely to be significant at the whole-tree level in affecting long-term growth of a tree or stand.

WATER RELATIONS AND PROCESS MODELING OF TREE GROWTH

In assessing the role of plant water relations in models of tree growth, it is useful to examine the context within which the models are developed and used. In empirical studies of tree growth, site productivity is often evaluated using an estimator of site quality such as site index (measured height growth of dominant or codominant trees during a specified period of time). Such indices are surrogates representing the effects of many factors such as soil water and nutrient availability and physiographic features of the site (slope, aspect, and elevation). However, the actual biophysical effects of these factors are not well understood or represented in an empirical growth equation or model.

An examination of the site factors affecting growth leads to three major lists, one including water availability and use, another nutrient availability, and a third energy availability (Grigal, this volume; Isebrands et al., this volume). Process models of growth that take into account the effects of these biophysical factors may lead to phenomenological expressions of site quality that are less empirical in nature. In addition, process models may account more directly for the effects of stand density and species composition on water, nutrient, and energy availability and use for individual trees and for aggregates of trees in stands.

Scientists who attempt to model growth in relation to the plant environment need to consider plant responses to two types of environmental conditions that involve plant water relations. First, models must reflect plant response to the environmental conditions prevailing through the physiologically active season. These conditions include not only the climatic variables, but also the effects of the edaphic and physiographic situation and stand structure. Second, responses to episodic or catastrophic events should be incorporated.

Few models take both types of conditions into account. The Drought Index for Southern Pines (DISP) (Zahner and Grier, this volume) is an example of a model that addresses average conditions having a scale of months, but not extreme events that may have a scale of several days or less. The Simple Whole Tree (SWT) model being developed by Ford and Kiester (this volume) includes phenological and morphological components. These components may provide the avenue through which episodic environmental effects on growth can be handled. The ECOPHYS model for irrigated and fertilized popular (*Populus*) described by Isebrands et al. (this volume) addresses light interception, photosynthesis, and allocation. It is proposed to incorporate the effects of water and nutrient stress through the use of an expert system, which presumably could account for both periodic cycles and episodic events. The canopy layer model described by Caldwell et al. (1986) focuses on canopy aspects of gas exchange processes over a wide range of climatic conditions, but it does not address carbon allocation and growth.

The season of physiological activity may extend well beyond the so-called growing season, which classically is limited to the period of shoot growth. Root growth, for example, may occur over a much longer season; and other processes, such as maintenance respiration, probably occur continuously except when temperatures are below freezing. Generally, plant water relations are thought to be most important during the periods of active shoot extension. However, it is during shoot growth that the effects of water deficits are simply the most obvious: Water stress reduces growth of all plant tissues, but effects on shoot growth are the most visible. Other effects, however, may be just as significant.

In the context of tree carbon balance, however, any effect of tree water relations on components of the carbon cycle or on regulation of carbohydrate use within trees any time during the year has the potential to affect growth. Several possible effects in addition to current shoot growth come to mind. First, photosynthesis may be reduced by water stress late in the season after shoot growth has terminated, thus directly affecting carbohydrate reserves that may be critical for growth (especially growth of roots) early in the following season (Kramer and Kozlowski, 1979). Second, root production may be restricted late in the season when soil water is depleted; this also may influence tree performance early the following season, when soil temperature is still low and restricting water uptake. Third, repeated drying cycles through the year could increase carbon demands for fine-root replacement because of dieback associated with each drying cycle (Santantonio and Hermann, 1985). Fourth, interaction with other factors such as tree nutrient status may alter tree growth.

Quite different effects may be associated with episodic or catastrophic environmental conditions. Except for trees growing in swamps or stable riparian sites, most trees experience drought periodically, even in regions characterized by ample precipitation. If drought occurs less frequently than annually, trees may grow most of the time under conditions of moderate to good supplies of water, which means that feedback pressures of water stress are minimal. Consequently, during severe drought, the trees have too much transpiring foliage in relation to the root system's capacity to absorb water. This may result not only in stomatal closure and loss of turgor leading to a complete cessation of growth, but even in dieback of shoots and roots. An example of this in radiata pine (*Pinus radiata* D. Don) is discussed later in this chapter (Linder et al., 1987). Whenever conditions are severe enough to cause shedding of plant parts (Kozlowski, 1976), or even when they are only severe enough to limit stomatal function for a few days after the water deficit has been removed (Rook et al., 1976; Sheriff and Whitehead, 1984), carbon balance and growth are affected adversely. In agroforestry situations, the application of fertilizer may increase water stress through rapid effects on the soil ionic environment, and it may make trees more susceptible to drought through reduction of the root-shoot ratio.

The degree of detail about tree water relations needed to describe adequately which processes are limiting growth and how those processes respond to environmental conditions depends largely on the type of tree growth model under consideration. For a detailed process (bottom-up) model, involving specific growth parameters for each type of tissue and time-steps of a day or less, extensive information may be needed to describe how cell expansion is affected by turgor, water transport across membranes, cell wall elasticity, the supply of structural materials, and so forth. While the development of such a model is useful for understanding individual tree growth, it is also likely that such a model will be unsuitable for modeling long-term growth of trees in stands because of the large amount of detail required, error propagation, and lack of utility. In purely empirical (top-down) models such as growth models based on site index, however, much knowledge about tree response to environmental conditions is ignored.

For modeling annual tree growth, perhaps a suitable approach is to attempt the development of phenomenological models which, though empirical in many respects, incorporate simplified functions that account for known physiological responses. Phenomenological expressions are those that describe a phenomenon or pattern of behavior at the level being considered; they do not take into account excessive mechanistic detail, nor are they so empirical that they ignore important information about processes (Thornley, 1976). Such models should be based upon considerable experimental evidence, and they should be consistent with theory or with known mechanistic detail. For example, a phenomenological model might not include information about the flux of water or carbon from one plant compartment to another, processes known to be involved in growth; but the model might well reflect the effects of daily water stress on carbon fixation and on cell enlargement through indices of weekly or seasonal water availability and atmospheric conditions. Such a model would satisfy the requirement that those phenomena known to be important in growth are addressed, yet the model does not become unwieldy by including detail not critical for estimating overall tree growth. The development of phenomenological functions or indices representing the effects of various environmental, stand, and physiographic parameters on growth deserves more attention, because it may be the most practical approach to modeling growth in many situations.

To summarize, we need a thorough understanding of tree growth in relation to a wide range of natural environmental conditions, and we need a wise selection and use of plant and environmental variables for estimating plant growth under those conditions. This requires information at several scales of space, time, and process detail to understand what is important and what is noise in the aggregate and how trees respond to stress in the aggregate.

The challenge to the researcher is to understand not only the effects of prevailing environmental conditions on growth through effects on water relations, but also how irregular and sometimes intense perturbations in the environment affect growth processes. Predicting the occurrence of random perturbations probably will remain outside the realm of tree growth models. Predicting their consequences may not, however, even though presently the consequences are poorly understood. Perhaps some research should be directed to elucidating tree responses that might be associated with severe conditions or events. Such research might involve, for example, experimentation with rain-out shelters and artificial heating of small forest stands. Appropriately designed experiments may well indicate how specific growth-related processes are influenced by some of the conditions associated with severe drought (and perhaps with global warming), and how models might take these effects into account.

WATER TRANSPORT AND THE DEVELOPMENT OF WATER STRESS IN TREES

In modeling tree growth, particularly in response to different environmental conditions of contrasting forest sites, we must bear in mind an important principle: An architectural balance among the water transport components of plants is required in each environment. The lack of such a balance places plants in a highly risky situation of having either too much foliage transpiring water compared with the capability of the plant to absorb it, or too much live biomass in the root system and bole requiring carbohydrate from a crown too small to provide it. Recalling the previous statement that plants do not have a strategy, it becomes clear that feedback mechanisms are responsible for regulating the balance of plant parts and growth to suit given environmental situations.

The two key feedback mechanisms involve water stress and internal carbon balance. Through water stress, the size of the transpiring surface is prevented from becoming too large relative to the size of the root absorbing surface and the xylem transport system delivering water to the foliage (see Zahner, 1968, for a review of water stress effects on growth of different tree tissues). The root and xylem conducting system are prevented from becoming excessively large relative to the foliage because the size of the foliage regulates the supply of carbohydrate for growth and maintenance of the lower parts of the tree. Less frequently, nutrient acquisition and use also may impose significant feedback effects (see Grigal, this volume, for a treatment of nutrient acquisition by trees).

The relationship between water balance and carbon balance and allocation is addressed later and in several other chapters (Bassow et al., Isebrands et al., Mäkelä, and Valentine, this volume). First, however, some of the principles of water transport and the development of tree water stress need to be considered, because it is through these principles that water relations feedback occurs.

A useful approach for examining how tree water stress is related to plant, atmospheric, and edaphic conditions is a rearrangement of the Van den Honert (1948) equation for water flux through the soil-plant-atmosphere continuum. The Van den Honert equation expresses flux as being equal to the water potential gradients divided by the relevant resistances for segments of the continuum. If an assumption is made that flux is quasi-steady state (Camacho et al., 1974), the gradients and resistances can be summed for several of the segments that exist in series. Then solving for leaf water potential (ψ_1) provides (Kaufmann, 1976):

$$\psi_1 = \psi_s + \psi_g - F \cdot r_{\text{soil to leaf}} \quad (1)$$

where ψ_s and ψ_g are soil and gravitational potentials, respectively; F is flux; and $r_{\text{soil to leaf}}$ is the resistance to liquid-phase flow from the soil to leaf. The flux term in Equation 1 may be taken as the transpiration rate:

$$F = g_s \cdot D \quad (2)$$

where g_s is leaf conductance and D is the vapor pressure difference from leaf to air.

Equation 1 is conceptual in the sense that expansion is required to account for the branched nature of the transport system and for the surface areas and volumes involved in absorption, internal transport, and transpiration (Richter, 1973; Tyree, 1988). The question arises, however, whether the expansion of Equation 1 is always necessary for addressing the effects of water stress on tree growth. Recalling remarks made earlier about the difficulty of using complex mechanistic models, we may conclude that the use of a simple model represented by Equation 1 is adequate in many situations for addressing the water relations phenomena affecting growth. The use of a simple, unbranched model would clearly not be adequate for incorporation into the Simple Whole Tree (SWT) model of Ford and Kiester (this volume), because their model includes a treatment of shoot branching. However, for less complex tree growth models intended for use at the level of stands, a simplified model predicting the development of water stress in relation to atmospheric, edaphic, and plant factors may be the most practical. Caution is in order, however, in view of the difficulty experienced by Edwards et al. (1986) in their attempt to develop an unbranched model for water transport.

Equation 1 is useful for describing in general terms how leaf water potential responds to changes in plant and environmental conditions and for determining several characteristics of the transport system (Elfving et al., 1972; Kaufmann, 1975, 1977; Kelliher et al., 1984; Sterne et al., 1977; Thompson and Hinckley, 1977). Figure 1 illustrates the types of responses of leaf water potential to transpiration rate obtained for several edaphic conditions. Transpiration is under the control of atmospheric conditions and stomatal function; and for non-limiting edaphic conditions, leaf water potential tracks along the upper curve in response to changes in transpiration.

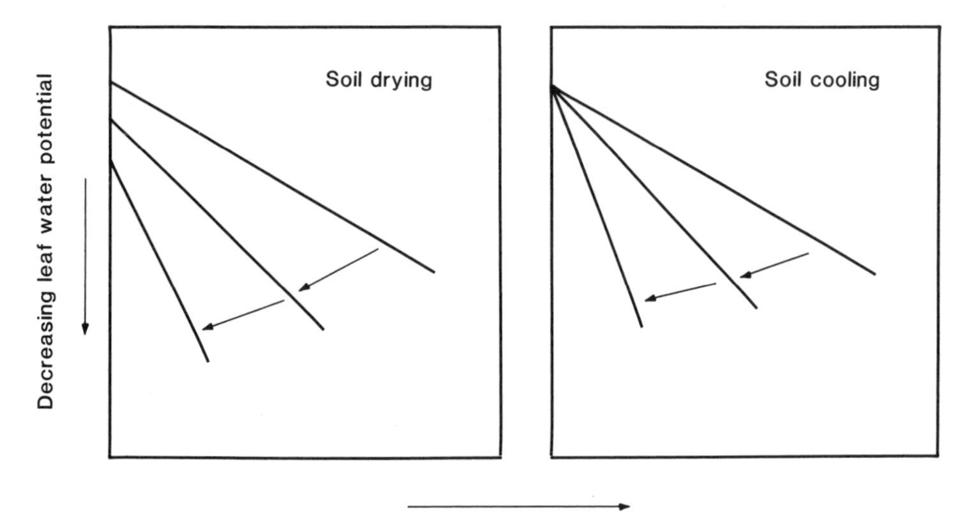

Figure 1. **Generalized relationship of leaf water potential to edaphic conditions over a range of transpiration rates. (*Left*) Soil drying, with changes in intercept occurring at very low transpiration rates (i.e., reduced pre-dawn leaf water potential) and slope resulting from increased flow resistances. (*Right*) Decreasing soil temperature, with changes in slope only, resulting from increased flow resistances. In both cases, the upper curve is for non-limiting soil water and temperature. While straight lines are shown, curvilinear relationships may be observed (Kaufmann, 1976).**

Effects of atmospheric and edaphic conditions on internal water transport and the development of water stress are the avenues through which water stress feedback is imposed on tree structure and development. Imagine, for example, that soil water supplies remain adequate during the growing season. Under these circumstances, shoot growth is normal; water flux depends primarily on leaf conductance and the vapor gradient, and it is not high enough to cause severe reductions in leaf water potential.

Water stress feedback occurs when something changes in Equation 1 causing leaf water potential to become low enough to cause stomatal closure. Reductions in leaf water potential can occur in several ways. First, high transpiration rates cause lower leaf water potentials, even if soil water potential is near zero. This is seen in Equation 1 because the flux-resistance term becomes larger; but it also is clear intuitively, because to accommodate a higher flux, the leaf water potential must decrease relative to soil water potential, thereby steepening the gradient for flow. A second common reason for leaf water potential reduction is a reduction in soil water potential. A third is an increase in the flow resistances from soil to leaf, for example, by root cooling, reduced root aeration, or by a restriction in the size of the root system. And finally, in tall trees, leaf water potential is reduced through the gravitational term at the rate of 0.1 MPa per 10 m of height. These effects also are obvious in Equation 1.

In the forest environment, reductions of leaf water potential may occur through each term in Equation 1. When the reductions are not severe enough to cause stomatal closure or loss of leaf turgor, photosynthesis and leaf expansion (or retention) may continue normally with no inhibition caused by water stress. Eventually, however, the leaf area of the crown becomes large enough that water stress occurs even when environmental conditions are normal. At this point, carbon gain and leaf expansion may become limited by water stress, and crown enlargement is restricted. In the annual cycle of physiological behavior, trees are probably subjected to feedback pressures from water stress and other factors a number of times, and the components of the water transport pathway of trees are fine-tuned to one another repeatedly.

Similar effects occur when irregular but often more adverse conditions result in severe water deficits. Very dry soil or hot, dry winds may result in drastic reductions in leaf water potential. This may result in the shedding of leaves and small branches as well as fine roots (Kozlowski, 1976). In fact, these episodic events may have stronger feedback effects on plant development than the more normal conditions, because the effects of shedding of plant parts persist longer than the effects of water stress on gas exchange, nutrient transport, and so on.

It follows, then, that forest sites differing in availability of soil water or evaporative demand will result in tree water stress occurring at different frequencies and intensities. Thus, through water stress feedback, trees on one site may never grow as fast or as large as those on another site characterized by less frequent or less intense water stress.

THE CONSEQUENCES OF WATER STRESS

Because trees are perennial and have relatively long growing seasons compared with annual plants, they are more resilient in their response to stressful conditions. An herbaceous plant subjected to water stress during leaf expansion is reduced in size for the duration of its life, whereas a tree, having larger reserves of water, nutrients, and carbohydrate in its sapwood, may recover almost fully and show little lasting effect of the stress. Miller (1965) demonstrated that stem elongation of loblolly pine (*Pinus taeda* L.) seedlings was rapid enough following a period of water stress to compensate for the reduction in growth during the stress period. While this response may not indicate that periods of water stress are insignificant if followed by a stress-free period (see Jackson et al., 1976, for evidence of prolonged effects on height growth), it does indicate that the consequences of drought may be hard to predict.

A wide variety of physiological processes in plants are affected by water stress. From the standpoint of understanding how water stress affects tree growth in forest stands, it is useful to focus on several types of responses that probably are most critical. Recalling the quote from Landsberg (1986) in the introduction to this chapter, there probably is no critical level of tree water stress at which adverse responses occur. Rather, as water stress becomes more severe, additional or more severe consequences of water stress become important.

The consequences of drought can be assessed by examining the changes that occur during drought, which might range from no change or damage (or even a benefit) to very severe damage and perhaps mortality. A positive effect of moderate water deficits is the hardening of plant tissues to make them more resistant to both water and temperature stress. Another response that may be positive is the triggering of reproduction. However, most other responses are negative.

Reduced Turgor and Stomatal Opening

While a number of cellular and tissue processes may be affected with very modest reductions in water potential (Bradford and Hsiao, 1982), perhaps the first critical effect of water stress on trees is loss of turgor followed by a reduction of stomatal opening. Both are gradual processes that do not occur at a single level of tissue water potential, but rather occur over a range of potentials. Loss of turgor in cells results from almost any reduction in leaf water potential, with turgor reaching a daily minimum near the time at which leaf water potential is lowest (depending upon changes in osmotic potential). The loss of turgor becomes critical for growth when turgor pressure is near zero (or, more specifically, below the threshold required for cell expansion) for a significant part of the day, or when it fails to recover substantially at night when plant water potential is highest because of decreased transpiration.

The diurnal range of turgor pressure differs for two types of conditions. The fluctuation of turgor pressures is highest for well-watered trees, especially when diurnal conditions favor high transpiration rates and nocturnal conditions favor complete recovery from water stress. The range of turgor pressures is lowest for trees under water stress severe enough to limit stomatal opening and transpiration. For these plants, leaf water potential changes much less through the course of a day-night cycle; the fluctuation in turgor pressure may be quite small, and the turgor pressure may remain low much of the time. Since tissue expansion occurs predominantly at night, growth of well-watered trees may be normal even when turgor pressure is reduced sharply during the day. In contrast, growth of stressed trees is restricted because turgor pressure is not restored at night.

A complicating feature of water absorption and transport in the soil-plant-atmosphere continuum that presents some difficulties in modeling growth is that water is not extracted uniformly from the soil profile. For wet soil, water uptake occurs primarily in the zone having the most roots. In dry soil, extraction occurs primarily from deep in the profile, usually at a more restricted rate that results in tree water stress. During the transition from wet to dry soil profiles, however, water may be extracted primarily from moderately dry soil having the highest root density during the daytime; but at low nocturnal absorption rates, the uptake may occur primarily from the less dense portion of the root system in deeper parts of the profile where water potential is higher. While the deeper roots in wetter soil may take up water rapidly during the day, there are too few such roots to keep significant tree water stress from occurring. Thus, during this transition period, daytime tree water stress may be substantial, but nighttime recovery from water stress may be nearly complete (see Sterne et al., 1977, for an example of this water absorption pattern in avocado (*Persea* sp.)). The effect of this phenomenon on the use of pre-dawn measurements of tree water potential as a measure of integrated water stress is discussed later.

Similarities exist between loss of turgor and reductions of stomatal opening, but significant differences also exist. Water stress causing loss of turgor in various plant tissues affects turgor in stomatal guard cells, but guard cell turgor also depends on dynamic osmotic effects unique to those cells. Consequently, the effects of water stress on stomatal function and carbon fixation may be different from those on cell turgor and tissue growth.

Examining the scenario of progressing from no water stress to severe water stress, we may assess how stomatal function and reduced photosynthesis are similar to or different from turgor effects on growth. Under well-watered conditions, diurnal stomatal opening is normal, with visible irradiance and atmospheric humidity the primary factors regulating variations in leaf conductance (Kaufmann, 1982a, 1982b). The maintenance of adequate guard-cell turgor for stomatal opening when reductions in daytime turgor occur in all other plant tissues illustrates the dynamic nature of turgor maintenance in guard cells. For well-watered trees, there may be relatively little daytime variation in leaf conductance, and both carbon fixation and tissue growth may occur at nearly optimum rates.

With progression to moderate restrictions in soil water availability, diurnal reductions in leaf water potential may become large enough for partial stomatal closure to occur. Under these conditions, stomata typically open in the morning because of overnight recovery of plant water content, but they close partially during midday. Afternoon recovery of stomatal opening (e.g., closure limited to a midday period) is uncommon because renewed transpiration prevents substantial recovery from water stress; and finally, reduced light conditions favor stomatal closure. In this situation, stomata respond primarily to irradiance and humidity, but leaf water stress reduces conductance below that predicted for the light and humidity encountered. Finally, under severe water stress, stomata fail to open even in the morning because nocturnal recovery is insufficient. Through this progression, both tissue growth and photosynthesis are reduced. There may be a stage at which growth has stopped but a positive net photosynthesis occurs, leading to modest increases in stored carbohydrate during the stress period. This may have contributed to the rapid recovery from water stress observed in seedlings subjected to drying cycles (Miller, 1965).

Pre-dawn leaf water potential has been used successfully as an index of integrated plant water stress (Donner and Running, 1986; Hinckley and Bruckerhoff, 1975; Running, 1984) because it provides a measure of a tree's ability to obtain adequate soil water for recovery from the previous day's water stress. It also is used for estimating the maximum leaf conductance during the day following measurement (Running, 1976). For modeling purposes, these uses are often appropriate, and they represent the types of phenomenological functions discussed earlier.

Pre-dawn measurements of water potential must be used cautiously, however. First, similar pre-dawn water potentials over a period of time suggest that soil water availability remains constant, but we saw in the discussion above that nocturnal recovery in leaf water potential and plant water content may be nearly complete even when only a portion of the root system is in moist soil (Sterne et al., 1977). Thus, during the daytime, far different levels of water stress and leaf conductance may develop in response to different levels of soil water availability, even though pre-dawn leaf water potentials are equal.

Second, it is fairly well agreed that stomata respond primarily to light and humidity and secondarily to water stress and temperature (Landsberg, 1986). Accordingly, pre-dawn water stress may be a useful observation for estimating the maximum leaf conductance, but it should not be used to the exclusion of light, humidity, and perhaps a measurement or prediction of daytime leaf water potential, because these terms determine the diurnal course of conductance.

Compensation Through Diebacks of Shoots and Roots

When moderate water stress occurs, fine root turnover is common, apparently occurring with almost every period during which the forest litter/mineral soil interface becomes dry (Deans, 1979). Santantonio and Hermann (1985) found that the mean standing crop of live fine roots in Douglas-fir (*Pseudotsuga menziesii* [Mirb.] Franco) turned over 1.7 times per year on a wet site, but 2.8 times per year on a dry site (Table 1). Similarly, early leaf senescence may occur when significant water stress develops during the season, particularly in deciduous species; and in conifers, annual patterns of soil water depletion may lead to periods of senescence of the oldest needles. The dieback of fine roots undoubtedly affects root water absorption immediately after precipitation rewets the forest litter and upper soil layers, but fine root production may rapidly increase the capacity for water uptake. In temperate climates, the senescence of foliage in response to

Table 1. Estimated live biomass of fine roots (< 1 mm diameter), annual root turnover, and the annual turnover rate for mature Douglas fir (*Pseudotsuga menziesii* [Mirb.] Franco) trees in dry, moderate, and wet sites in Oregon, U.S.A. (from Santantonio and Hermann, 1985).

Site	Live roots < 1 mm diameter (Mg/year)	Root turnover (Mg/ha/year)	Turnover rate (times/year)
Dry	2.53	7.2	2.8
Moderate	3.50	7.2	2.0
Wet	3.15	5.5	1.7

water stress is almost never offset by new foliage production during that growing season. Consequently, the photosynthetic capacity of the tree crown may be reduced, but transpiration and the likelihood of increased tree water stress also are reduced.

More severe water stress, especially that associated with episodic events such as infrequent drought, may be accompanied by dieback of larger roots and shoots. Rarely does such a drought lead directly to tree mortality, but mortality can occur in certain circumstances. Linder et al. (1987) reported on the effects of contrasting precipitation patterns during two years of an irrigation-fertilization experiment on 10-year-old radiata pine. The first year was characterized by a wet summer, but a severe drought occurred during the second. For unirrigated trees, canopy growth was extensive during the wet summer, especially in fertilized trees, a response consistent with minimal feedback effects of water stress discussed earlier. During the dry summer that followed, however, soil water was rapidly depleted because of the high leaf areas. In the unfertilized trees, this resulted in severe water stress, reduced growth of new foliage, and dieback of foliage older than one year; in the fertilized stands, some mortality occurred. This experiment inadvertently provided ideal circumstances for catastrophe: stimulated shoot growth by fertilizing during a wet growing season, followed by a severe drought the following season.

More commonly, a severe reduction in growth and dieback of shoots and roots might be expected during a very dry year, with mortality limited to conditions involving a combination of factors such as suppression in a dense stand or weakened resistance to insects or disease. Even in the absence of mortality, however, water stress severe enough to cause dieback of branches or larger roots undoubtedly has lasting effects on tree growth, because both water absorption and net photosynthesis must be reduced for at least part of the current and perhaps the next growing season.

Infrequent but recurring periods of fairly severe water deficits serve a useful purpose, however, in that they regulate the balance among plant parts and generally help avoid the vulnerability observed in the radiata pine experiment discussed above. From a modeling standpoint, these responses probably should be incorporated in process or phenomenological models of tree growth intended to project the growth of trees and stands over long periods of time. To ignore these effects could well result in models in which annual or long-term growth either is overestimated or is kept in check for the wrong reasons, for example, as a result of using predicted growth rates that are lower than expected or observed under more moderate conditions.

The irrigation-fertilization study of Linder et al. (1987) illustrates another point, namely, the interaction of water stress with other factors affecting growth. In their study, nutrient supply during a wet year favored shoot production and drastically altered the balance of the water transport pathway, resulting in trees unable to cope with severe drought. Obviously, growth and the control of growth are complex, and models representing growth must account for the significant interactions among factors.

Altered Carbon Allocation and Growth Efficiency

Several effects of water stress on internal carbon allocation have already been discussed, namely, effects on the balance of crowns and roots and effects on fine root and foliage turnover or, in severe cases, turnover of larger roots and branches. The degree to which internal allocation of fixed carbon is influenced by water stress is beyond the scope of this paper (see Mäkelä, this volume; Valentine, this volume), but some attention should be given to the question of growth efficiency.

One of the most important processes in plants is the interception of solar radiation and the partial conversion of radiant energy into photosynthetic products. Obviously, the potential for carbon fixation increases as the amount of leaf area increases or as the display of foliage is enhanced, because more radiant energy is intercepted by the foliage. Equally obvious, however, is that increased energy interception is accompanied by increased transpiration; and at the higher leaf areas, self-shading limits further gains in photosynthesis. One way to examine this trade-off is to evaluate the growth efficiency of trees.

Growth efficiency can be defined in several ways. Common expressions of growth efficiency relate some measure of growth (volume growth is typical) to available solar irradiance, nutrients, or water. Jarvis and Leverenz (1983) described how the upper bound of stand growth rate is related to intercepted photosynthetically active radiation, and actual growth reflects reductions caused by environmental stress, by structural, physiological, and biochemical properties of the canopy and leaves, and by respiratory losses, turnover of leaves and roots, and mortality (Figure 2). The ratio of growth to intercepted irradiance is a measure of growth efficiency, and it is clear that growth efficiency is reduced by any of the factors mentioned.

Kaufmann and Ryan (1986) examined growth efficiency, expressed as the annual volume increment per unit of absorbed irradiance, for three subalpine conifers. Their results indicated a decline in efficiency with age, but an increase in efficiency as site

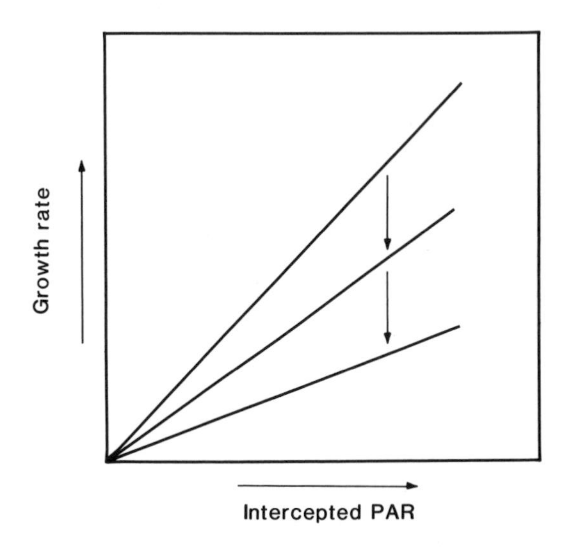

Figure 2. Relationship between growth rate and intercepted photosynthetically active radiation (PAR). (*Upper line*) Canopy performing at maximum efficiency. (*Middle line*) Reduction in efficiency attributed to structural, physiological, and biochemical properties of the canopy and leaves or to environmental stress. (*Lower line*) Additional losses of efficiency through respiration, turnover of leaves and roots, and mortality.

azimuth departed from due south. They suggested that the effect of azimuth may be related to higher temperatures (therefore higher respiration rates) and perhaps to reduced water supply on south slopes. They also suggested that the decline in efficiency with tree age may be related to increased maintenance respiration requirements and possibly increased carbon allocation to root production as well as reduced photosynthetic rates in older trees.

More recent studies by Ryan (1988) indicate that increased maintenance respiration rates cannot account for the entire decrease in growth efficiency. A reasonable hypothesis is that significant reductions may occur through steady declines in net photosynthesis as trees age. A nearly universal sign of maturity in forest trees is a flattening of the tops of crowns. While no clear explanation for this phenomenon has been developed, it is likely that as trees become taller, gravitational effects on water transport and tissue water potential become more influential. Eventually, water stress may become a significant factor limiting foliar gas exchange, and there may be implications regarding nutrient supply to the foliage and carbohydrate supply to the roots (Ford and Kiester, this volume). The height at which this occurs may be determined both by environmental conditions and by genetic effects on the structural characteristics of each species. The role of tree water stress in photosynthesis, growth efficiency, and tree growth as trees approach maturity deserves more research attention.

Water use efficiency, generally expressed as the amount of dry matter produced per amount of water transpired, typically is stated as a rate determined from instantaneous measurements of net photosynthesis and transpiration (e.g., Caldwell et al., 1986). While of some value for comparing the physiological performance of plants, this estimate of water use efficiency ignores factors that are important in long-term growth of trees. A major factor is annual respiration of the non-photosynthetic plant parts, and another is turnover of foliage, stem, and root dry matter.

Kaufmann (1985) attempted to evaluate water use efficiency over the course of a full rotation for several subalpine forest species (Table 2). Volume growth was estimated from an empirical growth and yield model developed for each species, and estimates of water use from a transpiration model (Kaufmann, 1984). During a stand rotation having a typical management scenario of thinning, tree water use efficiency was 0.0025 m^3 merchantable volume produced/m^3 transpired water for lodgepole pine (*Pinus contorta* var. *latifolia* Engelm.) compared with 0.0013–0.0014 m^3/m^3 for a stand of Engelmann spruce (*Picea engelmannii* Parry ex. Engelm.) and subalpine fir (*Abies lasiocarpa* [Hook.] Nutt.). While such calculations are subject to large error associated with a variety of causes, they illustrate an attempt to relate water use to annual measures of tree growth (specifically to merchantable volume) rather than to instantaneous measures of net photosynthesis. Growth models predicting annual or longer term growth may provide improved estimates of water use efficiency applicable to the forest stand.

Table 2. Total merchantable volume production, transpiration, and water use efficiency (ratio of volume production to transpiration) for lodgepole pine (*Pinus contorta* var. *latifolia* Engelm.) and Engelmann spruce (*Picea engelmannii* Parry ex. Engelm.) and subalpine fir (*Abies lasiocarpa* [Hook.] Nutt.) stands in Colorado, U.S.A. (from Kaufmann, 1985).

	Lodgepole pine	Engelmann spruce–subalpine fir	
	(120 years)	(120 years)	(180 years)
Volume production (m^3/ha)	505	448	762
Transpiration (m^3/ha)	205,000	323,000	602,000
Water use efficiency (m^3/m^3)	0.0025	0.0014	0.0013

SUMMARY

The representation of tree growth using process models requires an understanding of tree growth in relation to prevailing environmental conditions and to extreme environmental events or conditions. In all environments, trees develop a suitable architectural balance among the components of the transpiring system, and this balance is achieved through feedback processes involving water transport and carbohydrate availability for growth. There probably is no threshold level of water potential at which all adverse effects begin. Rather, as water stress becomes progressively more severe, the consequences also become more important. Many of the consequences or effects of environmental conditions may be taken into account in the context of continuous physiological adjustments in the structure and function of different parts of plants. Research is needed to further the development of phenomenological expressions of tree water relations useful for models of tree growth at the whole-tree or stand level. Research also is needed to determine how to account for the effects of episodic or catastrophic events in tree growth models.

LITERATURE CITED

Bassow, S. L., E. D. Ford, and R. Kiester. 1989. A critique of carbon-based tree growth models. In: Dixon, R. K., R. S. Meldahl, G. A. Ruark, and W. G. Warren, eds. *Process Modeling of Forest Growth Responses to Environmental Stress.* Timber Press, Portland, OR, U.S.A.

Bradford, K. J., and T. C. Hsiao. 1982. Physiological responses to moderate water stress. In: Lange, O. L., P. S. Nobel, C. B. Osmond, and H. Ziegler, eds. *Encyclopedia of Plant Physiology,* 12B. Springer-Verlag, Berlin, F.R.G. Pp. 263–324.

Caldwell, M. M., H.-P. Meister, J. D. Tenhunen, and O. L. Lange. 1986. Canopy structure, light microclimate and leaf gas exchange of *Quercus coccifera* L. in a Portuguese macchia: measurements in different canopy layers and simulations with a canopy model. *Trees* 1:25–41.

Camacho-B, S. E., M. R. Kaufmann, and A. E. Hall. 1974. Leaf water potential response to transpiration by citrus. *Physiologia Plantarum* 31:101–105.

Deans, J. D. 1979. Fluctuations of the soil environment and fine root growth in a young Sitka spruce plantation. *Plant and Soil* 52:195–208.

Dixon, R. K. 1989. Physiological processes and tree growth. In: Dixon, R. K., R. S. Meldahl, G. A. Ruark, and W. G. Warren, eds. *Process Modeling of Forest Growth Responses to Environmental Stress.* Timber Press, Portland, OR, U.S.A.

Donner, B. L., and S. W. Running. 1986. Water stress response after thinning *Pinus contorta* stands in Montana. *Forest Science* 32:614–625.

Edwards, W. R. N., P. G. Jarvis, J. J. Landsberg, and H. Talbot. 1986. A dynamic model for studying flow of water in single trees. *Tree Physiology* 1:309–324.

Elfving, D. C., M. R. Kaufmann, and A. E. Hall. 1972. Interpreting leaf water potential measurements with a model of the soil-plant-atmosphere continuum. *Physiologia Plantarum* 27:161–168.

Ford, E. D., and A. R. Kiester. 1989. Modeling the effects of pollutants on the processes of tree growth. In: Dixon, R. K., R. S. Meldahl, G. A. Ruark, and W. G. Warren, eds. *Process Modeling of Forest Growth Responses to Environmental Stress.* Timber Press, Portland, OR, U.S.A.

Grigal, D. F. 1989. Mechanistic modeling of nutrient acquisition by trees. In: Dixon, R. K., R. S. Meldahl, G. A. Ruark, and W. G. Warren, eds. *Process Modeling of Forest Growth Responses to Environmental Stress.* Timber Press, Portland, OR, U.S.A.

Hinckley, T. M., and D. N. Bruckerhoff. 1975. The effects of drought on water relations and stem shrinkage of *Quercus alba. Canadian Journal of Botany* 53:62–72.

Isebrands, J. G., H. M. Rauscher, T. R. Crow, and D. I. Dickmann. 1989. Whole-tree growth process models based on structural-functional relationships. In: Dixon, R. K., R. S. Meldahl, G. A. Ruark, and W. G. Warren, eds. *Process Modeling of Forest Growth Responses to Environmental Stress.* Timber Press, Portland, OR, U.S.A.

Jackson, D. S., H. H. Gifford, and J. Chittenden. 1976. Environmental variables influencing the increment of *Pinus radiata:* (2) effects of seasonal drought on height and diameter increment. *New Zealand Journal of Forestry Science* 5:265–286.

Jarvis, P. G., and J. W. Leverenz. 1983. Productivity of temperate, deciduous and evergreen forests.

In: Lange, O. L., P. S. Nobel, C. B. Osmond, and H. Ziegler, eds. *Encyclopedia of Plant Physiology,* New Series 12D. Springer-Verlag, Berlin, F.R.G. Pp. 233–280.
Kaufmann, M. R. 1975. Leaf water stress in Engelmann spruce—influence of the root and shoot environments. *Plant Physiology* 56:842–844.
Kaufmann, M. R. 1976. Water transport through plants: current perspectives. In: Wardlaw, I. F., and J. B. Passioura, eds. *Transport and Transfer Processes in Plants.* Academic Press, New York, NY, U.S.A. Pp. 313–327.
Kaufmann, M. R. 1977. Soil temperature and drying cycle effects on water relations of *Pinus radiata. Canadian Journal of Botany* 55:2413–2418.
Kaufmann, M. R. 1982a. Leaf conductance as a function of photosynthetic photon flux density and absolute humidity difference from leaf to air. *Plant Physiology* 69:1018–1022.
Kaufmann, M. R. 1982b. Evaluation of season, temperature, and water stress effects on stomata using a leaf conductance model. *Plant Physiology* 69:1023–1026.
Kaufmann, M. R. 1984. A canopy model (RM-CWU) for determining transpiration of subalpine forests. I. Model development. *Canadian Journal of Forest Research* 14:218–226.
Kaufmann, M. R. 1985. Species differences in stomatal behavior, transpiration, and water use efficiency in subalpine forests. In: Tigerstedt, P. M. A., P. Puttonen, and V. Koski, eds. *Crop Physiology of Forest Trees.* Helsinki University Press, Helsinki, Finland. Pp. 39–52.
Kaufmann, M. R., and M. G. Ryan. 1986. Physiographic, stand, and environmental effects on individual tree growth and growth efficiency in subalpine forests. *Tree Physiology* 2:47–59.
Kelliher, F. M., T. A. Black, and A. G. Barr. 1984. Estimation of twig xylem water potential in young Douglas-fir trees. *Canadian Journal of Forest Research* 14:481–487.
Kozlowski, T. T. 1976. Water supply and leaf shedding. In: Kozlowski, T. T., ed. *Water Deficits and Plant Growth,* Vol. 4. Academic Press, New York, NY, U.S.A. Pp. 191–231.
Kramer, P. J. 1980. Drought, stress, and the origin of adaptations. In: Turner, N. C., and P. J. Kramer, eds. *Adaptation of Plants to Water and High Temperature Stress.* John Wiley and Sons, New York, NY, U.S.A. Pp. 7–20.
Kramer, P. J., and T. T. Kozlowski. 1979. *Physiology of Trees.* Academic Press, New York, NY, U.S.A. 811 pp.
Landsberg, J. J. 1986. *Physiological Ecology of Forest Production.* Academic Press, London, U.K. 198 pp.
Linder, S., M. L. Benson, B. J. Myers, and R. J. Raison. 1987. Canopy dynamics and growth of *Pinus radiata.* I. Effects of irrigation and fertilization during a drought. *Canadian Journal of Forest Research* 17:1157–1165.
Mäkelä, A. 1989. Modeling structure-functional relationships in whole-tree growth: resource allocation. In: Dixon, R. K., R. S. Meldahl, G. A. Ruark, and W. G. Warren, eds. *Process Modeling of Forest Growth Responses to Environmental Stress.* Timber Press, Portland, OR, U.S.A.
Miller, L. N. 1965. Changes in radiosensitivity of pine seedlings subjected to water stress during chronic gamma irradiation. *Health Physics* 11:1653–1662.
Richter, H. 1973. Frictional potential losses and total water potential in plants: a re-evaluation. *Journal of Experimental Botany.* 24:983–994.
Rook, D. A., R. H. Swanson, and A. M. Cranswick. 1976. Reaction of radiata pine to drought. Reprint No. 1036, New Zealand Forest Service, Rotorua, New Zealand. Pp. 55–68.
Running, S. W. 1976. Environmental control of leaf water conductance in conifers. *Canadian Journal of Forest Research* 6:104–112.
Running, S. W. 1984. Documentation and preliminary validation of H_2OTRANS and DAYTRANS, two models for predicting transpiration and water stress in western coniferous forests. USDA Forest Service Rocky Mountain Forest and Range Experiment Station, Fort Collins, CO, U.S.A. Research Paper RM-252. 45 pp.
Ryan, M. G. 1988. The importance of maintenance respiration by the living cells in the sapwood of subalpine conifers. Ph.D. Thesis, Oregon State University, Corvallis, OR, U.S.A. 104 pp.
Santantonio, D., and R. K. Hermann. 1985. Standing crop, production, and turnover of fine roots on dry, moderate, and wet sites of mature Douglas-fir in western Oregon. *Annals Science Forestry* 42:113–142.
Sheriff, D. W., and D. Whitehead. 1984. Photosynthesis and wood structure in *Pinus radiata* D. Don during dehydration and immediately after rewatering. *Plant, Cell and Environment* 7:53–62.
Sterne, R. E., M. R. Kaufmann, and G. A. Zentmyer. 1977. Environmental effects on transpiration and leaf water potential in avocado. *Physiologia Plantarum* 41:1–6.
Thompson, D. R., and T. M. Hinckley. 1977. A simulation of water relations of white oak based on soil moisture and atmospheric evaporative demand. *Canadian Journal of Forest Research* 7:400–409.
Thornley, J. H. M. 1976. *Mathematical Models in Plant Physiology.* Academic Press, London, U.K. 318 pp.

Tyree, M. T. 1988. A dynamic model for water flow in a single tree: evidence that models must account for hydraulic architecture. *Tree Physiology* 4:195–217.

Valentine, H. 1989. A carbon-balance model of tree growth with a pipe-model framework. In: Dixon, R. K., R. S. Meldahl, G. A. Ruark, and W. G. Warren, eds. *Process Modeling of Forest Growth Responses to Environmental Stress.* Timber Press, Portland, OR, U.S.A.

Van den Honert, T. H. 1948. Water transport in plants as a catenary process. *Discussions of the Faraday Society* 3:146–153.

Zahner, R. 1968. Water deficit and growth of trees. In: Kozlowski, T. T., ed. *Water Deficits and Plant Growth,* Vol. 2. Academic Press, New York, NY, U.S.A. Pp. 191–254.

Zahner, R., and C. E. Grier. 1989. Concept for a model to assess the impact of climate on the growth of the southern pines. In: Dixon, R. K., R. S. Meldahl, G. A. Ruark, and W. G. Warren, eds. *Process Modeling of Forest Growth Responses to Environmental Stress.* Timber Press, Portland, OR, U.S.A.

SECTION II

Tree Structure and Function

The architecture of a tree affects the ways in which it can cope with environmental conditions and largely determines its ability to compete with other plants. The morphology of the shoot and root systems reflects the plants' functional need to obtain light, water, and nutrients in a balanced fashion. Process models may be used to describe the manner in which a tree integrates its responses to the above- and belowground environments as it grows.

The topics of tree root and crown system dynamics are emphasized in this section. Concomitant linkage of dynamic root and shoot system processes, a more complex topic, is also addressed. The authors discuss fundamental physiological processes such as photosynthesis, nutrient acquisition, and dormancy within the context of tree structure.

Several conceptual frameworks for modeling the growth of a whole tree or a selected tree component are presented. The common tack is to assume that the structure and size of the various plant parts are dictated by their functional roles. Thus mathematical relationships can be used to maintain proportional relationships among plant parts, such as foliage and fine roots, as the tree grows. Some models assume a constant proportioning among parts over time, while others allow carbon allocation to vary in response to changes in the stage of plant development and fluctuations in environmental conditions.

CHAPTER 7

MODELING STRUCTURAL-FUNCTIONAL RELATIONSHIPS IN WHOLE-TREE GROWTH: RESOURCE ALLOCATION

Annikki Mäkelä

Abstract. This paper reviews models of resource allocation and tree growth which derive the distribution pattern from balanced-growth considerations and optimality principles. The functional balance and the pipe-model theories are considered, and the analysis of height growth, treated as an evolutionary game, is described. The allocation principles are connected to a carbon-balance model of tree growth. The approach appears promising, but more empirical data are needed for testing and further development of the models. In particular, more information is needed on branch and height growth and heartwood formation.

INTRODUCTION

In the current development of process-oriented stand growth models, one of the most difficult questions is resource allocation. "Partitioning coefficients" are customarily used for assigning a share of total growth to each organ (Thornley, 1976; De Wit, 1978; Ågren et al., 1980; McMurtrie and Wolf, 1983; Mäkelä and Hari, 1986; Mohren, 1987), but no general procedure for determining these coefficients has emerged. In some applications, the straightforward assumption of constant allocation ratios has given reasonable predictions (McMurtrie and Wolf, 1983; Mäkelä and Hari, 1986); however, empirical data indicate that there is a temporal trend, especially in the partitioning between foliage and wood (Mohren, 1987). Also, the allocation ratios appear to vary with environmental conditions (White, 1937; Ingestad, 1979; Richards et al., 1979). To account for this variation, the allocation coefficients should be derivable from the physiological state and the environment of the tree.

One way of doing this is to analyze the physiological mechanisms of assimilate partitioning. The transport resistance model by Thornley (1972), where assimilated carbon and nitrogen are transported along concentration gradients and the partitioning pattern is a result of substrate-controlled growth rates thus achieved in the shoots and roots, seems to function reasonably well at least for crop plants; however, it does not account for the annual cycle important in trees. A "supply and demand" approach has been suggested to describe, e.g., seasonal variations in the growth activities at different locations (Landsberg, 1986). By removing the problem of allocation, however, this method appears to have created one of demand. To follow through the mechanistic approach eventually leads to

Dr. Mäkelä is in the Department of Silviculture at the University of Helsinki, Unioninkatu 40 B, 00170 Helsinki, Finland.

the analysis of the hormonal regulation of growth, which is likely to increase the dimension of the model more than is desirable in large-scale applications.

In contrast to the mechanistic approach, a number of recent studies derive the allocation patterns from different balanced-growth considerations (Reynolds and Thornley, 1982; Valentine, 1985; Mäkelä, 1986; Bossel, 1986; Ludlow et al., this volume). Among these are the principle of functional balance (Davidson, 1969), which states that the shoot-root ratios are adjusted so as to maintain a balanced carbon-nitrogen ratio in the plant, and the pipe-model theory (Shinozaki et al., 1964), which reasons that each unit of foliage requires a unit pipeline of wood to conduct water from the roots and to provide physical support. Based on empirical findings, these principles have been explained as evolutionary adaptations (Shinozaki et al., 1964; Davidson, 1969). Indeed, a balanced allocation to organs supplying mutually exclusive, vital elements, has been shown to maximize relative growth rate (e.g., Mäkelä and Sievänen, 1987).

As an extension of the idea that plant form is an instrument of adaptation, allocation patterns have also been treated as an optimization problem in which a function approximating the chance of survival is maximized under limitations set by the metabolic processes. For instance, this approach has been applied to the timing of reproductive growth (Vincent and Pulliam, 1980), shoot-root ratios (Iwasa and Roughgarden, 1984), crown shape (Iwasa et al., 1984), and height growth (Mäkelä, 1985).

The "evolutionary approach" appears attractive because of its relative simplicity, and the fact that it seems to provide a connection between the formation of structure and the metabolic process considered. To what extent can such derivations be applied to stand growth models at present, and what are the major problems concerning their further development? The aim of this chapter is to give insight into these questions by reviewing some of the allocation principles and their application to tree and stand growth models.

A CARBON-BALANCE FRAMEWORK

In order to focus on the basic principles of the allocation pattern yet simultaneously to look at their consequences for whole-tree or whole-stand growth, I use a simple carbon-balance framework. It incorporates the biomass of the organs, W_i; the total growth rate, G; the senescence rates of the parts, S_i; and the partitioning coefficients, λ_i ($i \varepsilon N$ = the set of indices of the parts). The latter are defined as the proportional share of total growth attributable to each biomass compartment. The variables can refer to either an individual tree or a whole stand.

The growth rates of the biomass compartments are

$$\frac{dW_i}{dt} = \lambda_i G - S_i \qquad \text{for all } i \varepsilon N \tag{1}$$

i.e., net growth is the difference between gross growth and senescence.

On an annual basis, it can be assumed that the carbon assimilated is totally consumed in growth and respiration. If annual average photosynthetic rate (kg C/year) is denoted by P, annual respiration (kg C/year) by R, and the carbon content of dry matter by f_C, annual growth, G (kg dry weight/year), is therefore

$$G = (P - R)/(fc) \tag{2}$$

Photosynthesis is proportional to foliage biomass, W_f, and to the annual average specific photosynthetic rate, σ_C, (kg C/year/kg dry weight/year)

$$P = \sigma_C W_f \tag{3}$$

σ_C depends on environmental and internal factors. In particular, it is a function of the light conditions within the stand. A simple way of accounting for this is provided by Beer's law, which states that the canopy-average specific photosynthetic activity depends upon the foliage biomass in the canopy:

$$\sigma_C = \sigma_{C0} \frac{1 - e^{-kB}}{kB} \quad (4)$$

where k is extinction coefficient, σ_{C0} is the unshaded specific photosynthetic rate, and B is canopy foliage biomass (Monsi and Saeki, 1953).

Respiration is divided into growth and maintenance respiration, R_g and R_m, respectively (McCree, 1970).

$$R = R_g + R_m \quad (5)$$

Growth respiration is proportional to growth rate, and maintenance respiration is proportional to the respiring biomass.

$$R_g = r_g G, \qquad R = \Sigma\, r_{mi} W_i \quad (6)$$

where r_g and r_{mi} are constant coefficients.

Senescence is customarily determined on the basis of specific senescence, s_i, which is the reciprocal of the average lifetime of the organ.

$$S_i = s_i W_i \quad (7)$$

Equations 1–7 represent a basic framework for a carbon-balance model (Thornley, 1976; De Wit, 1978; Ågren et al., 1980; McMurtrie and Wolf, 1983). In the following sections, the determination of the partitioning coefficients is considered.

FUNCTIONAL BALANCE

Principles

The principle of functional balance refers to the observation that the shoot-root ratios in plants appear to be adjusted so as to balance the uptake of carbon and nitrogen (e.g., White, 1937; Brouwer, 1962). The principle was first formulated mathematically by Davidson (1969), who introduced the following simple model of assimilate partitioning:

$$\sigma_N W_r = \pi \sigma_C W_s \quad (8)$$

where σ_C and σ_N are the specific carbon and nitrogen assimilation rates, W_s and W_r are the shoot and root dry weights, and π is an empirical parameter which is assumed constant over a range of shoot and root environments. Davidson (1969) argued that the bases for this relationship were the requirements that (1) carbon and nitrogen are used in a constant ratio for dry matter growth, and (2) the assimilation of the elements is in balance with their utilization. The coefficient of proportionality in Equation 8, π, can therefore be interpreted as the amount of nitrogen in dry matter growth per unit amount of carbon.

Equation 8 is based upon empirical observation and provides in certain cases a good approximate description of the variation in the shoot-root ratios with varying assimilation rates (e.g., Richards et al., 1979). However, the implicit assumption of a constant ratio of nitrogen to carbon, π, does not always hold true (e.g., Ingestad, 1979). To account for the variation in π and yet to describe the functional balance, Reynolds and Thornley (1982) developed a more detailed, dynamic model of assimilate partitioning to shoots and roots. This model allows for variation in the carbon and nitrogen substrate concentrations by treating the substrates as state variables, in addition to the structural biomass compartments. Two alternative ways of incorporating the idea of functional balance were considered, where the allocation coefficients were determined (1) so as to maintain a constant substrate nitrogen:carbon ratio, and (2) from the requirement of maximum relative growth rate.

It turns out that these models reproduce the condition of Equation 8, with π as a variable depending upon the substrate nitrogen and carbon concentrations (Reynolds and Thornley, 1982). Furthermore, it has been shown that in the optimal growth version of the model, the extent of environment-induced variation in π is determined by what has been assumed about substrate utilization in structural growth (Mäkelä and Sievänen, 1987). The less the carbon and nitrogen substrates can compensate for each other, the more constant the value of π (Figure 1a). Moreover, in spite of the difference in the variation of π, the corresponding variation in the ratio of shoot to total dry weight is almost identical for all substrate utilization patterns (Figure 1b).

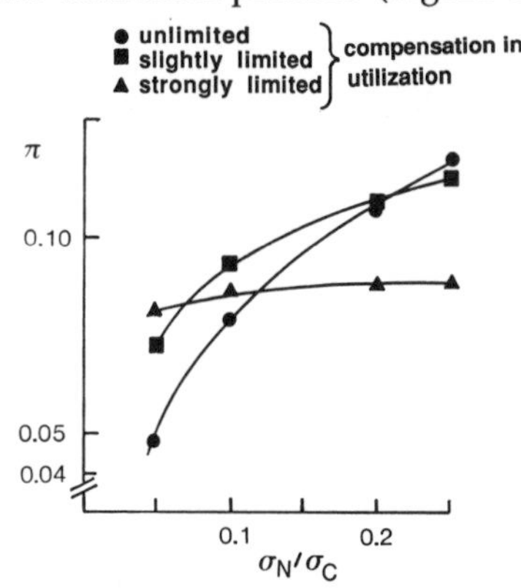

Figure 1a. **Ratio of nitrogen to carbon in the plant as a function of the environment, with different assumptions on substrate utilization in structural growth. ● Nitrogen and carbon compensate each other strongly; ■ nitrogen and carbon compensate each other slightly; ▲ nitrogen and carbon are utilized in a fixed ratio.**

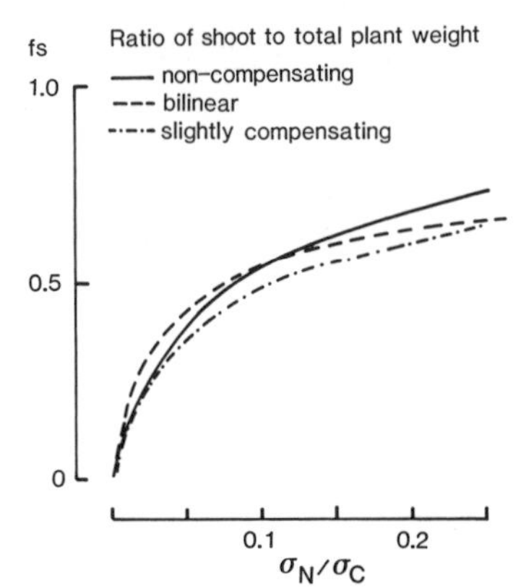

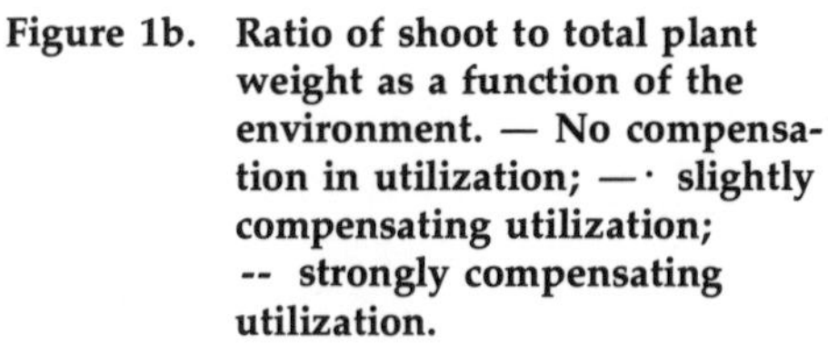

Figure 1b. **Ratio of shoot to total plant weight as a function of the environment. — No compensation in utilization; —· slightly compensating utilization; -- strongly compensating utilization.**

These results suggest that Equation 8 can be used as a good approximation of the more detailed assumptions of the Reynolds and Thornley (1982) model. However, the impact of some further physiological details on the formulation of the functional balance remains to be investigated. First, the above considerations do not take into account the possibility that the specific activities of the shoots and roots may depend on the substrate concentrations (cf. Ågren and Ingestad, 1987). Second, although the substrate concentrations vary, the structural contents of carbon and nitrogen are assumed constant. The general method of determining the functional balance by maximizing the relative growth rate at steady state is easily applied to such extensions (Mäkelä and Sievänen, 1987).

Application to Carbon Balance

In the carbon-balance model presented above, the partitioning of growth between foliage and feeder roots can be derived from the requirement that, in steady state, Davidson's functional balance (Equation 8) is satisfied. This yields the following relationship between the partitioning coefficients for foliage, λ_f, and roots, λ_r (Mäkelä, 1986):

$$\lambda_r = a_r\lambda_f + \beta_r \tag{9}$$

where

$$a_r = \pi\frac{\sigma_N}{\sigma_C}, \qquad \beta_r = \frac{S_r - a_rS_f}{G}$$

where the notation is as above. The partitioning between foliage and feeder roots is hence determined by (1) the relative specific activities of the two organs, and (2) their turnover rates relative to each other.

The significance of both roots and foliage for the growth of the tree in this model can be illustrated by defining the productive biomass as the sum of the foliage and feeder root biomass. A "specific activity" of this productive biomass can then be expressed as

$$\sigma = \frac{\sigma_C\sigma_N}{\sigma_N + \pi\sigma_C} \tag{10}$$

which is a Michaelis-Menten type equation in both σ_C and σ_N (Mäkelä, 1986). The dependence is such that if $\pi\sigma_C >> \sigma_N$, then root activity will restrict growth, and small changes in σ_C have little impact on the total productivity. If $\pi\sigma_C << \sigma_N$, then σ_C is limiting. If both terms are of the same order of magnitude, then respective changes in either one have equal effects. Environmental variations affecting root and foliage activity can be brought into the carbon-balance model through this relationship.

Results

It follows from Equation 10 that the greater the specific root activity, the less the share of root growth relative to foliage growth. If the specific root activity depends upon site quality, differences among growth sites may be analyzed with a carbon-balance model simply by varying the share of growth to feeder roots and foliage. McMurtrie and Wolf (1983) have carried out such an exercise using a model with the assumption that the partitioning coefficients are constant in time. Figure 2 shows the time development of foliage, feeder roots, and wood in a poor quality (2a) and good quality (2b) site according to this model. The decline in growth in this application follows from the decrease of foliage activity due to shading, while the differences in biomass accumulation of the two sites are attributable to the partitioning pattern.

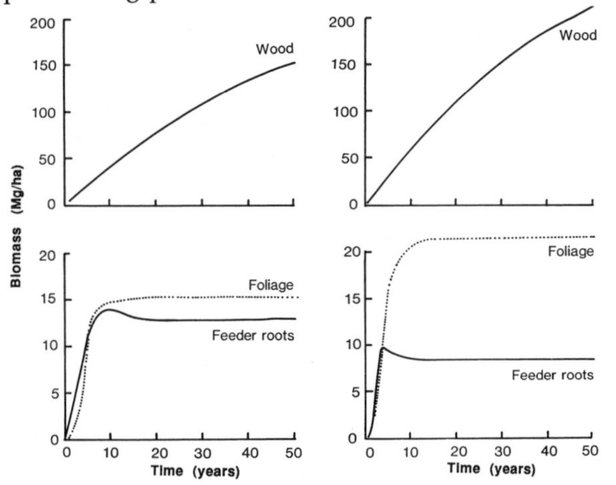

Figure 2a,b. Simulated pattern of growth of a coniferous forest on sites of poor and good quality. The only difference in the two simulations is the partitioning coefficients to foliage, wood, and fine roots, which are 20:20:60 (*a*) and 30:30:40 (*b*) (McMurtrie and Wolf, 1983).

Mäkelä (1988) applied the functional balance in a more complex stand growth model for Scots pine (*Pinus sylvestris* L.), with individual trees as the basic unit. The partitioning of dry matter between foliage and wood was derived using the pipe-model theory (see below). However, the basic result that the partitioning pattern can explain differences in aboveground growth between sites of different fertility was maintained. It was concluded that the specific nitrogen uptake rate, σ_N, and the fine-root specific senescence rate, s_r, can compensate each other to produce a certain level of growth, and that the levels correspond to different actual sites, when measured in terms of average diameter growth (Figure 3) (Mäkelä, 1988).

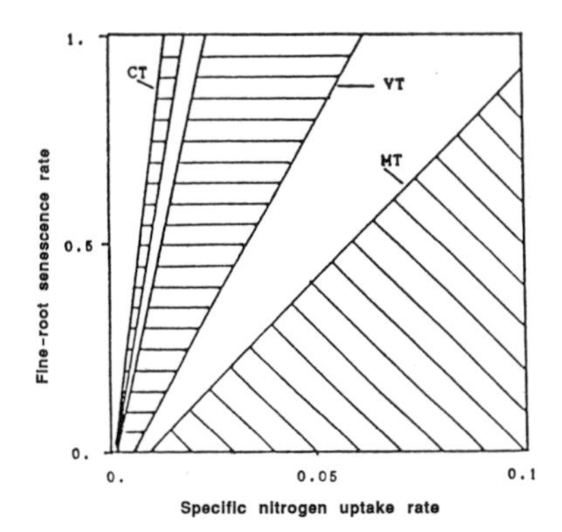

Figure 3. Combinations of root specific activity, σ_N, and root senescence rate, s_r, which lead to simulated stand growth corresponding to that described in yield tables for different site classes (Koivisto, 1959). MT = *Myrtillus* type growth sites, VT = *Vaccinium* type, CT = *Calluna* type (Mäkelä, 1988).

THE PIPE-MODEL THEORY

Principles

The pipe-model theory maintains that the sapwood area at height x and the foliage biomass above x are related through constant ratio (Shinozaki et al., 1964). More recent empirical results indicate that (1) the ratio may be different for stem and branches, and (2) the transport roots obey a similar relationship (Hari et al., 1986). With these supplementary notions, the basic observations can be elaborated to yield the following three relationships (Mäkelä, 1986):

(1) Stem sapwood area at crown base, A_s, is proportional to total foliage biomass, W_f.

$$A_s = \eta_s W_f \tag{11}$$

(2) The total sapwood area of primary branches (at foliage base), A_b, is proportional to total foliage biomass.

$$A_b = \eta_b W_f \tag{12}$$

(3) The total sapwood area of transport roots at the stump, A_c, is proportional to total foliage biomass.

$$A_c = \eta_c W_f \tag{13}$$

The constants η_i are species-specific parameters, which are related to the water-conducting capacity of sapwood (Hari et al., 1986), and probably also to the water economy at a particular growing site (Whitehead et al., 1984).

Application to Carbon Balance

In analogy with the shoot-root partitioning pattern, Equations 11–13 may be utilized in the carbon-balance model for deriving the partitioning of growth between foliage and the woody organs. This requires, however, that sapwood biomass be expressed in terms of sapwood area and the average length, h_i, of the sapwood "pipe":

$$W_i = \rho_i h_i A_i, \qquad i = \text{stem, branches, transport roots} \tag{14}$$

where ρ_i is the density of wood in sapwood compartment i (Mäkelä, 1988). Denoting the senescence rate of sapwood area by D_i (m^2/year), this yields (Mäkelä, 1988):

$$\lambda_i = (a_i\lambda_f + \beta_i)h_i \tag{15}$$

where

$$a_i = \frac{\rho_i}{\eta_i}, \qquad \beta_i = \frac{\eta_i D_i - S_f}{G}$$

This gives a general growth partitioning pattern between foliage and wood.

It is apparent from Equation 15 that the partitioning pattern is a function of the average sapwood lengths of the woody organs. However, the pipe-model theory does not provide us with any hypothesis concerning these quantities. For the time being, let us make the following simple assumption about the connection between height and foliage (or diameter; see Equations 11–13) growth:

$$\frac{dh}{dt} = a\,\frac{dW_f}{dt} \tag{16}$$

where a is an empirical parameter. This assumption is identical with that of Valentine (1987), stating that the extension and expansion growth of the pipes are partitioned at a constant ratio.

Results

Tree and stand development. Figure 4a shows the time development of the partitioning coefficients for needles, feeder roots, and wood, for an individual tree grown under constant light conditions (i.e., constant σ_C in Equation 3). The physiological parameter values have been evaluated from published information for Scots pine (Mäkelä, 1988), and the coefficient a has been chosen so as to produce reasonable height and diameter growth. Little information suitable for testing the result is available, but the dynamic pattern derived from the model very much resembles some empirical observations summarized for Douglas fir (Mohren, 1987).

With regard to model testing, more information is available about the time development of the state variables. Some of these are depicted in Figure 4b. The diameter at breast height (dbh) has been calculated by assuming that the tree lays a coating of constant basal-area increment over the part of the stem which is below the live crown. The model predicts a leveling off of growth for all the variables shown. Since no shading is assumed, this is now completely due to the allocation pattern, which tends to increase the relative share of the nonproductive part of the tree at the cost of the productive organs.

At the stand level, when mutual shading is included, the implications of the model are slightly different. Figure 5 shows the foliage biomass development in a Scots pine stand according to a stand growth model where individual tree growth follows the carbon-

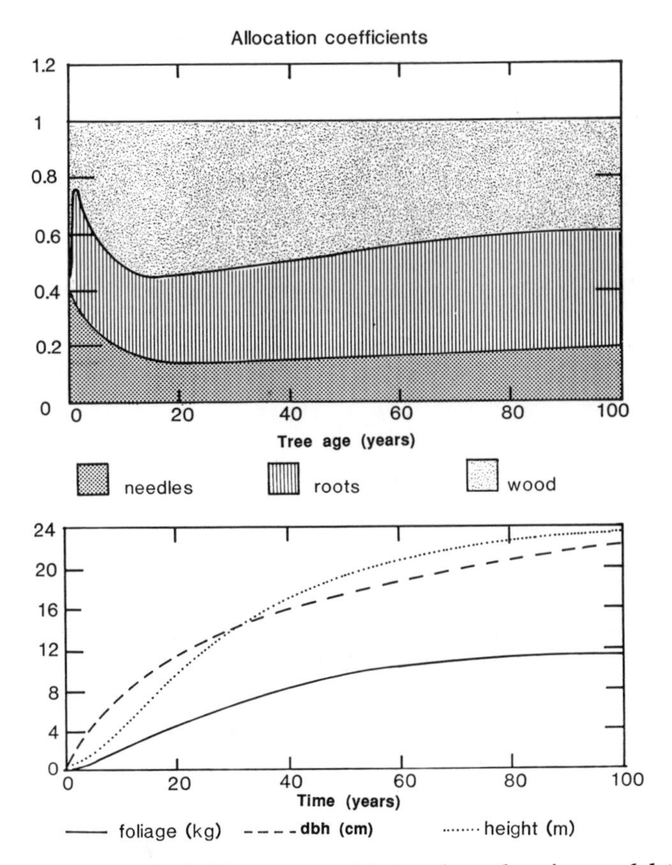

Figure 4. **Results from the individual-tree model based on the pipe-model theory, when no shading is assumed. (*a*) Time development of the allocation coefficients to needles, roots, and wood. (*b*) Time development of foliage, diameter, and height.**

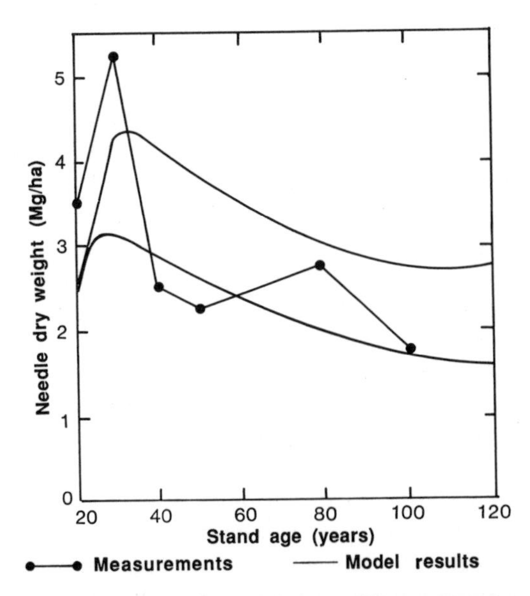

Figure 5. **Time development of needle dry weight according to a stand growth model which incorporates the pipe-model allocation pattern and shading in a horizontally homogeneous canopy (Mäkelä, 1988), compared with measurements (Albrektson, 1980).**

balance approach connected with the pipe model, and where mutual shading is described using Beer's law (Mäkelä, 1988). In this particular application, the parameter values were tuned so as to make the volume growth correspond to the growth and yield tables for Scots pine in southern Finland (Koivisto, 1959). The foliage biomass peaks at a fairly early stage then slowly declines to a rather constant level. This result is a direct consequence of combining the effects of light extension and the pipe-model partitioning pattern. Each of these mechanisms alone leads to leveling off of growth, but the two together yield a strengthened feedback effect. This pattern of foliage growth seems to be in close agreement with the empirical results given by Albrektson (1980) for Scots pine in central Sweden.

Implications for height growth. Let us now consider the effect of the height growth strategy on model behavior. Figure 6 shows what happens when the parameter a is modified. It illustrates the results of Mäkelä (1986) that (1) for each combination of parameters, e.g., corresponding to different growth sites, there is a maximum height which the tree can reach, regardless of the rate of height growth, and (2) simultaneously, the asymptotic level of the foliage biomass strongly depends on the rate of height growth, rather than the equilibrium height.

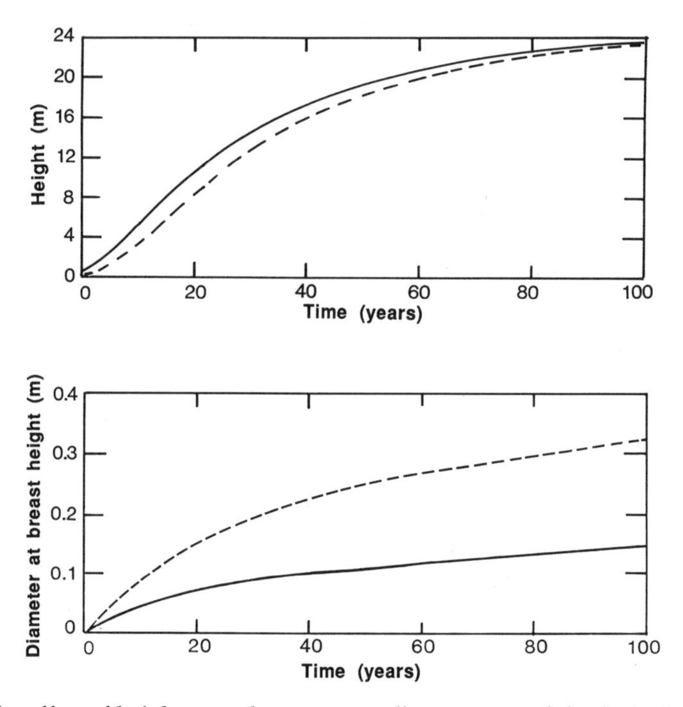

Figure 6. **The effect of height growth strategy on diameter growth in the individual tree model without shading. (*a*) Height growth of two trees as a result of varying the parameter a of Equation 16. (*b*) The corresponding diameter growth.**

The above example refers to the height growth strategies of an open-grown tree, as the rate of height growth is the only parameter varied. In a stand, however, height growth also affects other physiological parameters. The specific photosynthetic rate of the shorter trees will be reduced through shading, and they will probably also have a higher turnover rate of foliage, both factors decreasing their productivity relative to their taller neighbors. It therefore appears that a good strategy for survival in a stand is to keep up with the height growth of the rest of the canopy, a conclusion also supported empirically (Logan, 1966).

Figure 7 compares the growth of two trees with a small difference in initial height, which is assumed to result in a decrease in the specific photosynthetic activity, σ_C, of the shorter tree by 10% for all times. Simultaneously, the shorter tree is assumed to follow the height growth of the taller one as long as possible. Note that the asymptotic height is reduced because the productivity is smaller. The figure shows that while this strategy probably enhances the chances of survival of the tree, it will nevertheless lead to a decreasing share of growth to diameter, foliage, and roots. This result clearly contradicts the frequent model assumption of a constant diameter-height relationship (e.g., Shugart, 1984).

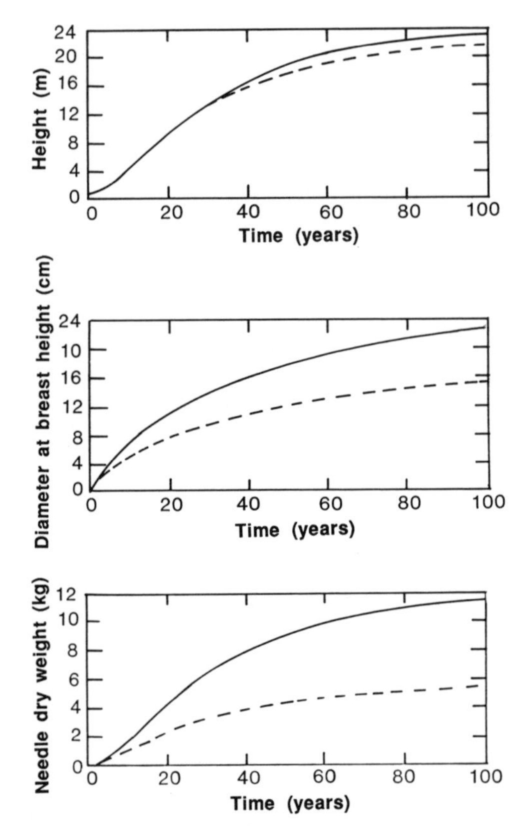

Figure 7. The development of two trees competing for light in a canopy. (*a*) Tree height. (*b*) Diameter at breast height. (*c*) Foliage dry weight.

The above is an example of the more general result that if trees with different photosynthetic capacities maintain the same length growth pattern, the ones with higher photosynthesis will have a larger share of growth to be allocated to foliage, roots, and diameter (Mäkelä, 1986). This applies to comparisons of allocation at different stocking densities, as height growth is not considerably affected but the availability of light per tree varies (Whitehead, 1978), and is consistent with the observation that higher stocking density increases allocation to stems (Bröms and Axelsson, 1984).

EVOLUTIONARILY STABLE HEIGHT GROWTH STRATEGIES

Significance of Tree Height

Although the pipe-model derivations result in reasonable growth patterns when height growth strategies are varied, we are left with the fact that a satisfactory dynamic connection between tree height and the other state variables is missing. In terms of the present approach, we need to understand how the parameter a in Equation 16 depends upon the environment and the state of the tree. Based on both model results and empirical observations (e.g., Whitehead, 1978; Logan, 1966), it does not appear that a balance principle similar to the functional balance or the pipe-model theory applies to height growth. On the contrary, tree height seems rather a stable variable in comparison with, for instance, diameter and foliage biomass. Therefore, retaining the evolutionary perspective, we have to go back to first principles and study the significance of tree height for survival.

From the point of view of production, rapid height growth is not preferable because it decreases the growth of the productive parts (see Figure 6 above). On the other hand, it has been suggested that height growth is required (1) for physical durability, (2) for decreasing self-shading by distributing the foliage vertically, and (3) for avoiding shading by other trees. The latter, in particular, has received attention in the recent literature. Indeed, at least in horizontally homogeneous canopies in high latitudes where the sun angle is low, the height of a tree relative to the whole canopy seems to be an important indicator of survivorship (Oker-Blom, 1986; Mäkelä and Hari, 1986).

Height Growth Game

Including interactions with the rest of the canopy, when calculating how the chance of survival is maximized, leads to the analysis of height growth as an evolutionary game (for evolutionary games, see Maynard Smith, 1982). Rose (1978) interpreted the height growth of trees as an example of the static "Scotch auction" game. Iwasa et al. (1984) considered tree height and crown shape as a result of a noncooperative game in the case of a time-independent equilibrium. Mäkelä (1985) used dynamic game theory for determining an evolutionarily stable height growth strategy (ESS) for an even-aged monoculture. This approach applied a carbon-balance model, involving a simplified version of light interactions in a homogeneous canopy, and an allocation pattern with increasing share to the trunk as the tree grows taller.

The results obtained by Mäkelä (1985) may be summarized as follows: If height growth produces a decrease in the fraction of growth allocated to the productive organs, and shading has a comparatively strong effect on photosynthesis, the maximum chance of survival in a dense population will be achieved through a twofold strategy that provides the maximum physiologically possible growth, in the beginning, and a complete cessation of growth at a certain height. The final height increases as the requirement of new sapwood for the support of new foliage decreases, and as the relative gain, with respect to light, of growing taller than the rest of the canopy increases. The time of the cessation of growth increases with decreasing maximum growth rate (Figure 8). Some more realistic assumptions about the restrictions of height growth would obviously smooth out the resulting curve, but the basic principle of diminishing growth remains.

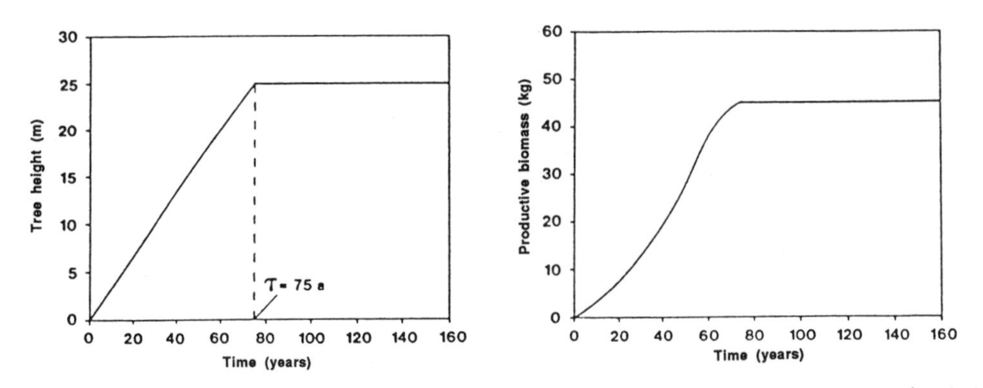

Figure 8. Solution to the "height growth game" for tree height and productive biomass (Mäkelä, 1985).

DISCUSSION

The results reviewed in this paper have given some insight into the applicability of evolutionary considerations to modeling the allocation of growth in stand growth models. The use of balance principles was demonstrated by deriving the partitioning coefficients of dry matter growth for a tree over its lifetime, and further applications of the evolutionary considerations were illustrated by means of the "height growth game."

The allocation patterns obtained seem realistic, at least as far as we can judge from our present, rather limited empirical information. The models derived resemble mechanistic models in the sense that they account for interactions of the different metabolic processes, as well as tree-environment relationships. Simultaneously, they are almost as simple in form as empirical partitioning functions (Mohren, 1987) and allometric ratios (see review by Landsberg, 1986). These characteristics assign high potential to the approach for applications in large-scale, process-oriented stand growth models, but a number of questions remain to be answered.

In a carbon-balance framework, the functional balance predicts that differences in productivity among site types is a consequence of the different allocation of growth among the sites. A similar approach has been presented by McMurtrie and Wolf (1983), while many other models attribute site differences directly to the growth rate (e.g., Ågren, 1983; see review by Shugart, 1984). The allocation to roots is greater if the specific uptake rate is lower, or if the root senescence rate is higher. In a performance analysis of a stand growth model (Mäkelä, 1988), this assumption did not seem critical for acceptable model results; however, it relies entirely upon the actual values of the specific nutrient uptake rate and specific senescence rate for each site type. Some empirical results (Keyes and Grier, 1981) seem to support the present conclusions, but the assumption cannot be properly validated until more accurate empirical information becomes available on the related parameters.

While the results of the foliage-root partitioning pattern focus on differences between growth sites, the essential implications of the pipe-model theory are related to the internal growth dynamics. The allocation between foliage and wood seems to produce a negative feedback between tree size and production, the dynamic implications of which, in a stand growth model, are similar to those of light interception. First, there is a direct effect through the annual investment on the buildup of nonproductive tissue. Second, wood production results in an increasing proportion of respiring tissue relative to the productive biomass. This means that information on the growth, senescence, and respiration of wood is critical for a better test of the partitioning pattern. As regards wood growth, the greatest uncertainty seems to be related to pipe length.

The role of pipe length in the dynamic interpretation of the pipe-model theory was already pinpointed above. For a more profound understanding of growth allocation between wood and foliage, studies are needed on the detailed branching structures of trees (e.g., Ford, 1985; Cannell et al., 1983; Kellomäki, 1986). Some insight into these questions has been obtained using evolutionary considerations similar to those mentioned in this paper (e.g., Nikinmaa and Hari, this volume).

Even if the present approach seems to function satisfactorily, one should always bear in mind that there is not straightforward, fail-safe way of determining the "goal" and the restrictions of the evolutionary process (Maynard Smith, 1982). Rather, it is a question of an educated guess which may or may not be realistic enough for our purposes. The approach should more realistically be understood as a means of finding relevant questions which are to be tested case by case. The best test would probably be the discovery of a physiological mechanism which performs the apparent "goal-seeking" function of the plant. The existence of such a linkage has already been demonstrated (Mäkelä and Sievänen, 1987) between the balanced-growth model by Reynolds and Thornley (1982) and the more mechanistic transport resistance models by Thornley (1972). When regarded as a source of new hypotheses, the evolutionary approach can serve as a useful tool for increasing our understanding of the development of tree form.

LITERATURE CITED

Ågren, G. I. 1983. Nitrogen productivity of some conifers. *Canadian Journal of Forest Research* 13:494–500.

Ågren, G. I., B. Axelsson, J. G. K. Flower-Ellis, S. Linder, H. Persson, H. Staaf, and E. Troeng. 1980. Annual carbon budget for a young Scots pine. In: Persson, T., ed. Structure and function of northern coniferous forests: an ecosystem study. *Ecological Bulletin* (Stockholm, Sweden) 32:307–313.

Ågren, G. I., and T. Ingestad. 1987. Root:shoot ratio as a balance between nitrogen productivity and photosynthesis. *Plant, Cell and Environment* 10:579–586.

Albrektson, A. 1980. Total tree production as compared to conventional forestry production. In: Persson, T. ed. Structure and function of northern coniferous forests: an ecosystem study. *Ecological Bulletin* (Stockholm, Sweden) 32:315–328.

Bossel, H. 1986. Dynamics of forest dieback: systems analysis and simulation. *Ecological Modelling* 34:259–288.

Bröms, E., and B. Axelsson. 1984. Variation of carbon allocation patterns as a base for selection in Scots pine. In: *Proceedings, an International Conference on Managing Forest Trees As Cultivated Plants,* July 23–28, 1984. Finland.

Brouwer, R. 1962. Distribution of dry matter in the plant. *Netherlands Journal of Agricultural Science* 10:361–376.

Cannell, M. G. R., L. J. Sheppard, E. D. Ford, and R. H. F. Wilson. 1983. Clonal differences in dry matter distribution, wood specific gravity and foliage efficiency in *Picea sitchensis* and *Pinus contorta. Silvae Genetica* 32:195–202.

Davidson, R. L. 1969. Effect of root/leaf temperature differentials on root/shoot ratios in some pasture grasses and clover. *Annals of Botany* 33:561–569.

De Wit, C. T. 1978. *Simulation of Assimilation, Respiration, and Transpiration of Crops.* Simulation Monograph, Pudoc, Wageningen, the Netherlands. 141 pp.

Ford, E. D. 1985. Branching, crown structure and the control of timber production. In: Cannell, M. G. R., and J. E. Jackson, eds. *Attributes of Trees As Crop Plants.* Institute of Terrestrial Ecology, Huntingdon, England, U.K. Pp. 228–252.

Hari, P., L. Kaipiainen, P. Heikinheimo, A. Mäkelä, E. Korpilahti, and J. Samela. 1986. Trees as a water transport system. *Silva Fennica* 20(3):205–210.

Ingestad, T. 1979. Nitrogen stress in birch seedlings. II. N, K, P, Ca and Mg nutrition. *Physiologia Plantarum* 45:149–157.

Iwasa, Y., D. Cohen, and J. A. Leon. 1984. Tree height and crown shape, as results of competitive games. *Journal of Theoretical Biology* 112:279–297.
Iwasa, Y., and J. Roughgarden. 1984. Shoot/root balance of plants: optimal growth of a system with many vegetative organs. *Theoretical Population Biology* 25:78–105.
Kellomäki, S. 1986. A model for the relationship between branch number and the effect of crown shape and stand density on branch and stem biomass. *Scandinavian Journal of Forest Research* 1:455–472.
Keyes, M. R., and C. C. Grier. 1981. Above- and below-ground net production in 40-year-old Douglas-fir stands on low and high productivity sites. *Canadian Journal of Forest Research* 11:599–605.
Koivisto, P. 1959. Growth and yield tables (in Finnish). *Communicationes Instituti Forestalis Fenniae* 51(8).
Landsberg, J. J. 1986. *Physiological Ecology of Forest Production.* Academic Press, New York, NY, U.S.A. 198 pp.
Logan, K. T. 1966. *Growth of Tree Seedlings As Affected by Light Intensity.* II. *Red Pine, White Pine, Jack Pine and Eastern Larch.* Canadian Forestry Branch, Department Publication 1160.
Ludlow, A. R., T. J. Randle, and J. C. Grace. 1989. Developing a process-based growth model for Sitka spruce. In: Dixon, R. K., R. S. Meldahl, G. A. Ruark, and W. G. Warren, eds. *Process Modeling of Forest Growth Responses to Environmental Stress.* Timber Press, Portland, OR, U.S.A.
Mäkelä, A. 1985. Differential games in evolutionary theory: height growth strategies of trees. *Theoretical Population Biology* 27:239–267.
Mäkelä, A. 1986. Implications of the pipe-model theory on dry matter partitioning and height growth in trees. *Journal of Theoretical Biology* 123:103–120.
Mäkelä, A. 1988. Performance analysis of a process-based stand growth model using Monte Carlo techniques. *Scandinavian Journal of Forest Research* 3:315–331.
Mäkelä, A., and P. Hari. 1986. Stand growth model based on carbon uptake and allocation in individual trees. *Ecological Modelling* 33:204–229.
Mäkelä, A., and R. Sievänen. 1987. Comparison of two shoot-root partitioning models with respect to substrate utilization and functional balance. *Annals of Botany* 59:129–140.
Maynard Smith, J. 1982. *Evolution and the Theory of Games.* Cambridge University Press, London, England, U.K.
McCree, K. J. 1970. An equation for the rate of respiration of white clover plants under controlled conditions. In: *Prediction and Measurement of Photosynthetic Productivity.* Proceedings of the IBP/PP technical meeting, September 14–21, 1969, Trebon. Pudoc, Wageningen, the Netherlands.
McMurtrie, R., and L. Wolf. 1983. Above- and below-ground growth of forest stands: a carbon budget model. *Annals of Botany* 52:437–448.
Mohren, G. M. J. 1987. *Simulation of Forest Growth, Applied to Douglas Fir Stands in the Netherlands.* Pudoc, Wageningen, the Netherlands. 83 pp.
Monsi, M., and T. Saeki. 1953. Ueber den Lichtfaktor in den Pflanzengesellschaften und Seine Bedeutung fuer die Stoffproduktion. *Japanese Journal of Botany* 14(1):22–52.
Nikinmaa, E., and P. Hari. 1989. A simplified carbon partitioning model for Scots pine to address the effects of altered needle longevity and nutrient uptake on stand development. In: Dixon, R. K., R. S. Meldahl, G. A. Ruark, and W. G. Warren, eds. *Process Modeling of Forest Growth Responses to Environmental Stress.* Timber Press, Portland, OR, U.S.A.
Oker-Blom, P. 1986. Photosynthetic radiation regime and canopy structure in modeled forest stands. *Acta Forestalia Fennica* 197. 44 pp.
Reynolds, J. F., and J. H. M. Thornley. 1982. A shoot:root partitioning model. *Annals of Botany* 49:585–597.
Richards, D., F. H. Goubran, and K. E. Collins. 1979. Root-shoot equilibria in fruiting tomato plants. *Annals of Botany* 43:401–404.
Rose, M. R. 1978. Cheating in evolutionary games. *Journal of Theoretical Biology* 75:21–34.
Shinozaki, K., K. Yoda, K. Hozumi, and T. Kira. 1964. A quantitative analysis of plant form: the pipe-model theory. I. Basic analyses. *Japanese Journal of Ecology* 14:97–105.
Shugart, H. H. 1984. *A Theory of Forest Dynamics: The Ecological Implications of Forest Succession Models.* Springer-Verlag, New York, NY, U.S.A. 278 pp.
Thornley, J. H. M. 1972. A balanced quantitative model for root:shoot ratios in vegetative plants. *Annals of Botany* 36:431–441.
Thornley, J. H. M. 1976. *Mathematical Models in Plant Physiology: A Quantitative Approach to Problems in Plant and Crop Physiology.* Academic Press, London, England, U.K.

Valentine, H. T. 1985. Tree-growth models: derivations employing the pipe-model theory. *Journal of Theoretical Biology* 117:579–585.
Valentine, H. T. 1987. A carbon-balance model of stand growth: a derivation employing the pipe-model theory and the self-thinning rule. IIASA WP-87-56:1–7.
Vincent, T. L., and H. R. Pulliam. 1980. Evolution of life history strategies for an asexual annual plant model. *Theoretical Population Biology* 17:215–231.
White, H. L. 1937. The interaction of factors in the growth of lemna. XII. The interaction of nitrogen in light intensity in relation to root length. *Annals of Botany* 1:649–654.
Whitehead, D. 1978. The estimation of foliage area from sapwood basal area in Scots pine. *Forestry* 51(2):137–149.
Whitehead, D., W. R. N. Edwards, and P. G. Jarvis. 1984. Conducting sapwood area, foliage area and permeability in mature trees of *Picea sitchensis* and *Pinus contorta*. *Canadian Journal of Forest Research* 14:940.

CHAPTER 8

WHOLE-TREE GROWTH PROCESS MODELS BASED ON STRUCTURAL-FUNCTIONAL RELATIONSHIPS

J. G. Isebrands, H. Michael Rauscher, Thomas R. Crow, and Donald I. Dickmann

Abstract. Concepts are presented for developing whole-tree growth process models based upon experimentally derived structural-functional relationships. Key aspects of this approach include recognition that (1) individual leaves are the primary regulating organs of the crown, (2) photosynthesis and respiration are central physiological processes, (3) carbon allocation is a major determinant of growth, (4) root-shoot interaction and feedback are essential model components, (5) individual branches are independent physiological units, and (6) tree growth is indirectly affected by the effects of stress on the physiological processes of the leaves.

ECOPHYS, an ecophysiological growth process model of a poplar tree, is presented as an example of this approach. Model structure includes a light interception and photosynthate production submodel, and a photosynthate allocation and growth submodel. In addition, a new approach is offered combining an "expert system" linked with ECOPHYS. The expert system serves as the regulatory control module providing feedback among external environmental factors, internal physiological factors, and plant developmental patterns. Its purpose is to allow the tree to adapt its growth and development to changing environmental, morphological, and physiological conditions.

INTRODUCTION

In recent years, emphasis on forest growth modeling has changed from empirical (i.e., descriptive and predictive) models to mechanistic (i.e., explanatory) process-based models as evidenced by this conference theme. Empirical models (also known as correlative models) describe relationships among variables without necessarily referring to the underlying biological principles of the system. These models are usually based almost entirely upon data from direct measurements (i.e., height, diameter, site index). Mechanistic process models, on the other hand, are based upon the underlying physiological or physical processes of the system, and usually attempt to simulate how the

---

Drs. Isebrands and Crow are associated with the USDA Forest Service, Forestry Sciences Laboratory, Rhinelander, WI, 54501 U.S.A. Dr. Rauscher is at the Forestry Sciences Laboratory, Grand Rapids, MN, 55744 U.S.A; and Dr. Dickmann is with the Department of Forestry, Michigan State University, East Lansing, MI, 48824 U.S.A. The authors acknowledge R. E. Dickson, T. M. Hinckley, G. E. Host, P. R. Larson, P. Layton, R. A. Leary, D. A. Michael, and K. D. Ware for their input and discussion regarding process modeling during the preparation of this manuscript. This research was performed under Interagency Agreement DE-A105-800R20763 with the U.S. Department of Energy, Oak Ridge, TN, U.S.A.

system reacts to environmental stimuli (Landsberg, 1986). In practice, the distinction between empirical and mechanistic process models is not so clear-cut, because a continuum of forest growth models exists (Figure 1). In fact, most successful plant growth models are combinations of empirical and mechanistic submodels, because all models become empirical at their lowest level (Whisler et al., 1986).

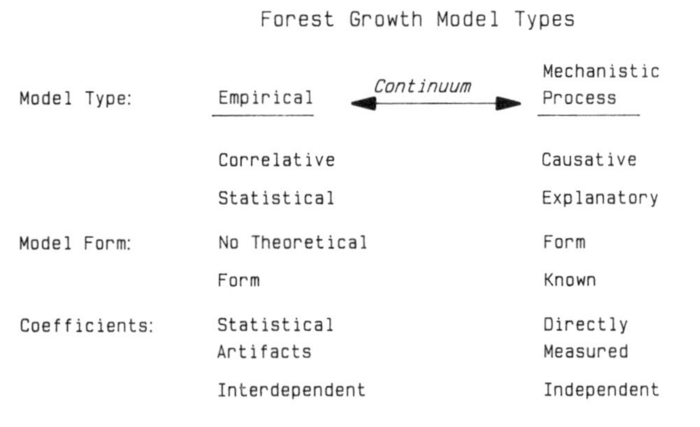

Figure 1. Schematic of the continuum of forest growth model types from empirical to mechanistic.

Forest scientists have lagged behind their agronomic and horticultural colleagues in developing mechanistic process models. Landsberg (1986) concluded that there are few, if any, mechanistic forest models in use. Contrast this situation with the multitude of mechanistic process models available in plant physiology and biochemistry (Thornley, 1976; Newman and Wilson, 1987), plant ecosystems (Hesketh and Jones, 1980; Wisiol and Hesketh, 1987), and agronomic crops (Whisler et al., 1986). The reasons for the lag in development of tree and forest growth process models are (1) there is a lack of knowledge about fundamental tree growth processes (Kramer, 1986), due in part to our national research funding priorities and our traditional forestry education, (2) trees are large, complex, multigenic, and perennial organisms (Larson, 1984), which makes whole-tree physiological experimentation very difficult, and (3) computers with sufficient power, speed, and parallel-processing capability for handling whole-tree and forest-stand process models have not been readily available to forest scientists until recently (Leary, 1988b). However, these limitations are not as great today as they have been in the past, because substantial progress has been made in our fundamental understanding of the structural-functional relationships of trees in several genera, such as spruce (*Picea*) and poplar (*Populus*). (See Larson et al., 1980, and Dickson and Isebrands, in press, for detailed explanations of structural-functional relationships.) Moreover, there have been major improvements in computer processing and storage capability, including a new generation of computers with parallel-processing units (Leary, 1988b; Rensberger, 1988). These technologic advances should help solve the computing problems associated with complex whole-tree growth process models.

There has also been a change in the kind and nature of scientific questions being asked about trees and forests (Leary, 1985). Formerly, many of the questions being asked of forest growth modelers were "what is" and "what if" questions (Figure 2). For example, What is the volume or biomass of a particular forest? or What would happen to diameter growth if a particular stand was thinned or fertilized to a specific level? These questions can be characterized as description and prediction questions with a limited scope of generality. Most early forest growth models fall into these categories (Fries, 1974; Chappell and Maguire, 1987; Ek et al., 1988). Today the public and our clients are more

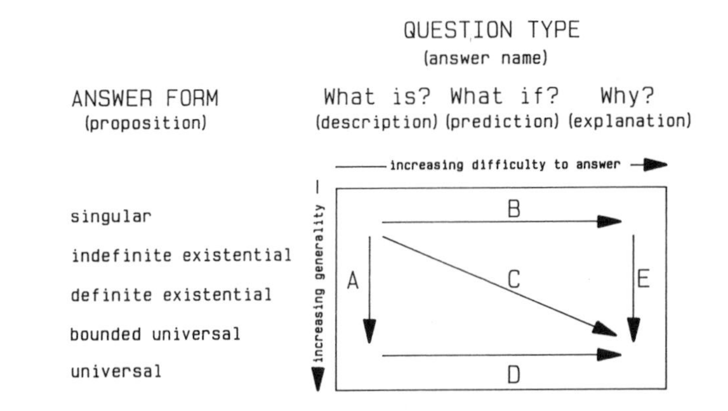

Figure 2. Framework for asking and answering scientific questions of increasing difficulty and increasing generality about a research problem (i.e., forest tree growth). Indefinite existential is "in at least one case," definite existential is "in n cases," bounded universal is "in all cases in the system," and universal is "in all cases" (adapted from Leary, 1985).

informed and are asking why or how a phenomenon in the forest occurs. How does acid deposition affect stand productivity? Why do oak seedling routinely die back after outplanting? Questions of this type are more difficult to answer and require an understanding of the underlying principles and processes (i.e., structural-functional relationships) involved in tree or forest growth in order to explain the phenomena.

Understanding why or how trees grow in stands may require that we first know how an individual tree grows, and perhaps how individual trees grow in pure stands on uniform sites, before we try to model mixtures of species on a variety of sites (Leary, 1988a). Moreover, increasing our understanding of trees and stands will probably require a multidisciplinary approach because the knowledge needed will likely exceed any one individual's scientific training. Paradoxically, in process modeling, investigators can increase their understanding of the overall system only if it is possible to break it into subsystems that are already understood.

In this contribution, we give the rationale and provide concepts for developing whole-tree growth process models based upon experimentally-derived structural-functional relationships. An example of this concept—an ecophysiological growth process model (ECOPHYS) for a juvenile poplar tree—is also given. In addition, we offer a new approach for evaluating the effects of environmental stresses on tree growth processes based upon linking the ECOPHYS process model with an "expert system" regulatory control module. Applications of our approach are presented, as well as the current limitations and research needs in whole-tree growth process models.

MODELING APPROACH AND RATIONALE

The goal of all forest growth process modelers is to better understand how trees and forests grow, and to be able to understand and predict the responses of physiological processes to changes in the environment (Landsberg, 1986). However, there is a hierarchy of process models and modeling approaches to be considered, based on different scales of time and space (Meentemeyer and Box, 1985; Whisler et al., 1986). The user's objectives and intended applications determine the scale and type of model that is appropriate. For example, a model of a forest biome with a response time of decades would require a different approach than a model of an individual tree with a response time of hours or days. The approach to modeling an uneven-aged, mixed conifer-deciduous forest would likely require a different approach than that of monoclonal plantation of a conifer or a hardwood growing under intensive culture. Moreover, an individual-tree growth process model of a conifer may differ somewhat from that of a hardwood because of differing morphology and physiology.

Mechanistic growth process models can be classified into two broad categories (Landsberg, 1986). Models of the first category are called "detailed physiological models," sometimes referred to as "bottom-up" models. (We do not use the term "bottom-up model" because of the somewhat negative connotation of this terminology.) This group of models simulate the physiological processes of growth and their interactions, and are largely research tools. It should be noted that there are detailed physiological models of individual trees and of forest stands (Sievänen, 1983; Caldwell et al., 1986). The second category consists of "top-down" models. This group of models provides a simplified representation of physiological processes that are usually based upon one or two major driving variables. Top-down models are usually simpler than detailed physiological models and are often used as management tools. No one model or modeling approach is likely to be suitable for all objectives and purposes. Rather, a variety of models and modeling approaches are required to meet different needs. The various modeling approaches available should be viewed as complementary rather than mutually exclusive. In fact, Landsberg (1986) suggests that the ideal modeling approach would be to link a "detailed physiological process model" with a more simplified "top-down" model in the same system.

Most models available in forestry today are of the top-down type. Emphasis has been on keeping models as simple as possible while still being realistic. Most top-down forest growth models have been developed at the canopy or stand levels (Reed, 1980), not at the individual tree level. However, the goal of most top-down modelers is to ultimately understand growth processes at the stand level.

There has also been some concern that detailed physiological models are too detailed and, therefore, too complex. However, with the advent of improved microcomputers and computers with parallel-processing capability, the concerns about too much detail should subside. One major advantage of the detailed physiological approach is that it allows the modeler to aggregate upward in the system from a lower-level model to a higher-level model (e.g., from the leaf to the branch to the crown to the canopy). This advantage is a result of the need to understand at least one or two levels below the level of prediction (Whisler et al., 1986). It is always easier to aggregate upward from a micro to the macro scale than to disaggregate downward. The payoff in the upward aggregation approach is an increased level of understanding of the system at all levels and the opportunity to incorporate biological and environmental variables directly. By comparison, many of the top-down models suffer from the inability to address the effect of changing environmental variables directly at the subprocess level.

Considerable insight can be gained by first considering the forest from an individual-tree perspective (Dale et al., 1985). Unfortunately, our knowledge is insufficient and based on too few forest tree species to allow us to adequately develop detailed physiological whole-tree growth process models. Moreover, it is difficult to develop a growth process model that is a function of environmental variables because of the complexity involved (Reed, 1980). Currently, few forest growth models include these variables because even the most simplistic model becomes exceedingly complex with an ever-changing microenviroment. For these reasons, we believe there is merit in developing a detailed whole-tree physiological process model of genetically uniform individuals growing under controlled field conditions before tackling more complex forest systems. However, tree genera of genetically uniform material are not always available, and rarely do we know enough about structural-functional relationships within those trees.

KEY GROWTH CONCEPTS

Through the years, our research program has focused on understanding the detailed structural-functional relationships within pine (*Pinus*), poplar (*Populus*), and oak (*Quercus*) trees. Our approach has been to conduct intensive anatomical and physiological experiments on genetically uniform material grown under controlled conditions (Larson, 1984; Isebrands, 1982; Isebrands et al., 1983; Isebrands et al., 1988). Emphasis has been on translation of this detailed knowledge obtained from trees growing in the growth chamber to those grown in the glasshouse and finally in the field. As a result of this experimentation, a number of key concepts (and hypotheses) have emerged based upon structural-functional relationships. These concepts can help us understand how trees grow and how we might build detailed whole-tree physiological growth process models. Moreover, these concepts are likely to hold for vascular plants in general (Dickson and Isebrands, in press).

The first and foremost concept is that leaves are major organs regulating growth and development in trees (Figure 3). Leaves have been shown to be the primary organizing unit of vascularization of the leaf-stem-root continuum in many plants. In fact, Larson (1980) has shown that in *Populus,* each wood element is intimately connected with specific vascular traces in the leaf. Leaves are also the major sites for CO_2 fixation. Leaves regulate carbon transport and allocation, and leaf surfaces are major sites of water loss. Moreover, leaves are centers of nitrogen metabolism and transport, and are the most important receptors of environmental stimuli (i.e., heat, drought, and anthropogenic pollutants), as well as sites of hormonal synthesis (Dickson and Isebrands, in press).

Second, photosynthesis and respiration are central physiological processes regulating tree growth and yield (Figure 4), as in agronomic crops (Wallace et al., 1976). Leaf size, shape, and orientation are important determinants of whole-leaf and whole-tree photosynthesis (Isebrands and Michael, 1986). When whole-leaf photosynthesis is integrated over the entire tree, over the day, and over the season (i.e., total photosynthesis), it is closely correlated with biomass yield (Isebrands et al., 1988).

Third, carbon allocation is a major determinant of within-tree growth components (Figure 4). Patterns of carbon allocation vary with plant ontogenetic stage and position within the plant, with season, and by genotype (Dickson, 1986), and are thought to be altered by anthropogenic and environmental stresses (Mooney et al., 1987). Understanding carbon allocation patterns coupled with knowledge of genotype × environment interactions has led to appreciable gains in crop yields in certain agronomic crops (Wallace et al., 1976; Gifford et al., 1984).

A fourth key concept is that a strong root-shoot interaction exists in trees, and relative root and shoot growth is closely controlled by multiple feedback systems (Borchert, 1975; Passioura, 1988). This feedback system is facilitated by an intimate and often direct connection between the leaves and roots (Dickson and Isebrands, in press).

A fifth concept is that branches are autonomous physiological units (Sprugel and Hinckley, 1988). Patterns of leaf development and carbon allocation in branches are similar to those in terminal shoots, and as a branch develops, it quickly becomes photosynthetically independent from the main shoot (Dickson, 1986). Moreover, in *Populus,* assimilates produced by branches are exported primarily to the stem and roots, and not to other branches or to the current terminal (Isebrands, 1982; Isebrands et al., 1983). Thus, the tree can be considered as an aggregation of branches that are rather independent physiological units, each of which is physiologically similar to other branches and to the terminal shoot. Moreover, once branch elongation is completed, the primary contribution of branches to tree growth is to the stem and roots.

Another concept is that environmental stress affects tree growth by directly affecting

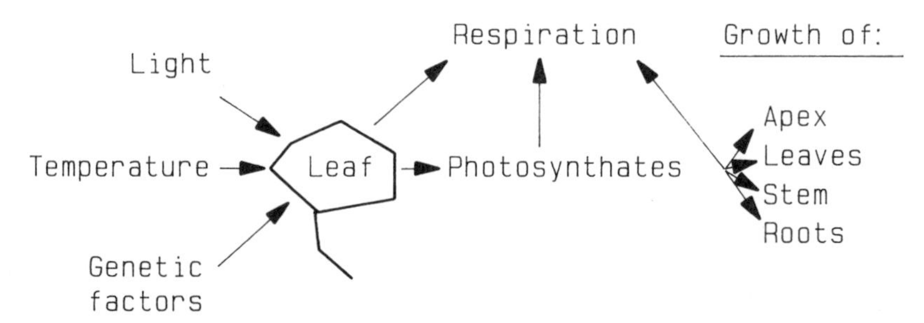

Figure 3. The leaf is a major organ regulating growth and development in plants, including anatomical and morphological development of the shoot and carbon allocation within the plant.

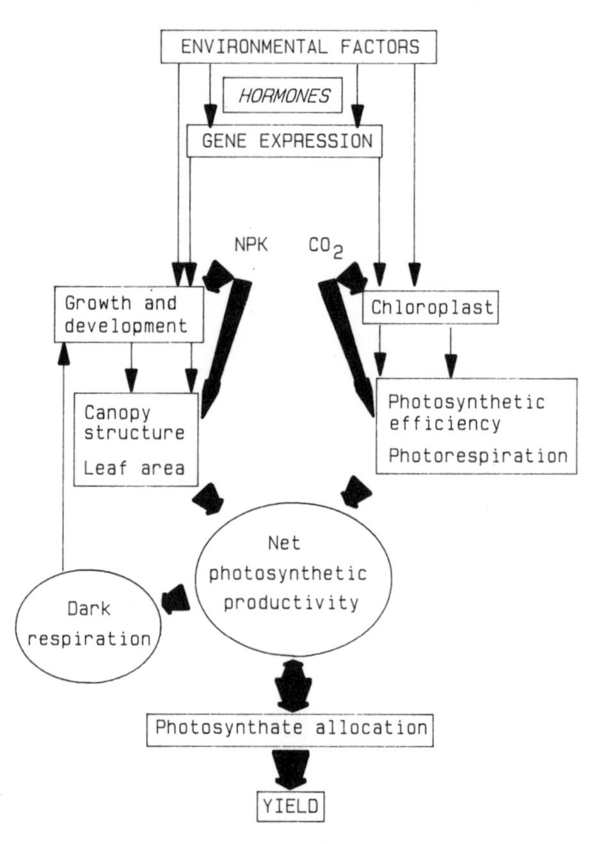

Figure 4. Schematic of the system of major physiological processes of growth and yield. The central processes in the yield diagram are photosynthesis and respiration coupled with photosynthate allocation (adapted from Wallace et al., 1976).

the physiological processes of leaves and roots. In the past, much confusion in the interpretation of tree growth has resulted from investigators trying to relate growth (both quantitatively and qualitatively) directly to environmental influences (Larson, 1984). Good examples are the attempts to determine the effect of acid rain on tree growth. A more unifying biological concept and approach is that environmental effects act directly on the physiological processes (i.e., photosynthesis and nutrient uptake) in leaves and roots and only indirectly on resultant growth (i.e., height, diameter, and biomass).

Lastly, from a modeling perspective, is the concept that as the biological processes of tree growth are understood, the mathematical representation of the system is simplified

(Bowerman and Glover, 1988). One reason for this simplicity is related to the inherent organizational structure that genetics brings to a tree system. Moreover, as we understand and conceptualize about how a small tree grows, we can often translate research results and concepts to larger trees (Larson, 1984).

A synopsis of these concepts and hypotheses suggests that detailed whole-tree physiological growth process models may contain many attributes in common with whole-plant growth process models in agronomic crops. For example, the basic physiological modeling unit in *Populus* should be the leaf, as in cotton (Norman, 1980). Also, predicting photosynthesis, respiration, and photosynthate production is a vital part of the modeling process (Ledig, 1969; Hesketh and Jones, 1980). In fact, photosynthate production can be considered the driving force of all plant growth; photosynthate allocation must be addressed as an important determinant of within-tree growth (McMurtrie and Wolf, 1983), and a root-shoot feedback system is an essential model component. Branches can be treated as independent physiological units contributing to the growth of the stem and root, and the effects of environmental stress must be dealt with through their effect on physiological processes in the leaf and root. Given this synopsis, a word of caution is necessary in that agronomic crop models should not be universally applied to trees without a critical examination of the underlying assumptions.

ECOPHYSIOLOGICAL GROWTH PROCESS MODEL OF POPLAR

We have combined the concepts outlined in the previous section to develop an ecophysiological growth process model of a juvenile poplar tree called ECOPHYS (Figure 5). Details of model development, model applications, and a user's manual are presented elsewhere (Crow et al., 1987; Rauscher et al., 1988; Host et al., in press). ECOPHYS is an individual tree model of a first-year poplar clone growing under the near-optimal field conditions of short rotation intensive culture (SRIC), including weed control, irrigation,

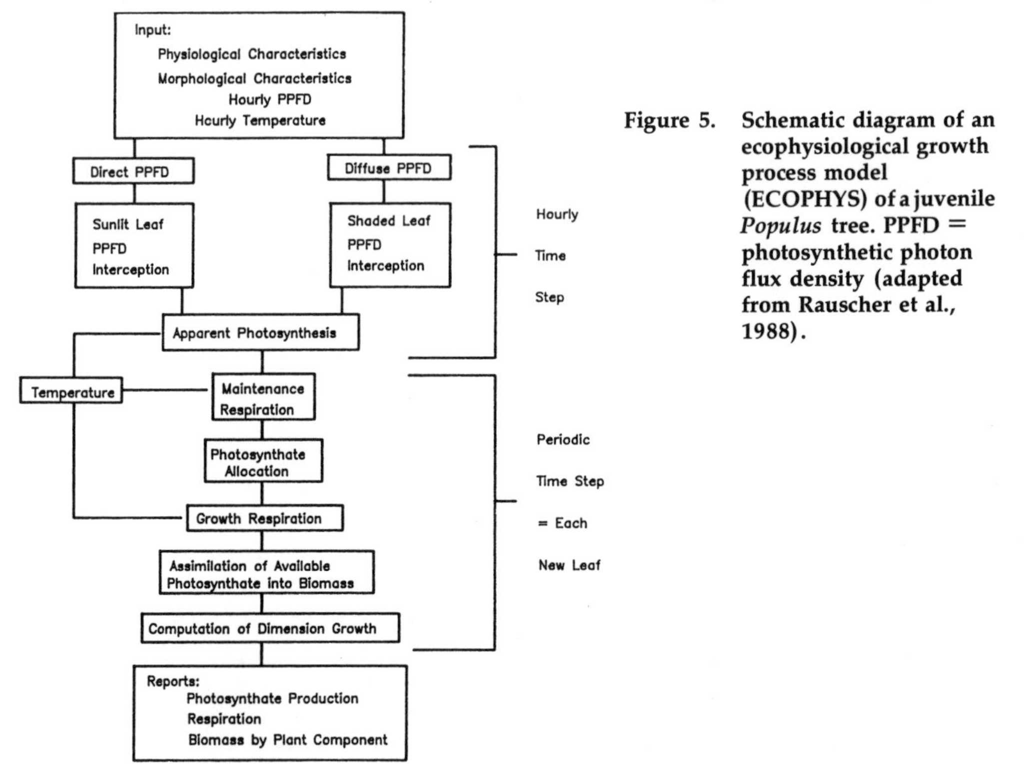

Figure 5. Schematic diagram of an ecophysiological growth process model (ECOPHYS) of a juvenile *Populus* tree. PPFD = photosynthetic photon flux density (adapted from Rauscher et al., 1988).

and fertilization (Isebrands, et al., 1983). The SRIC system approaches an agronomic system that provides a modeler with rapid feedback for model verification and validation. The SRIC system also provides a system where environmental, physiological, morphological, and yield data are readily available for modeling. Moreover, measurements are easy to make and are easily identified as critical or sensitive. The genus *Populus* was chosen because a wealth of fundamental knowledge is available about its structural-functional relations, perhaps more than for any other forest tree (Isebrands et al., 1988; Dickson and Isebrands, in press).

ECOPHYS simulates growth using the individual leaf as a primary biological unit and solar radiation, temperature, and clonal morphological factors as the major driving variables. The user provides input on clonal characteristics including morphological, phenological, and physiological traits, a summary of the seasonal environmental variables, and site variables. Clonal traits include leaf size and shape, leaf orientation, and budset date, as well as photosynthesis and carbon allocation coefficients. Environmental variables include hourly solar radiation and temperature. The user can also override all input variables to examine the effect of hypothesized changes in individual tree characteristics and/or environmental variables on photosynthetic production and growth (Crow et al., 1987; Rauscher et al., 1988). The model grows a poplar tree on an hourly basis over the entire season. Each leaf begins photosynthesizing each day when solar radiation begins in the morning and when temperature is favorable. The leaf receives both direct and diffuse light over the day according to solar position and shading, then photosynthesizes, respires, and produces photosynthate for transport to growth centers throughout the plant. The model gives an hourly photosynthesis report that includes an output of total leaf area, leaf area in sun and shade, light intensity (i.e., photosynthetic photon flux density [PPFD]), and total photosynthesis per leaf and integrated over the tree. The photosynthate is then allocated to the competing growth centers within the tree according to transport coefficients determined with radiotracers (Isebrands and Nelson, 1983; Dickson, 1986; Isebrands and Dickson, in press). The individual tree components (i.e., leaves, stem, and roots grow according to availability of photosynthate at the respective growth center after respiratory losses (i.e., growth and maintenance respiration) are subtracted (Amthor, 1986). The model gives a daily growth summary including height, diameter, number of leaves, biomass by components, and volume. In addition, there is an hourly graphical display of the tree so that the user can view the leaves and their relation to the sun at any time of the day. Moreover, the graphics capability is interactive so that the user can view the tree from any perspective desired.

ECOPHYS has been verified and validated as recommended by Jeffers (1982). Sensitivity analyses were run to ensure that the simulations of physiological and growth variables were within the expected range based on previous hourly, daily, and seasonal observations. A verification of ECOPHYS for stem height and diameter for a poplar tree growing under SRIC in Rhinelander, WI, U.S.A. (45°N, 89°W) in 1979 is shown in Figure 6. Leaf-by-leaf photosynthesis predictions were validated by comparing an independent photosynthesis data set on July 17, 1979, at Rhinelander (Figure 7). Number of leaves, leaf area, stem height, and biomass were also validated (Figure 8) on an independent data set from first-year poplar clones growing under SRIC near Seattle, WA, U.S.A. (47°N, 122°W), in 1985 (Isebrands et al., 1988). Figure 8 also shows a graphical display of the clone on Julian dates 183 and 250.

Sensitivity analyses of ECOPHYS were used to examine the influence of light and temperature on first-year growth of poplars (Rauscher et al., 1988) and the effect of changing leaf orientation on photosynthetic production (Crow et al., 1987). Light influenced growth more than temperature, but the interaction of both variables resulted in about a 50% increase or decrease in stem biomass of an individual tree compared to a base year. Moreover, there were distinct differences in photosynthetic production when leaf orientation was changed (manipulated) from "normal" to erect or to horizontal orientations. Observed differences could be related to mutual shading patterns of the leaves.

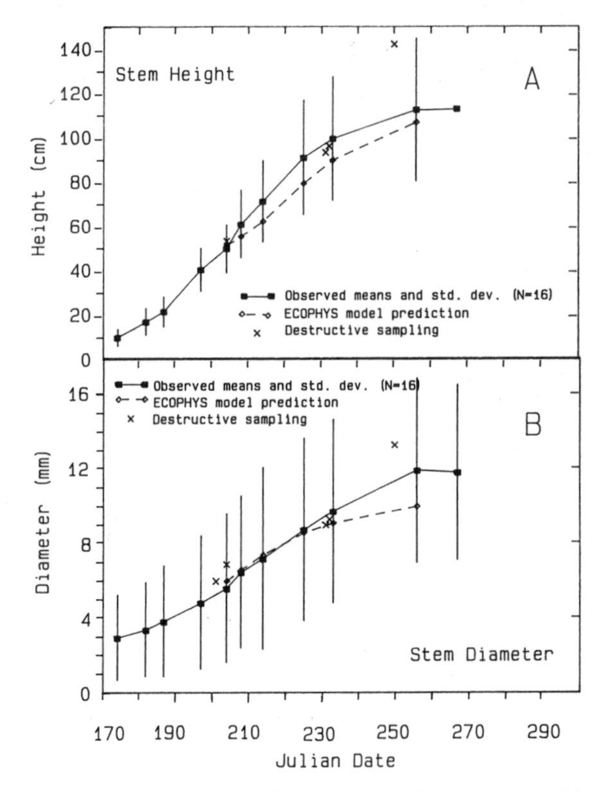

Figure 6. **Verification of ECOPHYS model prediction for one-year-old *Populus* clone growing at Rhinelander, WI, U.S.A. (45°N, 89°W) in 1979. Julian date 182 = July 1; Julian date 274 = October 1. Bars represent standard deviations of means of permanent growth plots, and x's indicate periodic destructive harvests. (*A*) Stem height. (*B*) Stem basal diameter. (Adapted from Dickson and Isebrands, in press.)**

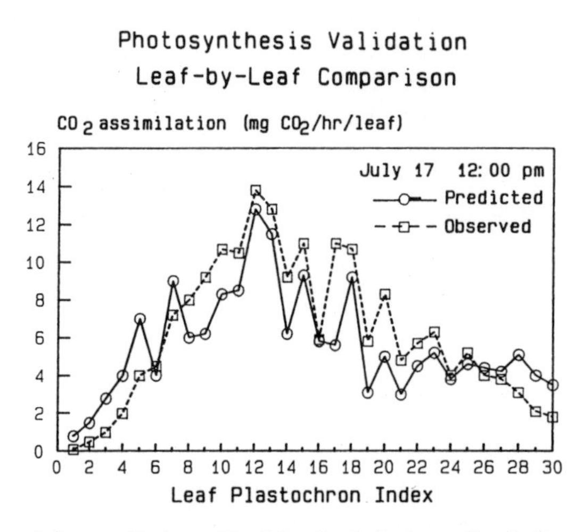

Figure 7. **Validation of the prediction of leaf-by-leaf photosynthesis (i.e., CO_2 assimilation) of a one-year-old *Populus* clone grown as in Figure 6 by the ECOPHYS model for July 17, 1979.**

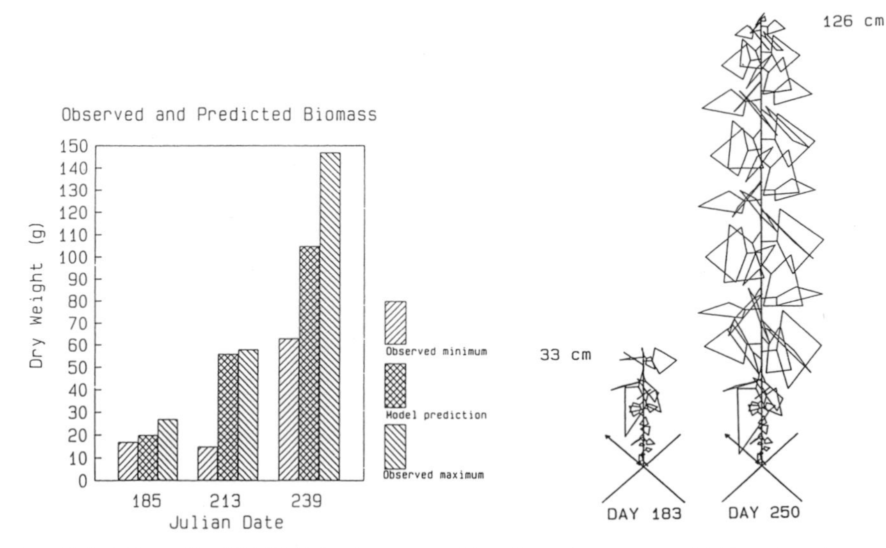

Figure 8. Validation of ECOPHYS for biomass estimation of one-year-old *Populus* clone grown near Seattle, WA, U.S.A. (47°N, 122°W), in 1985. Also, ECOPHYS graphical display of leaves on Julian dates 183 and 250.

ECOPHYS has several unique features when compared to most other whole-tree growth process models. First, the light interception submodel calculates (not estimates) the hourly sun/shade patterns of each individual leaf on the tree. Second, growth of each biomass component (i.e., leaf, stem, and root) is driven by available photosynthate at that growth center. The photosynthate is derived from the cumulative production over the whole leaf, the entire tree, and the entire season. The transport coefficients used to allocate photosynthate to these components were derived from a series of field studies employing radiotracers. ECOPHYS thereby allows one to determine a carbon budget for each biomass component and to quantify the exact sources of photosynthate for each competing growth center. We know of no other tree process model that uses this approach or that has this capability. The model was developed totally from experimentally derived data and is based upon our present knowledge of structural-functional relations in poplar. Although the model is detailed and complex in totality, each component is relatively simple, is straightforward mathematically, and has biological basis. Presently, an entire season run for the one-year-old tree takes approximately 3.5 hours on a 12 Mhz 80286 microcomputer.

EVALUATING STRESS EFFECTS—AN EXPERT SYSTEMS APPROACH

ECOPHYS was developed on poplar trees growing under near-optimal conditions of SRIC, where moisture and nutrient needs were broadly met. By contrast, trees growing under natural conditions are routinely subjected to a multitude of environmental stresses including physical and biotic stresses (Levitt, 1980). Whole-tree growth process models must be responsive to these stress factors to be useful in making research or applied forest management decisions. An expert system linked with an individual-tree growth model such as ECOPHYS can provide the integrated model structure necessary to be responsive to external factors and internal factors (Figure 9).

An expert system is a knowledge-based computer system that is capable of performing at the level of a human expert. The expert system usually consists of a set of rules

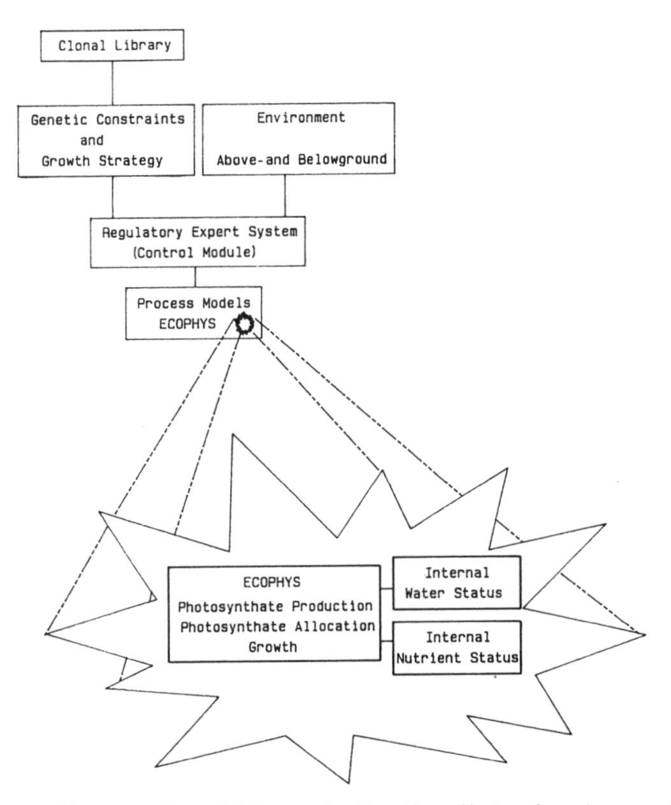

Figure 9. A proposed integrated model for evaluating the effects of environmental stresses on poplar tree growth processes consisting of a series of ECOPHYS simulation modules linked with an "expert system" regulatory control module.

from the general knowledge base connected to an "inference engine" that examines the rules one at a time (Whisler et al., 1986; Bowerman and Glover, 1988). Some of the advantages of an expert system are that it can (1) handle uncertainty, (2) treat complex systems as a whole, (3) handle hierarchical structures and their interactions, and (4) include expert intuition (i.e., heuristic knowledge) in the decision-making process (Wisiol and Hesketh, 1987). Recently, expert systems have been developed for use in natural resource management applications (Coulson et al., 1987; Rauscher, 1985; Schmoldt and Martin, 1986), but they have not been routinely applied in forest biology research or in process modeling. The advantage of an expert system used in conjunction with whole-tree physiological growth process models is that a set of decision-making rules can be developed based upon our fundamental knowledge of the structure and function that allow the tree to react to periodic fluctuations of internal and external factors. Expert system methodology also enables one to efficiently integrate processes understood qualitatively with quantitative processes.

The integrated model structure we propose for evaluating the effects of stresses on poplar tree growth processes is given in Figure 9. It consists of a series of ECOPHYS simulation modules linked with an expert system regulatory control module. The model has the following parts:

A. A genetic constraints and growth strategy subsystem that
   1. has a strategic morphological and physiological survival pattern, and
   2. permits a degree of morphological and physiological adaptation.

B. A regulatory expert system that controls
   1. carbohydrate translocation strategy,
   2. timing of new organ formation, i.e., leaves, branches, reproductive structures, etc.,
   3. root-shoot structural balance,
   4. timing of phenological events, i.e., budbreak, budset, etc., and
   5. reaction to physical, chemical, and biotic stress.

C. A physiological subsystem in ECOPHYS consisting of
   1. photosynthate production,
   2. respiration,
   3. photosynthate allocation,
   4. transpiration, and
   5. nutrient uptake and transport.

D. The morphological subsystem in ECOPHYS consisting of
   1. leaf size, shape, mass, and location in three-dimensional space,
   2. stem internode length, diameter, location of leaf and branch attachment, and
   3. root mass and distribution in three-dimensional space.

External to the tree is the above- and belowground environment with

E. An aboveground physical environment with inputs of
   1. light intensity (i.e., PPFD) and duration,
   2. temperature,
   3. humidity,
   4. precipitation,
   5. CO_2 concentration, and
   6. nutrient fluxes.

F. A belowground physical environment with inputs of
   1. soil horizon, texture, structure, nutrient and moisture status, and dynamics in three-dimensional space,
   2. nutrient availability to roots located in the three-dimensional mineral soil space,
   3. nutrient and moisture availability from the litter,
   4. soil temperature status and dynamics in the root space, and
   5. soil microbiological dynamics.

G. An aboveground biological environment with inputs of
   1. crown competition for light,
   2. changes in light quality,
   3. changes in temperature, and
   4. mechanical damage.

H. A belowground biological environment with inputs of
   1. root competition for water and nutrients,
   2. root grafting,
   3. mycorrhizal interactions, and
   4. allelopathic interactions.

The proposed model would run according to the following expert system scenario. The genetic constraints and growth strategy subsystem for the specific poplar clone being modeled would be in the clonal library and would provide the simulation model with a strategic goal for growth and development given favorable environmental conditions. The regulatory expert system would make day-to-day tactical decisions so that the tree would develop and grow according to design. But it would also acclimate to the real environment through feedback interactions with the tree's morphological and physiological state. The regulatory system would guide the structural-functional processes in the tree in such a way that growth is attained within the constraints imposed by the tree's above- and belowground environment as well as its genetic constitution. The physiological subsystem provided by the independent ECOPHYS simulation modules would be responsible for the functioning of the physiological growth processes under the guidance of the regulatory control module. Similarly, the morphological subsystem that is also provided by independent ECOPHYS modules would change the physical dimensions of the tree and their locations in three-dimensional space under the guidance of the regulatory control module.

In this model, the root system within ECOPHYS would be separated into both coarse and fine roots (i.e., < 1 mm in diameter), and the soil volume in the tree's rhizosphere would be divided into a matrix of discrete cubes. This approach would permit the profile of root architecture and biomass to be quantified in three-dimensional space; it would also facilitate the modeling of water and nutrient uptake by the roots and the modeling of temperature distribution in the root zone.

Branches on the tree would be treated as independent ECOPHYS simulation modules (i.e., independent central processing units) distributed in a three-dimensional coordinate system. Each branch would have carbon transport coefficients based upon radiotracer experiments (already conducted), and the cumulative carbon production of the branches would accumulate at the stem and roots with a root-shoot feedback loop. This integrated model structure would provide the framework for understanding the underlying principles and processes of tree growth needed to explain why or how a phenomenon occurs in the forest stand.

APPLICATIONS OF WHOLE-TREE GROWTH PROCESS MODELS

Whole-tree growth process models such as ECOPHYS have a wide variety of applications. They span a range of scientific disciplines from the fundamental to the applied sciences, and they have utility for geneticists, physiologists, entomologists, pathologists, and biometricians, as well as silviculturists. Some of the more important current applications of whole-tree growth process models are described below.

Early Selection and Breeding

With a growth process model, photosynthate production and growth of juvenile trees can be rapidly compared for clones with contrasting leaf morphology, crown architecture, and physiological traits (Isebrands and Michael, 1986; Ceulemans et al., 1987; Isebrands et al., 1988). The model can thereby be used for screening clones and also as a tool for testing selection and breeding objectives in the laboratory for subsequent comparison with field performance. This approach gives the breeder more confidence in choosing the morphological and physiological traits to look for in the populations. It also provides a range of physiological and morphological traits to work with, as compared to the more traditional mensurational traits of height and diameter. Such models also assist the breeder in construction of ideotypes, which are ideal conceptions (or targets) for selection and breeding (Dickmann, 1985).

Assessing the Impact of Environmental Stress

Growth process models provide a dynamic and sophisticated tool for assessing the impact of multiple stresses (Whisler et al., 1986). Moreover, different levels of a stress can be assessed. For example, the impact of leaf area reduction, decreased photosynthesis, increased respiration, and the associated compensatory effects of an anthropogenic stress such as ozone or elevated CO_2 can be assessed simultaneously. With ECOPHYS, a model developed under optimal conditions, one has the ability to create potential growth standards for comparison with growth and development under non-optimal conditions. These standards can be used as a phytometer for the detection of stresses. The impact of stress at the process level can also be assessed quantitatively at a specific growth center (i.e., leaf, stem internode, and root).

Identifying Areas for Further Research

One major research application of a growth process model is to identify gaps in our understanding of the underlying mechanisms of tree growth. Identifying these gaps is of much value in designing further experiments in a controlled environment or in the field. Such experiments are necessary to improve on the next version or generation of models. A growth process model also allows the investigator to pose "what if" questions and to generate hypotheses to be tested. For example, one can examine and compare the impact of different growth strategies, e.g., carbon allocation patterns, leaf initiation rate, leaf orientation (Crow et al., 1987), leaf area duration, leaf flushing patterns, and root/shoot ratios. Another research application of whole-tree process models is the ability to run sensitivity analyses on key driving variables of the growth process. An example of such an analysis was the study conducted by Rauscher et al. (1988) on the influence of temperature and light on juvenile poplar growth. Photosynthate production and growth can also be examined as a function of a matrix of environmental variables. In this way, the effect of fluctuations in the levels of the variables and the interactions among variables can be assessed. Growth process models also allow researchers to extend point estimates of physiological traits, such as photosynthetic rate per unit area, over the entire tree and over the season. As an example, few investigators have found that point estimates of photosynthetic rate correlate with growth and yield of trees (Isebrands and Michael, 1986). However, small differences in point estimates of photosynthetic rate translate to large differences in growth when integrated over the tree and over the season (Isebrands et al., 1988). There is also significant merit in understanding the physiological growth processes of a few representative tree species. For example, as Larson (1984) has stated, knowing the structural-functional relationships of poplar from the shoot tip to the root tip has many implications and analogies for understanding the growth of other important tree species.

Teaching

Whole-tree growth process models also have significant value in teaching tree physiology and forest biology. A model such as ECOPHYS provides an inexpensive, interactive, and user-friendly teaching tool for students studying the basics of tree and forest growth. A computer game format combined with graphics capability provides a stimulating approach for learning and for knowledge synthesis. The model can be used to supplement traditional laboratory exercises by allowing students to design, run, and analyze experiments of their own.

Future applications of whole-tree growth process models include the linkage of individual-tree growth process models with stand-level models. This approach shows promise for modeling stand growth as a function of environmental variables (Reed, 1980), for analyzing nutient cycles in stands, and for defining and quantifying above- and belowground competition among trees. Once these ecophysiological processes are understood, we should be better equipped to design silvicultural strategies, to engineer clone and species mixtures, and to predict the quantity and quality of stand growth and yield.

CURRENT LIMITATIONS AND RESEARCH NEEDS

The major limitation to developing detailed physiological growth process models in trees is still the lack of baseline structural-functional information (Dickson and Isebrands, in press). A great deal of anatomical and physiological information on trees has been generated through the years, but the knowledge synthesis is still lacking for many of our important forest species (Larson, 1984). Moreover, most of the physiological research has been conducted in controlled environments rather than in the field. Also, too few parallel physiological experiments have been conducted where controlled-environment research has been coupled with field studies. Unfortunately, the emphasis on indoor research has contributed to the lack of understanding of the underlying physiological mechanisms of tree growth and their interaction with environmental stresses. For example, there is not enough information about the effects of atmospheric pollutants and elevated CO_2 on basic physiological processes in field-grown trees. Sadly, there has also been a communication gap between researchers and field workers, so that even when information is available, it too often is not applied (Kramer, 1986).

Another limitation is that multidisciplinary research teams have rarely collaborated on the development of tree and forest growth process models. Moreover, such teams have not yet applied knowledge-based expert systems technology to basic physiological modeling problems.

Clearly, more multidisciplinary research on structural-functional relationships is needed for the major forest tree species. More specifically, research is needed on leaf initiation (Dale and Milthorpe, 1983), respiration (above- and belowground), carbon allocation within and from branches (Sprugel and Hinckley, 1988), root development, and root-shoot interactions and feedback (Dickson and Isebrands, in press). It is unfortunate (almost deplorable) that so little is known about the aforementioned processes in such a major forest species as loblolly pine, northern red oak, black walnut, and Douglas fir, to mention a few. There is also a need for more experimental studies to be conducted in conjunction with whole-tree growth process model simulations. As we pointed out earlier, growth process models can be used to suggest weaknesses or gaps in our knowledge base, and hypotheses on the effects of certain environmental stresses on key physiological processes can be tested rapidly with well-designed controlled-environment and/or phytotron experiments. In the end, if physiologists, geneticists, and modelers collaborate, our understanding of the underlying mechanisms of tree and forest growth should soon reach a higher level, and we should be better equipped to explain why various phenomena occur in the forest.

LITERATURE CITED

Amthor, J. S. 1986. Evolution and applicability of a whole plant respiration model. *Journal of Theoretical Biology* 122:473–490.

Borchert, R. 1975. Endogenous shoot growth rhythms and indeterminate shoot growth in oak. *Physiologia Plantarum* 35:152–157.

Bowerman, R. G., and D. E. Glover. 1988. *Putting Expert Systems into Practice.* Van Nostrand Reinhold, New York, NY, U.S.A.

Caldwell, M. M., H. P. Meister, J. D. Tenhunen, and O. L. Lange. 1986. Canopy structure, light microclimate and leaf gas exchange of *Quercus coccifera* L. in a Portuguese macchia: measurements in different canopy layers and simulations with a canopy model. *Trees* 1:25–41.

Ceulemans, R., I. Impens, and V. Steenackers. 1987. Variations in photosynthetic, anatomical, and enzymatic traits and correlations with growth in recently selected *Populus* hybrids. *Canadian Journal of Forest Research* 17:273–283.

Chappell, H. N., and D. A. Maguire. 1987. Predicting forest growth and yield: current issues and future prospects. Research Conference, No. 58. College of Forest Resources, University of Washington, Seattle, WA, U.S.A.

Coulson, R. N., J. Folse, and D. Loh. 1987. Artificial intelligence and natural resource management. *Science* 237:262–267.

Crow, T. R., J. G. Isebrands, H. M. Rauscher, and R. E. Dickson. 1987. Effects of canopy architecture on photosynthetic production in *Populus* seedlings: a simulation study. *Bulletin of the Ecological Society of America* 63(3):286. (Abstract.)

Dale, J. E., and F. L. Milthorpe. 1983. *The Growth and Functioning of Leaves.* Cambridge University Press, Cambridge, England, U.K.

Dale, V. H., T. W. Doyle, and H. H. Shugart. 1985. A comparison of tree growth models. *Ecological Modelling* 29:145–169.

Dickmann, D. I. 1985. The ideotype concept applied to forest trees. In: Cannell, M. G., and J. E. Jackson, eds. *Attributes of Trees As Crop Plants.* Institute of Terrestrial Ecology, Huntingdon, England, U.K. Pp. 89–101.

Dickson, R. E. 1986. Carbon fixation and distribution in young *Populus* trees. In: Fujimori, T., and D. Whitehead, eds. *Proceedings, IUFRO Conference on crown and canopy structure in relation to productivity.* Forestry and Forest Products Institute, Ibaraki, Japan. Pp. 409–426.

Dickson, R. E., and J. G. Isebrands. In press. Role of leaves in regulating structural-functional development in plant shoots. In: Mooney, H. A., W. E. Winner, and E. J. Pell, eds. *Integrated Response of Plants to Stress.* Academic Press, New York, NY, U.S.A.

Ek, A. R., S. R. Shifley, and T. E. Burk, eds. 1988. *Forest Growth Modelling and Prediction.* Vols. I and II. General Technical Report NC-120, USDA Forest Service North Central Forest Experiment Station, St. Paul, MN, U.S.A. 1149 pp.

Fries, J. 1974. Growth models for tree and stand simulation. Research Note No. 30, Royal College of Forestry, Stockholm, Sweden.

Gifford, R., J. Thorne, W. Hitz, and R. Giaquinta. 1984. Crop productivity and photoassimilate partitioning. *Science* 225:801–808.

Hesketh, J. D., and J. W. Jones. 1980. *Predicting Photosynthesis for Ecosystem Models,* Vols. I and II. CRC Press, Boca Raton, FL, U.S.A.

Host, G. E., H. M. Rauscher, J. G. Isebrands, D. I. Dickmann, R. E. Dickson, T. R. Crow, and D. A. Michael. In press. ECOPHYS: an ecophysiological growth process model of *Populus:* user's manual. General Technical Report, USDA Forest Service North Central Forest Experiment Station, St. Paul, MN, U.S.A.

Isebrands, J. G. 1982. Toward a physiological basis of intensive culture of poplar. In: *Proceedings, TAPPI R&D Conference,* August 29–September 1, 1982, Asheville, NC, U.S.A. Pp. 81–90.

Isebrands, J. G., R. Ceulemans, and B. Wiard. 1988. Genetic variation in photosynthetic traits among *Populus* clones in relation to yield. *Plant Physiology and Biochemistry* 26:427–437.

Isebrands, J. G., and R. E. Dickson. In press. Carbohydrate production and distribution—radiotracer techniques and application. In: Lassoie, J., and T. Hinckley, eds. *Ecophysiology of Forest Trees,* Vol. I. *Techniques and Methodologies.* CRC Press, Boca Raton, FL, U.S.A.

Isebrands, J. G., and D. A. Michael. 1986. Effects of leaf morphology and orientation on solar radiation interception and photosynthesis in *Populus.* In: Fujimori, T., and D. Whitehead, eds. *Proceedings, IUFRO Conference on crown and canopy structure in relation to productivity.* Forestry and Forest Products Institute, Ibaraki, Japan. Pp. 359–381.

Isebrands, J. G., and N. D. Nelson. 1983. Distribution of ^{14}C-labeled photosynthates within intensively cultured *Populus* clones during the establishment year. *Physiologia Plantarum* 59:9–18.

Isebrands, J. G., N. D. Nelson, D. I. Dickmann, and D. A. Michael. 1983. Yield physiology of short rotation intensively cultured poplars. General Technical Report NC-91, USDA Forest Service. Pp. 77–93.

Jeffers, J. N. R. 1982. *Modelling.* Chapman and Hall, London, England, U.K.

Kramer, P. J. 1986. The role of physiology in forestry. *Tree Physiology* 2:1–16.

Landsberg, J. J. 1986. *Physiological Ecology of Forest Production.* Academic Press, London, England, U.K. 198 pp.

Larson, P. R. 1980. Control of vasacularization by developing leaves. In: Little, C. H. A., ed. *Control of Shoot Growth in Trees.* Proceedings, IUFRO Workshop, Fredericton, N.B., Canada. Pp. 157–172.
Larson, P. R. 1984. Exploiting the tree as an experimental organism. In: Lanner, R., ed. *Proceedings, 8th North American Forest Biology Workshop,* July 30–August 1, 1984, Logan, UT, U.S.A. Pp. 1–10.
Larson, P. R., J. G. Isebrands, and R. E. Dickson. 1980. Sink to source transition of *Populus* leaves. *Berichte Deutsch Botanik Gesellschaft* 93:79–90.
Leary, R. A. 1985. A framework for assessing and rewarding a scientist's research productivity. *Scientometrics* 7(1–2):29–38.
Leary, R. A. 1988a. Some factors that will affect the next generation of growth models. In: Ek, A. R., S. R. Shifley, and T. E. Burk, eds. *Forest Growth Modelling and Prediction.* General Technical Report NC-120, USDA Forest Service North Central Forest Experiment Station, St. Paul, MN, U.S.A. Pp. 22–32.
Leary, R. 1988b. Parallel processing for spatio-temporal order of forest growing stock. *Proceedings, 18th IUFRO World Congress,* September 7–13, 1986, Ljubljana, Yugoslavia. P. 649. (Abstract.)
Ledig, F. T. 1969. A growth model for tree seedlings based upon the rate of photosynthesis and distribution of photosynthate. *Photosynthetica* 3:263–275.
Levitt, J. 1980. *Responses of Plants to Environmental Stresses,* Vol. I. Academic Press, New York, NY, U.S.A.
McMurtrie, R. M., and L. Wolf. 1983. Above- and below-ground growth of forest stands: a carbon budget model. *Annals of Botany* 52:437–448.
Meentemeyer, V., and E. O. Box. 1985. Scale effects in landscape studies. In: Turner, M. G., ed. *Landscape Heterogeneity and Disturbance.* Springer-Verlag, Berlin, F.R.G. Pp. 15–34.
Mooney, H. A., R. M. Pearcy, and J. Ehleringer. 1987. Plant physiological ecology today. *BioScience* 37:18–20.
Newman, D. W., and K. G. Wilson. 1987. *Models in Plant Physiology and Biochemistry,* Vols. I, II, and III. CRC Press, Boca Raton, FL, U.S.A.
Norman, J. 1980. Interfacing leaf and canopy light interception models. In: Hesketh, J. D., and J. W. Jones, eds. *Predicting Photosynthesis for Ecosystem Models,* Vol. II. CRC Press, Boca Raton, FL, U.S.A. Pp. 49–67.
Passioura, J. B. 1988. Water transport in and to roots. *Annual Review of Plant Physiology* 39:245–265.
Rauscher, H. M. 1985. The promise of expert systems for forest management. *FORS Compiler* 3:3 and 9.
Rauscher, H. M., J. G. Isebrands, T. R. Crow, R. E. Dickson, D. I. Dickman, and D. A. Michael. 1988. Simulating the influence of temperature and light on the growth of juvenile poplars in their establishment year. In: Ek, A. R., S. R. Shifley, and T. E. Burk, eds. *Forest Growth Modelling and Prediction.* General Technical Report NC-120, USDA Forest Service North Central Forest Experiment Station, St. Paul, MN, U.S.A. Pp. 331–339.
Reed, K. 1980. An ecological approach to modeling growth of forest trees. *Forest Science* 26:33–50.
Rensberger, B. 1988. New computer works 1000 times faster. *Washington Post* 105:A-12 and 16.
Schmoldt, D., and G. Martin. 1986. Expert systems in forestry: utilizing information and expertise in decision making. *Computer Electronics Agriculture* 1:233–250.
Sievänen, R. 1983. Growth model for mini-rotation plantations. *Communicationes Instituti Forestalis Finniae* 117:1–41.
Sprugel, D. G., and T. M. Hinckley. 1988. The branch autonomy theory. In: Winner, W. E., and L. G. Phelps, eds. *Response of Trees to Air Pollution: The Role of Branch Studies.* Workshop Proceedings, National Forest Response Program, November 5–6, 1987, Boulder, CO. Pp. 1–19.
Thornley, J. 1976. *Mathematical Models in Plant Physiology.* Academic Press, London, England, U.K.
Wallace, D. H., M. Peet, and J. L. Ozbun. 1976. Studies of CO_2 metabolism in *Phaseolus vulgaris* L. and applications in breeding. In: Burris, R., and C. Black, eds. *CO_2 Metabolism and Plant Productivity.* University Park Press, Baltimore, MD, U.S.A. Pp. 43–58.
Whisler, F. D., B. Acock, D. N. Baker, R. E. Fye, H. F. Hodges, J. R. Lambert, H. E. Lemmon, J. M. McKinion, and V. R. Reddy. 1986. Crop simulation models in agronomic systems. *Advances in Agronomy* 40:141–208.
Wisiol, K., and J. D. Hesketh. 1987. Plant growth modeling for resource management, Vol. I. *Current Models and Methods.* CRC Press, Boca Raton, FL, U.S.A.

CHAPTER 9

MECHANISTIC MODELING OF NUTRIENT ACQUISITION BY TREES

David F. Grigal

Abstract. A mechanistic model of nutrient acquisition by trees can be based on similar models developed for agronomic crops. Acquisition is considered to be a function of a relatively small number of parameters that can be grouped into four broad categories: the movement of nutrient ions through the soil to the plant root; their uptake by the root; the distribution of roots in the soil and its change with time; and the properties of the soil in supplying the ion. This formulation seems to be sufficiently general to serve as the basis for modeling nutrient uptake by tree roots. The difficulties faced by modelers of trees are lack of data for parameterization of the model and the significant differences between forested and agronomic systems that affect model conceptualization and numerical parameterization. Examples of these differences are the presence of forest floor as an important source of nutrients in forests, the prevalence of mycorrhizae in forests, the perennial nature of trees with multi-year storage of nutrients therein, and the markedly anisotropic soils in which forest trees grow. The latter characteristic leads to the necessity of substantially different parameterization of any model by soil horizon.

INTRODUCTION

The objective of this paper is to suggest approaches for mechanistically modeling the acquisition of nutrients by individual trees, and thereby to assess the effects of stress, including atmospheric pollution, on tree growth. Agronomists have made major strides in attempting to understand nutrient acquistion. Although their work requires modification to encompass features unique to forested ecosystems, the core of that work is applicable to forest trees.

MODELING GOALS OR ATTRIBUTES

The object of my discussion is a process-based or mechanistic model that can be extrapolated in space and time. All possible mechanisms, however, cannot be included in such a model. I will discuss a "conceptual model," one that lies between time-series analyses and physics-based approaches (Wheater et al., 1986). It is based on subjective assessment of the dominant processes and represents a compromise in complexity while

Dr. Grigal is a Professor in the Department of Soil Science, University of Minnesota, St. Paul, MN, 55108 U.S.A. This paper is the result of work under project MN-48-54 of the Minnesota Agricultural Experiment Station, and is published as Paper No. 16,082 of the Experiment Station. The author wishes to thank all who, through their reviews, suggested improvements and to assert full responsibility for not taking all the advice so offered.

retaining a form that can be considered physically. This is important in parameterization. Before I begin my detailed discussion, there are some characteristics that I would like to emphasize as necessary in achieving the goal of modeling nutrient acquisition by trees.

One of the attributes of a good mechanistic model is the use of measurable parameters and response variables in both model calibration and validation. Measurements must be routine and relatively easy to make. Too often, models use parameters that are theoretically measurable but are seldom routinely measured. In addition to measurability, an important consideration is the number of parameters and variables in the model. There is an urge to make the model as realistic as possible by including all factors thought to be important. Physical driving variables are often measured and used more accurately than is justified by knowledge of biology (Landsberg, 1981). As the number of variables that must be simultaneously fit to the observations increases, there is a decrease in the degrees of freedom for tuning the model to represent a local and not a general situation (Chen et al., 1983). Finally, the principles of conservation of mass and stability of plant stoichiometry are powerful constraints in developing a full model of tree nutrition.

A model may be developed for purposes ranging from a pure scientific interest to a management-related application, and most models fall somewhere between these extremes. Where management applications are among the goals, the accuracy of the model should be compatible with an acceptable risk level in managing the resource in question (Warrington, 1983). The risk level judged acceptable has to do with the loss of resource that can be tolerated at a given error level of the model. When is "good enough" good enough, and how do we decide? This is partially a policy question and must be considered in developing many models, especially those related to pollution effects.

A final issue is an appropriate time-step for the model. Most of the driving variables for nutrient acquisition are unlikely to be available at very short time-steps. Even temporally dynamic weather data are usually available on a daily basis, and many of the less dynamic parameters, such as soil nutrient status or plant phenology, have even longer time-steps. Conversely, too long a time-step may not have sufficient discrimination to simulate transitory phenomena, such as the effects of acute ozone exposures or a dry period of short duration (mini-drought).

The computational load introduced by small time-steps is less of a consideration today than even a decade ago. I therefore believe that a daily time-step is appropriate for simulating nutrient acquisition. The result of this time-step is simulation of the annual pattern of plant processes; this is important when such patterns for multiple-year phenomena are to be compared in order to detect changes induced by stress. This brings up another important feature of models of forest trees: In contrast to models of annual crops, where the slate is wiped clean at the end of the growing season, models of perennial woody species extend over many growing seasons, allowing minor errors in the simulation to propagate and eventually become dominant.

ALTERNATIVE MECHANISMS OF NUTRIENT SUPPLY TO TREES

Although my main topic is acquisition of nutrients by tree roots, it should be noted that trees satisfy their requirements for nutrients by other mechanisms as well. First, foliage is an organ of assimilation. Foliage is the path for entry of CO_2 into the plant, but it can also absorb other gases or nutrient ions, reducing the demand for nutrients from the root system. For example, in Tennessee forests, atmospheric deposition to the canopy supplied 40 to 100% of the nitrogen (N) and sulfur (S) requirements, respectively, for the annual woody increment (Lindberg et al., 1986). Many other examples of significant foliar uptake exist. The important point is that foliar uptake has implications with respect to the nutrient status of plants and must be included in any complete simulation model.

In addition, trees have developed mechanisms to conserve nutrient elements. Although trees usually are found on sites that are low in available nutrients, leaf fall and subsequent rapid leaching and mineralization can provide a flush of nutrients (Harley and Smith, 1983). Conversely, low soil water status can virtually halt the movement of nutrients to the roots. Probably the most common method of conserving nutrient elements within trees is their return from leaves into woody tissue prior to autumnal leaf fall. Redistribution also occurs among other plant organs. Elemental concentrations in heartwood, the metabolically inactive portion of the woody stem, are lower than those in the sapwood from which it forms; this infers translocation from heartwood (Miller, 1981).

Some investigators (e.g., Miller, 1981) have rejected the validity of the concept of luxury consumption in perennial plants. In annual agronomic species, continued accretion of a nutrient element above amounts necessary for physiologic functioning may be a luxury. In perennial plants, storage may provide reserves for periods when particular nutrients are in short supply. Such a period may be during budbreak and leaf expansion. At that time, maximum demand for nitrogen (N) and phosphorus (P) occurs within 3–4 weeks, and trees probably use reserves built up during the entire year. Miller et al. (1979) found that ⅔ of the N taken up during a single year of pine fertilization was used for growth in subsequent years by remobilization from intra-tree sources.

MODELING ION ACQUISITION

The remainder of my discussion will deal with modeling of nutrient acquisition from soil, the most important source to plants. Annual uptake of nutrients by forests is comparable to amounts measured at the same latitude for other vegetation, including row-crops (Miller, 1981). Concentration profiles of important nutrient ions near absorbing roots and their rates of uptake by agronomic crops can be satisfactorily explained by mechanistic models in which the ion moves to the root by mass flow and diffusion (Nye, 1984). These agronomic models are relatively similar. I will primarily use Barber's (1984) model as an example, although where appropriate I will also draw from the approach of Nye and Tinker (1977). Finally, specified units for parameters may differ from either approach in some cases because of dimensional considerations. Most of the parameters in the model are defined in terms of a specific nutrient.

Movement of ions to plant roots is described by

$$I_{out} = D_e \, (dC/dr) + v_o C_l \qquad (1)$$

and uptake by

$$I_{in} = I_{max} \times ((C_l - C_{min})/(K_m + C_l - C_{min})) \qquad (2)$$

where the parameters are defined in Table 1. I will discuss the parameters by categories that relate to the processes of ion movement and uptake.

Movement of Nutrients to Roots

Ions move to roots by two processes, diffusion and mass flow. The rate of diffusion is described by D_e, the diffusion coefficient, and by the diffusion gradient (Tables 1 and 2). The length of the diffusion path is a function of root distribution and growth rate, and the gradient is defined by the soil-supplying power and by uptake by the root. The gradient can often be assumed to approach near-zero concentration (C_{min}) at the root surface.

The diffusion coefficient can be more explicitly defined as a reduction in free-solution diffusion (D_l), or

$$D_e = D_l\, f_l\, \theta\, dC_l/dC \quad (3)$$

(Table 1) (Nye and Tinker, 1977). This equation adequately represents diffusion of simple inorganic ions, including those of nutrient elements, Ca^{2+}, Mg^{2+}, K^+, and $H_2PO_4^-$(Nye and Tinker, 1977). The impedance factor, f_l, encompasses effects of increased path-length for diffusion in soil, increased water viscosity near surfaces, and the negative adsorption of anions. It is strongly related to θ; an approximation based on spherical particles is

$$f_l = \theta^{0.5} \quad (4)$$

(Nye and Tinker, 1977). The product θC_l is the amount of solute per volume of soil. As a result of the relationships in Equations 3 and 4, ions that move predominantly by diffusion are strongly influenced by changes in θ.

The buffer power (b, or dC/dC_l; see Table 1) of the soil is the inverse of the change in concentration of ions in solution per unit change of those ions in the whole soil system (Nye and Tinker, 1977). It will be examined in more detail in the discussion of soil properties later in this chapter. It has a relatively large effect on diffusion of ions. For example, because nitrate is usually not absorbed by soil but only occurs in the solution phase, $b = \theta$, and the product $\theta\, dC_l/dC$ from Equation 3 = 1. Conversely, most of the phosphate in the soil system is present in the solid phase and not in solution, and so C can be approximated by C_s, the ion concentration in the solid phase. A resonable value for b may therefore be 100, although values up to 1000 have been reported. Because D_e is inversely proportional to b, at the same θ, f_l, and D_l, the computed D_e's are two orders of magnitude higher for NO_3^- than for $H_2PO_4^-$ (Table 2). Diffusing ions follow a gaussian distribution (in one dimension), with a root mean square displacement of

$$d = (2\, D_e\, t)^{0.5} \quad (5)$$

where t is time (Nye and Tinker, 1977). Because of differences in b, NO_3^-diffuses an order of magnitude or more farther than $H_2PO_4^-$in a given time. The buffer power therefore affects the volume of soil that is exploited by a root.

The parameter v_o (Tables 1 and 3) measures the flow of water moving toward the root in response to transpirational gradients. The product of this and C_l determines the nutrient supply via mass flow. There is some disagreement concerning the validity of separating ions moving in the transpiration stream from those moving by diffusion (Barley, 1970). For modeling, these disagreements are of limited consequence. Simulation of ion profiles near roots is important, but the relative value of the two processes matters little except under extreme conditions. Only if ion uptake into roots occurs at markedly different sites than does water uptake, or if I_{in} is less than v_oC_l, would separation of processes become important. In these cases, ions moved by mass flow would accumulate. This may occur if deep suberized roots with low I_{max} are important for water uptake when surface soils are dry.

The importance of the process of mass flow varies among systems. For example, in a beech forest in the Solling Mountains, F.R.G., elemental uptake by trees was much smaller than supply of aluminum (Al), chlorine (Cl), and sodium (Na) by mass flow; uptake and mass flow were nearly balanced for sulfur (S), iron (Fe), and magnesium (Mg); and mass flow supplied less than uptake in the case of calcium (Ca), potassium (K), nitrogen (N), and phosphorus (P) (Prenzel, 1979). Conversely, data from Oliver and Barber (1966a, b) indicate that Ca and Mg, but neither K nor Al, could be supplied by mass flow to soybeans. Although differences in species' requirements may partly explain these results, differences in soil solution concentrations are a more likely explanation. The soil solution has sharply contrasting chemistries at the two sites. Unless the soil chemistry is known, few generalizations can be made about the importance of mass flow for supplying nutrients.

An adequate simulation of mass flow of nutrients to plant roots depends on a good model of transpiration by trees. An apportionment of the amount of water reaching a tree from each soil horizon is necessary so that mass flow can be accurately estimated.

Table 1. Parameters used in modeling nutrient uptake by trees.[a]

Parameter	Units	Definition
b	dimensionless	buffer power of soil for supplying nutrients for diffusion, or dC/dC_l
C	μmol/L	ion concentration in the whole soil system, including solid phase and in solution
C_l	μmol/L	ion concentration in soil solution
C_{min}	μmol/L	solution concentration at which $I = 0$
C_s	μmol/L	ion concentration in soil solid phase
D_e	cm^2/sec	effective diffusion coefficient for the nutrient
D_l	cm^2/sec	diffusion coefficient of the solute in free solution
f_l	dimensionless	impedance factor
I_{in}	$\mu mol/cm^2/sec$	flux of ions into plant root
I_{max}	$\mu mol/cm^2/sec$	maximum influx at high concentrations
I_{out}	$\mu mol/cm^2/sec$	external flux of ions to plant root
k	cm/sec	rate of root growth
K_m	μmol/L	solution concentration at which $I = I_{max}/2$
L_a	1/cm	root length per area
L_v	$1/cm^2$	root length per volume
r	cm	radial distance from root axis
r_l	cm	half-distance between root axes
r_o	cm	mean root radius
v_o	L/sec	mean water influx
θ	dimensionless	volumetric water content of the soil (the cross-section for diffusion through solution)
t	seconds	time

[a]The parameters k, L_a, and v_o are expressed in units per cm^2 of land area, L_v in units per cm^3 of soil, and C and C_s in units per L of soil. The I's are expressed in units per cm^2 of root surface area.

Table 2. Order of magnitude estimates of important parameters varying with ion and used in modeling nutrient uptake by roots (10^x).

Parameter	$H_2PO_4^-$	K^+	NO_3^-	NH_4^+
D_e (cm^2/sec)	−8	−7	−6	−7
I_{max} ($\mu mol/cm^2/sec$)	−7	−6	−5	−6
K_m (μmol/L)	−1	0	1	1
C_{min} (μmol/L)	−2	−2	−1	?
C_l (μmol/L)	−2 to 0	0 to 2	−2 to 3	−1 to 1
b (dimensionless)	2	1	0	1

Table 3. Order of magnitude estimates of parameters varying with absorbing organ and used in modeling nutrient uptake by roots (10^x).

Parameter	Hyphae	Mycelial strand	Fine root
r_o (cm)	−4	−3	−2
L_a (1/cm)	4	3	1
k (cm/sec)	(−2)	(−3)	−5
r_1 (cm)	−3	−2	−1
θ (dimensionless)	——	−2 to −1	——
v_o (L/sec)	——	−9 to −8	——

Elemental Uptake by Roots

The process of nutrient uptake has been described by three parameters: I_{max}, K_m, and C_{min} (Tables 1 and 2). The first two parameters define a Michaelis-Menten kinetics model for uptake. Agronomic soils generally have higher levels of nutrients than do forest systems. As a result, I_{max} is often considered to be of lesser importance in forests; the limiting step in nutrient uptake is considered to be supply of ions from soil to root (Bowen, 1981). However, that may not be universally true. For example, some hypothesized anthropogenic stresses to trees are associated with situations in which toxic elements reach high concentrations, leading to rates approaching I_{max} (e.g., Al, in Huttermann and Ulrich, 1984; NO_3^-, in Nihlgard, 1985). This parameter could also be important in undisturbed systems if roots are active during periods of nutrient flush, such as at leaf fall.

Although elemental concentrations range over several orders of magnitude in the soil system, their range is narrow within the plant (Miller, 1981). Ion uptake is influenced by the nutrient status of the root or other absorbing organ, and both I_{max} and K_m may be considered to be functions of the internal concentration of the ion of interest (Glass and Siddiqi, 1984). A key to rapid and continuing uptake, therefore, is reduction in internal ion concentrations. For example, there is a close correlation between uptake of NH_4^+ and availability of carbohydrates as precursors for amino acids, reducing internal NH_4^+ (Plassard et al., 1985). Similarly, P can be precipitated as Ca polyphosphate (Harley and Smith, 1983), reducing internal phosphate concentrations. Interaction among ions also affects uptake. For example, Al interferes with uptake of other cations, especially Ca (Khanna and Ulrich, 1984). Conversely, P uptake can be enhanced because of protons in the rhizosphere (Nye, 1984).

The concentration C_{min} describes that level in solution at which plant uptake ceases (Equation 2). If C_l falls below C_{min}, the plant will not be able to absorb the particular nutrient. Sensitivity analysis using agronomic data has indicated that variation in C_{min} has little effect on estimated uptake because C_l is usually much larger (Barber, 1984). However, for those elements and situations where the two concentrations are of the same order of magnitude (see Table 2), C_{min} could become important in limiting uptake.

Mycorrhizae constitute an important fungi–higher plant symbiotic system in forests, involving uptake of nutrients by the fungi and supply of fixed carbon from the host. I will focus on ectomycorrhizae because they are present on the majority of commercially important forest trees in North America and Europe. Mycorrhizae affect uptake mechanisms and rates. The specific uptake rate of mycorrhizal tree roots is 2–9 times higher than that of non-mycorrhizal roots of the same species (Bowen, 1973; Harley and Smith, 1983). Although these differences are usually ascribed to an increased absorbing surface of the mycorrhizal system (Harley and Smith, 1983), there are indications that mycorrhizae can increase I_{max} (Reid, 1984) and lower C_{min} (Bowen and Cartwright, 1977). Uptake curves of excised mycorrhizae show active uptake and a tendency for rate to decrease with concentration (Harley and Smith, 1983), a relationship that can be modeled as Michaelis-Menten kinetics.

Tissue age and longevity also affect uptake. Nutrient uptake by non-mycorrhizal roots occurs mostly near the root tip, but uptake by mycorrhizae extends back from the tip (Harley and Smith, 1983). Mycorrhizae also maintain high absorbing rates for several months and are able to sustain absorption in one location rather than transiently through the soil as do uninfected roots (Bowen, 1973). This affects the volume of exploitation of nutrients. As described in Equation 5, nutrient movement by diffusion varies as the square root of time. An isolated root tip that is active in uptake for only 3 days exploits a volume of soil about 15% of that exploited over 3 weeks by a mycorrhizal root tip of the same size. In a natural system, overlapping zones of diffusion as affected by root distribution will have a bearing on such comparisons.

Root Distribution and Growth

The relevant parameters with respect to roots are L_a, r_o, r_1, and k (Tables 1 and 3). These four parameters define initial root distribution and its change with time. Estimates of root mass, absorbing area, spacing, and growth rate are bedeviling for parameterization in modeling tree uptake. First, there is the issue of root distribution. Although tree roots extend laterally for long distances, quantification of root extension is not an issue in a single-tree model. An individual tree shares its aboveground space with surrounding trees by intermingling of canopy branches and its belowground space with surrounding trees by intermingling of roots. We can assume that each tree has available a certain area of land, fit as an interlocking tessera (sensu Jenny, 1958) into the areas occupied by adjacent trees, with the size of the tessera a function of tree size. As trees compete, or as a tree dies and a tessera becomes vacant, then lateral extension of roots may become important. Although this competition enters into forest-stand models, it is not important in single-tree models.

The depth of rooting is important. Forest soils are markedly anisotropic, with high variation with depth in many properties related to nutrients. The majority of tree roots are located relatively near the surface. Gale and Grigal (1987) developed functional relationships describing root distribution with depth and concluded that rooting depth varies with successional status in northern forests. For example, 60% of the roots of early-successional species, 78% of the roots of mid-successional species, and 92% of the roots of late-successional species are located in the upper 30 cm of soil. Although the majority of roots are near the surface, roots can extend deeply (Schultz, 1972). During dry periods, the movement of ions by mass flow at depth may be a significant proportion of a tree's overall ion balance; diffusion, because of its dependence on θ, will be nearly nonexistent near the dry surface. This deep mass flow provides an explanation for the continual, albeit small, exploitation of nutrients from deep soil horizons by trees (Comerford et al., 1984). According to Comerford et al. (1984), two published studies of southern pines (*Pinus*) indicated that about 10% of the total P uptake came from deeper than 50 cm. Although data are limited, uptake from deep roots must ultimately be considered in any modeling effort.

Both root length and root radius, hence root surface area, are model parameters. Very little data are collected on the surface area of tree roots. Agronomists commonly measure either L_v or, if the entire rooting depth has been sampled, L_a (Table 1) (Barley, 1970). These parameters are appropriate surrogates for surface area because such roots vary little in diameter. The most common measure of tree roots is biomass. Because of the wide range in tree root sizes and the large mass but low surface area of large roots, estimates of total root mass are of limited use in model parameterization. Approximate L_v and L_a may be computed from biomass data, but such estimates can vary widely as a function of the assumed root diameter based on reported size classes. The use of either measures of actual root diameter or narrowly rather than broadly defined root diameter classes is important for parameterization.

Although root distribution is important, the distribution of mycorrhizae must also be considered in forests. While most tree roots are near the surface, the majority of mycorrhizae are even nearer the surface and primarily in the forest floor (Harvey et al., 1986). The forest floor is the site of the return of nutrients via throughfall and litterfall, and their release via mineralization. For mature stands, it is the major source of all elements, but especially of N and P (Jorgensen et al., 1980). The zone of elemental uptake can probably be approximated by the distribution of mycorrhizae. Mycorrhizae may markedly increase the nutrient-acquiring surface area of roots. Bowen (1973) stresses the abundance of mycelial strands (branching aggregates of hyphae) in radiata pine (*Pinus radiata* D.Don) forests with a relatively low density of roots. Occupancy of soil by mycorrhizae is difficult to estimate. Although fine roots (ca. 2 mm diameter) can be separated from soil, mycorrhizal hyphae (ca. 2 μm diameter) are impossible to routinely separate. Data concerning the total length or surface area of mycelial strands and hyphae in the field are therefore lacking. The number of mycorrhizae structures on the root is an inadequate measure, for each may have strands and hyphae protruding from it.

The distance between root axes (r_1) is also a model parameter, determining the competition among roots and the volume of soil that is exploited. From the perspective of parameterization, the most important issue is whether or not the total absorbing area changes markedly over time, and whether unexploited zones of soil thus become exploited. Because of the longevity of mycorrhizal roots and the close spacing of mycelial strands (Bowen, 1973), it appears that with reasonable mycorrhizal densities the soil is completely exploited, and a combination of low θ and the soil buffer power limits elemental influx to roots.

Rate of root growth, k, is also important. First, there is a long temporal scale. In newly established forest stands, the mineral soil is the primary zone for root and mycorrhizae distribution and elemental uptake. As a stand matures, the canopy closes and the forest floor fully develops; the active mycorrhizal roots and much of the nutrient uptake shift to the forest floor. On a shorter temporal scale (seasonal), root mortality and turnover must be considered (Joslin and Henderson, 1987). In contrast to annual plants, where the volume of exploited soil continually increases over the growing season, in perennial plants the soil is fully exploited. In plantations with moist soils in warm climates, there may be little fluctuation of the fine root population, but this may be accomplished through concurrent root death and initiation (Bowen, 1984). Root growth of woody plants apparently lacks a clear pattern but may be associated with suitable environmental conditions and with availability of carbohydrates (Lyford, 1980). For modeling nutrient uptake (not carbon allocation), can a certain constant density of roots per unit volume of soil be assumed? Or must a seasonal cycle be introduced?

An overwhelming uncertainty in these discussions is the presence and rate of growth of mycorrhizal hyphae or mycelial strands. For modeling, it may be necessary to simply adjust parameters dealing with mycorrhizal roots rather than to try to accurately measure these structures. In other words, because mycorrhizae are a sink for carbohydrates and a source of nutrients, it may be sufficient to consider them merely as an extension of the root system rather than differentiating them from the vascular plant. Depending on the data available, however, this adjustment deviates from a strictly mechanistic approach and leads to the development of a tuned model. It may be that some aspects of nutrient uptake are presently too ill-defined for a purely mechanistic approach, but that should surely be the goal toward which we strive.

Soil-Supplying Power

The parameters that deal with the supply of ions by the soil are C_l and b (Tables 1 and 2). In general, forest soils have much lower concentrations of most elements in solution than do agricultural systems. Soil solution concentrations in agricultural systems are commonly an order of magnitude higher than in forest systems, although forest soils with high levels of dissolved anions due to rapid mineralization, atmospheric deposition, or high respiration raising HCO_3^- may have concentrations that approach the levels in agricultural systems. Measurements are necessary for good parameterization of the model for specific sites.

The buffer capacity of the soil (b) for a given ion is operationally defined as the ability of the soil to supply that ion to solution for diffusion. Based on the constraint that model parameters must be measurable, presence in solution is a good index of availability. The buffer power approaches the slope of the adsorption isotherm for strongly adsorbed ions such as phosphate, and it assumes an equilibrium between ions in solution and in the solid phase (Nye and Tinker, 1977). A mechanistic representation of tree growth must also mechanistically represent nutrient supply. A model that deals with cation exchange and weathering reactions is necessary to define b for cations. For P, precipitation/solubilization reactions must be simulated. And for N and in part for P, a simulation of organic matter mineralization is necessary. Models exist that deal with soil chemistry or nutrient supply (e.g., Sposito and Mattigod, 1980; Reuss, 1983), but they may require modification. Mineralization or release of nutrient elements from the forest floor is critically important to forest tree nutrition. This process is not handled by existing models (e.g., Pastor and Post, 1986) at the daily time-step appropriate for simulation of tree growth.

Although mycorrhizae may alter mineralization and even rates of weathering, for modeling it is probably not efficient to attempt to separate their effects. However, if changes in mycorrhizae populations alter the availability of nutrients, then that information must be included in parameterization. For example, mycorrhizae may be present on one site and absent on another, or their activity may decline over time because of some anthropogenic effect, or participation in mineralization may shift from the tree-symbiont to soil microbes. All these changes can affect the simulation.

CONCLUSIONS

A number of conclusions may be drawn. Soils in forest systems are much more anisotropic than soils in most agricultural systems. Yet the latter systems have served as the basis for the development, parameterization, and validation of existing models of nutrient uptake. Many of the properties that affect uptake of nutrients, including the distribution of the tree roots themselves, vary markedly by horizon. For most forest systems, the forest floor or surface organic and organic-rich mineral horizons are most important in terms of fine-root density, mycorrhizae density, and rate of nutrient release via mineralization. The chemical properties of some soil horizons are such that they are not occupied by roots (Huttermann and Ulrich, 1984), while in other cases a physical property or properties (such as high bulk density) inhibits rooting. Most of the parameters in the model are soil horizon–specific.

Forest trees are large organisms with the ability to buffer environmental stress, including excessive or deficient nutrients or environmental conditions that inhibit nutrient movement or uptake. There is a high demand for nutrients by young trees before crown closure, with both less internal retranslocation and more dependence on the mineral soil. As stands mature, trees become more buffered against fluctuations in nutrient supply. Any model must deal with the biological buffering capacity in terms of quantity as well as seasonality.

I have discussed the two major processes involved in nutrient acquisition by tree roots: movement to roots, including release from the solid phase of the soil to solution; and the uptake process itself. Release from solid phase to solution should be addressed mechanistically, requiring a good submodel of soil chemistry. Most existing models are deficient in their treatment of forest floor/soil organic matter dynamics. Those dynamics deserve more attention because of their documented importance to the nutrition of forest trees. Movement of ions to roots, whether by mass flow or by diffusion, is highly dependent on soil water content. A good submodel of water balance of the soil, including transpirational demand by soil horizon, is essential for modeling of nutrient acquisition.

Uptake of ions depends on the rate per unit surface area of roots and on the area of absorbing surface. The rate may be approached either by Michaelis-Menten kinetics or by some other formulation. Although in most cases, movement to roots and not uptake is the limiting step, some hypotheses relating atmospheric pollution to tree growth involve excess ion accumulation at root surfaces. The amount and the spatial and temporal variability of the area of absorbing surface of roots are important parameters for which limited data exist. Quantitative estimates of fine root and mycorrhizae abundance, especially in near-surface horizons including the forest floor, are crucial. Unfortunately, we do not have a precise definition of the environmental factors that enhance or inhibit root development.

Mycorrhizae are very important in uptake of nutrients from soil: in their spatial exploitation; in their uptake characteristics; and in their potential role in solubilizing nutrients or otherwise making them more available for uptake. The difficulty of quantifying their presence may require any model to include them simply by modifying the standard parameters.

In spite of these obstacles, I am confident that by building on the foundation laid down by agronomists, we can create both detailed and more general models of nutrient acquisition by forest trees.

LITERATURE CITED

Barber, S. A. 1984. *Soil Nutrient Bioavailibility: A Mechanistic Approach.* John Wiley and Sons, New York, NY, U.S.A.

Barley, K. P. 1970. The configuration of the root system in relation to nutrient uptake. *Advances in Agronomy* 22:159–201.

Bowen, G. D. 1973. Mineral nutrition of ectomycorrhizae. In: Marks, G. C., and T. T. Kozlowski, eds. *Ectomycorrhizae: Their Ecology and Physiology.* Academic Press, New York, NY, U.S.A. Pp. 151–205.

Bowen, G. D. 1981. Approaches to nutritional physiology. In: *Proceedings, Australian Forest Nutrition Workshop: Productivity in Perpetuity,* August 10–14, 1981, Canberra, Australia. Pp. 79–91.

Bowen, G. D. 1984. Tree roots and the use of soil nutrients. In: Bowen, G. D., and E. K. S. Nambiar, eds. *Nutrition of Plantation Forests.* Academic Press, New York, NY, U.S.A. Pp. 147–179.

Bowen, G. D., and B. Cartwright. 1977. Mechanisms and models of plant nutrition. In: Russell, J. S., and E. L. Greacen, eds. *Soil Factors in Crop Production in a Semi-Arid Environment.* University of Queensland Press, St. Lucia, Queensland, in association with the Australian Society of Soil Science. Pp. 197–223.

Chen, C. W., S. A. Gherini, R. J. M. Hudson, and J. D. Dean. 1983. *The Integrated Lake-Watershed Acidification Study,* Vol. I. *Model Principles and Application Procedures.* EA-3221, Electric Power Research Institute, Palo Alto, CA, U.S.A. 214 pp.

Comerford, N. B., G. Kidder, and A. V. Mollitor. 1984. Importance of subsoil fertility to forest and non-forest plant nutrition. In: Stone, E. L., ed. *Forest Soils and Treatment Impacts. Proceedings of the Sixth North American Forest Soils Conference.* Department of Forestry, Wildlife, and Fisheries, University of Tennessee, Knoxville, TN, U.S.A. Pp. 381–403.

Gale, M. R., and D. F. Grigal. 1987. Vertical root distribution of northern tree species in relation to successional status. *Canadian Journal of Forest Research* 17(8):829–834.

Glass, A. D. M., and M. Y. Siddiqi. 1984. The control of nutrient uptake rates in relation to the inorganic composition of plants. In: Tinker, P. B., and A. Lauchli, eds. *Advances in Plant Nutrition,* Vol. 1. Praeger Special Studies–Praeger Scientific, New York, NY, U.S.A. Pp. 103–147.

Harley, J. L., and S. E. Smith. 1983. *Mycorrhizal Symbiosis.* Academic Press, London, England, U.K.

Harvey, A. E., M. F. Jurgensen, M. J. Larsen, and J. A. Schlieter. 1986. Distribution of active ectomycorrhizal short roots in forest soils of the inland Northwest: effects of site and disturbance. Research Paper INT-374, USDA Forest Service Intermountain Research Station, Ogden, UT, U.S.A. 8 pp.

Huttermann, A., and B. Ulrich. 1984. Solid phase–solution–root interactions in soils subjected to acid deposition. *Philosophical Transactions, Royal Society of London* B 305:353–368.

Jenny, H. 1958. Role of plant factor in the pedogenic functions. *Ecology* 39:5–16.

Jorgensen, J. R., C. G. Wells, and L. J. Metz. 1980. Nutrient changes in decomposing loblolly pine forest floor. *Soil Science Society of America Journal* 55:1307–1314.

Joslin, J. D., and G. S. Henderson. 1987. Organic matter and nutrients associated with fine root turnover in a white oak stand. *Forest Science* 33:330–346.

Khanna, P. H., and B. Ulrich. 1984. Soil characteristics influencing nutrient supply in forest soils. In: Bowen, G. D., and E. K. S. Nambiar, eds. *Nutrition of Plantation Forests.* Academic Press, London, England, U.K. Pp. 80–117.

Landsberg, J. J. 1981. The number and quality of the driving variables needed to model tree growth. *Studia Forestalia Suecica* 160:43–50.

Lindberg, S. E., G. M. Lovett, D. D. Richter, and D. W. Johnson. 1986. Atmospheric deposition and canopy interactions of major ions in a forest. *Science* 231:141–145.

Lyford, W. H. 1980. Development of the root system of northern red oak (*Quercus rubra* L.). Harvard Forest Paper 21, Harvard University, Harvard Forest, Petersham, MA, U.S.A. 30 pp.

Miller, H. G. 1981. Nutrient cycles in forest plantations, their change with age and the consequence for fertilizer practice. In: *Proceedings, Australian Forest Nutrition Workshop; Productivity in Perpetuity,* August 10–14, 1981, Canberra, Australia. Pp. 187–200.

Miller, H. G., J. M. Cooper, J. D. Miller, and O. J. L. Pauline. 1979. Nutrient cycles in pine and their adaptation to poor soils. *Canadian Journal of Forest Research* 9:19–26.

Nihlgard, B. 1985. The ammonium hypothesis—an additional explanation to the forest dieback in Europe. *Ambio* 14:1–8.

Nye, P. H. 1984. pH changes and phosphate solubilization near roots—an example of coupled diffusion processes. In: Barber, S. A., and D. R. Bouldin, eds. *Roots, Nutrient and Water Influx, and Plant Growth.* Special Publication 49, American Society of Agronomy, Madison, WI, U.S.A. Pp. 89–100.

Nye, P. H., and P. B. Tinker. 1977. *Solute Movement in the Soil-Root System.* Blackwell Scientific Publications, Oxford, England, U.K.

Oliver, S., and S. A. Barber. 1966a. An evaluation of the mechanisms governing the supply of Ca, Mg, K, and Na to soybean roots (*Glycine max*). *Soil Science Society of America Proceedings* 30:82–86.

Oliver, S., and S. A. Barber. 1966b. Mechanisms for the movement of Mn, Fe, B, Cu, Zn, Al, and Sr from one soil to the surface of soybean roots (*Glycine max*). *Soil Science Society of America Proceedings* 30:468–470.

Pastor, J., and W. M. Post. 1986. Influence of climate, soil moisture and succession on forest carbon and nitrogen cycles. *Biogeochemistry* 2:3–27.

Plassard, C., F. Martin, D. Mousain, and L. Salsac. 1985. Physiology of nitrogen assimilation by mycorrhiza. In: Gianinazzi-Pearson, V., and S. Gianinazzi, eds. *Physiological and Genetical Aspects of Mycorrhizae.* Institut National de la Recherche Agronomique, Nancy, France. Pp. 111–120.

Prenzel, J. 1979. Mass flow to the root system and mineral uptake of a beech stand calculated from 3-year field data. *Plant and Soil* 51:39–49.

Reid, C. P. P. 1984. Mycorrhizae: a root-soil interface in plant nutrition. In: Todd, R. L., and J. E. Giddens, eds. *Microbial-Plant Interactions.* Special Publication 47, American Society of Agronomy, Madison, WI, U.S.A. Pp. 29–50.

Reuss, J. O. 1983. Implications of the Ca-Al exchange system for the effect of acid precipitation on soils. *Journal of Environmental Quality* 12:591–595.

Schultz, R. P. 1972. Root development of intensively cultivated slash pine. *Soil Science Society of America Proceedings* 36:158–162.

Sposito, G., and S. V. Mattigod. 1980. GEOCHEM: a computer program for the calculation of chemical equilibria in soil solution and other natural water systems. Kearney Foundation of Soil Science, University of California, Riverside, CA, U.S.A.

Warrington, G. E. 1983. Modeling approaches to forest productivity. In: *IUFRO Symposium on Forest Site and Continuous Productivity.* General Technical Report PNW-163, USDA Forest Service Pacific Northwest Forest and Range Experiment Station. Pp. 58–60.

Wheater, H. S., K. H. Bishop, and M. B. Beck. 1986. The identification of conceptual hydrological models for surface water acidification. *Hydrological Processes* 1:89–109.

CHAPTER 10

MODELING GROWTH AND PRODUCTION OF TREE ROOTS

Dan Santantonio

Abstract. Several approaches to modeling root growth are described: (1) allometric relations, (2) allocation or partitioning fractions, (3) utilization of substrates, (4) specific activity, (5) functional balance between shoot and roots, (6) Thornley's mechanistic approach, and (7) indirect approaches based on carbon balance or root decomposition. These approaches were largely developed by plant and crop physiologists to apply to early growth of herbaceous plants in experimentally controlled environments. We lack sufficient understanding of the mechanisms controlling root growth to define mechanistic models except for limited, constant conditions, so nearly all models are more or less empirical. Phenomenological models attempt to represent biological and physiological processes in an elementary way. Several have been applied to trees or forests, although usually with much less sophistication. I briefly describe examples. Because coarse and fine roots have very different characteristics of growth, mortality, and function, different approaches are needed to model these two components separately. Allometric and pipe-model (a relation between conducting area and foliage mass) approaches appear best for coarse roots. For fine roots, the best approach may be a functional balance between the size and activity of fine-root and foliage systems. Phenomenological approaches are still in early stages of development and have not been tested widely.

INTRODUCTION

Biologically based models are useful and necessary tools to integrate understanding of processes underlying tree growth and forest production (Waring and Schlesinger, 1985; Landsberg and McMurtrie, 1985; Landsberg, 1986). But is it necessary to consider roots as a separate compartment in these models? If tree growth in stands has adapted to and is in balance with site conditions, and if these effects are consistently reflected in tree dimensions and stand characteristics, then roots should be a consistent and readily predictable portion of stand biomass. Growth would be the increase in mass with time. There is much truth in this, especially once trees have established a stand (Whittaker and Marks, 1975, p. 91; Santantonio et al., 1977; Landsberg, 1986, p. 101), but it applies only to one component of root production—the growth of coarse, or woody, roots. The other component that must also be considered is fine, or non-woody, roots.

Fine roots are an important carbon sink in temperate and boreal forests. Direct estimates of fine-root production in these forests range from 1–12 Mg/ha/year (Fogel, 1985; Gholz et al., 1986; Santantonio, 1989). Santantonio (1989) has compared results for 20

Dr. Santantonio is Guest Scientist at the Department of Ecology and Environmental Research, Swedish University of Agricultural Sciences, S-75007 Uppsala, Sweden. Support for the preparation of this paper came from the Swedish Council of Forestry and Agriculture.

coniferous stands, representing a wide range of sites and species, and found that production of fine roots of trees amounted to 5–68% of total net primary production (TNPP). Fine roots, however, constituted less than 5% of total tree biomass. Production of coarse roots ranged from 0.4–8 Mg/ha/year and accounted for 4–24% of TNPP. Coarse roots made up 15–25% of total tree biomass. These data indicate that the distribution of TNPP, or annual dry-matter partitioning, to fine roots may exceed the proportion they represent in total biomass by an order of magnitude or more. Dry-matter partitioning to coarse roots may equal, but was generally less than, their proportion of biomass. In a similar comparison of dry-matter partitioning for a broader range of forest types, though a fewer number of stands, Cannell (1985) found values which fall within the ranges given here. Variation in dry-matter partitioning does not appear to be as much the result of different species or forest types as the result of differences in site (Karizumi, 1974; Keyes and Grier, 1981; Linder and Axelsson, 1982; Kottke and Agerer, 1983) or differences in stand age and structure (Karizumi, 1974; Albrektson, 1980; Grier et al., 1981; Pearson et al., 1984; Beets and Pollock, 1987). Wide variation in the proportion of TNPP required by fine roots explains in part why net assimilation and forest yield can be poorly correlated (Jarvis and Leverenz, 1983; Linder and Rook, 1984; Briggs et al., 1986). Thus, the distribution of TNPP is not constant among tree components, nor is it easily predicted because production of fine roots varies so widely. It would appear to be largely under the control of environmental factors with the imposition of certain genetic constraints.

Little or no production of fine roots appears as an increment in the standing crop of live and dead fine roots once stands have achieved closed canopy (the maximum sustainable mass of foliage). Limited evidence indicates that the fine roots of trees achieve maximum biomass per hectare at or before the time of canopy closure (Karizumi, 1968, 1974; Vogt, et al., 1987). Subsequent growth of fine roots is offset mostly by mortality of individual root tips. Fine-root production and mortality apparently occur simultaneously within small microsites (Reynolds, 1970; Tippett, 1982; Santantonio and Hermann, 1985; Santantonio and Santantonio, 1987). In contrast, coarse-root mass continues to increase beyond the time of canopy closure in a manner similar to that of stems (Ovington, 1957; Karizumi, 1974; Albrektson, 1980). These roots are long-lived; mortality would appear to occur mainly through the death of diseased or suppressed trees. These differences in the production and mortality of fine and coarse roots result from the different functions they perform. Coarse roots structurally support the stem, anchor it to the soil, and provide the conducting network needed to transport water and nutrients, which are absorbed by fine roots in different soil microsites. (Fine roots are also the site of synthesis of growth regulators and other metabolites transported to the shoot.) Modeling growth and production of these two components of tree roots in forests requires fundamentally different approaches.

Although the main factors affecting root growth in forests are generally well known, the mechanisms controlling growth and longevity are poorly understood (Bowen, 1984; Coutts, 1987; Eis, 1986; Hermann, 1977; Kottke and Oberwinkler, 1986; Nylund, 1988; Orlov, 1980; Persson, 1983; Santantonio and Hermann, 1985; Vogt, et al., 1986; Wilcox, 1983). Our present knowledge of roots is fragmented and lags far behind that of shoots, largely because of technical difficulties in the study of root growth and mortality (Fogel, 1983, 1985; Persson, 1983, 1989; Vogt and Persson, in press; Santantonio, 1989). I know of no attempt to directly model root growth as a process of response to stimuli in forests. Roots have been included as a separate compartment in several tree and stand models; examples of these will be discussed in the course of this paper.

Since the late 1960s plant and crop physiologists have made considerable progress in modeling the growth of roots in interaction with shoots of herbaceous plants during early vegetative growth (Bastow Wilson, 1988; France and Thornley, 1984; Hesketh and Jones,

1980; Loomis et al., 1979; Luxmoore and Stolzy, 1987; Penning de Vries, 1983; Wareing and Patrick, 1975). Thornley (1976, 1977) has developed the most comprehensive and theoretical treatment of this subject. Reviews by Cannell (1985), Causton (1985), Landsberg (1986), and Ledig (1976) consider trees. Approaches range from completely empirical to mechanistic.

In this paper, I describe the approaches that have been developed to model root growth, or at least the utilization of photosynthate by roots, as a separate compartment in growth models. Where possible, I describe how they have been applied to forests and evaluate their appropriateness for modeling the growth and production of roots. Respiration, translocation to symbionts, and exudation are also important fates of photosynthate allocated to roots. I do not mean to diminish their importance by not specifically considering these processes within this paper. I will attempt to evaluate our present capability to develop process-level models of coarse-root and fine-root production in forests.

DEFINITIONS

A standard definition of fine roots has not been adopted. In fact, it is difficult to make an exact demarcation in function and morphology between fine and coarse roots for trees in forests (Hermann, 1977; Fitter, 1985; Fogel, 1985). From a practical standpoint, the distinction has usually been defined by a diameter size of <1 mm or <2 mm. These diameters generally coincide with a change in woodiness, branching, and longevity of roots (Lyford, 1975; Fogel, 1983; Santantonio and Santantonio, 1987; Persson, 1989). Fine roots include fungal mantle tissue in mycorrhizal root tips. Thus, mycorrhizal fungi should be considered as a functional component of fine roots. The distinction I have made for the purpose of this paper is that fine roots, sometimes called feeder roots, are non-woody, while coarse roots are woody.

Net photosynthetic production is the net carbon assimilated minus dark respiration of leaves. Carbon allocation is the process by which net photosynthetic production, or photosynthate, is distributed among different plant components and functions. Dry-matter partitioning is the resulting distribution of net primary production and does not include carbon lost through respiration, exudation, and translocation to symbionts. Dry-matter production, or net primary production, is usually estimated by harvest as the net increase in biomass plus losses of dry matter through mortality, grazing, and so on. Growth is the net increment of plant size or biomass. Steady-state growth is the condition where the concentration of elements within the plant remains constant, and the partitioning of growth among plant organs remains constant. If growth is also exponential, then the specific growth rates (i.e., the relative growth rates) of plant organs remain constant. Steady-state exponential growth means constant composition, constant partitioning, and constant specific growth rate. Steady-state growth has often been referred to as "balanced" growth, but in these cases it has not always been clear whether all three conditions were met.

Inasmuch as the conference that led to the present volume emphasized process modeling of forest growth in response to environmental stress, a definition of what constitutes a process-level model is appropriate. I will follow the definition of Landsberg (1986, p. 2), who describes a process-level model as "a formal and precise statement, or set of statements, embodying our current knowledge or hypotheses about the workings of a particular system and its responses to stimuli." I will not consider models in which plant growth is a function of time alone, as in classical growth equations (Causton and Venus, 1981; France and Thornley, 1984; and examples: Lohmann, 1983; Rose, 1983). Developmental models are needed; but from the standpoint of process-level models, chronological time must be converted to "physiological time" (Loomis et al., 1979; Penning de Vries, 1983).

APPROACHES TO MODELING ROOT GROWTH AND PRODUCTION

Approaches that have been developed to model root growth and production range from completely empirical to mechanistic. Completely empirical models may be simple statistical models that have a high predictive value when applied to a situation similar to that from which they were developed. They provide little or no biological explanation and may be applied beyond the range of data only with great risk. Mechanistic models are much more complex and generally apply to a much wider range of conditions, although they are probably less precise in any particular situation. These models provide an explanation of the biology but require an understanding of mechanisms controlling the processes involved.

Most current activity is in a broad area between these two extremes, in the development of what may be termed phenomenological models. These models are still empirical. They attempt to represent biological or physiological process in an elementary way, but generally they do not claim to define causal mechanisms. I will progress in these pages from completely empirical to mechanistic approaches. I will not consider sink strength or sink activity as a separate approach to modeling root growth because the use of source and sink concepts has many ambiguities and problems which cause confusion in the definition of models (see discussions by Warren Wilson, 1972; Wareing and Patrick, 1975; Thornley, 1976, pp. 38–42, 1977; Lang and Thorpe, 1983). While I believe the categories defined below represent different approaches to modeling root growth and production, I accept that they are not mutually exclusive and some overlap exists.

Allometric Relations

Allometric equations are not process-level models, but they are such an important, useful tool for estimating plant biomass by component in forests (see reviews by Whittaker and Marks, 1975; Pardé, 1980; and Satoo and Madgwick, 1982) that they must be included. The allometric growth formula (Pearsall, 1927; Huxley, 1932; Troughton, 1956) estimates the mass of one plant organ as a function of the mass of another organ, for example, roots (W_r) in relation to shoots (W_s):

$$W_r = a\ W_s^{\ b} \tag{1}$$

or

$$\log W_r = \log a + b \log W_s \tag{2}$$

where a and b are empirical constants.

Allometric relations may also be used to relate component weight to plant dimensions. Researchers working with trees have commonly used the equation in the form where weight is a function of diameter at breast height (dbh), dbh squared, or dbh squared times height:

$$\log W_r = \log a + b \log (\text{dbh}) \tag{3}$$

These equations have been used to indirectly estimate root biomass for many species over a broad range of tree sizes (Karizumi, 1974; Whittaker and Marks, 1975; Santantonio et al., 1977). They have also been used to estimate the increment of root biomass from root mass or from the stem increment (Kira and Ogawa, 1968; Yamakura et al., 1972). Ledig (1969) used allometric relations among organs as a basis for allocating photosynthate in a model for loblolly pine (*Pinus taeda* L.) seedlings.

Assumptions associated with the use of Equations 1–3 concern the appropriateness of the values of empirical constants for the species and site, and whether they remain constant over the interval of time and the range of tree sizes considered. If b in the equations is constant, then a constant ratio is maintained in the relative growth of root and shoot when no mortality occurs. This is a better index of a balance in growth than root:shoot ratio, which changes with plant size (except when b = 1.0).

Allocation or Partitioning Fractions

At the simplest level, the fraction of TNPP can be estimated empirically (Ledig and Botkin, 1974; Promnitz, 1975). These coefficients of allocation or partitioning can be estimated empirically from studies of carbon utilization or dry-matter production, respectively, depending on whether one starts with photosynthate or sums net primary production (NPP) by component as one might do from harvest data. Photosynthate or NPP may be distributed among tree components according to fixed fractions such that the sum of the fractions equals 1.0. These fractions must remain constant over the interval of estimation. This approach has been used in many recent stand-level growth models of trees where roots were designated as a separate compartment, though not explicitly modeled (e.g., McMurtrie and Wolf, 1983; Mohren et al., 1984; McMurtrie, 1985; Mäkelä and Hari, 1986; Running and Coughlan, 1988). These models typically operate at annual resolution with respect to net primary production. The production of roots ($Prod_r$) can be estimated, as in McMurtrie and Wolf (1983):

$$Prod_r = \eta_r \, Y \, P_n \, (1 - k_r) \qquad (4)$$

where η_r = the allocation fraction to roots, Y = the conversion efficiency of net photosynthate to dry matter, P_n = net photosynthetic production as weight of CO_2, and k_r = the fraction of net photosynthate consumed in the maintenance of woody tissue.

Root mortality has usually been estimated as a fractional loss of root biomass. A variation of this approach attempts to make the allocation fractions more phenomenological by defining a system of priorities in the utilization of net photosynthate or other substrate (Loomis et al., 1979). In such a system, one process or compartment is given absolute priority over another until its requirement or maximum rate of growth is achieved. Typically, the requirement of maintenance respiration is met first, with the remaining photosynthate allocated to a reserve pool. Growth and associated constructive respiration utilize this reserve based on priorities which may direct first the production of new foliage, then the production of fine roots to support the additional foliage. Remaining photosynthates in the reserve would finally be used to grow woody components. Growth of components is usually constrained by morphological characteristics and other controlling factors (e.g., Bassow et al., this volume; Weinstein and Beloin, this volume) or by response functions based on different situation scenarios (e.g., Bossel, 1986; Schäfer et al., 1988).

Utilization of Substrates

Thornley (1976, 1977) has explained the application of the Michaelis-Menten relations for enzyme kinetics to represent growth in response to substrate level. He proposes that they are also suited for modeling the response to growth regulators. These relations form curves which are rectangular hyperbolas. A simple form of the relation can be used to represent the utilization (U) of a single substrate (S) by one or more different organs, for example, the growth of roots (r) and shoots (s):

$$U_r = k_r \, S \, / \, (K_r + S) \quad \text{and} \quad U_s = k_s \, S \, / \, (K_s + S) \qquad (5)$$

where k and K are constants specific to the organs.

As the substrate density increases, growth increases rapidly and then levels off asymptotically at an upper limit, much as photosynthetic rate does in response to increasing photon flux density. The initial slope is k/K, and the asymptote is k. The equations do not describe the details of these complex processes, but only the overall outcome. When responses of different organs are compared, allocation priorities can be set in relation to substrate density.

The Michaelis-Menten relation can also represent the effect of a second substrate or growth factor on the growth of an organ:

$$U = k' S_1 S_2 / (1 + K_1 S_1 + K_2 S_2 + K_{12} S_1 S_2) \tag{6}$$

where k' and K are constants specific to the organs.

Up to three substrates or three compartments can be handled in this manner. This approach can be taken one step further to represent the interaction of two organs in the utilization of two substrates limiting growth with different levels of substrates in each organ, as will be discussed later with regard to functional relations between roots and shoots.

Michaelis-Menten equations have proved to be very useful in phenomenological models. Assumptions associated with the use of Michaelis-Menten relations include the following: (1) Utilization follows Michaelis-Menten kinetics, (2) the supply of substrate or other vital substance is the main factor limiting utilization, (3) substrate is utilized directly in growth, i.e., there is no long-term storage, and (4) substrate is converted to dry matter at a known conversion efficiency.

Specific Activity

In the most general sense, specific activity is an empirical coefficient by which a particular process or rate is related to a state variable. Growth of roots (dW_r/dt), for example, can be a function of root mass (W_r):

$$dW_r/dt = \sigma W_r (t_2 - t_1) \tag{7}$$

where σ = the specific activity of root growth, and t = time.

For growth, the specific activity is the specific growth rate, or relative growth rate. During exponential growth, the specific growth rate remains constant. The specific activity is assumed to remain constant during the time interval. This simple approach has often been applied to plants during steady-state exponential growth in a constant environment (e.g., Ingestad, 1979; Ingestad and Lund, 1979).

A more complex phenomenological approach, however, consists of multiplying the "potential" specific growth rate by the current effects of environmental and internal factors affecting the status of carbohydrates, water, and nutrients within the plant (Brouwer and De Wit, 1969; Loomis et al., 1979; Penning de Vries, 1983; Hunt and Nicholls, 1986). Ågren and Axelsson (1980) applied this approach to model growth under natural conditions in Sweden of a 15-year-old pine with eight different organs, including fine and coarse roots. Run at a time-step of one day, the model modified specific growth rates each day by multiplicative functions of temperature, available carbohydrates, water potential in the tree, nutrient status of the tree, and a function explicitly containing the time of year. Root mortality was a constant fraction of fine-root biomass.

The specific-activity approach can be seen in the method used by Nadelhoffer et al. (1985) to estimate fine-root production in forests from budgets of nitrogen uptake and utilization by using nitrogen as a "conservative tracer" to estimate the turnover of fine roots. The specific turnover rate equals the nitrogen allocated to fine roots divided by the amount of nitrogen in fine roots. Nadelhoffer et al. (1985) estimated annual fine-root production as the specific turnover rate, based on annual estimates, times the mean annual biomass of fine roots.

Functional Balance

Following a series of experiments with pondweed (*Lemna*), White (1937) proposed a simple mechanism where growth of leaf and root meristems required carbohydrate and nitrogenous substrates in balanced concentrations corresponding to the amounts used in development. He assumed that the supply of carbohydrates to roots was determined by the diffusion gradient between leaves and roots. Factors which reduced photosynthesis would decrease the steepness of the gradient, as would those which promoted leaf-area development and high rates of utilization within the leaf (e.g., when nitrogen acquisition was high). Factors which limited utilization of carbohydrates in leaves (e.g., when nitrogen acquisition was low) would lead to a steepening of the gradient and increased transport of carbohydrates to roots. More roots would grow and nitrogen acquisition would increase, thus permitting more leaf growth and decreasing the carbohydrate gradient.

Brouwer (1963) and Davidson (1969) contributed further to the development of the idea of a functional balance between the size and activity of root and shoot systems:

$$\sigma_r W_r \propto \sigma_s W_s \tag{8}$$

where σ_r = the specific activity of absorption by roots, and σ_s = net photosynthetic production per unit of shoot (or leaf) mass.

Uptake of nutrients, especially nitrogen, has been the main area of interest in the application of functional equilibrium, but it applies equally to absorption of water (De Wit, 1978; Huck and Hillel, 1983; Cowan, 1986; Fiscus, 1986). The issue of immediate concern has been the definition of the proportionality factor, more in relation to the specific activities of uptake and photosynthetic production than in relation to weights of roots and shoots (Brouwer, 1966; Davidson, 1969). If the proportion is assumed constant, then the nitrogen and carbon composition will remain constant fractions of plant dry weight during growth—there will be a fixed N:C ratio. This approach supposes that uptake of elements is in balance with their utilization and that storage is relatively small. This will hold for steady-state growth in a constant environment, but it also implies that the same plants grown under different constant environments will have the same N:C ratio (Thornley, 1977). Yet it is well known that plants will differ in nitrogen content when grown under constant conditions but with different rates of nitrogen supply (e.g., Ingestad, 1979; also see discussion by Mäkelä and Sievänen, 1987).

There has been considerable theoretical development in the formulation of this approach to allow its application with less restrictive assumptions to a wider range of conditions. The entire plant mass may not be involved in photosynthetic production or absorption. Charles-Edwards (1979, 1981) and Thornley (1977) defined specific activities in relation to the mass involved in photosynthesis or uptake and allowed for substrates in storage. In addition to roots and shoots, they included a third compartment—reproductive structures; but the argument should apply to woody structures as well. Reynolds and Thornley (1982) defined an adaptive partitioning function which enabled specific activities to vary in time such that the partitioning between roots and shoots adapts to changes in the environment to keep the substrate ratio N:C in the storage pool at a given value. Johnson (1985) found that this value "always settles to a fixed value, and this value is not influenced by the external environment." He modified the approach so that changes in partitioning produce a constant ratio between carbon in the shoot and nitrogen in roots, thus making the root:shoot ratio a function of the state of the plant, and not a function of rates or activities.

Hirose (1987) developed empirical partitioning functions based on the nitrogen content (state) of the plant. Ågren and Ingestad (1987) applied the nitrogen productivity concept (Ingestad, 1979) to estimate root:shoot ratio in relation to nitrogen content of the plant. Mäkelä (1986a) modified the partitioning function of Reynolds and Thornley (1982) to allow for non-zero and unequal rates of root and shoot mortality. Brugge (1985) introduced vertical diffusion of a mobile root pool as a means to stimulate vertical penetration of roots into uncolonized soil layers. Iwasa and Roughgarden (1984) and Johnson and Thornley (1987) used optimization of photosynthetic production and specific growth rate, respectively, to determine partitioning between root and shoot. The application of these models has usually been limited to conditions of steady-state exponential growth. Johnson and Thornley (1987) and Mäkelä and Sievänen (1987) have compared and discussed many of these models in greater detail.

Generally, the concept of a functional balance between roots and shoots has been applied to trees or forests with less sophistication. Borchert (1973) developed a model of episodic extension growth of tropical trees under constant conditions. Here shoot growth expands transpirational surface until limited by water deficit. The water status of the shoot system is determined by the balance between transpiration and absorption of water by roots. Sollins et al. (1981) hypothesized that root growth occurred in response to transpiration by foliage and in proportion to leaf growth in spring, and to support the stem in proportion to shoot growth throughout the year:

$$dW_r/dt = c_l\, dW_l/dt + c_b\, dW_b/dt \tag{9}$$

where c's are empirical constants, and subscripts r, l, and b refer to root, leaf, and bole.

Sievänen et al. (1988) made net photosynthetic production of young willow plants proportional to the dry weight and nitrogen content of the shoot, and nitrogen uptake proportional to soil nitrogen concentration and the mass of roots. Carbon and nitrogen substrates were immediately used in exponential growth, and the allocation fraction between roots and shoots was held constant. Following Reynolds and Thornley (1982), they optimized the allocation fraction such that increase in plant weight would be maximized at different levels of soil nitrogen during the first 14 weeks of growth.

Recent developments have brought the pipe-model theory (Shinozaki et al., 1964a) into the problem to account for the development of woody, conducting and supportive structures of forest trees. Functional equilibrium accounts for partitioning to fine roots. These models use a time resolution of one year and a time span of the lifetime of the tree. Mäkelä (1986b) developed a model where the partitioning of growth between foliage and wood maintained a constant sapwood:foliage ratio. Production of branches, stem and transport roots depended on the change in sapwood area and the change in tree height, radius of crown, or radius of transport root system with coefficients for sapwood per unit of foliage, wood density, and form factor. Partitioning of growth to fine roots maintained a functional equilibrium with foliage mass and depended on the specific activities of foliage and fine roots, and an empirical coefficient which was the N:C ratio of dry weight, or, possibly, the efficiency of water use. Senescence of sapwood and mortality was accounted as a constant fractional loss of the state variables. The specific activity of tree growth was expressed as a Michaelis-Menten function of root and shoot activities and the N:C ratio of dry weight. Responses to environmental changes such as shading and fertilization during the lifetime of the tree acted by changing the specific activities, coefficients of sapwood per unit of foliage, and eventually the partitioning coefficients. Valentine (1988) developed a similar model but assumed that fine-root mass was proportional to foliage mass without the influence of root and shoot specific activities. He used the self-thinning rule (Yoda et al., 1963; White, 1981; Westoby, 1984) to account for tree mortality in his model of an even-aged stand following canopy closure.

Thornley's Mechanistic Approach

Thornley (1972a, 1972b) developed two models which formalized the mechanisms proposed by White (1937) and which illustrate the hypothesis of a functional balance between the size and activity of root and shoot systems. Allocation and growth in these two models depend on the processes of transport and utilization, with the assumptions that transport is proportional to the difference in substrate concentration, and utilization is proportional to substrate level and follows a Michaelis-Menten relation. The first model (1972a) consists of three compartments and one substrate: root, stem, and leaf, with the supply of photosynthate from leaves as the factor limiting growth. The second model (1972b) consists of two compartments and two substrates: root and shoot, with the supply of nitrogen from root and photosynthate from leaves interacting to limit growth. These models apply to steady-state exponential vegetative growth in a constant environment. All substrate is used directly for growth and in constant proportion throughout the plant—there is a constant N:C ratio. Maintenance and wastage are not included as processes but are part of an empirical constant for the conversion of net photosynthate to dry matter. Response to different environmental conditions (as in successive experiments) results in changes in partitioning between root and shoot through changes in the specific activities of roots and shoots.

Cooper and Thornley (1976) applied the second model to tomato (*Lycopersicon*) with "encouraging results"; but further use of this model has been discouraged by criticism that the approach is too complex and that resistances to substrate transport need to be measured or estimated. This led Reynolds and Thornley (1982) to develop a functional equilibrium model as "a useful compromise between the complex mechanistic approach and the simple, highly empirical approach." Bastow Wilson (1988) has reviewed evidence on the control root:shoot ratio in relation to Thornley's (1927b) mechanistic model. Mäkelä and Sievänen (1987) have compared this model to the Reynolds and Thornley (1982) model in detail.

Indirect Approaches

It may be possible to estimate tree root production by difference in carbon balance models (Linder and Axelsson, 1982; Jarvis and Leverenz, 1983; Briggs et al., 1986). This approach, however, will be of little value without independent data, unless the error in the estimates for other compartments and processes of the carbon balance is known and reasonably small. Otherwise, the uncertainty of estimates impacts the estimate for roots to an unknown extent.

Simultaneous production, mortality, and decomposition in many forests cause serious problems for the estimation of fine-root production and mortality. Santantonio and Grace (1987) developed a compartment—flow model to estimate fine-root production and mortality at monthly resolution in a radiata pine (*Pinus radiata* D. Don) plantation on a moist, fertile site. In the model run, enough fine roots were produced and shed to maintain the observed standing crops of live and dead fine roots given losses through decomposition each month. Decomposition was represented by an exponential decay equation in which the coefficient (k) was a function of monthly soil temperature. Although not directly modeling production and mortality, this model is a process-level model of fine-root decomposition and estimates the other two processes in a conceptually sound way. This approach is useful because it adequately handles simultaneous process, it is driven by true driving variables, and it can be run at different time frames. It may also serve as an independent way to confirm estimates developed by other means.

DISCUSSION

A General Assessment

Growth as an output of models, has been considered almost exclusively as dry weight/time or dry weight/area/time. Although impacts to net production or yield may be the main interest, production of dry matter alone is a crude characterization of plant response to environmental stimuli. The expansion or loss of surface area and length is more functionally meaningful but requires a much greater understanding of the process which control elongation, branching, cell expansion, physiological activity, senescence, abscission, and so on. Thus, modeling dry-matter production is only a first step in the development of process-level models of tree growth in response to environmental stress.

All biological models are constrained in some way by the purpose and situation for which they were designed. For root growth, technical problems with access and observation make matters more difficult than for other tree components. A range of modeling approaches exists, and each has advantages and disadvantages. No single model or approach will meet all objectives or be applicable to all situations. Usually what dictates the approach taken is the purpose of the model and level of explanation sought, the amount and kind of supporting data available, the level of understanding of fundamental processes involved, assumptions which must be met in the formulation of the model, and so on. Because our understanding of partitioning and growth is inadequate to define mechanistic models (except Thornley's for steady-state exponential growth in a constant environment), we must continue to develop and use empirical models. Phenomenological models will be the most useful. They are needed to develop and interpret descriptive information, integrate system processes, and explore new theories. But, in future modeling, we should keep Thornley's (1976, p. 153) warning in mind:

> If a model is to be used to try to assess the relative importance of processes such as transport and chemical conversion or utilization to the pattern of plant growth, then it is desirable that these processes should be described in mechanistic terms. Often these processes are dealt with empirically or semi-empirically; such an approach has the disadvantage that when the performance of the simulated plant is compared with that of the real plant, it is usually not clear how much of the agreement that may exist is due to correct mechanistic or structural relationships having been built into the model, and how much is due to empirical adjustment. This severely limits the use of the model for predictive purposes, or for examining the consequences of particular assumptions about mechanisms.

Errors and bias enter models at several steps in their development, evaluation, and use. Kremer (1983), Loehle (1987), and Sala (1988) provide useful discussions of these problems in relation to ecological models.

Phenomenological models are in early stages of development. Typically, they have been developed by plant and crop physiologists for herbaceous plants during early vegetative growth under experimental conditions, and they usually describe the behavior of the average plant without mortality of plant parts and without competition. Without these effects, dry-matter production equals growth, and partitioning has not been modified by competition. Roots have been considered as a single compartment without any distinction between fine and coarse roots or between differences in their function.

Most models rely heavily on empirical data (see reviews cited in the introduction to this chapter). Those models that are more theoretical, as in functional-balance and mechanistic approaches, generally pertain to conditions of steady-state exponential growth in a constant, controlled environment. Under these conditions, the rate and partitioning of growth, the amount of growth in relation to plant weight, and the chemical composition of plant parts remain constant at values specific to the conditions; this greatly simplifies the analytical formulation of these models. Implications of these simplifications must be considered when the models, or the approaches they represent, are applied to long-lived woody perennials in natural environments. Phenomenological models have not been tested widely, and descriptive information on the production and function of roots in relation to shoots is extremely limited. For these reasons, I feel that it is premature to judge the models or approaches with regard to their ability to accurately represent root growth and production. Rather, one should ask for what situations they are appropriate and for what purposes they are useful.

For reasons discussed in the introduction above, root growth in forests must be considered as two separate components: coarse roots and fine roots. The approaches I have described are not equally appropriate for modeling both components. Many developing process-level models are too empirical because they contain parameters that (1) have little or no biological meaning or (2) lump together effects of several processes that are not constant. All phenomenological approaches are not equally suited to a particular situation; they may require substantial modification, for example, for experimental conditions as opposed to natural environments, or for herbaceous as opposed to woody plants. Application of all approaches with regard to root production in forests has apparently been limited to annual resolution. Development of growth models for trees has mainly been oriented toward changes from one year to the next; seasonal periodicity of carbon allocation and root production has not been addressed. This ignores much of the physiology of growth and major factors that control it.

Modeling Coarse Roots

Although allometric equations are completely empirical devices developed from harvest data, they do reflect a structural continuity in the growth and dimensions of tree components. Since the coefficients in these simple equations come from regression analysis, statistical tests can be used to determine objectively whether coefficients may be considered the same at a given level of confidence. Deviations about the log-log line, however, are not arithmetic, but are in orders of magnitude. Error in the estimate for a single tree can be quite large, but the errors tend to balance out when summed over many trees in a stand. For nine sample plots representing seven tree species, Madgwick (1983) found the average bias was less than 2% for leaves, branches, and stems in a comparison of known plot weights to estimates based on allometric equations developed from small samples of trees in the same plots.

Results of Shinozaki (1964b) and others (Benecke and Nordmeyer, 1982; Kaipiainen and Hari, 1985; Carlson and Harrington, 1987) indicate that coarse roots are simple pipe assemblages without the complications created by heartwood formation in the stem below the lowest living whorl of branches. The sapwood or cross-sectional area of coarse roots is highly correlated to that of living branches and to foliage mass. Allometric equations based on sapwood area may be an effective way to estimate the mass and increment of coarse roots. More descriptive data are needed to determine how robust such relations might be with respect to tree, stand, and site factors.

Allometric and pipe-model approaches pertain to individual trees. Stands must be created based on plot data or statistical distributions. Mortality of individual trees can be based on the self-thinning rule (e.g., Valentine, 1988) or on net loss of productive biomass as a result of negative carbon balance (e.g., Mäkelä, 1986a). Mortality of coarse roots can be directly linked to tree mortality, with the assumption that survival through root grafts and coarse-root mortality (other than that which occurs with tree mortality) are insignificant.

Modeling Fine Roots

Present approaches circumvent modeling fine-root growth and production directly as a response to stimuli; instead, they treat this compartment as a "black box" and attempt to allocate carbon or partition production based on the size of the box and/or output from it. Models of absorption at the root level are very sensitive to root morphology and elongation rate but lack linkage to tissue growth, longevity, and activity in relation to shoot growth and environmental influences (Clarkson and Hanson, 1980; Clarkson, 1985; Luxmoore and Stolzy, 1987; Moorby, 1987; Ingestad and Ågren, 1988). Fine-root mortality has been neglected completely and appears in nearly all models only as a constant fractional loss of fine-root biomass. Potentially high mortality and widely varying partitioning to fine roots make this component important to phenomenological modeling.

Thornley's mechanistic approach does not appear possible given its present assumptions and level of development. Partitioning-fraction approaches remain useful not only in developing descriptive information and integrating system processes, but also in "what if" gaming, given that an effective carbon balance can be applied. In gaming with these models, one may see how factors changing partitioning to fine roots might be expected to affect production of other tree components (McMurtrie and Wolf, 1983; McMurtrie, 1985). Models based on limiting factors reducing the potential specific growth rate have serious problems with interacting factors (Loomis et al., 1979; Ågren and Axelsson, 1980). Their use is probably limited to situations with a high level of experimental control. Further development of process-level models based on a functional-balance approach should prove worthwhile. Since interaction and feedback are fundamental to their conception, these models would appear to have the most flexibility to evolve from more to less empiricism as our understanding of controlling factors grows.

Implementation of most functional-balance approaches require considerable descriptive information about the size and activity of fine-root and foliage systems, as well as information about their substrate composition—all of which needs to be known in relation to tree, stand, and site factors. We need ways to quantify tree status and site conditions in terms which are directly applicable to modeling growth in response to stimuli. Evidence from experiments of controlled nutrition in controlled environments shows that nitrogen content and starch accumulation are consistent indicators of partitioning between root and shoot and of relative growth rate in small birch (*Betula pendula* Roth.) plants at different levels of nitrogen stress (Ingestad, 1979; Ingestad and Lund, 1979; McDonald et al., 1986a, 1986b). However, very little information exists for forests, and what does exist often appears to be contradictory. For example, compare the results of Vogt et al. (1987) with Santantonio (1989) with regard to the effect of site quality on the relation of foliage biomass to the standing crop of fine roots, or compare those of Aber et al. (1985) with Keyes and Grier (1981) with regard to fine-root turnover versus nitrogen availability. The fundamental relations needed to use the functional-balance approach in forests remain to be defined.

CONCLUDING REMARKS

Conceptual and technical problems present obstacles to developing process-level models of root growth and production in forests. We lack sufficient understanding of the mechanisms controlling production, senescence, and mortality of roots to define mechanistic models. Progress in the near future will rely on empirical models which incorporate biological and physiological processes in an elementary way. They are needed to develop and interpret descriptive information, integrate system processes, and explore new theories. I believe functional-balance approaches are the key to developing process-level system models. They will require considerable descriptive information. Technical problems, however, limit the amount and quality of supporting data. Concepts of the models appear to be far ahead of the data base needed to test them. While not feeling the disappointment of Loomis et al. (1979, p. 361–362), I come to the same conclusion:

> [The] unbridled enthusiasm that many of us displayed during our early euphoria with [systems analysis in plant physiology] must now be tempered. A great deal of hard work remains, and "grand" models are not about to substitute for real plants and real experiments. Still, in many ways the modeling is ahead of the information base, and is likely to remain there . . . Their *raison d'être* is that the problem is there. No other means exists [that is] as powerful for the integrative physiology of plants.

LITERATURE CITED

Aber, J. D., J. M. Melillo, K. J. Nadelhoffer, C. A. McClaugherty, and J. A. Pastor. 1985. Fine root turnover in forest ecosystems in relation to quantity and form of nitrogen availability: a comparison of two methods. *Oecologia* 66:317–321.

Ågren, G. I., and B. Axelsson. 1980. PT: a tree growth model. In: Persson, T., ed. *Structure and Function of Nothern Coniferous Forests: An Ecosystem Study. Ecological Bulletin* (Stockholm, Sweden) 32:525–536.

Ågren, G. I., and T. Ingestad. 1987. Root:shoot ratio as a balance between nitrogen productivity and photosynthesis. *Plant, Cell and Environment* 10:579–586.

Albrektson, A. 1980. Relations between tree biomass fractions and conventional silvicultural measurements. In: Persson, T., ed. *Structure and Function of Northern Coniferous Forests: An Ecosystem Study. Ecological Bulletin* (Stockholm, Sweden) 32:315–327.

Bassow, S. L., E. D. Ford, and R. Kiester. 1989. A critique of carbon-based tree growth models. In: Dixon, R. K., R. S. Meldahl, G. A. Ruark, and W. G. Warren, eds. *Process Modeling of Forest Growth Responses to Environmental Stress.* Timber Press, Portland, OR, U.S.A.

Bastow Wilson, J. 1988. A review of evidence on the control of shoot:root ratio, in relation to models. *Annals of Botany* 61:433–449.

Beets, P. B., and D. S. Pollock. 1987. Accumulation and partitioning of dry matter in *Pinus radiata* as related to stand age and thinning. *New Zealand Journal of Forestry Science* 17:246–271.

Benecke, U., and A. H. Nordmeyer. 1982. Carbon uptake and allocation by *Notofagus solandri* and *Pinus contorta* at montane and subalpine altitudes. In: Waring, R. H., ed. *Carbon Uptake and Allocation in Subalpine Ecosystems As a Key to Management.* Forest Research Laboratory, Oregon State University, Corvallis, OR, U.S.A. Pp. 9–21.

Borchert, R. 1973. Simulation of rhythmic growth under constant conditions. *Physiologia Plantarum* 29:173–180.

Bossel, H. 1986. Dynamics of forest dieback: systems analysis and simulation. *Ecological Modelling* 34:259–288.

Bowen, G. D. 1984. Tree roots and the use of soil nutrients. In: Bowen, G. D., and E. K. S. Nambiar, eds. *Nutrition of Plantation Forests.* Academic Press, New York, NY, U.S.A. Pp. 147–179.

Briggs, G. M., T. W. Jurik, and D. M. Gates. 1986. A comparison of rates of aboveground growth and carbon dioxide assimilation by aspen on sites of high and low quality. *Tree Physiology* 2:29–34.

Brouwer, R. 1963. Some aspects of the equilibrium between overground and underground plant parts. *Jaarboek 1963.* Instituut voor Biologisch en Scheikundig Onderzoek van Landbouwgewassen, Wageningen, the Netherlands. Pp. 31–39.

Brouwer, R. 1966. Root growth of grasses and cereals. In: Milthorpe, F. L., and D. J. Ivins, eds. *The Growth of Cereals and Grasses.* Butterworth's, London, England, U.K. Pp. 153–166.

Brouwer, R., and C. T. de Wit. 1969. A simulation model of plant growth with special attention to root growth and its consequences. In: Whittington, W. J., ed. *Root Growth.* Butterworth's, London, England, U.K. Pp. 224–244.

Brugge, R. 1985. A mechanistic model of grass root growth and development dependent upon photosynthesis and nitrogen uptake. *Journal of Theoretical Biology* 116:443–467.

Cannell, M. G. R. 1985. Dry matter partitioning in tree crops. In: Cannell, M. G. R., and J. E. Jackson, eds. *Attributes of Trees As Crop Plants.* Institute of Terrestrial Ecology, Huntingdon, England, U.K. Pp. 194–207.

Carlson, W. C., and C. A. Harrington. 1987. Cross-sectional area relationships in root systems of loblolly and shortleaf pine. *Canadian Journal of Forest Research* 17:556–558.
Causton, D. R. 1985. Relationships among tree parts. In: Cannell, M. G. R., and J. E. Jackson, eds. *Attributes of Trees As Crop Plants.* Institute of Terrestrial Ecology, Huntingdon, England, U.K. Pp. 137–159.
Causton, D. R., and J. C. Venus. 1981. *The Biometry of Plant Growth.* Edward Arnold, London, England, U.K. 307 pp.
Charles-Edwards, D. A. 1979. Shoot and root activities during steady-state plant growth. *Annals of Botany* 40:767–772.
Charles-Edwards, D. A. 1981. *The Mathematics of Photosynthesis and Productivity.* Academic Press, London, England, U.K.
Clarkson, D. T. 1985. Factors affecting mineral nutrient acquisition by plants. *Annual Review of Plant Physiology* 36:77–115.
Clarkson, D. T., and J. B. Hanson. 1980. The mineral nutrition of plants. *Annual Review of Plant Physiology* 31:239–298.
Cooper, A. J., and J. H. M. Thornley. 1976. Response of dry matter partitioning, and carbon and nitrogen levels in the tomato plant to changes in root temperature: experiment and theory. *Annals of Botany* 40:1139–1142.
Coutts, M. P. 1987. Developmental processes in tree root systems. *Canadian Journal of Forest Research* 17:761–767.
Cowan, I. R. 1986. Economics of carbon fixation in plants. In: Givnish, T. J., ed. *On the Economy of Plant Form and Function.* Cambridge University Press, Cambridge, England, U.K. Pp. 133–170.
Davidson, R. L. 1969. Effect of root/leaf temperature differentials on root/shoot ratios in some pasture grasses and clover. *Annals of Botany* 33:561–569.
De Wit, C. T. 1978. *Simulation of Assimilation, Respiration and Transpiration of Crops.* Simulation Monographs, Pudoc, Wageningen, the Netherlands. 141 pp.
Eis, S. 1986. Differential growth of individual components of trees and their interrelationships. *Canadian Journal of Forest Research* 16:352–359.
Fiscus, E. L. 1986. Belowground costs: hydraulic conductance. In: Givnish, T. J., ed. *On the Economy of Plant Form and Function.* Cambridge University Press, Cambridge, England, U.K. Pp. 275–298.
Fitter, A. H. 1985. Functional significance of root morphology and root architecture. In: Fitter, A. H., D. Atkinson, D. J. Read, and M. B. Usher, eds. *Ecological Interactions in Soil.* Blackwell Scientific Publications, Oxford, England, U.K. Pp. 87–106.
Fogel, R. 1983. Root turnover and productivity in coniferous forests. *Plant and Soil* 71:75–85.
Fogel, R. 1985. Roots as primary producers in belowground ecosystems. In: Fitter, A. H., D. Atkinson, D. J. Read, and M. B. Usher, eds. *Ecological Interactions in Soil.* Blackwell Scientific Publications, Oxford, England, U.K. Pp. 23–36.
France, J., and J. H. M. Thornley. 1984. *Mathematical Models in Agriculture.* Butterworth's, London, England, U.K.
Gholz, H. L., L. C. Hendry, and W. P. Cropper, Jr. 1986. Organic matter dynamics of fine roots in plantations of slash pine (*Pinus elliottii*) in north Florida. *Canadian Journal of Forest Research* 16:529–538.
Grier, C. C., K. A. Vogt, M. R. Keyes, and R. L. Edmonds. 1981. Biomass distribution and above- and below-ground production in young and mature *Abies amabilis* zone ecosystems of the Washington Cascades. *Canadian Journal of Forest Research* 11:155–167.
Hermann, R. K. 1977. Growth and production of tree roots: a review. In: Marshall, J., ed. *The Belowground Ecosystem: A Synthesis of Plant-Associated Processes.* Range Science Department, Series 26, Colorado State University, Ft. Collins, CO, U.S.A. Pp. 7–28.
Hesketh, J. D., and J. W. Jones. 1980. Integrating traditional growth analysis techniques with recent modeling of carbon and nitrogen metabolism In: Hesketh, J. D., and J. W. Jones, eds. *Predicting Photosynthesis for Ecosystem Models.* CRC Press, Boca Raton, FL, U.S.A. Pp. 51–92.
Hirose, T. 1987. A vegetative plant growth model: adaptive significance of phenotypic plasticity in matter partitioning. *Functional Ecology* 1:195–202.
Huck, M. G., and D. Hillel. 1983. A model of root growth and water uptake accounting for photosynthesis, respiration, transpiration, and soil hydraulics. *Advances in Irrigation* 2:273–333.
Hunt, R., and A. O. Nicholls. 1986. Stress and coarse control of growth and root-shoot partitioning in herbaceous plants. *Oikos* 47:149–158.
Huxley, J. S. 1932. *Problems of Relative Growth.* Methuen and Company, London, England, U.K.
Ingestad, T. 1979. Nitrogen stress in birch seedlings. II. N, K, P, Ca and Mg nutrition. *Physiologia Plantarum* 45:149–157.
Ingestad, T., and G. Ågren. 1988. Nutrient uptake and allocation at steady-state nutrition. *Physiologia*

Plantarum 72:450–459.
Ingestad, T., and A. Lund. 1979. Nitrogen stress in birch seedlings. I. Growth technique and growth. *Physiologia Plantarum* 45:137–148.
Iwasa, Y., and J. Roughgarden. 1984. Shoot/root balance of plants: optimal growth of a system with many vegetative organs. *Theoretical Population Biology* 25:78–105.
Jarvis, P. G., and J. W. Leverenz. 1983. Productivity of temperate, deciduous and evergreen forests. In: Lange, O. L., P. S. Nobel, C. B. Osmond, and H. Ziegler, eds. *Encyclopedia of Plant Physiology* (New Series) Springer-Verlag, New York, NY, U.S.A. 12D:233–280.
Johnson, I. R. 1985. A model of partitioning of growth between the shoots and roots of vegetative plants. *Annals of Botany* 55:421–431.
Johnson, I. R., and J. H. M. Thornley. 1987. A model of shoot:root partitioning optimal growth. *Annals of Botany* 60:133–142.
Kaipiainen, L., and P. Hari. 1985. Consistencies in the structure of Scots pine. In: Tigerstedt, P., A. P. Puttonen, and V. Koski, eds. *Crop Physiology of Forest Trees.* Helsinki University Press, Helsinki, Finland. Pp. 31–37.
Karizumi, N. 1968. Estimating root biomass in forests by the soil block sampling. In: Ghilarov, M. S., et al., eds. *Methods of Productivity Studies in Root Systems and Rhizosphere Organisms.* Nauka, Leningrad, U.S.S.R. Pp. 79–86.
Karizumi, N. 1974. Mechanisms and function of tree root in process of forest production. II. Root biomass and distribution in stands. *Bulletin of Government Forest Experiment Station* (Tokyo, Japan). No. 267:1–88.
Keyes, M. R., and C. C. Grier. 1981. Above- and below-ground net production in 40-year-old Douglas-fir stands on low and high productivity sites. *Canadian Journal of Forest Research* 11:599–605.
Kira, T., and H. Ogawa. 1968. Indirect estimation of root biomass increment in trees. In: Ghilarov, M. S., et al., eds. *Methods of Productivity Studies in Root Sytems and Rhizosphere Organisms.* Nauka, Leningrad, U.S.S.R. Pp. 96–101.
Kottke, I., and R. Agerer. 1983. Cited from Kottke and Oberwinkler, 1986.
Kottke, I., and F. Oberwinkler. 1986. Mycorrhiza of forest trees—structure and function. *Trees* 1:1–24.
Kremer, J. N. 1983. Ecological implications of parameter uncertainty in stochastic simulation. *Ecological Modelling* 18:187–207.
Landsberg, J. J. 1986. *Physiological Ecology of Forest Production.* Academic Press, London, England, U.K.
Landsberg, J. J., and R. McMurtrie. 1985. Models based on physiology as tools for research and forest management. In: Landsberg, J. J., and W. Parsons, eds. *Research for Forest Management.* CSIRO, Melbourne, Australia. Pp. 214–228.
Lang, A., and M. R. Thorpe. 1983. Analysing partitioning in plants. *Plant, Cell, and Environment* 6:267–274.
Ledig, F. T. 1969. A growth model for tree seedlings based on the rate of photosynthesis and the distribution of photosynthate. *Photosynthetica* 3:263–275.
Ledig, F. T. 1976. Physiological genetics, photosynthesis, and growth models. In: Cannell, M. G. R., and F. T. Last, eds. *Tree Physiology and Yield Improvement.* Academic Press, London, England, U.K. Pp. 21–54.
Ledig, F. T., and D. B. Botkin. 1974. Photosynthetic CO_2-uptake and the distribution of photosynthate as related to growth of larch and sycamore progenies. *Silvae Genetica* 23:188–192.
Linder, S., and B. Axelsson. 1982. Changes in carbon uptake and allocation as a result of irrigation and fertilization in a young *Pinus sylvestris* stand. In: Waring, R. H., ed. *Carbon Uptake and Allocation in Subalpine Ecosystems As a Key to Management.* Forest Research Laboratory, Oregon State University, Corvallis, OR, U.S.A. Pp. 38–44.
Linder, S., and D. A. Rook. 1984. Effects of mineral nutrition on carbon dioxide exchange and partitioning of carbon in trees. In: Bowen, G. D., and E. K. S. Nambiar, eds. *Nutrition of Plantation Forests.* Academic Press, New York, NY, U.S.A. Pp. 211–236.
Loehle, C. 1987. Errors of construction, evaluation, and inference: a classification of sources of error in ecological models. *Ecological Modelling* 36:297–314.
Lohmann, H. 1983. Stand growth as a system of accumulation and circulation. *Mitteilungen der Forestlichen Bundesversuchsanstalt Wien* 147:189–197.
Loomis, R. S., R. Rabbinge, and E. Ng. 1979. Explanatory models in crop physiology. *Annual Review of Plant Physiology* 30:339–367.
Luxmoore, R. J., and L. H. Stolzy. 1987. Modeling belowground processes of roots, the rhizosphere, and soil communities. In: Wisiol, K., and J. D. Hesketh, eds. *Plant Growth Modeling for Resource Management,* Vol. II. CRC Press, Boca Raton, FL, U.S.A. Pp. 129–153.
Lyford, W. H. 1975. Rhizography of non-woody roots of trees in the forest floor. In: Torrey, J. G., and

D. T. Clarkson, eds. *The Development and Function of Roots.* Academic Press, London, England, U.K. Pp. 179–196.

Madgwick, H. A. I. 1983. Above-ground weight of forest plots—comparison of seven methods of estimation. *New Zealand Journal of Forestry Science* 13:100–107.

Mäkelä, A. 1986a. Partitioning coefficients in plant models with turnover. *Annals of Botany* 57:291–297.

Mäkelä, A. 1986b. Implications of the pipe-model theory on dry matter partitioning and height growth in trees. *Journal of Theoretical Biology* 123:103–120.

Mäkelä, A., and P. Hari. 1986. Stand growth model based on carbon uptake and allocation in individual trees. *Ecological Modeling* 33:205–229.

Mäkelä, A., and R. P. Sievänen. 1987. Comparison of two shoot-root partitioning models with respect to substrate utilization and functional balance. *Annals of Botany* 59:129–140.

McDonald, A. J. S., T. Lohammar, and A. Ericsson. 1986a. Growth response to a step-decrease in nutrient availability in small birch (*Betula pendula* Roth.). *Plant, Cell and Environment* 9:427–432.

McDonald, A. J. S., T. Lohammar, and A. Ericsson. 1986b. Uptake of carbon and nitrogen at decreased nutrient availability in small birch (*Betula pendula* Roth.) plants. *Tree Physiology* 2:61–71.

McMurtrie, R. 1985. Forest productivity in relation to carbon partitioning and nutrient cycling: a mathematical model. In: Cannell, M. G. R., and J. E. Jackson, eds. *Attributes of Trees As Crop Plants.* Institute of Terrestrial Ecology, Huntingdon, England, U.K. Pp. 194–207.

McMurtrie, R., and L. Wolf. 1983. Above- and below-ground growth of forest stands: a carbon budget model. *Annals of Botany* 52:437–448.

Mohren, G. M. J., C. P. Van Gerwen, and C. J. T. Spitters. 1984. Simulation of primary production in even-aged stands of Douglas-fir. *Forest Ecology and Management* 9:27–49.

Moorby, J. 1987. Can models hope to guide change? *Annals of Botany* 60:175–188.

Nadelhoffer, K. J., J. D. Aber, and J. M. Melillo. 1985. Fine roots, net primary production and soil nitrogen availability: a new hypothesis. *Ecology* 66:1377–1390.

Nylund, J.-E. 1988. The regulation of mycorrhiza formation: carbohydrate and hormone theories reviewed. *Scandinavian Journal of Forest Research* 3:465–479.

Orlov, A. Y. 1980. Cyclic development of roots of conifers and their relation to environmental factors. In: Sen, D. N., ed. *Environment and Root Behavior.* Geobios International, Jodhpur, India. Pp. 43–61.

Ovington, J. D. 1957. Dry-matter production of *Pinus sylvestris* L. *Annals of Botany* 21:287–314.

Pardé, D. R. 1980. Forest biomass. *Forestry Abstracts* 41:343–362.

Pearsall, W. H. 1927. Growth studies. VI. On the relative sizes of growing plant organs. *Annals of Botany* 41:549–556.

Pearson, J. A., T. J. Fahey, and D. H. Knight. 1984. Biomass and leaf area in contrasting lodgepole pine forests. *Canadian Journal of Forest Research* 14:259–265.

Penning de Vries, F. W. T. 1983. Modeling of growth and production. In: Lange, O. L., P. S. Nobel, C. B. Osmond, and H. Ziegler, eds. *Encyclopedia of Plant Physiology* (New Series) 12D:117–150.

Persson, H. 1983. The distribution and productivity of fine roots in boreal forests. *Plant and Soil* 71:87–101.

Persson, H. 1989. Methods of studying root dynamics in relation to nutrient cycling. In: Harrison, A. F., et al., eds. *Field Methods in Terrestrial Ecosystem Nutrient Cycling.* Elsevier Science Publishers, Amsterdam, the Netherlands.

Promnitz, L. C. 1975. A photosynthate allocation model for plant growth. *Photosynthetica* 9:1–15.

Reynolds, E. R. C. 1970. Root distribution and the cause of its spatial variability in *Pseudotsuga taxifolia* (Poir.) Brit. *Plant and Soil* 32:501–517.

Reynolds, J. F., and J. H. M. Thornley. 1982. A root:shoot partitioning model. *Annals of Botany* 49:585–597.

Rose, D. A. 1983. The description of the growth of root systems. *Plant and Soil* 75:405–415.

Running, S. W., and J. C. Coughlan. 1988. A general model of forest ecosystem processes for regional applications. I. Hydrologic balance, canopy gas exchange and primary production processes. *Ecological Modelling* 42:125–154.

Sala, O. E. 1988. Bias in estimates of primary production: an analytical solution. *Ecological Modelling* 44:43–55.

Santantonio, D. 1989. Dry-matter partitioning and production of fine roots in forests—new approaches to a difficult problem. In: Pereira, J. S., and J. J. Landsberg, eds. *Biomass Production by Fast-Growing Trees.* Pp. 57–72.

Santantonio, D., and J. Grace. 1987. Estimating fine-root production and turnover from biomass and decomposition data: a compartment-flow model. *Canadian Journal of Forest Research* 17:900–908.

Santantonio, D., and R. K. Hermann. 1985. Standing crop, production, and turnover of fine roots on dry, moderate, and wet sites of mature Douglas-fir in western Oregon. *Annales des Sciences Forestières* 42:113–142.

Santantonio, D., and E. Santantonio. 1987. Effects of thinning on production and mortality of fine roots in a *Pinus radiata* plantation on a fertile site in New Zealand. *Canadian Journal of Forest Research* 17:919–928.

Santantonio, D., R. K. Hermann, and W. S. Overton. 1977. Root biomass studies in forest ecosystems. *Pedobiologia* 17:1–31.

Satoo, T., and H. A. I. Madgwick. 1982. *Forest Biomass.* Nijhoff/Junk, the Hague, the Netherlands.

Schäfer, H., H. Bossel, H. Krieger, and N. Trost. 1988. Modelling the responses of mature forest trees to air pollution. *GeoJournal* 17:279–287.

Shinozaki, K., K. Yoda, K. Hozumi, and T. Kira. 1964a. A quantitative analysis of plant form: the pipe-model theory. I. Basic analyses. *Japanese Journal of Ecology* 14:97–105.

Shinozaki, K., K. Yoda, K. Hozumi, and T. Kira. 1964b. A quantitative analysis of plant form: the pipe-model theory. II. Further evidence of the theory and its application to forest ecology. *Japanese Journal of Ecology* 14:133–139.

Sievänen, R., P. Hari, P. J. Orava, and P. Pelonen. 1988. A model for the effect of photosynthetic allocation and soil nitrogen on plant growth. *Ecological Modelling* 41:55–65.

Sollins, P., R. A. Goldstein, J. B. Mankin, C. E. Murphy, and G. L. Swartzman. 1981. Analysis of forest growth and water balance using complex ecosystem models. In: Reichle, D. E., ed. *Dynamic Properties of Forest Ecosystems.* Cambridge University Press, Cambridge, England, U.K. Pp. 537–565.

Thornley, J. H. M. 1972a. A model to describe the partitioning of photosynthate during vegetative plant growth. *Annals of Botany* 36:419–430.

Thornley, J. H. M. 1972b. A balanced quantitative model for root:shoot ratios in vegetative plants. *Annals of Botany* 36:431–441.

Thornley, J. H. M. 1976. *Mathematical Models in Plant Physiology.* Academic Press, London, England, U.K. 318 pp.

Thornley, J. H. M. 1977. Root:shoot interactions. *Symposia of the Society for Experimental Biology* 31:367–389.

Tippett, J. T. 1982. Shedding of ephemeral roots in gymnosperms. *Canadian Journal of Botany* 60:2295–2302.

Troughton, A. 1956. Studies on the growth of young grass plants with special reference to the relationship between root and shoot systems. *Journal of British Grassland Society* 11:56–65.

Valentine, H. T. 1988. A carbon balance model of stand growth: a derivation employing pipe-model theory and the self-thinning rule. *Annals of Botany* 62:389–396.

Vogt, K. A., and H. Persson. In press. Root methods. In: Hinckley, T. M., ed. *Techniques and Approaches in Forest Tree Physiology.* CRC Press, Boca Raton, FL, U.S.A.

Vogt, K. A., C. C. Grier, and D. J. Vogt. 1986. Production, turnover and nutrient dynamics of above- and belowground detritus of world forests. *Advances in Ecological Research* 15:303–377.

Vogt, K. A., D. J. Vogt, E. E. Moore, B. A. Fatuga, M. R. Redlin, and R. L. Edmonds. 1987. Conifer and angiosperm fine-root biomass in relation to stand age and site productivity in Douglas-fir forests. *Journal of Ecology* 75:857–870.

Wareing, P. F., and J. Patrick. 1975. Source-sink relations and partition of assimilates in the plant. In: Cooper, J. P., ed. *Photosynthesis and Productivity in Different Environments.* Cambridge University Press, Cambridge, England, U.K. Pp. 481–499.

Waring, R. H., and W. H. Schlesinger. 1985. *Forest Ecosystems: Concepts and Management.* Academic Press, Orlando, FL, U.S.A.

Warren Wilson, J. 1972. Control of crop processes. In: Rees, A. R., K. E. Cockshull, D. W. Hand, and R. G. Hurd, eds. *Crop Processes in Controlled Environments.* Academic Press, London, England, U.K. Pp. 7–30.

Weinstein, D., and R. Beloin. 1989. Evaluating effects of pollutants on integrated tree processes: a model of carbon, water, and nutrient balances. In: Dixon, R. K., and R. S. Meldahl, G. A. Ruark, and W. G. Warren, eds. *Process Modeling of Forest Growth Response to Environmental Stress.* Timber Press, Portland, OR, U.S.A.

Westoby, M. 1984. The self-thinning rule. *Advances in Ecological Research* 14:167–225.

White, H. L. 1937. The interaction of factors in the growth of *Lemna.* XII. The interaction of nitrogen and light intensity in relation to root length. *Annals of Botany* 1:649–654.

White, J. 1981. The allometric interpretation of the self-thinning rule. *Journal of Theoretical Biology* 89:475–500.

Whittaker, R. H., and P. L. Marks. 1975. Methods of assessing terrestrial productivity. In: Lieth, H., and R. H. Whittaker, eds. *Primary Productivity of the Biosphere.* Springer-Verlag, New York, NY, U.S.A. Pp. 55–118.

Wilcox, H. E. 1983. Fungal parasitism of woody plant roots from mycorrhizal relationships to plant disease. *Annual Review of Phytopathology* 21:221–242.

Yamakura, T., H. Saito, and T. Shidei. 1972. Production and structure of under-ground part of Hinoki (*Chamaecyparis obtusa*) stand. I. Estimation of root production by means of root analysis. *Journal Japanese Forestry Society* 54:118–125.

Yoda, K., T. Kira, H. Ogawa, and K. Hozumi. 1963. Self-thinning in over-crowded pure stands under cultivated and natural conditions (inter-specific competition among higher plants XI). *Journal of Biology, Osaka City University* (Osaka, Japan) 14:107–129.

CHAPTER 11

MODELING THE INTERCEPTION OF SOLAR RADIANT ENERGY AND NET PHOTOSYNTHESIS

Jennifer C. Grace

Abstract. The influence of canopy structure, crown shape, and amount and distribution of foliage within tree crowns on the interception of solar radiant energy is outlined. Alternative approaches for modeling net photosynthesis are outlined. A process-level model that simulates the interception of solar radiant energy and net photosynthesis for an array of tree crowns indicates that for a forest stand where tree crowns touch, annual canopy net photosynthesis is only slightly affected by crown shape, and tends to a maximum as leaf area index increases. The simulations also indicate that when tree crowns do not touch, the shape of the tree crown has a large influence on annual canopy net photosynthesis. For isolated trees, simulations indicate that annual net photosynthesis depends on crown shape and leaf area within the crown. The results of the simulation study were used to generate a simple model to predict annual net photosynthesis from leaf area, crown shape, tree size, and number of stems per ha (i.e., per 2.471 U.S. acres).

INTRODUCTION

Many different modeling approaches may be used to obtain estimates of tree and/or forest growth, ranging from simple empirical models developed from large data bases to detailed models which simulate the processes controlling tree growth, termed *process-level models* in this chapter. Each approach has advantages and disadvantages. Models which simulate the processes controlling tree growth are likely to give more realistic estimates of the likely consequences of increased levels of atmospheric carbon dioxide, pollution, or new forest management practices than empirical models because they are based on our understanding of how trees grow. However, a major disadvantage is the long execution time for this type of model. Consequently, if one has to pay for computer time, cost will be a disadvantage. By using process-level models to determine how easily measured tree and stand variables influence tree growth, it should be possible to develop models of forest growth with a short execution time that give realistic estimates of the effects of new forest management practices, pollution, and climate changes.

Essential components for a process-level model of tree growth are modules which simulate the interception of solar radiant energy, photosynthesis, respiration, allocation of net carbon fixed to tree components, and the consequent changes in leaf area and tree dimensions.

Dr. Grace is a Scientist, Forest Research Institute, Ministry of Forestry, Rotorua, New Zealand.
The author wishes to thank the organizers of the conference "Forest Growth: Process Modeling of Responses to Environmental Stress," April 18–22, 1988, Gulf Shores, AL, U.S.A., for financial assistance enabling her to attend the conference.

This chapter outlines important factors influencing the interception of solar radiant energy. Mention is made of different approaches to modeling photosynthesis. The process-level model of Grace et al. (1987a,b) is used to show important tree and stand variables affecting annual net photosynthesis of individual trees and stands. The results of these simulation runs are then used to develop a simple model for predicting annual net photosynthesis for forest stands.

INTERCEPTION OF SOLAR RADIANT ENERGY

The interception of solar radiant energy by a forest canopy depends on the shape of the canopy, the amount and distribution of foliage, branches, and stems within the canopy, and the amount of incoming solar radiant energy. Ross (1981) gives a comprehensive account of how plant architecture influences the radiation regime within plant canopies. Models that have been developed for predicting the interception of solar radiant energy may be classified into four main groups:

1. Models which assume that the canopy is made up of a number of discrete geometrical shapes which are opaque (e.g., Terjung and Louie, 1972; Jackson and Palmer, 1972).
2. Models which calculate the penetration of solar radiant energy through a canopy which is assumed to be continuous in the horizontal plane, and in which the foliage elements are assumed to be distributed according to a probability distribution. These foliage elements are assumed to absorb, transmit, and reflect solar radiant energy. The solar radiant energy intercepted by stems and branches is usually neglected. Such models are termed *statistical models* by Lemeur and Blad (1974) and *turbid layer models* by Ross (1981).
3. Models where the canopy is made up of a number of discrete geometrical shapes, but where the penetration of solar radiant energy is taken into account as in group 2. Such models are termed *weighted-random models* by Norman and Welles (1983).
4. Models where the canopy is assumed to be made up of individually placed foliage elements. This approach, developed by Myneni and Impens (1985), is called a *procedural approach.*

The model execution time and the information needed on canopy structure increase from group 1 to group 4. I consider that models from groups 2 or 3 are likely to be most suitable for developing process-level models of tree and/or forest growth. The following discussion describes assumptions which need to be made when choosing or developing a group 2 or 3 model for predicting the interception of solar radiant energy.

Canopy Structure

The first assumption that needs to be made is whether the canopy is continuous in the horizontal plane (e.g., Norman and Jarvis, 1975; Oker-Blom, 1986) or whether it is made up of discrete blocks of foliage (e.g., Norman and Welles, 1983; Oker-Blom, 1986). If one is interested in estimating growth for individual trees, however, it will be more appropriate to assume that the crown of each tree forms one block of foliage, even when the canopy could be considered to be continuous in the horizontal plane.

Shape of Tree Crowns

The shape of tree crowns varies widely among species and in many cases can be described by three parameters: the absolute size, the ratio of height to width, and the convexity (Horn, 1971). The modeled crown shape needs to be a realistic representation of the actual shape since the shape affects the interception of solar radiant energy. For example, Terjung and Louie (1972) estimated the proportion of solar radiant energy absorbed by four different shapes: a sphere, an upright cylinder, an upright cone, and a reverse cone, each having the same surface area. Assuming perfect absorption by the surface, they found that an upright cone with a small ratio of height (h) to radius (r) intercepted the most direct solar radiant energy at almost all sun zenith angles. Over a year, the shape that intercepted the most solar radiant energy varied with latitude.

Amount of Foliage within Tree Crowns

There are several alternative techniques for estimating the surface area of foliage within tree crowns, which is the most important tree variable for determining the penetration of solar radiant energy through tree crowns (Jarvis and Leverenz, 1983). Techniques may be divided into direct and indirect methods (see Norman and Campbell, in press, for a comprehensive review). Direct methods involve direct measurements of selected plant components. They are labor-intensive and either disturb or destroy the canopy. The general strategy is to make detailed measurements on selected components on a small representative sample of trees and then to derive equations to predict these components from more easily measured components. One example is estimating foliage surface area from sapwood basal area. However, the ratio between these two components does not appear to be independent of site factors (e.g., Dean and Long, 1986). Also, sapwood area is unlikely to change as rapidly as foliage surface area when trees are partially defoliated by insects or pathogens. While direct methods can give the foliage surface area for individual trees within a stand, it appears that direct measurements may be needed at each site for an acceptable level of accuracy.

Indirect methods utilize the relationship between canopy structure and the interception of solar radiant energy. Although the collection of field data is generally quick, complex algorithms are needed to estimate foliage surface area from the data. Indirect methods include spectral methods and gap fraction methods (e.g., hemispherical photographs and the technique of Lang and Xiang (1986), which estimates foliage surface area from the transmittance of direct solar radiant energy). Each method is suitable only in a limited range of conditions. Spectral methods are not suitable for leaf area indices greater than 2. Hemispherical photographs require calm days with overcast conditions. The technique of Lang and Xiang (1986) requires clear sunny days; moreover, although it is suitable for estimating the foliage surface area of isolated trees, it will not give the foliage surface area for individual trees within a stand. Obviously further research is required to develop a method for estimating foliage surface area which is suitable for all conditions.

Distribution of Foliage Within Tree Crowns

When calculating the penetration of solar radiant energy through tree crowns, one needs to know whether the foliage distribution within tree crowns is regular, random, or clumped; and one also needs to know the inclination angles of foliage with respect to the vertical and azimuth. Only a few field studies have examined the structure of tree crowns in detail (e.g., Norman and Jarvis, 1974; Kellomäki and Oker-Blom, 1983; Hutchison et al., 1986). While these studies indicate that the foliage distribution within tree crowns is unlikely to be random, reasonable agreement has been obtained between field measurements and model estimates of the penetration of solar radiant energy through tree crowns assuming that foliage is randomly distributed (e.g., Norman and Jarvis, 1975; Grace et al., 1987a). Though Baldocchi and Hutchison (1986) found that assuming the foliage was clumped gave a better prediction of direct photosynthetically active radiant energy in an oak-hickory (*Quercus – Carya* sp.) forest than the assumption of random distribution, they do not discount the random assumption, as it gives estimates within the typical range of experimental errors and is easier to use. The random assumption appears reasonable due to compensating errors. For example, there is less direct solar radiant energy intercepted and less scattered radiant energy in canopies with clumped as opposed to random foliage (e.g., Norman and Jarvis, 1975; Oker-Blom and Kellomäki, 1983). In addition, ignoring branches and stems helps to compensate for the overestimation of intercepted direct solar radiant energy in a random canopy (Norman and Jarvis, 1975).

The distribution of foliage inclination angles appears to have little effect on the amount of solar radiant energy absorbed over a prolonged period (Oker-Blom and Kellomäki, 1982). However, it does affect the distribution of solar radiant energy within the forest canopy over a short period (Kuruoiwa, 1970).

The distribution of leaf azimuthal angles is the least important structural property of a canopy which influences the interception of solar radiant energy (Jarvis and Leverenz, 1983). For many species, a uniform distribution of leaf azimuth angles can generally be assumed (Norman, 1979). If this is not the case, equations presented by Lemeur (1973) can be used to specify the azimuth distribution.

Distribution of Incoming Solar Radiant Energy

As foliage has different transmittance and reflectance coefficients for light in different wavebands, incoming solar radiant energy needs to be split into a minimum of four components: diffuse and direct-beam photosynthetically active radiant (PAR) energy (400–700 nm); and diffuse and direct-beam near-infrared radiant (NIR) energy (700–3000 nm). Incoming solar radiant energy is usually split only into the above four components. Equations are available to determine the proportions of each component in total incoming solar radiant energy (e.g., Weiss and Norman, 1985).

Equations for Calculating the Interception of Solar Radiant Energy

An appropriate equation to predict the interception of solar radiant energy will depend on the assumptions made on the distribution of foliage within the forest canopy. A poisson model is appropriate if foliage is randomly distributed, a negative binomial model if the foliage is clumped, and a positive binomial model if the foliage is regularly distributed (Nilson, 1971).

When foliage is randomly distributed within individual tree crowns, with no preferred azimuth direction, the probability that direct-beam radiant energy is not intercepted is given by Norman and Welles, (1983):

$$p = \exp(-k\,\rho\,s) \tag{1}$$

where p is the probability of non-interception, s is the distance the beam passes through tree crowns within the stand, ρ is the one-sided foliage surface area/crown volume, and k is the fraction of one-sided foliage area projected on a plane normal to the solar beam.

The probability that diffuse radiant energy is not intercepted is calculated by numerically averaging Equation 1 over a hemisphere (e.g., Norman, 1979; Norman and Welles, 1983).

The scattering of solar radiant energy can also be predicted by considering the amount of solar radiant energy intercepted, transmitted, and reflected (e.g., Norman and Jarvis, 1975; Norman and Welles, 1983).

Penumbral Effects

Many models predicting the interception of solar radiant energy assume that the sun is a point source. In fact, the sun has an angular radius of approximately 0.5° when viewed from the earth. Thus individual foliage elements do not necessarily intercept the whole solar beam. This creates partial shade (penumbral shade) and hence a more uniform distribution of solar radiant energy within the canopy. Including penumbral effects adds to the complexity of models. Oker-Blom (1985) found that the inclusion of penumbral effects increased the net photosynthesis at specified points within a modeled Scots pine (*Pinus sylvestris* L.) canopy. The percentage increase in net photosynthesis increased with depth in the canopy, reaching 69% at the base of the canopy. However, it is the outer foliage which contributes most to the photosynthesis of an individual tree (e.g., Rook et al., 1985), hence the change in net photosynthesis for a whole tree should not be large. For a deciduous forest, the inclusion of penumbral effects increases canopy photosynthesis by 10% (Baldocchi and Hutchinson, 1986), while Denholm (1981) suggests that penumbral effects are unlikely to be of practical significance.

ESTIMATION OF NET PHOTOSYNTHESIS

While PAR is essential for photosynthesis, other factors also influence the rate of net photosynthesis (see Jarvis and Sandford, 1986, for a review). Models simulating net photosynthesis can be developed at varying levels of complexity ranging from empirical models (e.g., Thornley, 1976) to biochemical models (e.g., Von Caemmerer and Farquhar, 1981). Models can also be developed at several different structural levels, such as chloroplast, foliage element, or canopy. An advantage of models predicting net photosynthesis at the level of individual foliage elements is that they can be used to simulate the effects on net photosynthesis of changing foliage distributions within the crown caused by disease, forest management practices, and other factors. Hence such models are likely to be most suitable for developing process-level models of forest growth. Methods to link equations predicting rates of net photosynthesis with models predicting the interception of solar radiant energy are discussed by Norman (1980) and Grace et al. (1987b).

A PROCESS-LEVEL MODEL TO PREDICT THE INTERCEPTION OF SOLAR RADIANT ENERGY AND NET PHOTOSYNTHESIS

Grace et al. (1987a,b) presented a process-level model which will predict the interception of solar radiant energy at any site, and net photosynthesis for radiata pine (*Pinus radiata* D. Don) at sites not limited by water and nutrient stress. As the model is modular, equations for predicting rates of net photosynthesis for stressed trees or even other species can easily be included.

In the model, the location of each tree within a given area is specified by x, y, and z coordinates, allowing different spacing patterns to be simulated. The crown of each tree is assumed to be ellipsoidal in shape. The ellipsoid may be truncated at the base. Within the ellipsoidal shape, three other ellipsoids can be defined (Figure 1) forming four shells. The shells allow four ages of foliage to be considered, and the amount of foliage in each shell is specified. The shells also allow for non-random distributions of foliage within the tree crown, which is particularly useful for investigating the effects of partial defoliation. The foliage within each shell is assumed to be randomly distributed, with foliage inclination angles distributed according to the spherical leaf angle distribution. There is assumed to be no preferred azimuth direction.

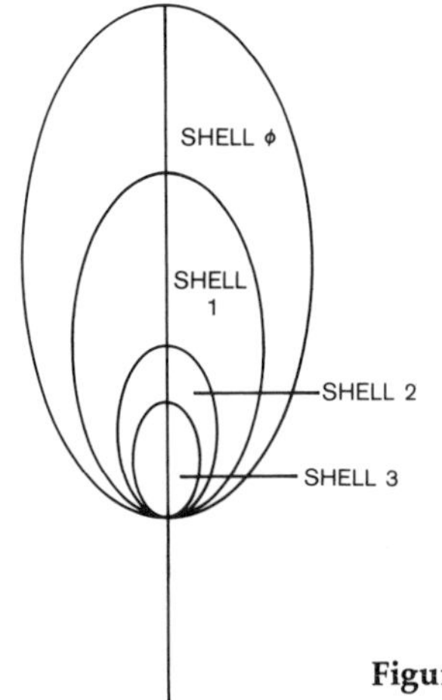

Figure 1. Diagram showing how ellipsoids are arranged to form four shells.

The sun is assumed to be a point source. Given the incoming solar radiant energy and canopy structure, the model of Grace et al. (1987a,b) calculates the amount of diffuse and direct solar radiant energy in both PAR and NIR wavebands at specified points within or below the forest canopy, using Equation 1 to calculate the penetration of direct and diffuse solar radiant energy as described above, and using the method of Norman and Welles (1983) to calculate the scattering of solar radiant energy.

Since acceptable agreement between model estimates and measurements of PAR reaching the forest floor was obtained in a stand of radiata pine using the model of Grace et al. (1987a), it is considered that the model assumptions are acceptable when predicting the interception of solar radiant energy in stands of radiata pine.

To enable net photosynthesis to be predicted, the crown is split into six levels as well as the four shells. Each level is also split into four quadrants, giving a maximum of 52 segments (Grace et al., 1987b). The incident PAR at a given point within each segment is calculated, and this value of PAR is input to the equation predicting the rate of net photosynthesis at that point. This rate of net photosynthesis is assumed to apply to all foliage within that segment. The estimates of net photosynthesis for each segment are summed to give net photosynthesis for a tree.

Because the model of Grace et al. (1987a,b) calculates the interception of solar radiant energy and net photosynthesis on an hourly basis, it is time-consuming and also costly (if computer time is not free) to run a simulation for a whole year. Annual intercepted solar radiant energy and annual net photosynthesis are therefore estimated by running the model on an hourly basis for 28 representative days throughout the year covering the range of climatic conditions. Regression equations are then derived between daily intercepted solar radiant energy and daily incoming solar radiant energy, and between daily net photosynthesis and daily incoming PAR. These equations are used to estimate annual totals (see Grace et al., 1987a,b). Built into this simplification is the assumption that crown shape and leaf area remain constant during a one-year period. A similar procedure could be used if one wished to change crown shape and leaf area more often.

Since it is not possible to measure canopy photosynthesis directly, the validity of this model was checked by comparing model estimates of canopy net photosynthesis with field estimates of dry matter production and likely values for respiration. The model was considered to give realistic estimates of annual net photosynthesis for radiata pine stands (Grace et al., 1987b). Hence the model is considered suitable for investigating the effects of different canopy structures on annual net photosynthesis.

SIMULATION STUDY

Methods

In this study, the model of Grace et al. (1987a,b) was used to investigate how annual net photosynthesis changes with number of stems per ha, crown shape, and leaf area within the crown. The simulations were limited to one location: Puruki, New Zealand (38°26'S, 176°13'E), a forested catchment without water or nutrient stress, and where meteorological data needed to drive the model had been collected (Beets and Brownlie, 1987). Meteorological data used in the simulations are summarized in Table 1. Equation 2 was used to predict the rate of net photosynthesis. It assumes that the rate of net photosynthesis depends only on incident PAR and was chosen so that the effects of foliage distribution within the stand were easily discernible.

$$Pn = \frac{7.67 \times 0.012 \times I}{7.67 + 0.012 \times I} \qquad (2)$$

where Pn is the rate of net photosynthesis in μmol $CO_2/m^2/s$, and I is the incident PAR in $\mu mol/m^2/s$.

Equation 2 was derived from field data collected at the Forest Research Institute, Rotorua, New Zealand (38°9'S, 176°16'E), on rates of net photosynthesis for radiata pine shoots with the needle fascicles left in their natural position (Grace et al., 1987b). Unless otherwise stated, all trees within the stand have been assumed to be the same size with the foliage randomly distributed throughout the whole crown. Square spacing has been assumed.

Table 1. Average daily incoming shortwave solar radiant energy measured at Puruki, New Zealand (38°26'S, 176°13'E), during 1985.

Shortwave solar radiant energy (MJ/m^2)					
Month	Mean	Standard deviation	Month	Mean	Standard deviation
January	21.2	7.6	July	5.9	2.3
February	19.4	6.0	August	9.1	3.7
March	15.3	4.8	September	12.6	3.4
April	10.8	4.2	October	18.3	5.6
May	8.5	2.7	November	22.2	6.1
June	4.4	2.7	December	20.7	7.3

In order to determine how the number of stems per ha, crown shape, and leaf area within the crown interact to affect annual net photosynthesis, simulations were run for different combinations of the above variables. The simulations were based on data collected from a 5-year-old stand of radiata pine with a closed canopy near Rotorua, New Zealand (Table 2). Some of the combinations simulated may not occur in the field but are included to give a better understanding of how the variables interact to affect net photosynthesis in extreme situations. In this paper, a closed canopy is assumed to occur once tree crowns touch. Some simulations (see Appendix 1 at the close of this chapter) were designed to show how annual net photosynthesis is affected by the number of stems per ha, crown shape, and leaf area within the crown. Another set of simulations (see Appendix 2 at the close of this chapter) was designed to show how annual net photosynthesis for an individual tree changes with crown shape and leaf area within the crown. A third set of simulations (see Appendix 3 at the close of this chapter) was designed to show how annual net photosynthesis changes when alternate trees are partially defoliated by removing foliage from the center of the crown utilizing the shells (Figure 1).

Table 2. Structure of a 5-year-old stand of radiata pine growing near Rotorua, New Zealand (38°17'S, 176°7'E).

Stems per ha	2163.0
One-sided leaf area index	3.1
Average crown width	2.4 m
Average crown length	7.36 m
Average tree height	8.26 m

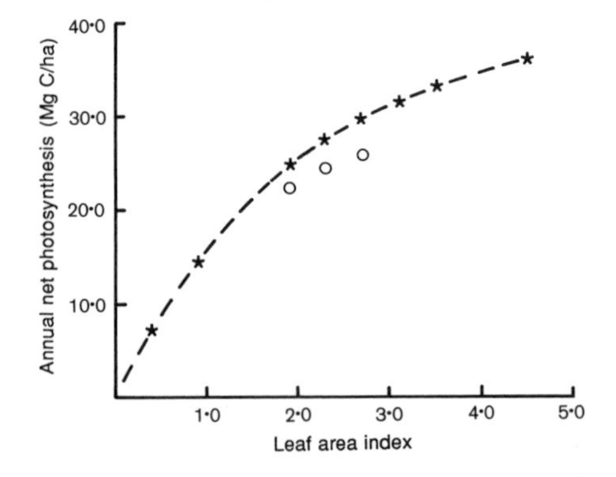

Figure 2. Annual canopy net photosynthesis as related to leaf area for stands with 2163 stems per ha (model estimates). * = closed canopy; o = open canopy, crown width = 1.2 m.

Results

For a given crown shape and number of stems per ha, the model predicts that for forests with a closed canopy, annual net photosynthesis increases to a maximum as leaf area index increases (Figure 2). When crowns do not touch, annual net photosynthesis for a given leaf area index is lower (Figure 2).

Annual net photosynthesis for a closed forest canopy with a given leaf area index increases only slightly with an increasing number of stems per ha (Appendix 1, group c). The increase in annual net photosynthesis is considered to be due to the increase in crown surface area within the stand (Figure 3), which allows more foliage to be exposed to the sun.

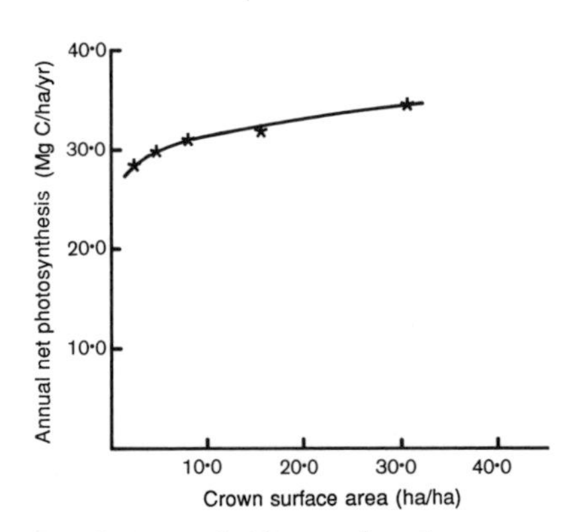

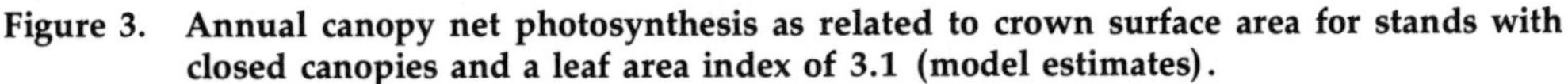

Figure 3. Annual canopy net photosynthesis as related to crown surface area for stands with closed canopies and a leaf area index of 3.1 (model estimates).

For a given leaf area index and number of stems per ha, the model predicts that crown length (the maximum vertical extent of foliage) has little influence on annual net photosynthesis when the canopy is closed. However, as the ratio of crown width (the maximum horizontal spread of foliage) to spacing decreases (i.e., as the canopy becomes more open), changes in crown shape become far more important (Figure 4). Various functions incorporating crown shape, tree size, and spacing were investigated to see whether the variability shown in Figure 4 could be explained by one function. The most appropriate function, R (Figure 5), was given by

$$G = \max(S - W, 0.0)$$
$$r = 1.0 - \min(G^{1.5}/H, 1.0) \qquad (3)$$
$$R = r \times r$$

where S is the spacing between trees, W is the crown width, and H is the tree height.

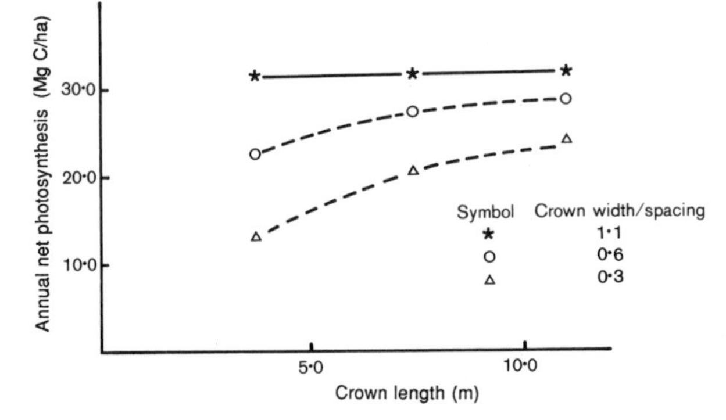

Figure 4. Annual canopy net photosynthesis as related to crown shape for stands with a leaf area index of 3.1 (model estimates).

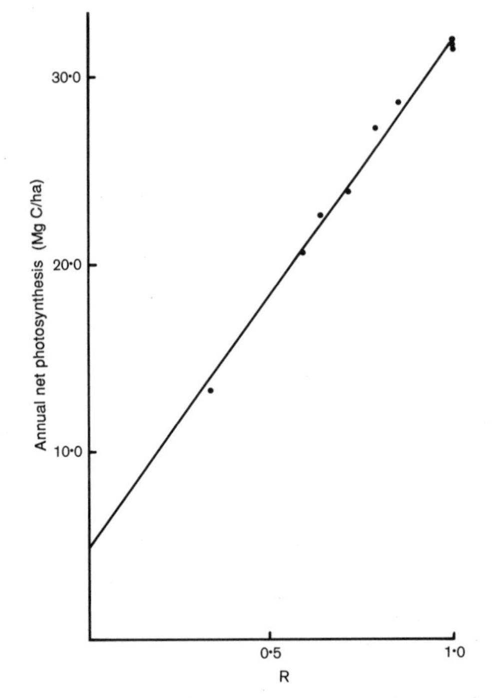

Figure 5. Annual canopy net photosynthesis as related to R for stands with a leaf area index of 3.1 (model estimates). R is defined in Equation 3 in the text.

To develop a simple model for predicting annual net photosynthesis, one needs to know when trees can be treated as isolated individuals and when trees are competing with one another. Comparing the estimates of tree net photosynthesis for trees within a stand (Appendix 1, groups i and j) with the corresponding estimates of net photosynthesis for isolated trees (starred results in Appendix 2), one can see that the estimates of tree net photosynthesis for trees within the stand are close to those for the isolated trees, indicating that there is little or no competition among the trees within the simulated stands. For the simulations (Appendix 1, groups i and j), the tree spacing was such that

$$S = H + W \tag{4}$$

where S is the spacing between trees, W is the crown width, and H is the tree height.

The assumption that competition begins when $S = H + W$ is therefore considered realistic for the location considered in this study. The spacing at which competition starts may differ from location to location due to variations in the path of the sun causing different patterns of shade.

When there is no competition among individual trees, annual net photosynthesis for an individual tree depends on the leaf area within the crown and the crown shape (Figure 6). Most of the variation among individual trees can be accounted for by expressing both annual net photosynthesis and foliage surface area as a ratio with crown surface area (Figure 7).

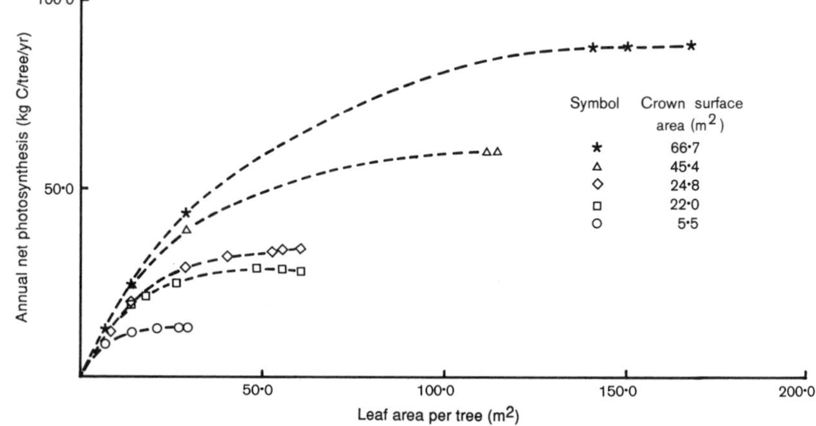

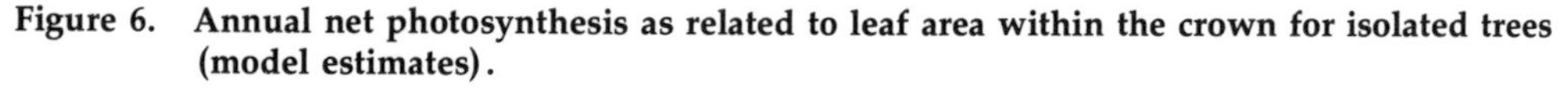

Figure 6. Annual net photosynthesis as related to leaf area within the crown for isolated trees (model estimates).

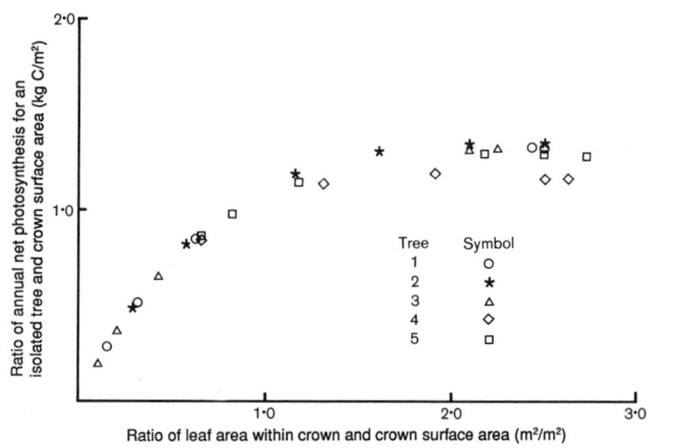

Figure 7. For isolated trees, annual net photosynthesis per unit of crown surface area versus leaf area per unit of crown surface area (model estimates).

The simulations shown in Appendix 3 indicate that annual net photosynthesis for the forest canopy in which some trees have been partially defoliated is close to but slightly less than that for an unaffected stand with the same leaf area. In this case, however, annual net photosynthesis for the individual trees varied widely. The value of annual net photosynthesis for individual trees relative to the canopy average was found to depend on the amount of leaf area within the tree crown relative to the canopy average (Figure 8).

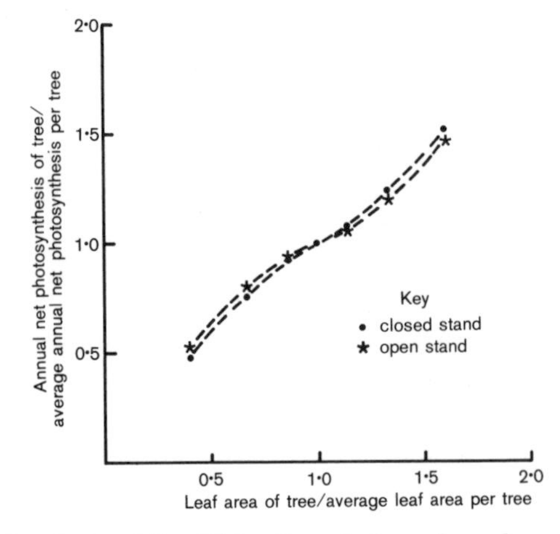

Figure 8. For a simulated stand in which alternate trees have been partially defoliated, ratio of annual net photosynthesis for an individual tree and canopy average versus ratio of foliage surface area for an individual tree and canopy average.

DEVELOPMENT OF A SIMPLE MODEL FOR PREDICTING ANNUAL NET PHOTOSYNTHESIS

Methods

Results from the above simulations were used to generate a model for predicting annual net photosynthesis for a forest canopy from easily measured variables.

While the results (Appendix 1, group c) indicate that annual net photosynthesis is a function of the number of stems per ha at canopy closure, plantation stands of radiata pine are unlikely to contain more than 7000 stems per ha. Between 100 and 7000 stems per ha, canopy annual net photosynthesis varies by only 11%. This may not be significant. Hence I make the assumption that at canopy closure, annual canopy net photosynthesis depends only on the leaf area index.

Figure 2 indicates that for a closed canopy, annual net photosynthesis tends to an asymptote with increasing leaf area index. Using the data shown in Figure 2, Equation 5 was derived to predict annual net photosynthesis for a closed forest canopy.

$$Pn_c = 40.2 \times (1.0 - \exp(-0.50 \times L)) \quad (5)$$

where Pn_c is the annual net photosynthesis for the forest canopy in Mg C/ha, and L is the leaf area index of the forest canopy.

While Figure 7 indicates that the ratio of annual net photosynthesis/crown surface area (Pn_t/A) to foliage area/crown surface area (L_t/A) may be a function of crown shape, the ratio of crown length to crown width in radiata pine stands is likely to be in the range of 1.5–4.0 (Grace, unpublished data), i.e., comparable to trees 1–3 in Appendix 2. For these three trees, Pn_t/A appears to depend only on L_t/A. Hence Equation 6 was derived using the simulation results for trees 1–3 to provide a means of estimating net photosynthesis for isolated trees.

$$Pn_t/A = 1.34 \times (1.0 - \exp(-1.65 \times L_t/A)) \quad (6)$$

where Pn_t is the annual net photosynthesis for an isolated tree in kg C, L_t is the one-sided foliage surface area for the tree in m^2, and A is the crown surface area of the tree in m^2.

To develop a model for predicting annual canopy net photosynthesis, assumptions are needed to tell us when trees start competing with one another and how annual canopy net photosynthesis changes from the time when trees start competing with one another to canopy closure.

As shown in Equation 4 above, trees start competing when $S = H + W$, where S is the spacing between trees, W is the crown width, and H is the tree height. At this spacing, the number of stems per ha, N, is given by

$$N = 10000.0/S^2 \quad (7)$$

When the number of stems per ha, N*, is less than or equal to N, annual canopy net photosynthesis is given by

$$Pn_c = Pn_t \times N^* \quad (8)$$

The function R (Figure 5) gives a way of calculating changes in annual canopy net photosynthesis due to the openness of the stand. If R is zero, this suggests that there is no competition among trees. However, the spacing at which this occurs is

$$S = H^{2/3} + W \quad (9)$$

Also, Figure 5 suggests that annual canopy net photosynthesis is constant when R is zero. However, the values of annual canopy net photosynthesis when competition begins (Appendix 1, groups i and j) are not constant. These results indicate that R is not a suitable variable to predict the change in canopy net photosynthesis due to the openness of the canopy. Yet openness of the canopy is clearly important (Figure 4), and further research is necessary to determine the role of crown shape in competition.

To account for the influence of canopy openness in the simple model, the following equation, to predict annual canopy net photosynthesis, was included.

$$Pn_c = Pn_{cl} + (Pn_{cm} - Pn_{cl} \times R^* \quad (10)$$

where R* is given by

$$G = \max(S - W, 0.0)$$
$$R^* = \min(G/H, 1.0)$$

and where Pn_{cl} is annual canopy net photosynthesis with N stems per ha, and Pn_{cm} is annual canopy net photosynthesis when the canopy is closed.

These assumptions were made to allow annual net photosynthesis to vary at the time when competition began, and to prevent any discontinuity in the model at this point.

Equations 5–10 form the simple model allowing annual canopy net photosynthesis to be predicted from stocking, leaf area, crown dimensions, and tree height. This model was used to predict annual canopy net photosynthesis for the stands shown in Appendix 1, groups a and b.

Results

The results (Table 3) indicate that one can predict annual canopy net photosynthesis for a forest stand from stocking, leaf area, crown dimensions, and tree height.

Table 3. Estimates of annual canopy net photosynthesis from the process-level model and from a simple model developed from the process-level model.

				Annual canopy net photosynthesis	
Crown width (m)	Crown length (m)	Spacing (m)	Leaf area index (LAI)	Process-level (Mg C/ha)	Simple (Mg C/ha)
2.4	7.36	2.15	3.1	31.6	31.7
		2.4	2.5	28.0	28.7
		2.5	2.3	26.7	27.2
		4.0	0.9	13.2	12.2
		6.0	0.4	6.3	5.1
		10.13	0.1	2.3	2.2
		10.66	0.1	2.1	2.2
		11.00	0.1	1.9	2.0
1.2	7.36	2.15	2.7	25.9	26.6
			2.3	24.3	24.5
			1.9	22.3	22.0

DISCUSSION

Process-level models of forest growth can be used to estimate the likely consequences for tree growth under conditions not previously experienced by the tree species, such as a new site, a new silvicultural regime, increased air pollution, or increased atmospheric carbon dioxide. However, to develop reliable process-level models requires detailed information on the ecophysiology of the species, namely, foliage distribution, rates of net photosynthesis, rates of respiration, how carbon is allocated to various parts of the tree, how leaf area and tree dimensions change, and how the various processes vary with changes in nutrition, water stress, air pollution, increased carbon dioxide, and so on. Even when processes are not fully understood, sensitivity analyses using process-level models are valuable for indicating where small changes in a parameter will cause large variations in growth, thus pinpointing the most important areas for further research.

Because process-level models have a long execution time, it is advantageous to try to simplify the model when predicting forest growth over a rotation. However, the simplification needs to be done in such a way as to retain the advantages of process-level models. Since many variables influence tree growth, it is best to develop a simplified model for a particular scenario. If one tried to build a comprehensive model, it is likely that it would be as complicated as the model being simplified.

In this paper, I have investigated the possibility of simplifying a process-level model simulating annual net photosynthesis on a site where water and nutrients are not limiting by determining the important, easily measured tree and stand variables which influence annual net photosynthesis.

When the forest canopy is closed, the simulation study indicates that within the range of realistic crown shapes or number of stems per ha, these variables are relatively unimportant for determining canopy net photosynthesis (Figures 3 and 4 and Appendix 1, group c). Since, in this study, the rate of net photosynthesis was assumed to depend only on incident PAR, the above result suggests that it is acceptable to use models from group 2 (as defined near the beginning of this paper) to predict the interception of solar radiant energy when the forest canopy is closed and one is only interested in determining canopy values. Indeed, group 2 models have been found to give acceptable estimates of transmitted solar radiant energy when compared to field measurements (e.g., Norman and Jarvis, 1975).

When the canopy is closed, the simulation study indicates that canopy net photosynthesis tends to an asymptote with increasing leaf area index (Figure 2). This result appears reasonable; however, it is difficult to validate since it is not possible to measure canopy net photosynthesis directly, and one tends to get only a limited range of leaf area indices at canopy closure.

The simulation study indicates that as the forest canopy becomes more open, the shape of the crown becomes more important (Figure 4); and for isolated trees, the important factors are the crown surface area and the amount of foliage within the crown (Figure 6).

While Figure 5 indicates that it is feasible to express the reduction in canopy net photosynthesis due to the openness of the canopy by a single function, it is clear that further research is necessary to fully understand the role of gaps within the canopy, as the function R was not suitable for inclusion in the simple model developed from the process-level model.

A comparison of the estimates of net photosynthesis from the simple and process-level models (Table 3) indicates that one can use the process-level model described in this paper to develop a model to predict annual net photosynthesis from easily measured variables. This is not surprising, but the advantage of this approach is that the simple model includes an understanding of how trees function, which is not necessarily the case with growth models developed from field data on tree size.

The simple model was not so successful at predicting annual net photosynthesis for some of the unrealistic crown shapes and leaf areas, indicating that further research is necessary to determine the range of realistic crown shapes and leaf areas within tree crowns.

Due to the variation in the path of the sun with latitude, the simplified model presented in this paper is unlikely to be applicable on all sites. A study of the interaction between gap size and latitude would therefore be beneficial.

The simulation results shown in Appendix 3 indicate that for stands in which some trees have been partially defoliated, the actual amount of foliage within the stand is of more importance than the distribution of foliage among trees for determining canopy net photosynthesis. On an individual tree basis, the amount of foliage within the crown compared to the canopy average is important for determining tree net photosynthesis (Figure 8). These results can easily be built into the simple model presented above by assuming that canopy net photosynthesis depends only on leaf area index and openness of the canopy. Net photosynthesis for individual trees can also be estimated by parameterizing Figure 8.

To summarize, process-level models have many advantages. But to make full use of their potential requires a commitment to ecophysiological research, as many of the processes affecting tree growth are not fully understood.

LITERATURE CITED

Baldocchi, D. D., and B. A. Hutchison. 1986. On estimating canopy photosynthesis and stomatal conductance in a deciduous forest with clumped foliage. *Tree Physiology* 2:155–168.

Beets, P. N., and R. K. Brownlie. 1987. Puruki experimental catchment: site, climate, forest management and research. *New Zealand Journal of Forestry Science* 17:137–160.

Dean, T. J., and J. N. Long. 1986. Variation in sapwood area–leaf area relations within two stands of lodgepole pine. *Forest Science* 32:749–758.

Denholm, J. V. 1981. The influence of penumbra on canopy photosynthesis. II. Canopy of horizontal circular leaves. *Agricultural Meteorology* 25:167–194.

Grace, J. C., P. G. Jarvis, and J. M. Norman. 1987a. Modelling the interception of solar radiant energy in intensively managed forests. *New Zealand Journal of Forestry Science* 17:193–209.

Grace, J. C., D. A. Rook, and P. M. Lane. 1987b. Modelling canopy photosynthesis in *Pinus radiata* stands. *New Zealand Journal of Forestry Science* 17:210–228.

Horn, H. S. 1971. *The Adaptive Geometry of Trees.* Princeton University Press, Princeton, NJ, U.S.A. 144 pp.

Hutchison, B. A., D. R. Matt, R. T. McMillen, L. J. Gross, S. J. Tajchman, and J. M. Norman. 1986. The architecture of a deciduous forest canopy in eastern Tennessee, U.S.A. *Journal of Ecology* 74:635–646.

Jackson, J. E., and J. W. Palmer, 1972. Interception of light by model hedgerow orchards in relation to latitude, time of year and hedgerow configuration and orientation. *Journal of Applied Ecology* 9:341–358.

Jarvis, P. G., and J. W. Leverenz. 1983. Productivity of temperate, deciduous and evergreen forest. In: Lange, O. L., P. S. Nobel, C. B. Osmond, and H. Ziegler, eds. *Encyclopedia of Plant Physiology,* New Series, Vol. 12d. *Physiological Plant Ecology.* Springer-Verlag, Berlin, F.R.G. Pp. 234–280.

Jarvis, P. G., and A. P. Sandford. 1986. Temperate forests. In: Baker, N. R., and S. P. Long, eds. *Photosynthesis in Contrasting Environments.* Elsevier Science Publishers, Amsterdam, the Netherlands. Pp. 199–236.

Kellomäki, S., and P. Oker-Blom. 1983. Canopy structure and light climate in a young Scots pine stand. *Silva Fennica* 17:1–21.

Kuruoiwa, S. 1970. Total photosynthesis of foliage in relation to inclination of leaves. In: Setlik, I., ed. *Prediction and Measurement of Photosynthetic Productivity.* Pudoc, Wageningen, the Netherlands. Pp. 79–89.

Lang, A. R. G., and X. Yueqin. 1986. Estimation of leaf area index from transmission of direct sunlight in discontinuous canopies. *Agricultural and Forest Meteorology* 37:229–243.

Lemeur, R. 1973. A method for simulating the direct solar radiation regime in sunflower, Jerusalem artichoke, corn and soybean canopies using actual stand structure data. *Agricultural Meteorology* 12:229–247.

Lemeur, R., and B. L. Blad. 1974. A critical review of light models for estimating the shortwave radiation regime of plant canopies. *Agricultural Meteorology* 14:255–286.

Myneni, R. B., and I. Impens. 1985. A procedural approach for studying the radiation regime of infinite and truncated foliage spaces. Part I. Theoretical considerations. *Agricultural and Forest Meteorology* 33:323–337.

Nilson, T. 1971. A theoretical analysis of the frequency of gaps in plant stands. *Agricultural Meteorology* 8:25–38.

Norman, J. M. 1979. Modeling the complete crop canopy. In: Barfield, B. J., and J. F. Gerber, eds. *Modification of the Aerial Environment of Plants.* Monograph No. 2, American Society of Agricultural Engineers, St. Joseph, MI, U.S.A. Pp. 249–277.

Norman, J. M. 1980. Interfacing leaf and canopy light interception models. In: Hesketh, J. D., and J. W. Jones, eds. *Predicting Photosynthesis for Ecosystem Models,* Vol. II. CRC Press, Boca Raton, FL, U.S.A. Pp. 49–67.

Norman, J. M., and G. S. Campbell. In press. Canopy structure. In: Pearcy, R. W., et al., eds. *Plant Physiological Ecology: Field Methods and Instrumentation.*

Norman, J. M., and P. G. Jarvis. 1974. Photosynthesis in Sitka spruce (*Picea sitchensis* [Bong.] Carr.). III. Measurements of canopy structure and interception of radiation. *Journal of Applied Ecology* 11:375–398.

Norman, J. M., and P. G. Jarvis. 1975. Photosynthesis in Sitka spruce (*Picea sitchensis* [Bong.] Carr.). V. Radiation penetration theory and a test case. *Journal of Applied Ecology* 12:839–878.

Norman, J. M., and J. M. Welles. 1983. Radiative transfer in an array of canopies. *Agronomy Journal* 75:481–488.

Oker-Blom, P. 1985. The influence of penumbra on the distribution of direct solar radiation in a canopy of Scots pine. *Photosynthetica* 19:312–317.

Oker-Blom, P. 1986. Photosynthetic radiation regime and canopy structure in modeled forest stands. *Acta Forestalia Fennica.* 197. 44 pp.

Oker-Blom, P., and S. Kellomäki. 1982. Effect of angular distribution of foliage on light absorption and photosynthesis in the plant canopy: theoretical computations. *Agricultural Meteorology* 26:105–116.

Oker-Blom, P., and S. Kellomäki. 1983. Effect of grouping of foliage on within-stand and within-crown light regime: comparison of random and grouping canopy models. *Agricultural Meteorology* 28:143–155.

Rook, D. A., J. C. Grace, P. N. Beets, D. Whitehead, D. Santantonio, and H. A. I. Madgwick. 1985. Forest canopy design: biological models and management implications. In: Cannell, M. G. R., and J. E. Jackson, eds. *Atrributes of Trees As Crop Plants.* Institute of Terrestrial Ecology, Hunting-

don, England, U.K. Pp. 507–524.
Ross, J. 1981. The radiation regime and architecture of plant stands. Dr. W. Junk Publishers, Dordrecht, the Netherlands. 391 pp.
Terjung, W. H., and S. S. F. Louie. 1972. Potential solar radiation on plant shapes. *International Journal of Biometeorology* 16:25–43.
Thornley, J. H. M. 1976. *Mathematical Models in Plant Physiology.* Academic Press, London, England, U.K.
Von Caemmerer, S., and G. D. Farquhar. 1981. Some relationships between the biochemistry of photosynthesis and the gas exchange of leaves. *Planta* 153:376–387.
Weiss, A., and J. M. Norman. 1985. Partitioning solar radiation into direct and diffuse, visible and near-infrared components. *Agricultural and Forest Meteorology* 34:205–213.

Appendix 1. Number of stems per ha, crown shape, and leaf area related to annual net photosynthesis for simulation runs.

Group	Crown width (m)	Crown length (m)	Spacing (m)	Stems per ha	Leaf area per tree (m^2)	Leaf area index (LAI)	Annual net photosynthesis (kg C/tree)	(Mg C/ha)
a	2.4	7.36	2.15	2163	20.80	4.5	16.6	36.0
					16.18	3.5	15.4	33.2
					14.33	3.1	14.6	31.6
					12.54	2.7	13.8	29.8
					10.75	2.3	12.8	27.6
					8.96	1.9	11.5	24.9
					4.14	0.9	6.7	14.4
					1.84	0.4	3.3	7.2
b	2.4	3.68	2.15	2163	14.33	3.1	14.5	31.5
		7.36					14.6	31.6
		11.04					14.8	32.0
c	9.6	7.36	9.6	108	285.7	3.1	263.2	28.6
	4.8		4.8	434	71.42		69.5	30.1
	2.4		2.4	1736	17.86		17.8	30.9
	1.2		1.2	6944	4.46		4.6	31.9
	0.6		0.6	27778	1.1		1.2	34.5
d	2.4	7.36	2.15	2163	14.33	3.1	14.6	31.6
	1.2						12.6	27.3
	0.6						9.5	20.6
e	0.6	3.68	2.15	2163	14.33	3.1	6.1	13.3
		11.04					11.0	23.9
f	1.2	3.68	2.15	2163	14.33	3.1	10.5	22.7
		11.04					13.3	28.7
g	2.4	7.36	2.15	2163	14.33	3.1	14.6	31.6
			2.4	1736		2.5	16.1	28.0
			2.5	1600		2.3	16.9	26.7
			4.0	625		0.9	21.1	13.2
			6.0	278		0.4	22.7	6.3
			10.13	97		0.1	23.4	2.3
			10.66	88		0.1	23.4	2.1
			11.00	83		0.1	23.4	1.9
h	1.2	7.36	2.15	2163	12.54	2.7	12.0	25.9
					10.75	2.3	11.2	24.3
					8.96	1.9	10.3	22.3
i	2.4	3.68	6.98	205	14.33	0.3	20.2	4.1b[1]
		7.36	10.66	88		0.1	23.4	2.1a
		11.04	14.34	48		0.1	24.9	1.2c
j	1.2	7.36	9.46	112	14.33	0.2	18.9	2.1e
	0.6		8.86	127		0.2	12.9	1.6d

[1]Letters correspond to isolated trees with similar characteristics listed in Appendix 2.

Appendix 2. Crown shape, leaf area, and modeled estimates of annual net photosynthesis for isolated trees.

Tree	Crown width (m)	Crown length (m)	Crown surface area (m^2)	Foliage surface area (m^2)	Annual net photosynthesis (kg C/tree)
1	2.4	7.36	45.4	7.16	13.0
				14.33	23.6 *a[1]
				28.66	39.0
				111.00	60.5
				113.50	60.5
				244.17	51.8
2	2.4	3.68	24.78	7.16	12.1
				14.33	20.4 *b
				28.66	29.3
				40.0	32.2
				52.0	33.4
				61.95	33.6
3	2.4	11.04	66.68	7.16	13.4
				14.33	25.0 *c
				28.66	43.8
				140.0	87.9
				150.0	88.3
				166.7	88.5
4	0.6	7.36	10.9	7.16	9.3
				14.33	12.4 *d
				20.80	13.0
				27.3	12.8
				28.66	12.7
5	1.2	7.36	22.0	14.3	19.0 *e
				18.0	21.6
				26.0	25.3
				48.0	28.5
				55.1	28.5
				60.0	28.4

[1]Letters correspond to trees with similar characteristics within forest stands reported in Appendix 1.

Appendix 3. Model estimates of annual canopy net photosynthesis for stands where 50% of the trees have been defoliated by a given percentage, and for stands with the same leaf area index where all trees have normal foliage levels.

Crown width (m)	Crown length (m)	Spacing (m)	Leaf area index (LAI)	Defoliation of crown (%)	Annual net photosynthesis: Defoliated stand (Mg C/ha)	Annual net photosynthesis: Normal stand (Mg C/ha)
2.4	7.36	2.15	2.7	25	29.7	29.8
			2.3	50	27.2	27.6
			1.9	75	24.2	24.9
1.2	7.36	2.15	2.7	25	25.7	25.9
			2.3	50	23.7	24.3
			1.9	75	20.8	22.3

CHAPTER 12

MODELING DORMANCY RELEASE IN TREES FROM COOL AND TEMPERATE REGIONS

Heikki Hänninen

Abstract. Principles and examples of a modeling approach to dormancy release are presented for trees from cool and temperate regions. The models are based on accumulation of developmental units, which in turn are based on temperature data. Chilling units are used to simulate the effect of low temperatures on rest break, and forcing units to simulate the effect of high temperatures on dormancy release. The predicted dates of rest break and dormancy release are obtained as output. According to the simulations, both genetic properties of the trees and temperature conditions during autumn, winter, and spring affect the timing of dormancy release and the subsequent risk of frost damage. The predicted climatic warming is found to drastically increase the risk of frost damage.

INTRODUCTION

The timing of dormancy release is an essential part of the climatic adaptation of trees in cool and temperate regions. Early dormancy release is accompanied by a high risk of frost damage, while late dormancy release entails the loss of part of the season favorable for growth. The timing of dormancy release is determined as an interaction between the genetic properties of the tree and the environmental factors prevailing at the growing site. This interaction has given rise to various models (e.g., Sarvas, 1972, 1974; Richardson et al., 1974; Landsberg, 1974). This chapter presents principles and applications of the modeling approach to the dormancy release of trees in cool and temperate regions.

THE MODEL

In models of dormancy release, the developmental stage of the tree is usually described by one cumulative variable. The daily increment in this variable depends on the environmental factors prevalent during the day (Hänninen, 1987). The present study uses a slightly modified version of the model by Sarvas (1972, 1974) (Figures 1 and 2).

In the annual ontogenetic development of the buds, three point events are distinguished, dividing the cycle into three segment events (Figure 1). During the rest period, the buds remain in a state of dormancy regardless of the prevailing environmental factors.

Dr. Hänninen is a Research Scientist, Faculty of Forestry, University of Joensuu, P.O. Box 111, SF-80101 Joensuu, Finland. The author wishes to thank Marja-Leena Jalkanen for drawing the figures and Joann von Weissenberg for linguistic revision of the text.

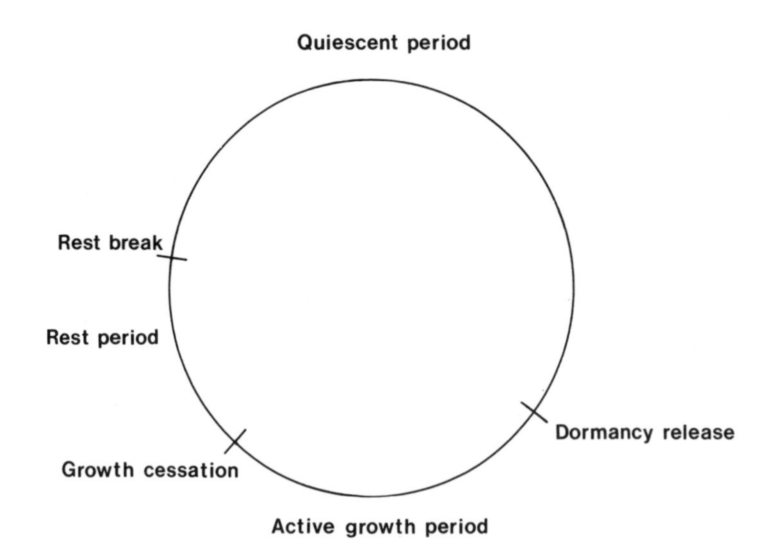

Figure 1. Point events and segment events in the annual developmental cycle of trees in cool and temperate regions.

During the quiescent period, the buds are dormant due to low ambient temperature.

Rest is broken by prolonged exposure to chilling temperatures. This phenomenon is simulated by the accumulation of a chilling unit sum (CU sum). Starting annually on September 1, the daily increment in the CU sum is calculated as follows (Figure 2a):

$$\text{CU/day} = \begin{cases} 0, & T \leq -3.4 \\ 0.159 \times T + 0.506, & -3.4 < T \leq 3.5 \\ -0.159 \times T + 1.621, & 3.5 < T \leq 10.4 \\ 0, & T > 10.4 \end{cases} \quad (1)$$

where T is daily mean temperature. When the CU sum attains a genotype-specific critical value, CU_{crit}, rest is broken. After that, during the quiescent period, high temperatures are required for dormancy release. This part of the cycle is simulated by accumulating a forcing unit sum (FU sum). The daily increment in the sum is calculated as follows (Figure 2b):

$$\text{FU/day} = \begin{cases} 0, & T \leq 0 \\ \dfrac{28.361}{1 + e^{-0.185(T-18.431)}}, & T > 0 \end{cases} \quad (2)$$

where T is the mean daily temperature. Dormancy release takes place when the FU sum attains a genotype-specific critical value, FU_{crit}.

TEMPERATURE DATA

Daily mean and minimum temperatures measured in standard meteorological screens in Luonetjärvi, central Finland (62°55′N, 25°39′E), were used in the simulations. The data covered the period 1961–1978.

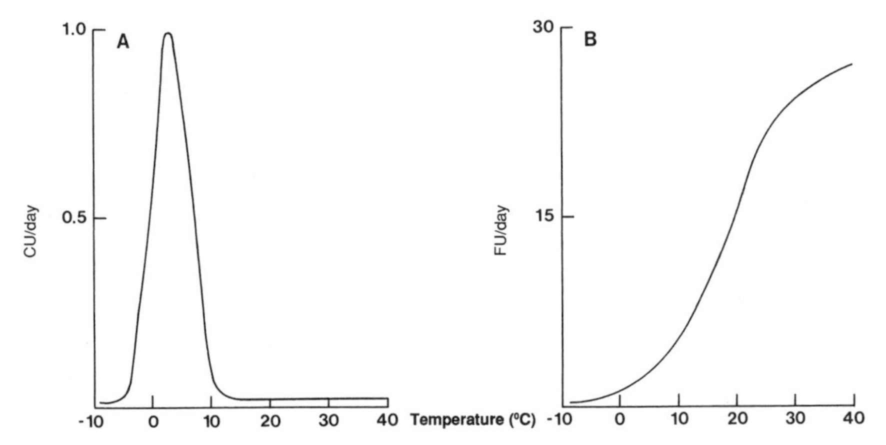

Figure 2. Dependence of the accumulation rate of developmental units on daily mean temperature during rest period (*a*) and quiescent period (*b*). CU = chilling unit, FU = forcing unit. Curves are fitted to the data of Sarvas (1972, 1974) (Equations 1 and 2).

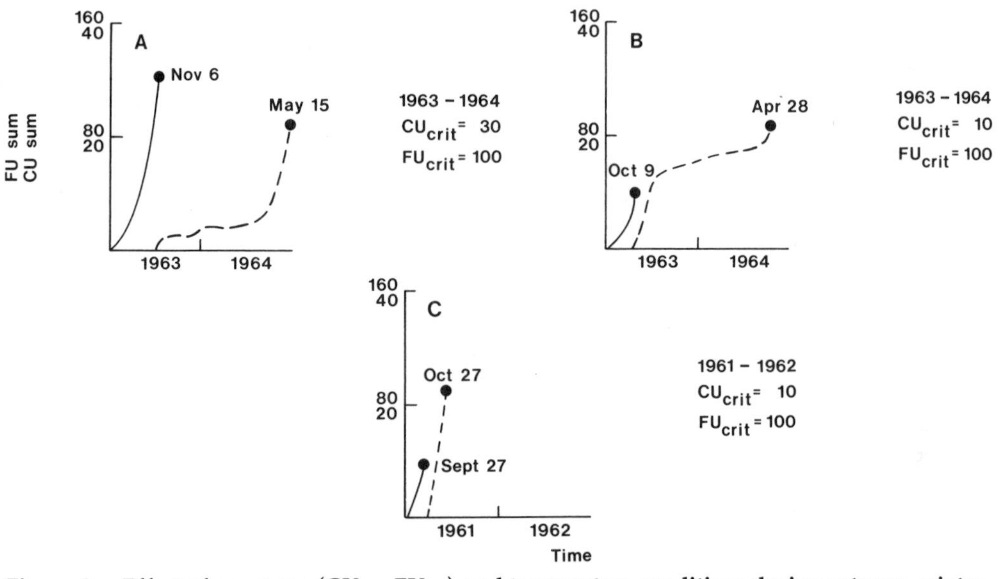

Figure 3. Effect of genotype (CU_{crit}, FU_{crit}) and temperature conditions during autumn, winter, and spring on the timing of dormancy release. *Solid line* (—): accumulation of chilling unit sum. *Broken line* (---): accumulation of forcing unit sum. In each figure the earlier date indicates the timing of the rest break, and the later date the timing of dormancy release.

SIMULATIONS

Timing of Dormancy Release

The timing of dormancy release depends on (1) the genetic properties of the trees (CU_{crit}, FU_{crit}) and (2) the temperature conditions at the growing site (temperature data).

The value of CU_{crit} regulates the starting date for accumulation of the FU sum. Consider the simulated development during 1963–1964 with $FU_{crit} = 100$. With a relatively high value of CU_{crit}, rest is broken so late that little subsequent accumulation of FU sum takes place during the autumn (Figure 3a). With a low value of CU_{crit}, rest is broken relatively early (Figure 3b). In that case, considerable accumulation of FU sum takes place during the autumn, due to the relatively high temperatures still prevailing after the rest break. As a result, dormancy release takes place more than 2 weeks earlier with the low value of CU_{crit} (Figure 3b) than with the high value (Figure 3a).

The year-to-year variation in temperature conditions during autumn, winter, and spring causes variation in the development of a given genotype growing at a given site. Consider the simulated development with parameter values $CU_{crit} = 10$ and $FU_{crit} = 100$. During 1961–1962, dormancy release takes place during the autumn (Figure 3c). This is because the relatively early attainment of the critical chilling unit sum is followed by a warm period, which in turn causes the attainment of the critical forcing unit sum. In contrast, during 1963–1964, the critical forcing unit sum is not attained until spring (Figure 3b).

Risk of Frost Damage

In any given temperature conditions, the genetic properties of the trees (CU_{crit}, FU_{crit}) determine the timing of dormancy release (Figure 3). In that way, they also determine the minimum temperature to which the trees are exposed after dormancy release. This makes it possible to assess, with the aid of the model, the risk of post-dormant frost damage.

The date of dormancy release was calculated using 48 combinations of CU_{crit} and FU_{crit} for each of the 17 years represented in the temperature data. After that, the minimum temperature during the period between the date of dormancy release and August 1 was determined for each parameter combination and year. Thus, the distribution of the annual minimum temperatures was obtained for each parameter combination. The minimum value for this distribution indicates the overall minimum temperature (for the whole simulation period) to which the tree is exposed after dormancy release (minimum post-dormant temperature).

The minimum post-dormant temperature increases with increasing values of either of the parameters CU_{crit} or FU_{crit} (Figure 4), indicating a decreased risk of frost damage. The effects of the parameters are, however, interdependent. If one of the parameters has a high value, then the minimum post-dormant temperature is relatively insensitive to the value of the other parameter (Figure 4).

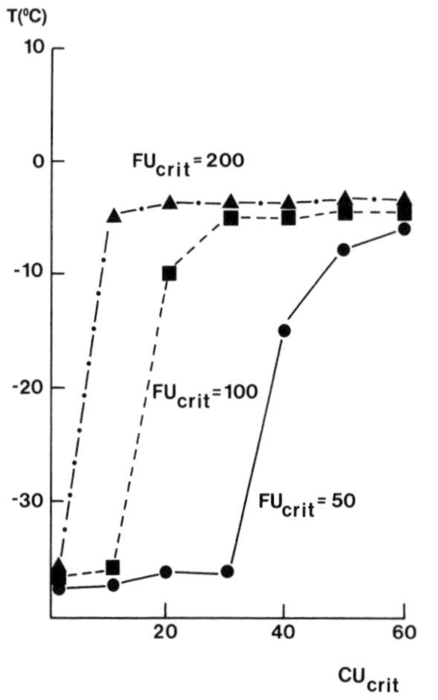

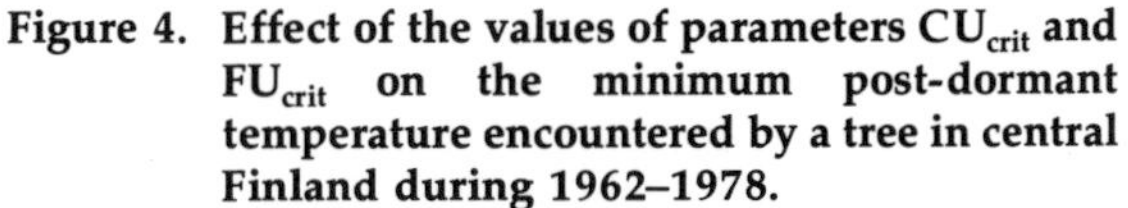
Figure 4. Effect of the values of parameters CU_{crit} and FU_{crit} on the minimum post-dormant temperature encountered by a tree in central Finland during 1962–1978.

Climatic Warming and Risk of Frost Damage

It has been predicted that the temperature of the earth will increase substantially during the next decades. At high latitudes, it has been predicted that the warming will occur especially during winter (Kettunen et al., 1987; Figure 5). The effect of this climatic warming on the risk of post-dormant frost damage was assessed by running another set of calculations with the 48 combinations of CU_{crit} and FU_{crit}, using modified temperature data. The values for both daily mean and minimum temperatures were increased according to the scenario shown in Figure 5.

According to Repo and Pelkonen (1986), the frost hardiness of central Finnish Scots pine (*Pinus sylvestris* L.) seedlings is −10°C at the time of dormancy release. When the results of the two sets of calculations were analyzed, this value was used as the threshold for acceptance of the minimum post-dormant temperature (Figure 6). In the simulation with measured temperature data, most of the parameter combinations tested produced an acceptably high value for post-dormant minimum temperature (Figure 6a); while in the simulation with the scenario temperature data, this was the case with only two combinations of the highest parameter values (Figure 6b).

The value of CU_{crit} is 30–50 units for central Finnish Scots pine seedlings (Hänninen and Pelkonen, 1988). In the conditions of central Finland, dormancy for Scots pine is usually released in mid-May (Repo, 1986). At that time the FU sum accumulation is about 100 units (Figure 3a). Thus the parameter combinations of real trees clearly are among the accepted parameter combinations in the present temperature conditions (Figure 6a), but not in the scenario conditions (Figure 6b). Accordingly, we may assume that the risk of post-dormant frost damage is drastically increased by the predicted climatic warming.

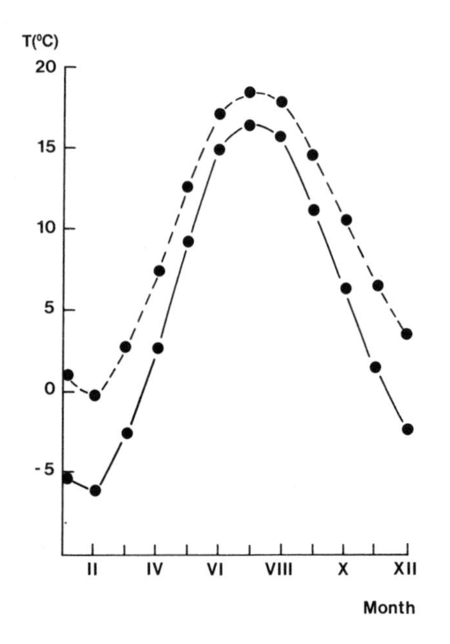

Figure 5. Mean monthly temperatures in Helsinki, southern Finland. ***Solid line*** **(—): measured for 1951–1980.** ***Broken line*** **(---): predicted under doubled concentration of atmospheric CO_2 (Kettunen et al., 1987).**

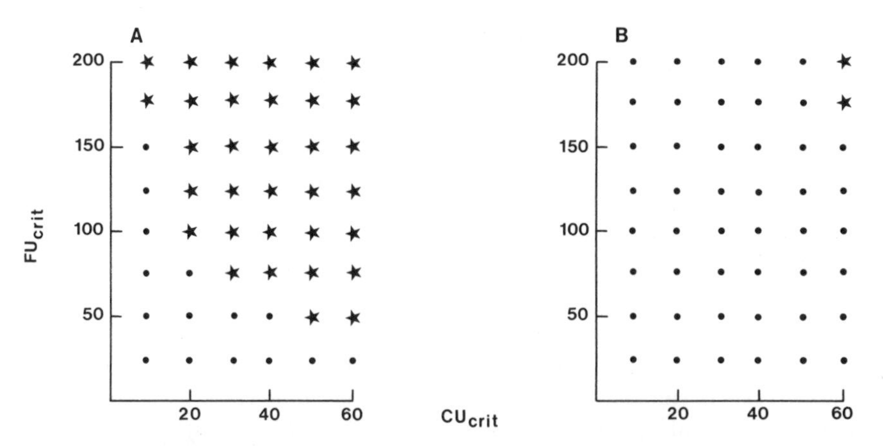

Figure 6. Combinations of parameter values producing a minimum post-dormant temperature greater (*stars*) and lower (*dots*) than −10°C in the conditions of central Finland during the years 1962–1978. (*a*) Simulation with measured temperature data. (*b*) Simulation with modified temperature data (the scenario shown in Figure 5).

DISCUSSION

In the present study, the model of Sarvas (1972, 1974) was used as an example of the modeling approach to dormancy release in trees of cool and temperate regions. In the literature, several other models are found, all in some respects contradictory to one another. All models, however, share the common approach of accumulating chilling and forcing units.

None of the models considered in this study takes photoperiod into account, although photoperiod is known, in some cases, to affect dormancy release (e.g., Worrall and Mergen, 1967). Nevertheless, the effect of photoperiod can be added into the models without changing the basic approach.

At present, the models of dormancy release serve as a useful framework for designing experiments (Hänninen, 1987). If a rigorously tested model is available, together with long-term temperature data, then the modeling approach can be used to solve many practical problems. These involve, for instance, provenance transfers (Campbell, 1974), tree breeding (Cannell et al., 1985), and consequences of climatic change (Cannell and Smith, 1986). In all these applications, the interaction between the genetic properties of the trees and the climatic factors prevailing at the growing site is of crucial importance.

The present results concerning the effects of climatic warming are not conclusive, because not all the assumptions in the underlying model have been tested. It is possible, for instance, that the FU sum accumulated during autumn (Figure 3) is no longer effective during spring. The model, however, contains many of the features regulating dormancy release. Thus, climatic warming will obviously cause marked changes in the timing of dormancy release, resulting in an increased risk of frost damage (cf. Cannell and Smith, 1986). In this new situation, the factors regulating the dormancy release of various species and provenances should be known. Simulation models provide a useful quantitative tool for understanding this process of regulation.

LITERATURE CITED

Campbell, R. K. 1974. Use of phenology for examining provenance transfers in reforestation of Douglas-fir. *Journal of Applied Ecology* 11:1069–1080.

Cannell, M. G. R., M. B. Murray, and L. J. Sheppard. 1985. Frost avoidance by selection for late budburst in *Picea sitchensis. Journal of Applied Ecology* 22:931–941.

Cannell, M. G. R., and R. I. Smith. 1986. Climatic warming, spring budburst and frost damage on trees. *Journal of Applied Ecology* 23:177–191.

Hänninen, H. 1987. Effects of temperature on dormancy release in woody plants: implications of prevailing models. *Silva Fennica* 21(3):279–299.

Hänninen, H., and P. Pelkonen. 1988. Effects of temperature on dormancy release in Norway spruce and Scots pine seedlings. *Silva Fennica* 22(3):241–248.

Kettunen, L., J. Mukula, V. Pohjonen, O. Rantanen, and U. Varjo. 1987. The effect of climatic variations on agriculture in Finland. In: Parry, M. L., T. R. Carter, and N. T. Konijn, eds. *The Impact of Climatic Variations on Agriculture,* Vol. 1. *Assessments in Cool Temperate and Cold Regions.* Reidel, Dordrecht, the Netherlands. Pp. 1–90.

Landsberg, J. J. 1974. Apple fruit bud development and growth: analysis and an empirical model. *Annals of Botany* 38:1013–1023.

Repo, T. 1986. Impedance-estimated frost resistance of pine and spruce during one year. *Acta Universitatis Ouluensis* A 179:183–186.

Repo, T., and P. Pelkonen. 1986. Temperature step response of dehardening in *Pinus sylvestris* seedlings. *Scandinavian Journal of Forest Research* 1:271–284.

Richardson, E. A., S. D. Seeley, and D. R. Walker. 1974. A model for estimating the completion of rest for 'Redhaven' and 'Elberta' peach trees. *HortScience* 9(4):331–332.

Sarvas, R. 1972. Investigations on the annual cycle of development of forest trees. Active period. *Communicationes Instituti Forestalia Fenniae* 76(3):1–110.

Sarvas, R. 1974. Investigations on the annual cycle of development of forest trees. II. Autumn dormancy and winter dormancy. *Communicationes Instituti Forestalia Fenniae* 84(1):1–101.

Worrall, J., and F. Mergen. 1967. Environmental and genetic control of dormancy in *Picea abies. Physiologia Plantarum* 20:733–745.

SECTION III

Model Structure and Evaluation

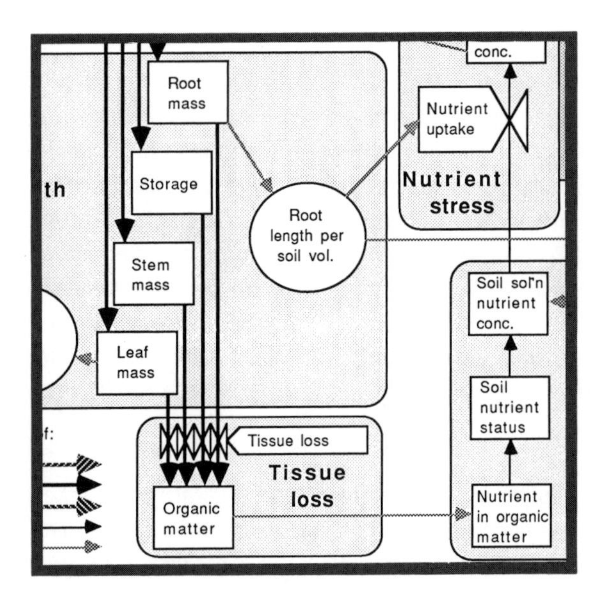

Scientists and policy-makers define process modeling differently, as reflected in the contributions to this section. At one end of the spectrum is a formal definition involving levels of organization (e.g., a process model simulates processes occurring at a more detailed level than the information to be output). In contrast, there is the notion of an empirical model in which the parameters can be given a biological or physical interpretation which is viewed as process-based. Since process modeling of biological phenomena is still in its formative stages, the lack of a recognized, rigorous definition is not surprising, and perhaps at this point not even desirable. Authors with divergent backgrounds have written the chapters in this section. Collectively, their work helps to define and clarify the issues of model structure and evaluation.

Models are typically criticized for failing to reflect reality. If they are to be used for decision-making, in the broadest sense of the term, their limitations must be properly understood. This requires identifying potential sources of error and quantifying how these errors can influence model output. The methodology for error quantification and propagation has much in common with that developed for controlling quality in manufacturing processes. These and other questions concerning model validation and internal consistency are addressed in Section III.

CHAPTER 13

MODEL STRUCTURE AND DATA BASE DEVELOPMENT

Basil Acock and James F. Reynolds

Abstract. We need to find ways of involving more researchers in model development and of improving communication among them. Most of the plant models in existence today have been developed by individuals or small groups of people. The developers have necessarily had a limited understanding of the processes involved, limited time, and a limited amount of data for building and testing their models. One method to increase participation in model development would be to identify the key processes that need to be incorporated in our models, group related processes into modules to provide a generic modular structure, and specify what the output from each of the modules should be. Given agreement on a structure, any researcher could concentrate on those modules within his area of specialization, collect the data needed to develop or validate those modules, and write improved versions. Through the involvement of specialists, our best understanding of each process would be recorded in code. As we learn more, existing models could easily be updated.

It is also highly desirable to agree on a minimum data set for model validation. The data set should include all variables used by any model, except those variables that are model-dependent. Specifying the format would also make sharing the validation data easier. Data sets for model development cannot be specified in the same way because the data collected depend on the hypotheses being tested.

Publishing large data bases or large amounts of code in scientific journals is not possible because of space limitations. This makes it difficult to share code and data and to benefit from peer comment. One solution is to publish models in a peer-reviewed electronic journal, including the full code, documentation, and validation data sets. Efforts to start such a journal are under way.

INTRODUCTION

In the last five years, modeling has become an acceptable and even fashionable activity among agricultural scientists. Prior to that, the field-plot experimenters (or mensurationists) held sway, and conventional wisdom was that all questions about the management of farm, range, and forest could and should be answered by direct observation in field experiments. Modeling, with some justification, was equated with mindlessly fitting multiple regression equations to field data, and was considered to be a passing fad. Modelers huddled together in small groups like medieval monks and fended off a hostile world. To justify their activities, many made extravagant claims based more on their belief in the potential of models than on actual achievements.

All this has changed. There is now a realization that some experiments are too expen-

Dr. Acock is Research Leader, USDA, ARS, Systems Research Laboratory, BARC-W, Beltsville, MD, 20705 U.S.A. Dr. Reynolds is Director, Systems Ecology Research Group, San Diego State University, San Diego, CA, 92182 U.S.A.

sive or dangerous in terms of their consequences for society. For example, we cannot afford to experiment with groundwater pollution or the effect of CO_2 and other greenhouse gases on climate. Models can be used as surrogates for plants in such experiments. They can also be used to capture our fundamental knowledge about plant behavior in computer code, a form that makes it available to lay users. In this new environment, modelers are communicating more freely about their successes and failures, sharing ideas, and exchanging code. Other scientists, who are not necessarily very comfortable with computers, mathematics, and code, want to know what hypotheses the models contain and what they can do to improve them. However, we are all hampered by model incompatibility and the lack of a formal medium for publishing models. In this chapter, we discuss these and other current problems from the perspectives of a horticulturist (B.A.) and an ecologist (J.F.R.) and offer some possible solutions.

THE PRESENT TECHNOLOGY TRANSFER SYSTEM

At present, most research results end up as articles in peer-reviewed journals (Figure 1). If these are read at all, they are read by other scientists. Occasionally, results of the more applied research go into journals which are read by extension specialists and agents and others concerned with advising the forest manager. These specialists reinterpret the information and pass it on to the manager in the form of general advice, which usually gives good results for the climate and soils of the area where the manager operates. Some research produces new cultivars or "silver bullet" solutions to specific problems, and these results are taken straight to the forest and applied there. Examples of such direct application of research are exceptions to the rule and account for about 2% of research effort. With the present system of communicating research results, very little of what we know about plant behavior is communicated in a useful way to the end user. Process-level models can be used to capture some of this information and make it available.

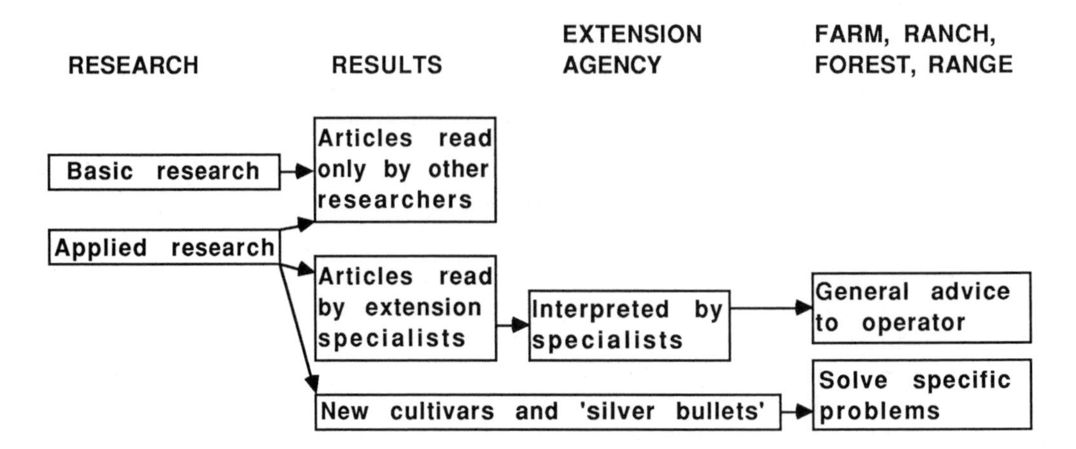

Figure 1. Present technology transfer system for conveying research results to forest managers.

In addition, we now have expert systems to capture heuristic knowledge: understanding that comes from long experience of the behavior of the forest system even though we do not understand all the processes that cause that behavior. Expert systems are computer programs that use knowledge in the form of rules to mimic the reasoning processes of a human expert in a limited domain. With process models to capture our hard knowledge about the system behavior and expert systems to capture our soft, more heuristic, knowledge about that system, we can put together a package that can be run on a computer in the forest manager's office and give specific advice for the species grown on the soil type in his forest. In effect, the process model becomes one of the rules in the expert system.

Such a package has been developed for the cotton crop with COtton MAnagement eXpert/*Gossypium* simulation (COMAX/GOSSYM) (Lemmon, 1986). GOSSYM (Baker et al., 1983) is a cotton-crop process model, and COMAX is an expert system. COMAX runs GOSSYM up to the current day of the season, using actual soil, weather, and management data for the site. It then uses several sets of historic weather data to project forward in time to final harvest. COMAX optimizes the amounts and timing of irrigation and fertilizer applications to maximize yield without vegetative regrowth at the end of the season. It also predicts the best time to apply the harvest aid chemicals that start the harvesting process. This software package has been quite successful and is now in use on about 200 farms across the U.S.A. cotton belt.

PROBLEMS WITH CURRENT METHODS OF MODEL DEVELOPMENT

Given the great potential of process models for capturing our understanding of plant behavior, why are so few researchers involved in building them? Most of the models in existence today have been built by individuals or small groups. This is because building a process model is a complex and difficult task. The builders have to assume that they know enough about soil physics, plant physiology, celestial mechanics, and so on, to describe all the processes correctly. Model development takes a long time—in some cases, the whole of a person's working life. Modeling results in few peer-reviewed articles because journals do not publish large amounts of code or the necessary documentation. As a result, most models are published in experiment station bulletins, with only a superficial description of the model appearing in a glossy journal.

Even when other researchers want to get involved in improving a model, they find that the models are monolithic and idiosyncratic. It is almost impossible for them to disentangle the part of interest to them so that they can work on that part alone. Thus, because so few researchers are involved in their construction, existing models are limited by their author's understanding of the processes and the availability of data for development and validation. In order to realize the full potential of process-level models to capture our knowledge of plant behavior in a useful form, we must change the way we build them so that more people participate in the building.

Here is an analogy to explain why change is needed and what the nature of that change might be. The small town of Harper's Ferry, West Virginia, U.S.A., stands at the confluence of the Shenandoah and Potomac Rivers. In 1801, George Washington established an armory at Harper's Ferry to take advantage of the abundant water-power; and in a few years, production reached 10,000 muskets per year. Every musket was made by a master craftsman, by hand, from beginning to end. The master craftsmen served a 10-year apprenticeship. Since the guns were hand-made, each was slightly different; when a part broke in use, the guns had to be returned to the armory for repair.

Then, in 1819, John Hall proposed making rifles from standardized, interchangeable components. The advantages were that the components could be made by less skilled laborers, and that broken rifles could be repaired in the field from a stock of standard spare parts. Gauges and government standards ensured that the components were completely interchangeable. This was the first known instance in the U.S.A. of a standard being adopted to assure interchangeability of parts. Today we take it for granted that, for instance, whichever make of lamp we buy, its plug will fit the sockets in our house.

A GENERIC MODULAR STRUCTURE

Process modeling is still at the master craftsman stage. One way of getting more people involved in building the models would be to identify standard interchangeable components or modules. In this case, of course, we are not interested in using unskilled labor to build the components, but rather in capturing the expertise of people who specialize in those limited areas.

Criteria for Separating the Modules

We suggest that suitable criteria for developing the generic modular structure for plant models are the following:

1. that modules should separate along disciplinary lines;
2. that modifying one module should not necessitate changes in any other modules;
3. that modules should have a minimum number of input and output variables;
4. that these input and output variables should be measurable properties of the system;
5. that each module should be organized internally in a hierarchical structure so that each module can be as simple or as complex as desired; and
6. that the structure should be suitable for modeling all plant species.

Proposed Generic Modular Structure

On the basis of these criteria and after examining existing models for common elements, we propose the modular structure shown in Table 1. The components of the structure are divided between two major categories: environmental processes and plant physiology. Apart from light interception, which is a function of plant geometry, the environmental process modules are completely independent of the plant growing in the system. Once we have satisfactory modules for aerial environment and soil environment, these can be used for all plants without amendment. This will enable us to save a good deal of time in generating new plant models. In the category of plant physiology, some modules, such as potential growth of organs, will probably differ for each species, although there will undoubtedly be similarities in the code for such modules. Other modules, such as carbon fixation, need exist only in versions describing the main photosynthetic pathways: C_3, C_4, and crassulacean acid metabolism (CAM).

For each module, it will be necessary to specify the exact function of the module and the essential output. Table 2 shows the possible function, input, and output variables for the light interception module. Similar specifications have been prepared for all the modules shown in Table 1. In fact, it is not necessary to specify the input variables. A modeler can use as input any variables that occur as output from any other module or variables that we can reasonably expect to have measured.

The hierarchical structure of the modules is best seen under the heading "Soil Environment" in Table 1. The system of indentation shows the different levels in the hierarchy. For many forest soils, it would be desirable to have the full complexity shown in Table 1. However, in, say, a tropical rain forest, it may not be necessary to deal with soil water in great detail, and that portion of the proposed modular structure could be greatly simplified or even eliminated.

Table 1. A proposed generic modular structure for plant growth models. (From Reynolds, J. F., B. Acock, R. L. Dougherty, and J. D. Tenhunen. In press. A modular structure for plant growth simulation models. In: Pereira, J. S., ed. *Forest Biomass for Fiber and Energy.* Reproduced with permission from Kluwer Academic, Dordrecht, the Netherlands.)

ENVIRONMENTAL PROCESSES	PLANT PHYSIOLOGY
INITIAL SOIL ENVIRONMENT	DEVELOPMENT (or PHENOLOGY)
AERIAL ENVIRONMENT	CARBON FIXATION
Celestial geometry	Photosynthesis
PAR & diffuse radiation	Respiration
Water vapor pressure	POTENTIAL GROWTH OF ORGANS
LIGHT INTERCEPTION	Vegetative shoot & root
SOIL ENVIRONMENT	Reproductive organs
Soil Water	Storage organs
Plant uptake	CARBON LIMITATIONS TO GROWTH
Evaporation	Initial carbon partitioning
Profile recharge	WATER LIMITATIONS TO GROWTH
Soil water potential	Potential transpiration
Soil nutrients	Actual water uptake
Plant uptake	Leaf water potential
Fertilizer additions	NUTRIENT LIMITATIONS TO GROWTH
Chemical transformations	Plant nutrient supply & demand
Leaching	Distribution of nutrients
Soil mechanical impedance	ACTUAL GROWTH OF ORGANS
Soil temperature	Vegetative shoot & root
Soil oxygen	Reproductive organs
	Storage organs
	TISSUE LOSS
	MORPHOLOGY (or ARCHITECTURE)
	Plant geometry

Table 2. The possible function, input, and output variables for a light interception module.

	LIGHT INTERCEPTION
Function	Calculates radiation interception by crop
Input	Row spacing and orientation
	Crop height
	Fractional cloud cover
	Solar altitude and azimuth
	Total radiation and PAR flux density
	Proportion of radiation that is diffuse
Output	Proportion of total radiation intercepted
	Proportion of PAR intercepted

Potential Benefits of Adopting a Generic Modular Structure

The expected benefits of adopting such a structure are that

1. the models would be more intelligible;
2. specialists could modify modules of interest to them without having to understand how the rest of the model worked;
3. it would be easier to validate individual modules;
4. it would be easier to locate faults in the model;

5. it would be easier to interchange code between models;
6. it would be easier to maintain existing models and continue efforts started by other people;
7. the individual modules could be used for teaching; and,
8. eventually, new crops could be modeled using existing modules developed for plants with similar growth and development.

To illustrate these benefits, we will examine the diagram for an existing model and show how it can be adapted to a modular structure. Landsberg's (1986) diagram for a tree model is reproduced in Figure 2. This diagram differs only slightly from his original version. Figure 2 is typical of such diagrams produced by modelers (including the authors) and suggests a monolithic structure with many interconnections among the processes. Essentially the same model elements are shown in Figure 3, rearranged so that materials flow in almost straight lines up and down the diagram and using the Forrester symbols as originally defined (Forrester, 1961). Also included on Figure 3 are shaded backgrounds showing the various processes grouped into modules. These correspond approximately with the modules given in Table 1. Figure 3 clearly defines the boundaries of each module and shows where the module can be detached from the rest of the model for improvement.

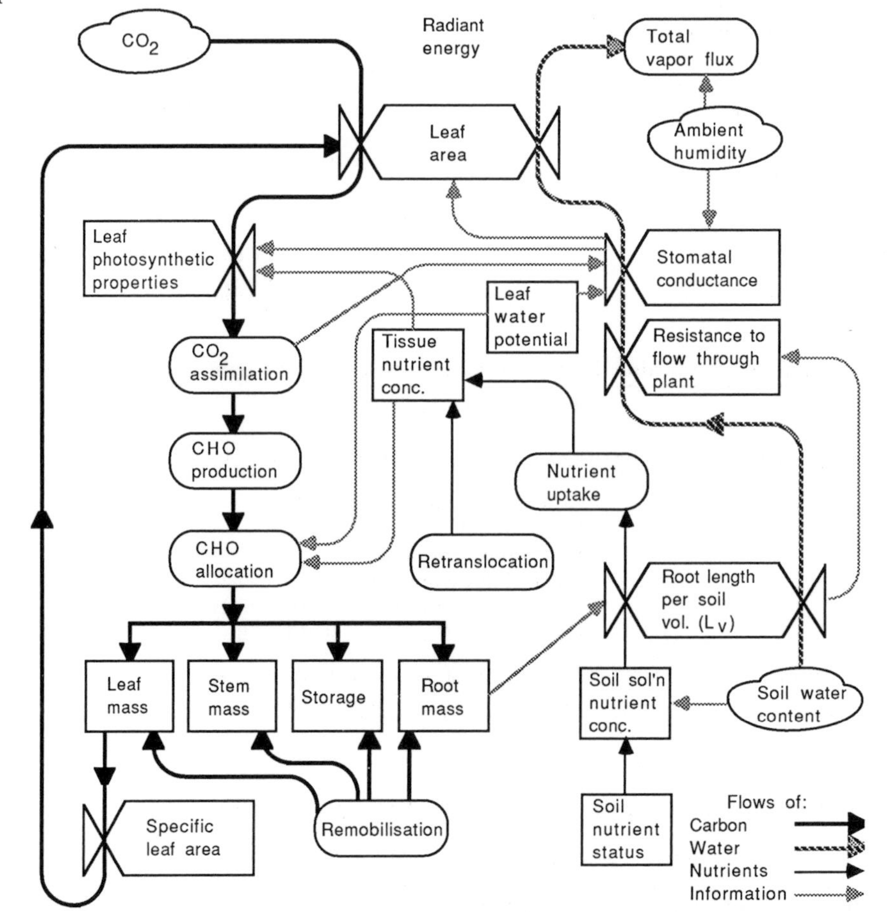

Figure 2. Diagram of Landsberg's model of tree growth (after Landsberg, 1986). (From Reynolds, J. F., B. Acock, R. L. Dougherty, and J. D. Tenhunen. In press. A modular structure for plant growth simulation models. In: Pereira, J. S., ed. *Forest Biomass for Fiber and Energy.* Reproduced with permission from Kluwer Academic, Dordrecht, the Netherlands.)

Although we have placed great emphasis on the importance of process-level or mechanistic modeling, we have also observed that the necessary information to build such models may not be available. This should not deter us from developing our modular structure. However, having developed a structure, we then have to encode each module. For a given module, there may be a good deal of empirical data but no real understanding of the processes involved. In such circumstances, an empirical module may be better than a process-level module—better, that is, until more information about the processes becomes available.

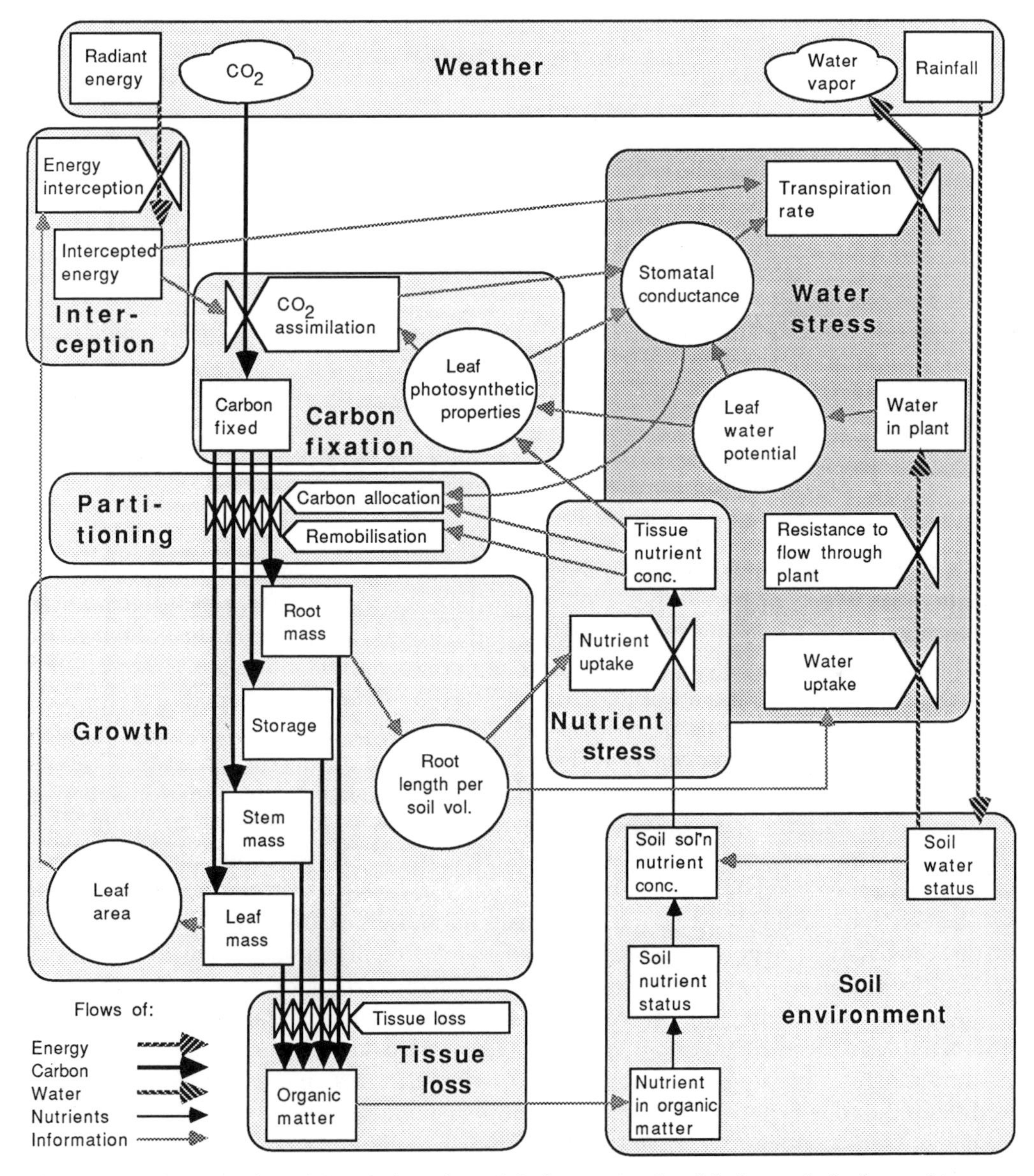

Figure 3. **Diagram of Landsberg's (1986) model of tree growth, with the symbols changed to more closely follow Forrester's (1961) notation, and the elements rearranged to emphasize how they fit the proposed generic modular structure. The shaded areas show which elements are in each module. (From Reynolds, J. F., B. Acock, R. L. Dougherty, and J. D. Tenhunen. In press. A modular structure for plant growth simulation models. In: Pereira, J. S., ed. *Forest Biomass for Fiber and Energy*. Reproduced with permission from Kluwer Academic, Dordrecht, the Netherlands.)**

A FUTURE TECHNOLOGY TRANSFER SYSTEM

Given agreement on a generic modular structure for plant models, researchers can package their experimental results as new modules or algorithms. We envisage the development of a technology transfer system such as that shown in Figure 4. In addition to journal publication and extension system advice, new knowledge will be packaged in process-level models and expert systems for use in computerized crop management packages similar to COMAX/GOSSYM. However, in order for this to happen, we need to amend the reward system.

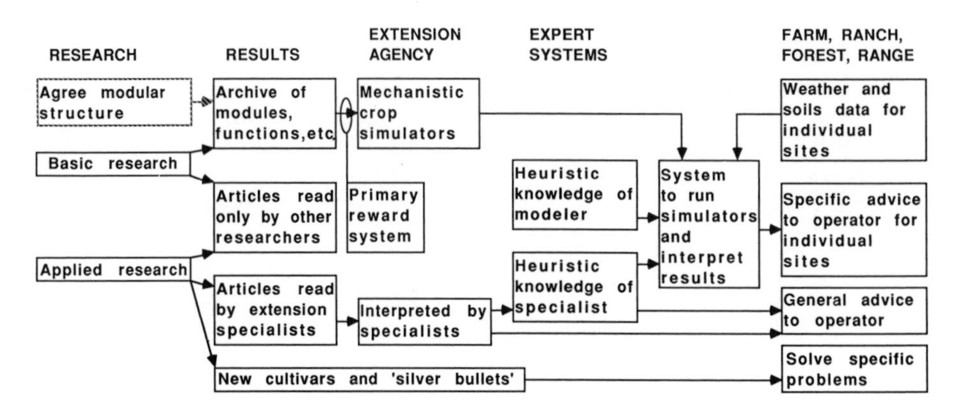

Figure 4. Enhanced technology transfer system for conveying research results to forest managers, using modular plant simulators and expert systems.

Whatever agricultural agencies may say about solving problems for growers, the fact remains that scientists are rewarded for publishing refereed journal articles. Scientists, like most laboratory animals, concentrate on doing things that are rewarding. At present, building models is not very rewarding. Conventional scientific journals will not publish models. However, the mission of the United States Department of Agriculture, Agricultural Research Service (USDA, ARS) Systems Research Laboratory in the Natural Resources Institute at Beltsville, MD, U.S.A., is to "devise means for integrating knowledge into practical technologies that can be transferred to users such as action agencies and state organizations to solve problems in the area of agricultural production, processing and marketing." In partial fulfillment of this mission, we are starting an electronic peer-reviewed journal to publish agricultural system software. The new journal will be electronic so that we do not have to place limitations on the length of an article and so that the code can be down-loaded and used readily. It will be peer-reviewed in order to maintain standards and to ensure that the authors receive credit for their contributions.

A publication in this journal would consist of an abstract, the model code liberally annotated, complete documentation of the ideas and data used, a statement of critical assumptions and model limitations, a dictionary with units, and a validation data set. Although the code is necessarily the most complete definition of the model, there should be enough explanatory material in standard English to allow anyone to understand what hypotheses have been incorporated into the model. This is essential in order to make the model accessible to specialists who do not read code. Only in this way will our models benefit from their criticism.

DATA BASES

Large amounts of data are needed to develop, run, and validate process-level models. Since data collection is expensive, the question immediately arises as to whether we can use existing data bases. Certainly there are weather and soils data bases for running the models, but even here the data are incomplete.

Class A weather stations do not collect solar radiation, which is a vital input to process models. However, the United States National Oceanic and Atmospheric Administration (NOAA) has a program for estimating solar radiation from satellite observations. Also, recent work suggests that it may be possible to estimate solar radiation above the earth's atmosphere from a knowledge of time, day of the year, and latitude of the site, and hence solar radiation at the earth's surface, aided by measurements of cloud cover and visibility (Pat Lestrade, pers. comm., 1988).

The most comprehensive soils data base is that assembled by the USDA Soil Conservation Service. It contains primarily textural information, but statistical correlations have been developed for calculating soil-water release curves and soil hydraulic conductivity from this data base (Saxton et al., 1986).

A few complete soil, weather, and plant data sets have been collected for validating crop models. Participants in the International Benchmark Sites Network for Agroforestry Transfer (IBSNAT) project have agreed on a format for recording such data (IBSNAT, 1986). However, many of the variables in the IBSNAT files are specific to certain models. Moves are under way to modify the file specifications so that the files contain only variables that are properties of the system and that contain all the variables used by existing models. (See Acock and Allen, 1986 for a complete listing of the soil and weather data needed to develop and validate process-level models.)

Only rarely is it possible to use existing data bases to develop process-level models. This is because model development involves testing new hypotheses, and rarely will another researcher have already collected all the data needed to test a new hypothesis.

When using existing data bases, we need to be concerned about the quality of the data. What techniques and equipment were used to make the measurements? Were the researcher and his helpers careful and honest? Did they estimate the values of any missing data? Were there any unusual events during the experiment which might have affected the result? At the moment, we have no method of evaluating the quality of such data. The best we can do is to assess the reliability of the researcher who collected them. Perhaps, in the future, it will be possible to publish these data bases in the electronic journal mentioned earlier. Peer review could then provide some degree of quality assurance.

Fortunately, the technology for capturing and distributing large data bases cheaply is already available. CD-ROM's use the same technology as the familiar music compact discs. Each will hold 670 million characters, the equivalent of 1,200 floppy discs or several large encyclopedias. The cost of CD-ROM's is now down to $2000 for mastering and the first 100 copies, and $5 for each additional copy. CD-ROM readers should eventually be no more expensive than CD music players.

AGGREGATING OVER TIME AND SPACE

Most of the discussion above has dealt with modeling the processes in individual plants. It is appropriate that we should deal with representative individuals in the system. This is because organisms react individually to their environment. They also modify that environment for themselves and other organisms. The reaction of a forest to its environment is the sum of the reactions of all the trees and the ways in which they modify the environment for one another.

Some ecologists believe that as you aggregate up from the tree to the forest level, additional variables that are properties of the forest but not of individual trees need to be considered. They call these "emergent" properties. In the authors' opinion, these are artifacts of approaches that are too empirical.

It is clearly impractical to run a process-level model for every tree in the forest. If the forest is very uniform, we may get by with running the model for one representative tree. If, more typically, the forest contains many species in various stages of growth, models for representative groupings of plants must be run simultaneously. Modeling competition for light, nutrients, and water under these conditions is very difficult, and we still have much to learn. In addition, forests often contain different soil types, slopes, and aspects of terrain.

The process of using single-plant models to simulate a variable ecosystem is called "aggregation" and has been discussed recently by Acock and Reynolds (in press). Essentially, the forest is mapped as a mosaic of cells that differ in the important variables affecting tree growth or differ in species and size groupings. Such maps can be generated using geographical information systems, parts of which come from satellite observations. The variables used to separate the cells in the mosaic can then be subjected to uncertainty analysis. In uncertainty analysis, the range and frequency distribution of each variable is used along with a process-level model to generate a multidimensional response surface. These data are then analyzed to see how much of the variability in response can be attributed to each of the variables. Cells separated only by variables that have little effect on the response may be combined to form a simpler mosaic. In a similar way, uncertainty analysis can be used on the process-level models to identify those model variables that have little effect on the response. These can then be eliminated to simplify the model. Another way of simplifying the process model is to generate a response data set covering the full range of input variables likely to be encountered. These data can then be used to generate an empirical model which will have most of the predictive capability of the original process-level model.

When both the plant model and the mosaic have been simplified to the extent considered permissible, the model is run for each cell type in the mosaic, and the results are weighted for the relative area occupied by each cell type in the forest.

RESOURCES IN THE U.S.A.

The liveliest medium presently available for the exchange of ideas in process modeling is the crop simulation workshop hosted by the Biological Systems Simulation Group at a major university in the U.S.A. in early March each year. Another useful medium is the Agricultural Systems Research Resource, an electronic conference hosted by the USDA, ARS, Systems Research Laboratory at Beltsville, MD, U.S.A. Further information about either of these organizations can be obtained from the senior author.

LITERATURE CITED

Acock, B., and L. H. Allen, Jr. 1986. Crop responses to elevated carbon dioxide concentrations. In: Strain, B. R. and J. D. Cure, eds. *Direct Effects of CO_2 on Vegetation.* U.S. Department of Energy, Carbon Dioxide Research Division, Washington, DC, U.S.A.

Acock, B., and J. F. Reynolds. In press. Notes on aggregation of plant process models. In: *Biological Response to Environmental Change.* U.S. Department of Energy, Carbon Dioxide Research Division, Washington, DC, U.S.A.

Baker, D. N., J. R. Lambert, and J. M. McKinion. 1983. GOSSYM: a simulator of cotton growth and yield. Bulletin 1089, South Carolina Agricultural Experiment Station, Clemson, SC, U.S.A. 134 pp.

Forrester, J. W. 1961. *Industrial Dynamics.* MIT Press, Cambridge, MA, U.S.A.

IBSNAT. 1986. Decision support system for agrotechnology transfer (DSSAT). IBSNAT Technical Report 5, University of Hawaii, Honolulu, HI, U.S.A.

Landsberg, J. J. 1986. Experimental approaches to the study of the effects of nutrients and water on carbon assimilation by trees. *Tree Physiology* 2:427–444.

Lemmon, H. 1986. COMAX—an expert system for cotton crop management. *Science* 233:29–33.

Saxton, K. E., W. J. Rawls, J. S. Romberger, and R. I. Papendick. 1986. Estimating generalized soil-water characteristics from texture. *Soil Science Society of America Journal* 50:1031–1036.

CHAPTER 14

FOREST MODELING APPROACHES: COMPROMISES BETWEEN GENERALITY AND PRECISION

Peter J. H. Sharpe

Abstract. Forest models are currently based on procedures that have separate roots in statistics, mathematics, scientific method, and process engineering. Each of these methodologies places a different emphasis on generality, precision, and reality. The statistical approach fuses precision and reality, the mathematical approach couples generality and precision, and the scientific method links generality and reality. Engineering approaches aim to combine generality, precision, and reality with varying degrees of success. Associated with each methodology is a distinct criterion for model validation: goodness-of-fit for statistics, rigor for mathematics, refutation for the scientific method, and performance for engineering. These approaches are described and discussed within the context of forest modeling objectives and priorities.

INTRODUCTION

The forestry modeling community is divided as to which of several modeling methodologies is most suitable for any particular objective. Forest modelers can currently choose among procedures that have separate roots in statistics, mathematics, biological science, and process engineering. Because these four methodologies have different emphases and traditions, conflicts among the proponents of the four approaches have been inevitable. Nowhere are these conflicts more evident than in the shift in emphasis from the prediction of forest yields based upon past productivities to the assessment of potential declines in future productivity due to air pollution effects and to the likely consequences of global changes in carbon dioxide levels, temperature, and rainfall.

MODELING APPROACHES

Although it is obvious to most observers that different philosophical approaches to modeling exist, the underlying rationale for these separate approaches has not been clarified. For example, Hall and DeAngelis (1985) refer to "a curious taxonomy of modelers,"

Dr. Sharpe is a Professor at the Biosystems Research Group, Department of Industrial Engineering, Texas A & M University, College Station, TX, 77843 U.S.A. The concepts presented in this manuscript were developed as part of ongoing theoretical ecosystems modeling studies supported by NSF grant BSR-86-14911. The author wishes to thank A. Ross Kiester, of the Corvallis Environmental Research Laboratory, Corvallis, OR, U.S.A., for stimulating his interest in modeling compromises. The author also thanks W. Michael Childress of Biosystems and Craig Loehle of Savannah River Laboratory, Aiken, SC, U.S.A., for their comments on the manuscript.

but do not elaborate on their classification, nor consider the reasons for the separate evolution of modeling approaches. In this section, causes for the emergence of separate modeling approaches are addressed.

The development of different methodologies within the modeling community stems from: (1) the complex nature of biological systems, (2) fundamentally different modeling priorities, (3) methodological trade-offs, (4) stopgap solutions to missing theory, and (5) academic traditions. Different approaches to modeling are a natural consequence of the limited number of ways for handling *precision, generality,* and *reality.* In this paper, I define precision as the degree of exactness in a measurement or a prediction, generality as the applicability of a concept to a whole range of instances, and reality as the collection of true causes and effects which underlie superficial appearances.

Levins (1966) suggests that it is not possible to maximize precision, generality, and reality in a single model; two of these attributes can be maximized, but not all three. This constraint has been recognized, but its implicit consequences have not yet been addressed. I believe Levins's insight can be used as a stimulus to understand: (1) the evolution of separate modeling approaches, (2) the nature of the philosophical conflicts among modeling groups (Hall and DeAngelis, 1985), (3) the underlying structure of the different model formats, and (4) the criteria employed for model validation.

At the risk of oversimplification, let us consider a single classification scheme. I follow the postulate of Levins (1966) and represent the three attributes (generality, precision, and reality) as edges of a triangle (Figure 1). Maximization of the three attributes taken two at a time leads naturally to three methodologies, GP, PR, and GR, shown at the corners of the triangle in Figure 1. I believe that it can be loosely argued that the intersection of precision and reality can be identified with *statistical method,* generality and precision with *mathematical method,* and generality and reality with *scientific method.* In the present context, these intersections are more aptly termed *biostatistics, biomathematics,* and *forest science* (Figure 2).

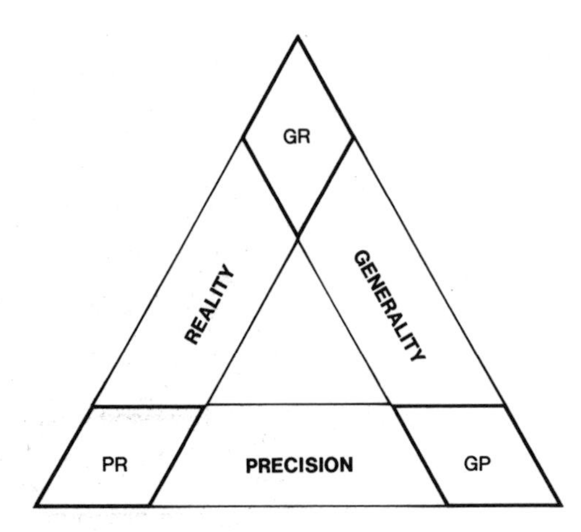

Figure 1. Triangular representation of intersection of knowledge attributes.

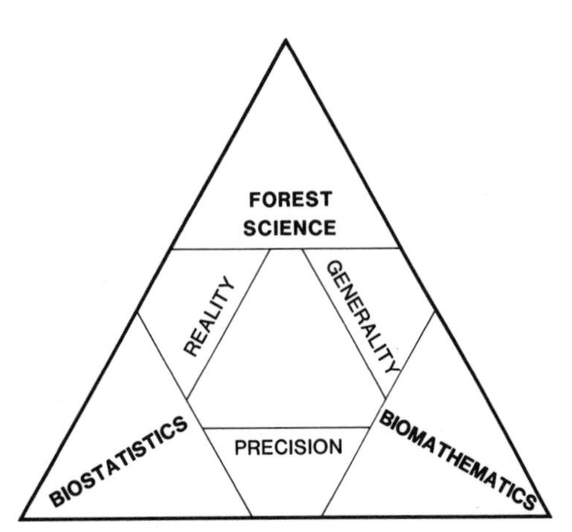

Figure 2. Biological academic disciplines defined by maximization of knowledge attributes taken in pairs.

Biostatistics

Precision and reality are coupled in the biostatistical or biometric school of forest modeling by emphasizing the precise estimation of empirical relationships derived from data. Regression is one of the principal techniques employed in this approach. Examples of biostatistical methodologies are analysis of variance (ANOVA) and regression-based growth and yield estimates of forest productivity. However, regression analysis does not accommodate the development of cause-and-effect concepts. Further examples of this approach are provided in companion chapters in the present volume.

Biostatistical modeling is the traditional methodology of choice in forestry and most other areas of structural resource management. Other modeling approaches are viewed with suspicion and will probably have to demonstrate substantial benefits before regression is superseded. The drawbacks of biostatistical models in general are often ignored in forestry.

Regression models cannot, by definition, be modified or improved. For every data set, a unique best-fit empirical relationship exists. When additional data are introduced, a new, unique regression must be formulated. This is both a strength and a weakness. The high precision by which regressions may be fit to the data is a considerable strength. The weakness lies in the model's inability to adapt the original formulation to new situations, locations, and time periods; to incorporate new concepts; and to provide explanation and interpretation. Precise prediction is usually the only goal of this approach.

There is little doubt that biometrics is more accurate than other approaches for predicting yield based on projection of previous growth data. However, forest response to multiple environmental stresses, as well as many other emerging management problems, requires extrapolation beyond existing data and investigation of alternative resource management options.

Precise, narrow-scope, empirical simulation models also fall into this precision-reality (PR) classification. Although these models may not use statistical procedures to fit data, they have a similar objective, i.e., to directly represent observed reality with one or more mathematical equations.

Biomathematics

The premise of the biomathematical approach is that idealized ecological systems can be adequately represented by models simple enough for closed-form mathematical analysis (Hall and DeAngelis, 1985). In this approach, generality and precision are merged by proposing rigorous definitions and idealized ecological relationships with little reference to observed reality. An example is the Blackman growth law (see Assmann, 1970), which predicts the annual height increment Z of a tree at age T by

$$\log Z = k \times \log^2 T \qquad (1)$$

The integral of this equation is a transformation of a standard gaussian integral, a typical mathematical model.

The study of complexity and stability by means of idealized models has helped to distinguish among different types of ecosystem stability (Hall and DeAngelis, 1985). From these idealized studies, the effects of species interaction, species numbers, and spatial heterogeneity on different types of stability have been determined. The approach of the mathematician is to start with the simplest functional model that will explain the phenomenon in question.

Biomathematics is considered by many practicing silviculturists to have little value to forestry. The ideal approach, however, is defended by Levin (1981), who argues that mathematicians play a larger role than that of mere participants in the induction-deduction-verification process. The mathematician aims at a general understanding, not at precise prediction or representation of biophysical process dynamics. Thus, fundamental properties of the forest system may be revealed that otherwise might be obscured in model representations that are broader in scope. Mathematics provides other benefits in the form of analytical solutions, proofs that one model may be a special case of another, and generalization of results.

Forest Science

Science, particularly forest science, is viewed by many land managers and forest extension administrators as largely irrelevant. This is because basic science in forestry has focused on fragmented, short-term studies of small-scale processes. Long-term testing of stand and tree population hypotheses is facilitated by the availability of historical data in the form of tree rings, fire scars, and stand populations of different ages (Loehle, 1988). Therefore, a long-term data base exists for the development and refutation of scientific theory in forest science.

The methodological emphasis on statistical procedures in forest modeling has inhibited the development of hypotheses with a high risk of refutation. Loehle (1987) points out that "significant progress can be made by following Popper (1963) and making *risky* predictions rather than attempting to use only goodness-of-fit or rejection of null hypotheses." Although most foresters would agree to the logical necessity of hypothesis formulation and refutation as a means of inferring the truth of universal statements from observations (Popper, 1978), the practical, short-term needs of the wood and paper industries for precise estimates of growth and yield limit the risky hypotheses that can be tested.

If science is defined as the intersection of generality and reality, then for science to have explanatory power, it must formulate its hypotheses at a mechanistic level. However, the mechanistic conjectures underlying one conceptual framework may be empirical in another framework. The important distinction is the relationships among hierarchical levels (Rykiel et al., 1988). Hierarchy theory (Allen and Starr, 1982; O'Neill et al., 1986) in an applied or management context identifies three levels: management, observation, and mechanism. The management level is the level of consequence. Below this is the observation level, where system responses are measurable. For biometric or regression hypotheses, only the observation and possibly the consequence levels are defined. For mechanistic hypotheses to be formulated, a focus on still lower-level processes is required; any model without a focus at this level is empirically based. It is the importance attached to this third level that determines the unique nature of mechanistic conjectures. To maintain mechanistic consistency, each testable hypothesis in the model must be described in terms of lower-level causal processes which are falsifiable.

One of the challenges in scientific and other mechanistic modeling is the avoidance of a mechanistic regress, where each mechanism is in turn defined in terms of more primitive processes. Choice of an appropriate level of mechanism early in the model formulation phase is an important component of the art of ecological modeling.

ENGINEERING: OPTIMIZATION OF ATTRIBUTES

What alternatives exist to the maximization of model attributes proposed by Levins (1966) and represented in Figure 1? In the physical realm, engineering can be viewed as an optimization of generality, precision, and reality. The foundation of engineering is process analysis. Typically, the engineering approach represents processes by equations which predict transfers of material, energy, and information within and between systems. Engineering attempts to use *general* concepts to formulate *precise* numerical solutions to difference or differential equations which hopefully describe *real* systems. This combination of all three knowledge attributes, as shown in the Venn diagram (Figure 3), is widely applied as a fundamental approach to almost all aspects of engineering analysis and design.

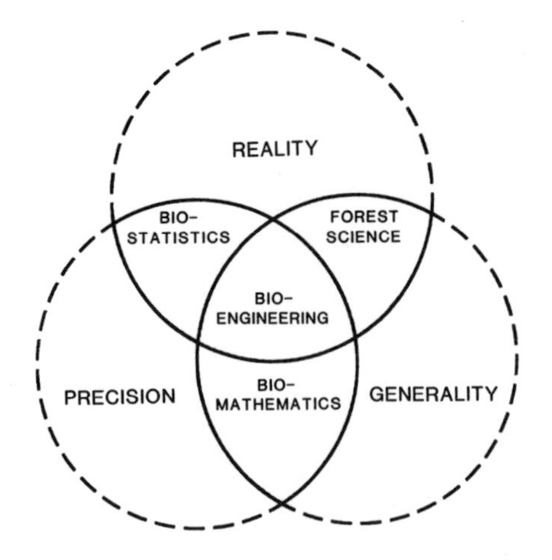

Figure 3. Venn diagram of biological academic disciplines including combinational optimization of knowledge attributes.

Engineering approaches work well when dealing with isolated components such as automobile design; but when applied to hierarchical systems such as automobile manufacturing plants, the full power of the optimization approach can rarely be realized. Physical laws, although the basis for all lower-level behavior, are insufficient to describe higher-level behaviors, particularly when these include human personality and management styles. A simulation of a large manufacturing plant has many of the same limitations as a large-scale ecosystem simulation. These systems typically exhibit stochastic behavior with non-linear feedbacks and hierarchy-dependent constraints.

Because engineering simulations aim to describe real rather than idealized systems behavior, new methodologies in artificial intelligence (AI) have a high priority. These methodologies place greater emphasis on explanation than on prediction. The rationale behind the AI approach is that when managers understand what the dominating factors are and can trace how these factors affect the final outcome, they feel more comfortable making their own predictions, and explaining and defending them to higher authorities. They are also in a position to assess risk based on the probabilities of underlying processes rather than variabilities in observed data. This is important since the data may reflect errors in experimental measurement rather than the risk of occurrence of events not included in the experimental study. The application of artificial intelligence in forestry is in a beginning stage (e.g., Kourtz, 1987; Thieme et al., 1987; White and Morse, 1987; Rauscher, 1988) and therefore is not included in this review. However, it is rapidly becoming an area of interest (Rykiel, 1988) and will contribute to the scope of forest modeling methodologies in the future.

Engineering process modeling is added to the three approaches previously identified; but because of the special characteristics of ecological systems, engineering methods must be specially adapted to the biological realm. Specifically, process engineering methods must be expanded to include the ramifications of the genome. Genetic variability has the potential to make individual trees in the forest unique in their response to the environment and to neighbors. Quantum mechanical approaches therefore appear more relevant than classical mechanics. The term bioengineering is used for engineering procedures to maintain consistency with the terminology used in Figure 2.

CONFLICTS IN VALIDATION CRITERIA

Associated with each of the four modeling methodologies (statistics, mathematics, biological science, and process engineering) are four different approaches to model validation. It should be recognized, however, that not all models can be falsified. Criteria for concept falsification are highlighted by Popper (1959), who is appalled that many explanatory concepts in science have been structured to be unfalsifiable. (Examples he cites include the theories of Freud, Adler, and Marx.) Unfalsifiable concepts can explain any observed behavior; because these theories cannot fail, they have no testable scientific content.

Popper (1959, 1978) studied the general problem of logically inferring the truth of universal statements from observations of particular phenomena. He concluded that in fact there is no logical method to demonstrate the truth of a theory or any universal statement via induction from data. A scientist's energies should therefore be devoted to attempts to invalidate theories. Invalidation is logically possible because one can deduce which observations disagree with projections of the theory. Thus Popper maintains that science is a sequence of "conjectures and refutations." Invalidation is followed by revision of theories and by proposing alternate hypotheses arising from the insights gained from previous refutations. This process of theory maturation is outlined by Loehle (1987).

Modeling methodologies, however, are not based upon hypothesis-testing in the strict scientific sense. Each modeling approach has its own criterion for validation. Each criterion relates specifically to the associated methodology itself and not necessarily to the scientific concepts embedded in the model. In mathematics, models are judged by their rigor (strict adherence to mathematical formalism), in statistics by their goodness-of-fit, in science by the validity of underlying conjectures (hypotheses or concepts), and in engineering by their performance. These four approaches to validation are shown diagrammatically in Figure 4 and discussed below.

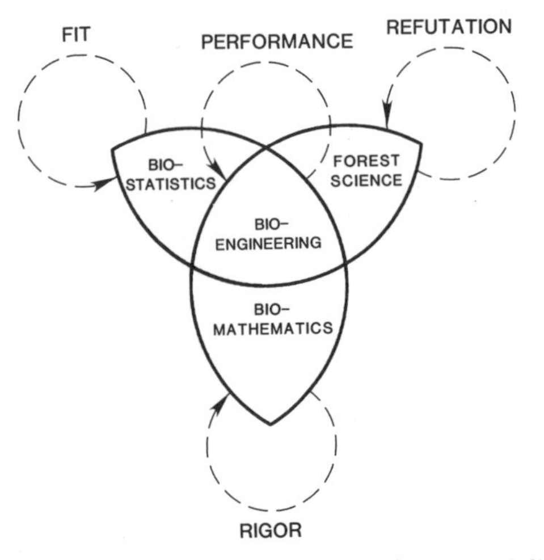

Figure 4. Validation criteria used by each of four forest modeling methodologies.

Mathematical Rigor

Give a mathematician a forest growth model, and he will begin by challenging the validity and consistency of the mathematical operations and manipulations. For mathematicians, this aspect is all-important in judging ecological models. Many of the breakthroughs in the physical sciences, however, have occurred in spite of mathematicians, rather than with their help. Examples include Newton's (1686) development of calculus, Dirac's delta function (Marsden, 1974), and many aspects of statistical mechanics (Kittel, 1958). Thus, although ecological modelers need to maintain mathematical rigor wherever possible, they must be prepared to step outside the narrow confines of current mathematical theory where this becomes an impediment to progress. On the other hand, mathematics can often convert a real but intractable problem into a solvable form.

Goodness-of-Fit

Regression models are typically built and validated from one or more data sets. In some cases, the original data set is split, with half the data used to develop the regression and the other half used to validate the statistical fit within some confidence limit. Generality cannot be expected because such models are specific to a given data set and cannot be structurally altered to include new knowledge or insight without deriving a new model. A statistical goodness-of-fit criterion can only reject or fail to reject a statistical model. When tested beyond the domain of the measurements from which they have been built and validated, statistical models frequently fail.

Refutation

In forest science, the refutation process is confounded as to whether the model or the ecological concepts underlying the model are on trial. The role of data is further confused as data are often used to derive the model, validate the model, and validate the concepts of the model. In my opinion, for scientific modeling to establish itself on a strong foundation of reputable hypotheses, the role of falsifiable concepts and the role of data must be clearly separated. Further, I advocate that only falsifiable hypotheses be included as components of ecological models where the objective is the testing of hypotheses and the explanation of cause-and-effect dependencies. For example, consider two propositions of Loehle (1988) for explaining tree-life strategies: (1) If investment in chemical and physical defenses slows down growth rate but increases tree longevity, then for sites of similar quality (i.e., site index), growth rate and longevity are hypothesized to be inversely related. (2) The counter-hypothesis of Waring and Franklin (1979) proposes that longevity is reduced by conditions where respiration demand exceeds photosynthesis. It can, therefore, be hypothesized that on poor sites, trees have short life spans. Hypotheses 1 and 2 are instances of concepts that can be refuted using experimental data. Readers are referred to Loehle (1988) to determine the fate of these hypotheses.

For scientific models, data should be used only for calibration and refutation. This is currently a minority view, although few foresters would go so far as Peters (1980a,b) and advocate the exclusion of all explanatory concepts from models.

Performance

Validation criteria for process models are rarely defined, because in engineering extensively tested principles from the physical sciences are used as the models' foundation. In the ecological sciences, this broad-based, mechanistic foundation is missing. To fill this gap, Mankin et al. (1977) have proposed the use of performance to objectively evaluate ecological models. Their model evaluation scheme uses two criteria, which they define as "validity" and "usefulness." According to their scheme, all ecological models fall into the invalid-useful category. They suggest that no "valid" model is possible, and no useless model will be retained. Unfortunately, as Loehle (1987) points out, models that are demonstrably wrong are retained in ecology. The logistic growth equation, for example, has been falsified in the overwhelming number of cases where it has been tested, yet it continues to be "taught in every elementary ecology course around the world" (Fagerstrom, 1987).

The relative usefulness of an engineering model is defined by Mankin et al. (1977) in terms of the "adequacy" and "reliability" of the model. Here adequacy is defined as the fraction of system behaviors which is duplicated by the model, and reliability as the fraction of predicted behaviors which duplicates observed behavior. The logical conclusion of the argument of Mankin et al. (1977) is that no valid process model has yet been or is ever likely to be developed in ecology. I interpret this gloomy conclusion as the logical result of attempting to build engineering models without first establishing a scientific foundation for such models.

Of the four approaches to model validation, I favor the conjecture-refutation approach of Popper (1963) because it is both the most risky and the most severe. It has the ability to unambiguously discriminate among models with different conceptual foundations. As a consequence, models subject to this degree of harshness in their validation are more likely to be elegant, simple, and coherent, as well as robust and powerful.

DISCUSSION

A much-debated issue among both modelers and experimentalists is the proper role and importance of data (Fagerstrom, 1987; Caswell, 1988). There has been something of a double standard with respect to data and models, because some have considered data as the ultimate reality (e.g., Simberloff, 1981), and therefore above the level of criticism aimed at models. A more objective assessment recognizes that data embody only a small subsample of the appearance of reality. Data may in fact represent the form of knowledge that is *least* real because of its conceptual limitations (Gardner, pers. comm.). There is a bias toward evaluating data solely in terms of its precision (i.e., measurement exactness), rather than in terms of its reliability, meaning, and conceptual significance. In addition, to obtain highly precise measurements or predictions, ecological systems must often be distorted from their natural state or behavior. A biological uncertainty principle enters in to cast doubt on the natural validity of very precise data. The distortions caused by the demands for precise measurements can be recognized in the physiological differences between plants grown under natural field conditions and those grown in precisely controlled environments.

The generality of any set of data relates to whether it has particular or broad significance. Do the data reflect some underlying universal principle? This issue cannot be assessed by examining the data in isolation from other sources of knowledge. In retrospect, it is difficult to justify the proposition that data are the ultimate reality because data are variable, depending on location, time, and prevailing environmental factors, as well as mechanistic detail and universal applicability.

This argument does not imply that data in the form of "hard facts" are not a reliable guide for model refutation. It does, however, question the factual foundation of much ecological data, because ecologists, and foresters, are often forced to measure what is measurable (e.g., fertilization rates) rather than what directly contributes to forest productivity (e.g., competitive and discriminatory nutrient uptake mechanisms).

Models similarly should not be confused with reality. In fact, modeling approaches based upon ideal representations (Levin, 1981) do not claim absolute reality. Where models attempt to mimic reality, they tend to do so either in terms of precision or generality. When they attempt to combine all three attributes, as in the case of engineering simulation, the final result tends to move towards an ideal rather than a real representation.

The major difference between complex engineering systems and ecological systems lies in their degree of coupling. Biosystems are so tightly coupled and interdependent that they can rarely be decoupled without destroying function. They span a wide range of spatial and temporal scales as well as hierarchical levels. Within-system interactions are so pervasive that it is extremely difficult to capture their essence in mechanistic forms that are both general and realistic. A one-to-one correspondence between cause and effect for all mechanistic details leads to hypotheses of unmanageable complexity. On the other hand, the exclusion of relevant factors results in systems concepts that are incomplete.

Two approaches to the problem of tightly coupled complexity are evident. Jorgensen (1986) speaks for many modelers when he claims that it is practically impossible to capture the complete mechanisms of biosystems. Therefore, he argues, numerous alternative mathematical representations are likely to be equally applicable. On the other hand, attempting to include known physiological mechanisms in higher-level models leads to a circular situation. To construct a forest growth simulator using physiological mechanisms requires parameter values that do not exist or that are extremely difficult to obtain. If uncertain parameter values are adjusted specifically to give reasonable predictions, the model may be dismissed as simply speculation (Landsberg, 1987). Alternatively, if the model is formulated at a hierarchical level for which growth parameters are directly measurable, physiological detail must be sacrificed (O'Neill et al., 1986), which is also unacceptable (Landsberg, 1987). This is a "damned if you do and damned if you don't" situation.

The most detailed long-term (0–500 years) forest growth models to date have been simulators for multiple-species stands of individual trees pioneered by Botkin et al. (1972) and extended by Shugart (1984) and colleagues. These simulations have extended the scope of forest growth modeling by introducing conceptual elements and comparing which changes in functions or parameters are required to apply the basic model structure to different forest systems.

The practical problem remains that there are insufficient valid components that can be used to bridge the chasm between tree physiology and silviculture. Most foresters focus their activities at one or the other end of this functional range, and for good reason. But if forestry is to develop as a unified science, methodologies which transcend hierarchical levels of forest organization must be developed.

A major requirement for modeling in forestry is its role in management decision-making, which far overshadows its role as a tool for scientific study. Empirical methodologies must be deemed a necessary evil as long as various causal mechanisms remain black boxes. Management decisions cannot be postponed until the relevant cause-effect relationships are defined.

In my judgment, there is a serious underestimation of forest-land managers' need for explanation of model predictions. Davis and Whigham (1988) suggest that "extrapolation is a particularly important requirement for management models since managers need to justify their decisions to interest groups and political representatives of the general public." Further, they point out that although both regression and process models can be used for prediction, only process models are capable of supplying an explanation of predictions beyond the "domain of observation." In retrospect, it is interesting to speculate why forest managers have been willing to accept prediction without explanation. Why has there been no demand for an explanation of the underlying rationale of forest growth and yield models? Maybe the specter of forest decline in the wake of projected changes in the global environment will lead to a recognition of the need for explanatory models in forestry.

The thrust of this article is that each of the four modeling approaches has its own criterion for validation. In traditional forestry, however, there has been a tendency to focus on only one validation criterion (e.g., goodness-of-fit), regardless of the type of model formulated. Forest modeling activities may be strengthened by the use of multiple modeling approaches and validation criteria. Satisfaction of more than one criterion lends credibility to the robustness of the modeling analysis. It can also partially compensate for inadequacies in any one approach taken in isolation.

In conclusion, it is apparent that numerous approaches exist for the development of growth models in forestry. These approaches represent different compromises with respect to generality, precision, and reality. The important message is that when forest modelers have a flexible mind-set and take advantage of the opportunities that these different approaches and criteria offer, then hopefully they will develop models which can bridge the gap between forest science and applied forestry.

LITERATURE CITED

Allen, T. F. H., and T. B. Starr. 1982. *Hierarchy: Perspectives for Ecological Complexity.* University of Chicago Press, Chicago, IL, U.S.A. 310 pp.

Assmann, E. 1970. *The Principles of Forest Yield Study.* Pergamon Press, New York, NY, U.S.A. 506 pp.

Botkin, D. B., J. F. Janak, and J. R. Willis. 1972. Some ecological consequences of a computer model of forest growth. *Journal of Ecology* 60:849–873.

Caswell, H. 1988. Theory and models in ecology: a different perspective. *Bulletin of the Ecological Society of America* 69:102.

Davis, J. R., and P. A. Whigham. 1988. Integrated systems for managing natural areas. Paper presented at the Resource Technology '88 Symposium, June, 1988, Fort Collins, CO, U.S.A.

Fagerstrom, T. 1987. On theory, data and mathematics in ecology. *Oikos* 50:258–261.

Hall, C. A. S., and D. L. DeAngelis. 1985. Models in ecology: paradigms found or paradigms lost? *Bulletin of the Ecological Society of America* 66:339–346.

Jorgensen, S. E. 1986. *Fundamentals of Ecological Modelling.* Elsevier Science Publishing, Amsterdam, the Netherlands. 389 pp.

Kittel, C. 1958. *Elementary Statistical Physics.* John Wiley and Sons, New York, NY, U.S.A. 228 pp.

Kourtz, P. 1987. Expert system dispatch of forest fire control resources. *AI Applications in Natural Resource Management* 1:25–33.

Landsberg, J. J. 1987. Forest dynamics: a book review. *Forest Ecology and Management* 18:161–163.

Levin, S. 1981. Mathematics, ecology, ornithology. *Auk* 97:422–425.

Levins, R. 1966. The strategy of model building in population biology. *American Scientist* 54:421–431.

Loehle, C. 1987. Hypothesis testing in ecology: psychological aspects and the importance of theory maturation. *Quarterly Review of Biology* 62:397–409.

Loehle, C. 1988. Tree life history strategies: the role of defenses. *Canadian Journal of Forestry Research* 18:209–222.

Mankin, J. B., R. V. O'Neill, H. H. Shugart, and B. W. Rust. 1977. The importance of validation in ecosystems analysis. In: Innis, G. S., ed. *New Directions in the Analysis of Ecological Systems,* Part 1. Simulation Councils Proceedings Series, Vol. 5, No. 1. Simulation Councils, La Jolla, CA, U.S.A.

Marsden, J. E. 1974. *Elementary Classical Physics.* W. H. Freeman, San Francisco, CA, U.S.A. 549 pp.

Newton, I. 1686. *Philosophiae Naturalis Principia Mathematica.* Royal Society, London, England.

O'Neill, R. V., D. L. DeAngelis, J. B. Waide, and T. F. H. Allen. 1986. *A Hierarchical Concept of Ecosystems.* Princeton University Press, Princeton, NJ, U.S.A.

Peters, R. 1980a. From natural history to ecology. *Perspectives in Biology and Medicine,* Winter, 1980:191–203.

Peters, R. 1980b. Useful concepts for predicting ecology. *Synthese* 43:257–269.

Popper, K. R. 1959. *The Logic of Scientific Discovery.* Hutchinson, London, England, U.K. 480 pp.

Popper, K. R. 1963. *Conjectures and Refutations: The Growth of Scientific Knowledge.* Harper & Row, New York, NY, U.S.A. 417 pp.

Popper, K. R. 1978. The myth of inductive hypothesis generation. In: Tweney, R. D., M. E. Dougherty, and R. C. Mynatt, eds. *On Scientific Thinking.* Columbia University Press, New York, NY, U.S.A. Pp 72–76.

Rauscher, M. 1988. Using AI methodology to advance forestry science. *AI Applications in Natural Resources Management* 2:58–59.

Rykiel, E. J., Jr. 1988. *Expert Systems in Ecology: Forestry Applications.* Symposium, International Society for Ecological Modelling, 39th Annual Meeting, August, 1988. American Institute of Biological Sciences, University of California, Davis, CA, U.S.A.

Rykiel, E. J., Jr., R. N. Coulson, P. J. H. Sharpe, T. F. H. Allen, and R. O. Flamm. 1988. Disturbance propagation by bark beetles as an episodic landscape phenomenon. *Landscape Ecology* 1:129–139.

Shugart, H. H. 1984. *A Theory of Forest Dynamics.* Springer-Verlag, New York, NY, U.S.A. 278 pp.

Simberloff, D. 1981. The sick science of ecology. *Eidema* 1:1.

Thieme, R. H., O. D. Jones, H. G. Gibson, J. O. Fricker, and T. W. Reisinger. 1987. Knowledge-based forest road planning. *AI Applications in Natural Resources Management* 1:25–33.

Waring, R. H., and J. F. Franklin. 1979. Evergreen coniferous forests of the Pacific Northwest. *Science* 204:1380–1386.

White, W. B., and B. W. Morse. 1987. Aspenex: an expert system interface to a geographic information system for aspen management. *AI Applications in Natural Resource Management* 1:49–53.

CHAPTER 15

DEVELOPMENT OF EMPIRICAL FOREST GROWTH MODELS

David Bruce

Abstract. Empirical models predicting periodic growth of trees and forests are based on measurements in the forest at two or more times and are developed without direct reference to the biological processes involved. Process models describing changes in trees and forests may be more complicated than empirical models; although they emphasize processes, those that predict quantities necessarily include some empirical elements. Growth models programmed to run on computers can be listed in order of increasing complexity, both forming progressions of information content and changing best uses for the models. A recent growth simulator that adds measures of the effects of nutrient cycling to an empirical model shows promise of increasing versatility without improving reliability of growth estimates under conditions currently measurable. Growth models differ greatly in how they handle competition among trees, and it has been found that competition measures at the stand level often work as well as those at the tree level. Models are evaluated objectively by predicting and observing growth in stands that sample the conditions modeled. Simple empirical models are accurate enough to guide short-term forest management decisions, because prudent managers monitor growth of their stands. Mixed models including both empirical and process elements are likely to be needed to make reliable long-term forecasts and credible growth predictions for large heterogeneous areas or for management conditions that cannot currently be observed.

INTRODUCTION

A few years after I helped developed Douglas-Fir Interim Tables (DFIT), a computer program that simulated managed yields of Douglas-fir (*Pseudotsuga menziesii* [Mirb.] Franco) stands (Bruce et al., 1977), I reviewed a lengthy position paper claiming that research on such stand growth models should be abandoned, and the money saved should be devoted to single-tree growth models, which were more sophisticated and potentially more accurate. I did not entirely or heartily agree with this. Shortly after this, I was told that neither stand nor tree models were the best way to simulate forest growth, but that both would be replaced by better models describing the growth processes of trees. Knowing a good deal about the shortcomings of both stand and tree models, I had to agree that better models were possible; knowing little about these process models, I wondered when they would be readily accepted.

The proceedings of a 1979 workshop in Sweden on "understanding and predicting tree growth" (Linder, 1981) helped me to understand the structure of models based on various levels of information about physiological processes. But, like many other foresters, I considered these models entirely different from the familiar empirical forest growth

Mr. Bruce is a Consulting Mensurationist and Volunteer, USDA Forest Service Pacific Northwest Research Station, Portland, OR, 97208 U.S.A.

models. In 1987, I described some of these differences and suggested that a link between the two kinds of models might be made, but not soon (Bruce and Wensel, 1988). This view of the differences between the model types was reinforced by the topic first assigned me for this chapter: "The interface between growth and yield methodologies and process modeling."

This emphasis on differences now seems less important to me. Papers in a report on a 1971 British symposium on mathematical models in ecology (Jeffers, 1972) as well as other, more recent literature led me to the idea of a sequence of models. The sequence began with a simple and totally empirical growth percentage for one species in a given region that no one would program separately, and it ended with a fantastically complicated and probably unprogrammable model. This model described how cells divided, forming all parts of all trees that were major components of a forest growing for a whole rotation on a sizable area. Between these extremes were many kinds of models that simulated forest growth well. Under current growing conditions, outputs of all these models could be reduced to the same growth percentage. More detailed models would respond better to factors that influenced growth and would answer more questions about how various management actions or other changes in forest conditions would affect growth and stand structure.

Although empirical and process models are different, elements of both can be and usually are included in a single simulator program. Landsberg (1981) suggested that this distinction between types of models is unimportant, and that all models should include as much information about processes as possible and resort to empiricism only when we know too little about underlying mechanisms to program them. I had similar thoughts when I was assembling algorithms for DFIT: With just a few more details, the program would explain all useful things about Douglas-fir growth. But now I believe that the planned use of a model should partly determine its structure, because it costs too much to assemble detailed input data and to develop and run a complicated computer program producing volumes of valuable and interesting information that will not help decide how to reach some specific goal.

I will discuss the development of empirical models based on studies of growth and yield that use remeasured plots in forested areas and some differences between such models and process models, which may suggest how to improve growth models. But first I will mention some names commonly used for growth models.

CLASSIFICATION AND NAMES

Growth models have many names. Sometimes two different kinds of models have the same name, but more often the same kind of model has several different names. Although some models are small copies of large objects, and some are maps of large regions, what I call empirical models are systems of algorithms used to relate tree and forest growth to measured variables. These models estimate growth under a range of conditions, and these estimates can be used as predictions.

Empirical forest growth models were classified by Munro (1974) as whole-stand or single-tree, with the latter divided into distance-independent and distance-dependent varieties. Another kind of model has become quite common since 1974; diameter-distribution models, which are between whole-stand and single-tree models in complexity but are considered to be stand models. It is no accident that Munro's three kinds of models correspond to long-established methods of recording information taken on forest growth plots. Whole-stand models can be based on untagged remeasured plots, distance-independent single-tree models on plots with tagged trees, and distance-dependent models on mapped plots. Diameter-distribution models are closely related to stand tables used to summarize data from uniform stands.

Other writers will discuss process models in this volume. I will use the term "mixed model" to identify a model that has elements of both process and empirical modeling. To categorize names used by other authors for the various models, I have listed contrasting pairs in Table 1. This incomplete list is limited to the names given to dynamic models and does not include static or state models, which may have quite similar names. The list is included to demonstrate the diversity of names, not to analyze types of models. Some names emphasize prospective uses of models, and some would be paired differently by others.

Table 1. Designations of growth models used in forestry.

Empirical	Process	
Managerial	Mechanistic	(Linder, 1981)
Management	Explanatory	(Landsberg, 1981)
Predictive	Functional	(Draper and Smith, 1981)
Simulation	Analytic	(Pielou, 1972)
Inductive	Deductive	(Pielou, 1972)

STRUCTURE OF EMPIRICAL MODELS

A purely empirical growth model is developed by measuring trees at intervals of a year or more and determining relations between change in size and certain other observable variables, with no direct reference to underlying processes. As the names listed in Table 1 imply, most empirical models are developed to simulate or predict forest growth as a guide to management. Basic variables included in the simple models for even-aged stands are species, site index, an index of stand density, and initial size, which is usually based on age. Most models are necessarily more complicated than this and use other variables, which I will mention later, but I will begin with these simple models.

In an even-aged stand, site index is defined as the average height of larger trees of a specified age. Stand density indices are combinations of stand basal area with one of the following: stand age, stand average diameter, or number of trees. These two indices can be objectively measured and continuously quantified, which is not true of crown class, which theoretically might be a better indicator of growth potential. Basal area alone is not a useful competition index because its apparent effect on growth varies with average tree height and crown depth.

Stand density and site indices express the results of growth processes and factors influencing them, without delving into how or why the processes modify growth. The indices are examples of using stands and trees as phytometers. For this reason, when these indices can be used in relatively uniform stands to indicate growth conditions, adding other information about site quality or competition seldom improves estimates. Use of these two indices together with age for a plantation is as close to pure empiricism as growth models get. Some individual-tree models for irregular stands use social structure (crown classes) to predict how each tree will grow. This comes close to process modeling, which considers both growth processes and the factors affecting them.

Despite many obvious differences in the design of stand and tree growth models, they have many similarities. They estimate changes with time in tree size and in tree numbers, which may include tree diameter, height, form, volume, or all these, as well as regeneration, ingrowth, and mortality. I like to call the equations describing these changes the driving functions. Driving variables include weather and other rapidly changing factors influencing tree growth. Factors such as species, age, and soil also influence growth and are included in these equations, but these factors change either regularly or slowly if at all. Driving functions include age or time as a variable. Housekeeping functions, such as volume based on height and diameter, do not include time but have coefficients specific to the model. Structural functions, such as stand basal area based on a list of tree diameters, are the same for all models.

Some factors influencing growth are included by definition in all kinds of models. Models for a single species or for a limited geographic area are called species-specific or site-specific models—other species and areas are excluded by definition. Empirical models may include by default such factors as prevalent soils and current climate; in contrast, process models may be based on careful measurement of soil or weather conditions.

Most empirical models separate increment into two parts, which simplifies equations and clarifies model operation. The first part is potential growth estimated by an equation well suited for describing biological development. This estimate is then modified by one or more functions limiting or accelerating growth. These modifiers are sometimes based on special studies separate from those in which growth potential was determined. There are two kinds of modifiers; internal, determined by stand development and built into models; and external, resulting from management decisions or other factors operating outside the models. When the results of management decisions are described by models, they require special signals to make them work. Postulated changes in weather can be handled in the same way. Submodels of biological processes are good modifiers to use in growth models.

Many forest growth simulators use submodels for juvenile stands. These submodels operate differently from the main model. The purpose of the main model is to describe the growth and development of stands that have passed the sapling stage. The end of the sapling stage is defined in different countries as the time the stand reaches an average dbh of 9–15 cm. Suitable models for young stands are entirely different for natural and artificial regeneration because method of regeneration controls species composition and influences spacing. Often there are important differences between natural and planted stands with respect to competition from low vegetation, which can retard seedling growth. A major problem in modeling the transition from this early stage of stand growth to the main growth model is that dbh is not used to describe the young stand but must be estimated to start the main model, and dbh distributions are sensitive to apparently minor variations in stand density and spacing.

GROWTH POTENTIAL AND COMPETITION

With regular stands (relatively uniform in species, size, and spacing), growth potential of trees can be assigned by tree size alone, and stand density is often the only index of competition needed (Dahms, 1983). With irregular stands, the growth potential of each tree must be estimated individually. No single tree characteristic will serve well for all trees, and often each tree is listed showing elements that determine its potential. Measures of competition are but one of several controlling elements. When stands are extremely irregular, no single measure of stand density adequately simulates competition. It has been claimed (Clutter et al., 1983) that this is best done by using a distance-dependent single-tree model where species, size, vigor, and proximity of trees and their neighbors are evaluated. Consideration of competition suggests some problems with this approach.

Usual descriptions of tree size, crown size, and vigor can account for some but not all elements of growth potential and competition. Elements not visible and not easily measured may be just as important; examples include roots, soil, stand history, and genetic diversity within species of trees in the stand. Roots determine the ability of a tree to take up water and nutrients. Root growth consumes part of a tree's energy. Roots occupy vacant space faster than do branches, and in some stands, root grafting occurs (Smith, 1986). Sometimes competition of understory vegetation reduces the growth of the overstory, particularly on sandy soils and when rainfall is low. For these and related reasons, comparisons of individual trees with stem and crown sizes and distance to neighbors are sometimes not used, even in irregular stands. Instead, individual tree characteristics and comparisons with stand average tree size and stand density are used in estimating growth potential.

In studies where measures of stand density and competition have been compared, expensive point measures for individual trees often have added no useful information to easily derived stand measures (Martin and Ek, 1984). Some studies have ignored stand measures and compared only point measures; others merely compared the two kinds of measures without testing how much information point measures might add to that provided by stand measures (Alemdag, 1978). Even the latest comparisons should not be accepted as definitive for all stands, because most such comparisons were made in plantations or even-aged stands where competition usually is reasonably uniform (Daniels et al., 1986). More critical tests eventually will be made in stands with mixed ages and several species.

Process modeling could improve individual tree competition indices by increasing understanding of the relative importance of all elements of competition, and by devising simple ways to quantify them. This promises greater progress than does additional tinkering with ways to express the relations among crown or stem sizes and distances separating the trees.

DEVELOPING AND TESTING GROWTH MODELS

There were forest growth plots long before there were computers, and there were forest yield tables before formal biometric methods were developed. But these two later developments, coupled with a growing interest in intensive forest management, have stimulated and assisted the recent development of forest growth simulators. Although the best data used in developing these simulators come from plots installed in the forest to observe the effects of treatments such as thinning and fertilizing, some information comes from plots or other tests that do not meet the standards for yield studies recently described by Curtis (1983). Plots tracing stand development for complete rotations exist only for southern pines (*Pinus*) and a few tropical species. A few plots have been remeasured for 50 years in natural stands of northern species, but managed stand records from standard plots dating back 30 years are rare.

An important reason for managed forests not matching yields on research plots is the relative degree of uniformity of the stand (Bruce, 1977). Research plots generally are much more uniform than managed stands, so that treatment effects can be detected over background noise or unmeasured sources of variation in forest plots. There is no way to generalize about differences in growth between irregular stands and uniform stands, which makes it difficult to calibrate growth responses observed on research plots in order to apply them correctly in a growth simulator. One persons's calibration is another's fudge factor! Fortunately for those who want to predict growth, many forest management activities or treatments make the forest more uniform and a little more like a research plot.

One of the biggest headaches in forest growth research comes from nonstandard yield plots. A common example is an attempt to use information from forest inventory plots, which are well suited for their primary purpose of determining tree volume on large areas but often have little or no value as yield plots. Their almost universal failing is that they are small and include no information about competition from the surrounding area. Compensation for this can sometimes be made by measuring past growth on all inventory trees, which reflects the effects of past competition and other factors influencing growth.

Another research problem is that detailed studies of tree growth are usually conducted in greenhouses or growth chambers. Greenhouses are usually small and often have growth conditions radically different from those of forest yield plots. Growth chambers are even smaller and have a still less natural ambience. Major allowances for these differences are usually necessary before results can be safely transferred. Another barrier to direct use of growth information from either of these environments is that the trees are usually young and have growth habits different from those of older trees. Despite these difficulties, much of what we know about tree growth has come from these sources.

Once suitable records of measurements of acceptable plots are assembled, they are analyzed. The analysis begins with plotting graphs and running ordinary least squares regressions to evaluate the data. Then more complicated analyses are made, but analysts still rely mainly on standard errors of estimate and coefficients of determination to judge goodness-of-fit and precision of estimate. Direct regression estimates for a single batch of data always have lower standard errors than estimates made in two or three steps, but sometimes information gained from the intermediate steps is well worth this loss in precision. Some complicated analytic methods make outsiders wonder if the extra effort is really productive, but most people are quite willing to accept results as long as precision is confirmed by the usual comparisons of actual and estimated values. The only analysis methods or regression equations generally avoided are those that have been demonstrated to produce irrational results.

Good modelers do not rely entirely on statistical measures in selecting regressions to use in growth models. Often several estimators have nearly the same low standard error of estimate. One that makes good biological sense and that does not produce ridiculous estimates near the limits of observed variables is favored. Only exceptional considerations would cause the choice of an estimator with a standard error much higher than the minimum level characteristic of the data set. This examination of standard errors of several estimators invalidates most formal tests of hypotheses.

Predictions are validated by history only after the fact. Thus the potential accuracy of growth estimates is determined by comparing recent growth on a check area with estimates based on older growth measurements on some original study area. This should be repeated frequently to detect unanticipated changes in growth conditions, because predictions based on empirical models are good only so long as such conditions do not change much. This same kind of test can be made for a process model where validation within the range of observable factors affecting growth increases credibility of its estimates outside this range. These validations are often incomplete because data sufficient for validation are frequently used to develop a new model. Such comparisons are the only objective way to test an estimator that will be used as a predictor. At some stage, other tests involve acts of faith and appear somewhat less convincing.

There may be little reason to improve the potential accuracy of models based on observed growth and used mainly to guide short-term forest management decisions. No matter how reliable and well-tuned a model, stands whose management it guides will vary more in growth than the plots measured to develop it. Prudent managers monitor their stands to determine actual growth and to detect events not included in the model development that may reduce growth, such as seed crops or insect defoliation. Under these circumstances, small improvements in the precision of estimates are of low value to the manager, although even small biases should be avoided whenever possible. Improved knowledge of the relative effects on growth of various treatments is more important than great accuracy of any single growth estimate.

WHAT ABOUT THE WEATHER?

Some empirical growth models are criticized because they make no allowance for changes in weather. Many growth studies on which models are based have been installed for short periods in limited geographic areas and include little or no information about effects of a range of weather conditions. This deficiency can be overcome by extending the period or area observed. Short-term weather forecasts are often disappointing, and long-term forecasts are notoriously inaccurate. The simpler model, without allowance for weather change, will do a good job of predicting growth if future weather is like that of the recent past. Improved models will quantify the effects of changes in the weather, which can be useful if weather patterns change from those of the base period, and also in speculative long-range planning. (Here I use speculative to mean conjectural, not risk-taking.)

Another problem with most growth studies is that any weather records used were taken at an airport near a city, not in the forest where the trees grow. It is possible, at some expense, to install automatic recording weather stations adjacent to growth plots and include instruments not used at most airports. Now some uncertainties about how weather influences growth can be removed, and correlations of cumulative weather measurements and observed growth should be higher. But automatic weather stations cannot be installed in all parts of a forest, nor can weather measurements be summarized before a growing season ends; thus growth cannot be predicted any better than weather in the forest can be predicted. Greater understanding of the effects of weather on growth processes may help improve computer simulators in other ways, and new models may be used for purposes other than guiding current management.

Weather and attacks by pests are almost the only rapidly changing and hard-to-predict environmental factors that alter growth patterns. These or other agents may kill so many trees that the stand falls below the stocking levels that could be carried in a managed forest, which may end the simulation.

PUBLISHED EXAMPLES

Two recent publications described growth models based on studies comparing the effects of several treatments on growth in 15- to 36-year-old coniferous plantations. Supported by forest plots with similar experimental designs and with similar considerations of the effects of nutrient cycling on growth, the models and their proposed uses are quite different.

A good example of a process model developed for use in forestry is SHAWN (Shawnigan Lake, B.C., Canada) (Barclay and Hall, 1986). The model is based on measurements of a factorial study of thinning and fertilizing in a 24-year-old Douglas-fir plantation. The purpose of the model is to provide a framework for increasing understanding of basic processes, emphasizing those involved in cycling nitrogen through the stand. Among other things, this will identify areas of inadequate understanding, suggest need for more research, and identify factors to which the model is sensitive so that they may be more completely measured or manipulated in other studies. A first comparison of rankings of estimates produced by the model shows general agreement with quantitative results from the experiment, with one anomaly ascribed to inadequacy of some of the process algorithms.

The other model is the Integrated Forest Process Model (IFPM) (Rennolls and Blackwell, 1987), which is based on measurements of a factorial study of thinning intensity and thinning cycle in a 15-year-old Sitka spruce (*Picea sitchensis* [Bond.] Carr.) plantation. It was developed to qualitatively expand an empirical distance-dependent tree model by following an unnamed resource through soil, litter, and foliage. For this revised model, one method of rating competition that was tested involved quantifying competition for light by calculating vertical angles from a plant tip to its neighbors' tips (Ford and Diggle, 1981). The other two competition indices were based on diameters and distances to neighbors. In this first trial, amounts of the resource in canopy and litter were set at arbitrary initial values in the hope that final relative values would be meaningful. Two methods were suggested for getting coefficients of four equations involved in resource cycling: using trials external to the growth study or setting coefficients at plausible values. Apparently, these equations were successfully linked to observations of diameter growth because the model was tested on three hypothetical examples that could best be run by starting with different amounts of reserves.

Two things may be typical in the development of such models. First, heights estimated from the model were adjusted so that top height would follow an already well-accepted height curve. This height adjustment makes it obvious which height-over-age relation the authors trust and really use to predict height growth. Second, the measure of competition based on biological concepts, after substitution of diameter for height, which made it just another index based on distance and diameter, was improved by averaging it with an empirical measure. This combination, however, was no better than yet another mostly empirical measure. This is not surprising because before alteration, the process-based measure was developed for well-watered marigolds growing for 8 weeks in a greenhouse.

MIXED MODELS AND THE FUTURE

Few models are purely empirical or strictly process operated. Empirical models are supposed to be developed without reference to underlying biological processes, yet most good ones use mathematical functions well suited for describing tree growth. Models using measures of individual-tree competition are closer to being process operated than are whole-stand models. Perhaps the former are nearly as much mixed models as are stand models with superimposed information about nitrogen cycling. To identify the true type of a model, determine the driving function; if growth estimates from one source are adjusted to agree with those from another source, the second source controls and is the driver. The adjusted function, however, may supply useful information about the effects of changing growth conditions.

Growth modifiers that can be added to empirical models include not only competition and nitrogen cycling but also water or carbon cycling, submodels of pests, radiant energy, and, of course, atmospheric pollution. Linkages can be made in several ways, such as using submodels for special events, using parallel driving functions, and using multiple conditional growth modifiers. How this is done depends on the information available, how the driver or modifier operates, and the planned use of the model. If only minor changes occur in growing conditions, none of these mixed models seems likely to produce more useful estimates of growth for short-term management planning than simpler growth models based on remeasured plots.

Long-range predicting is another ball game, as are predicting for large and inevitably heterogeneous areas, and estimating growth under conditions that are changing in major but currently unobservable ways. No existing empirical model will always track growth within 50% for 100 years. Only patchworks of observations of forest stands in various stages of development are available now to develop such estimates, mostly from plots barely adequate for use in growth studies. To get a reliable long-term model, or one suitable for heterogeneous or changing conditions, one alternative would be to develop a super-process model; but perhaps the best choice is to opt for a mixed model developed by a group including experts in both empirical and process modeling.

LITERATURE CITED

Alemdag, I. S. 1978. *Evaluation of Some Competition Indexes for the Prediction of Diameter Increment in Planted White Spruce.* Information Report FMR-X-108, Forest Management Institute, Ottawa, Ont., Canada. 39 pp.

Barclay, H. J., and T. H. Hall. 1986. *SHAWN: a model of Douglas-fir ecosystem response to nitrogen fertilizing and thinning: a preliminary approach.* Information Report BC-X-280, Canadian Forest Service, Pacific Forestry Centre, Victoria, B.C., Canada. 30 pp.

Bruce, D. 1977. Yield differences between research plots and managed forests. *Journal of Forestry* 75:14–17.

Bruce, D., D. J. DeMars, and D. L. Reukema. 1977. *Douglas-Fir Managed Yield Simulator—DFIT User's Guide.* General Technical Report PNW-57, USDA Forest Service. 26 pp.

Bruce, D., and L. P. Wensel. 1988. Modeling forest growth: approaches, definitions, and problems. In: Ek, A. R., S. R. Shifley, and T. E. Burk, eds. *Forest Growth Modelling and Prediction.* General Technical Report NC-120, USDA Forest Service North Central Forest Experiment Station, St. Paul, MN, U.S.A. Pp. 1–8.

Clutter J. L., J. C. Fortson, L. V. Pienaar, G. H. Brister, and R. L. Bailey. 1983. *Timber Management: A Quantitative Approach.* John Wiley and Sons, New York, NY, U.S.A. 333 pp.

Curtis, R. O. 1983. *Procedures for Establishing and Maintaining Permanent Plots for Silvicultural and Yield Research.* General Technical Report PNW-155, USDA Forest Service. 56 pp.

Dahms, W. G. 1983. *Growth-Simulation Model for Lodgepole Pine in Central Oregon.* Research Paper PNW-302, USDA Forest Service. 22 pp.

Daniels, R. F., H. E. Burkhart, and T. R. Clason. 1986. A comparison of competition measures for predicting growth of loblolly pine trees. *Canadian Journal of Forest Research* 16:1130–1237.

Draper, N. R., and H. Smith. 1981. *Applied Regression Analysis,* 2nd Ed. John Wiley and Sons, New York, NY, U.S.A. 709 pp.

Ford, E. D., and P. J. Diggle. 1981. Competition for light in a plant monoculture modelled as a spatial stochastic process. *Annals of Botany* 48:481–500.

Jeffers, J. N. R. 1972. *Mathematical Models in Ecology: The 12th Symposium of The British Ecological Society.* March 23–26, 1971, Grange-over-Sands, Lancashire, England. Blackwell Scientific Publications, Oxford, England, U.K. 398 pp.

Landsberg, J. J. 1981. *The Number and Quality of Driving Variables Needed to Model Tree Growth. Studia Forestalia Suecica* 160:43–50.

Linder, S. 1981. *Understanding and Predicting Tree Growth.* Proceedings of the SWECON workshop at Jädraäs, Sweden, September, 1979. *Studia Forestalia Suecica* 160. 86 pp.

Martin, G. L., and A. R. Ek. 1984. A comparison of competition measures and growth models for predicting plantation red pine diameter and height growth. *Forest Science* 30:731–743.

Munroe, D. D. 1974. Forest growth models: a prognosis. In: Fries, J., ed. *Growth Models for Tree and Stand Simulation.* Royal College of Forestry, Stockholm, Sweden. Pp. 7–21.

Pielou, E. C. 1972. On kinds of models (review of Jeffers, J. R., ed. *Mathematical Models in Ecology.* Blackwell Scientific Publications, Oxford, England, U.K. 398 pp. *Science,* September 15, 1972. Pp. 981–982.

Rennolls, K., and P. Blackwell. 1987. *An Integrated Forest Process Model: Its Calibration and Predictive Performance.* Forestry Commission Research and Development Paper 148. Great Britain Forestry Commission, London, England, U.K. 31 pp.

Smith, D. M. 1986. *The Practice of Silviculture,* 8th Ed. John Wiley and Sons, New York, NY, U.S.A. 527 pp.

CHAPTER 16

A METHODOLOGY FOR CONSTRUCTING, ESTIMATING COEFFICIENTS OF, AND TESTING PROCESS MODELS

Richard G. Oderwald and Richard P. Hans

Abstract. Process modeling provides an opportunity to develop theories of biological relationships, especially the complex relationships encountered in forestry. Without theories of tree growth, departures from the norm due to pollution, insects, and disease cannot be detected or predicted. However, our means of constructing and testing these models must change since current methods cannot use all the information provided by process specification. A methodology that can accomplish this task is described and demonstrated using forest growth and yield data.

INTRODUCTION

A process model attempts to describe a biological relationship by specifying rates, distribution, and proportionality among the variables involved in the relationship. These specifications can be used not only to design a model, but also to determine whether the model is workable as a theory. The properties of the specification can be used to determine its internal and external consistency and to establish all coefficient values. By specifying properties that are known to be true and others that we desire to test, we can construct and test biological theories. Hence, the methods described here will be referred to as "property fit." This proposed methodology is similar to methods described by Dolby (1982) and Mäkelä and Hari (1986). Much of the rationale for this methodology can be found in Popper (1968).

Variable selection and coefficient estimation for models are usually performed by regression techniques. While sufficient for many purposes, regression has characteristics harmful to successful process modeling, among them:

1. specification of a dependent variable, leading to one-way equations that cannot be rearranged in another context and obviating numerical compatibility between equations within a model;
2. reliability of a model over only the range of the data used to fit the model, allowing no extrapolation for experimental purposes and leaving regression-based models always one data set behind needs;
3. inherent nonconformity of biological data to necessary regression assumptions, causing appeals to regression robustness or use of biased-estimation techniques;

Dr. Oderwald is an Associate Professor in the Forestry Department, VPISU, Blacksburg, VA 24061, U.S.A. Mr. Hans was graduate Assistant in the Forestry Department, VPISU, at the time of this study.

4. requirement of a single minimized sum of squared differences, which cannot fit individual relationships in a model; and
5. inability to use known properties in an explicit manner that cannot be changed by the regression analysis, making summability and compatibility a matter of force or happenstance.

As an alternative to regression, the methodology of property fit is proposed, in which the processes to be examined are expressed as equations and considered as hypotheses to be tested. Equation coefficients may be set by hypothesis or as consequences of the hypothesized equations and coefficients. Tests of hypotheses are constructed from the properties and consequences of the system. In brief, once a system has been expressed, all the statements that can be made have been made. The only remaining task is to find out what statements were made and whether the statements can be accepted.

PROPERTY FIT METHODS

Specifying Hypotheses

The process description is considered as a system of equations that describe the relationships among the selected variables. This system must have the following characteristics:

1. Each equation in the system and the system itself must be considered as hypotheses to be tested.
2. Coefficients in the equations are necessarily a part of the equations containing them, and coefficient values must be determined as an integral part of the system of equations.

The equations and system must embody the relationships among the variables and equations as possible explanatory statements. Properties of the system and of the equations comprising it must be unaffected by transformation and rearrangement of the equations.

Specified properties may concern connection to a higher or lower level of detail than the process at hand. For example, the sum of masses of individuals may be required to total the population mass, or the constituents of mass of an individual may be required to total the individual's mass. The property of connection to a higher or lower order of detail is the most powerful property that can be specified.

Estimating Coefficients

Coefficient values are set in two ways: first, by direct hypothesis concerning the value; second, by consequence of the system and equation properties. Coefficient values must be independent of transformation and rearrangement of the equations. Conflict between a coefficient specified by hypothesis and a solution for that coefficient based on system properties means the hypothesis represented by the coefficient or some hypotheses represented by one or more equations must be rejected.

Each coefficient must have one and only one solution. If a coefficient value has no solution, then the system is incompletely specified. More than one distinct solution for a coefficient means a conflict exists in the system, and one or more hypotheses must be rejected.

Testing Hypotheses

The system hypotheses are tested by determining if the system is both internally and externally consistent. The system is checked for internal consistency by comparing the properties of each equation with the properties of the other equations and of the combined system. Conflicting properties mean one or more hypotheses must be rejected. All derivatives, integrals, limits, maximums, minimums, inflection points, and so on, of the equations also represent properties, many of which may be unknown when the equations are specified, that must be checked for internal consistency.

External consistency is determined by comparing properties and consequences of the system, including coefficient values, with desired properties and with data collected to examine those consequences. A property of the system must agree with the behavior of the biological process the system is intended to emulate. A property must also hold over the entire possible range of the variables concerned. If a property is untenable at the fringes of the range, then the property is externally inconsistent.

Comparison with data must be confined to specific tests that will concern only the consequence being considered. As compared to regression analysis, property fit methodology includes data at the opposite end of the model development process. Instead of collecting data over a broad range of conditions for which the model is to be used, and then selecting models and coefficients that best fit that data, relationships and coefficients that are hypothesized to be true are tested by comparison with specific data collected to test only those hypotheses.

In many instances, internal or external inconsistency in a system may be accepted if other features of the system are sufficiently attractive or if the inconsistency lies in a currently unimportant region of the system. Identifying an inconsistency for later revision or simply knowing it exists is sufficient in many instances. Also, a system may be internally consistent but not externally consistent.

Advantages of the Approach

The property fit approach can lead to theories of development. Past work can be used as a basis of future work, without having to begin anew each time. Successful models can be tested with other species and growth situations to determine common growth processes. In this way, it will be possible to develop general theories into which hypotheses concerning environmental stress can be incorporated and tested.

Theories and models for different aspects and different levels of detail for a process can be combined. Indeed, the combinations of different aspects or different levels represent new hypotheses that can advance the work in any one process description. Theories, however simple or complex, will provide a structure that can be worked toward, with, or against. For example, a stand-level model may be used as an upper limit for an individual-tree model, or a photosynthesis model may be used as a lower limit.

Data requirements are small and specific. A consequence or property is tested with data specifically collected to test that consequence. No large-scale data collection across the range of all variables is needed. In many instances, data already in hand may be sufficient, even if the process model is principally concerned with situations for which no data exists.

It is not to be imagined that there will be one and only one theoretical structure for any process or combination of processes, complete in all areas and difficult to dislodge. There should be many such structures, each with its own area of emphasis, strengths and weaknesses, and levels of detail. The more structures proposed the better, since each will help clarify the characteristics of the others. Examining differences among theories and trying to incorporate the best features of each, particularly as the level of detail becomes smaller and closer to the chemical reactions that occur in plants, will increase our knowledge and help to explain the effects of pollutants, disease, and insects.

EXAMPLE APPLICATIONS

The examples used here are forest growth and yield models. Growth and yield models are process models in that they are attempts to describe the relationships among the site and stand factors that go together to produce wood. These models need not be only attempts to find a function or functions that predict growth and yield. They can be viewed as attempts to develop theories of how forest variables interact.

As applied to growth and yield models, the property fit approach can help determine the differences among species in model specification and selection, what relationships can be used in many species, and the effects on the models of choosing one functional form over another. In this way, we can begin to develop a theory of forest growth and yield that can be tested for adequacy in many species and sites, find common factors among species and sites, and identify those areas in which further research is needed.

Growth and Yield Model of Clutter (1963)

Setting coefficients and testing hypotheses will be demonstrated using the volume and basal area growth and yield equations of Clutter (1963). The system will be taken as given, with no attempt to justify or respecify the hypotheses the system represents. As presented by Clutter, the equations are

$$\ln V = b_0 + b_1 S + b_2(\ln B) + b_3/A \quad (1)$$
$$\ln B = c_0 + c_1 S + c_2/A + c_3(\ln D)/A + c_4 S/A \quad (2)$$
$$dV/dA = V[(b_2/B) \times (dB/dA) - b_3/A^2] \quad (3)$$
$$dB/dA = B/A \times (c_0 + c_1 S - \ln B) \quad (4)$$

where ln is the natural logarithm, V is main stem cubic foot volume inside bark on a per acre basis, B is basal area per acre in square feet, S is site index in feet at index age 25, A is stand age in years, D is basal area per acre in square feet at age 20, and c's and b's are coefficients to be estimated. (To convert English units to metric units, please refer to Conversion Tables at the close of this volume.)

In the original paper, Clutter estimated the coefficients c_0, c_1, b_0, b_1, b_2, and b_3 by regression techniques involving several different dependent variables. Some coefficients were estimated independently several times as a result. The remaining coefficients $c_2 - c_4$, were not estimated, but can be recovered by rearranging the equations in which coefficients are estimated.

An example of developing system consequences for testing hypotheses can be shown by taking the limit of Equation 2 as age goes to infinity. The resulting ln(B) will be the maximum at the maximum value of B, therefore

$$\ln(\text{MaxB}) = c_0 + c_1 S \quad (5)$$

from Equations 2 and 4, represented as the culmination of mean annual increment (MAI) in B,

$$\ln(\text{Max MAI in B}) = c_0 + c_1 S - 1 \quad (6)$$

Two possible tests are thus generated:

1. Is site the sole controlling factor in these two instances?
2. Is the relationship linear in the logs?

These questions can be examined by comparison with data or with the expectations of the researcher. It is quite possible that no new data would be required to examine these questions since the questions are specific to basal area and site, without reference to age, volume, or other considerations. Further, from Equations 5 and 6, an additional property of the system can be educed, namely,

$$\ln(\text{Max MAI in B}) = \ln(\text{Max B}) - 1 \tag{7}$$

This property can also be tested, and may be even more revealing than the two separate questions above.

The properties represented by Equations 5, 6, and 7 are inherent in the system of equations. Even if they are ignored or set aside from immediate consideration, any solution for the system that contradicts these properties means the solution and the system are fundamentally at odds.

The coefficients c_0 and c_1 may be estimated using the properties represented by Equations 5 and 6. The only datum required is maximum basal area by site for the species being considered (or maximum mean annual increment in basal area), since the solution for c_0 and c_1 is the solution of two equations in two unknowns. For example, if maximum basal areas for site index 50 and site index 90 (15.2 and 27.4 m) are 120 and 170 square feet per acre (27.5 and 39 m^2/ha), respectively, the value of c_0 is 4.35 and the value of c_1 is .00875. This solution is species-specific only to the extent that the maximum basal area differs by species. Any species with the same maximums will have the same solutions for c_0 and c_1.

A property concerning relative rates of basal area and volume growth can be derived by taking the derivative of V in Equation 1 with respect to B and rearranging. The result is

$$b_2 = (dV/V) \ / \ (dB/B) \tag{8}$$

This requires that the ratio of relative change in volume to relative change in basal area be a constant, and that constant is b_2. The constancy of the ratio in Equation 8 can be tested by tree species, site, and age. This constancy also depends on the type of volume; it may be appropriate for cubic foot volume but not for board foot volume. The value of b_2 may be set using data specific to the ratio in Equation 8, or by hypothesis (e.g., $b_2 = 1$) with data used to test this hypothesis.

A conflict in the system can be brought to light by examining the point of maximum mean annual increment in volume, i.e., where $dV/dA = V/A$:

$$(b_2/B) \times (dB/dA) - (b_3/A^2) = 1/A \tag{9}$$

In many equations of this type, the analog of b_3 is equal to the negative of the age of MAI in volume. This is the case here if basal area is a maximum ($dB/dA = 0$) near the age of maximum MAI in volume. Therefore b_3 would be approximately -18 to -22. However if $dB/dA = B/A$ when $dV/dA = V/A$, then b_3 must be zero. Therefore, two widely different values exist for b_3, showing that the system contains an internal inconsistency. This contradiction can be seen in Clutter's original work; his fit of Equation 1 gave $b_3 = -22$, and his fit of Equation 3 gave $b_3 = -6$. The source of the contradiction lies in the commonality of the volume (1) and basal area (2) equations. If b_2 has a non-zero value, the terms b_0, b_1S, and b_3/A in Equation 1 become redundant, as exemplified by the possible value of zero for b_3, since their analogs are present in Equation 2. Avoiding this redundancy would require that basal area be removed from the volume equation (1) by setting b_2 to zero, thereby severing the connection between volume and basal area.

All coefficients in Equations 1–4 can be set using the procedures shown above.

Stand Density Index

Although Zeide (1987) has shown some difficulties with stand density index (SDI) (Reineke, 1933) by examining its implications, the intent here is to specify SDI and a mortality function as hypotheses to demonstrate the ways in which a system can be specified and tested. Stand density index is described as a limiting relation between stems per acre and average stem diameter. Presumably, stands grow to the SDI limit and then develop in number and size along the limit until they reach senescence or become old enough to drop below the line. The relation to the "3/2 power law of self thinning" is shown by Zeide.

The SDI equation is

$$\log_{10}N = \log_{10}c + d \times \log_{10}D \text{ or } N = c'D^d, \qquad c>0,\ d < 0 \tag{10}$$

where N is number of trees per acre, D is quadratic mean diameter per tree, c is number of trees at D equal one, c' is 10^c, and d controls rate of descent. In N and D, the function is concave upward, decreases at a decreasing rate, and N approaches zero as D becomes large.

The hypothesized mortality function is

$$N = a \times \exp(b \times A), \qquad a > 0,\ b < 0 \tag{11}$$

where N is number of trees per acre, A is age in years, a is initial number of trees per acre, and b controls rate of descent. This function also is concave upward, decreases at a decreasing rate, and approaches zero as a limit as A becomes large. This is a potential mortality function in that it begins at an initial number of trees at year 0, and decreases more quickly at small ages than at large ages.

While the properties of each equation individually are reasonable, the system as a whole must be tested. One possible test of the system is to derive a function for diameter development by age. This function should increase at a decreasing rate and approach a maximum as an asymptote or as an absolute. An additional desirable property is an inflection point at some low age, e.g., 5 to 15 years.

A diameter by age function from Equations 10 and 11 is

$$D = (a/c')^{1/d} \times \exp[(b/d)A], \qquad a/c' \ \&\ b/d > 0,\ 1/d < 0 \tag{12}$$

This function is concave upward with no upper limit, with a lower limit at $(a/c')^{1/d}$. None of the desirable properties is fulfilled. Therefore the combination of stand density index as a development line with the mortality function is untenable, and either the stand density index function or the mortality function or both must be reconsidered.

Further consideration can be given by determining the circumstances under which the diameter by age function can show the property of increasing at a decreasing rate, i.e., be concave down, when stand density index is used as a development line. To be concave down, the second derivative of D with respect to A must be less than zero, and by the chain rule,

$$d^2D/dA^2 = (dD/dN) \times (d^{2N/dA^2}) + (dN/dA)^2 \times (d^2/dN^2) < 0 \tag{13}$$

The slope dD/dN is negative, d^2D/dN^2 is positive, and $(dN/dA)^2$ is positive. The sign of d^2N/dA^2 depends on the shape of the mortality function. To achieve downward concavity,

$$(d^2N/dA^2)/(dN/dA)^2 > -(d^2D/dN^2)/(dD/dN) \tag{14}$$

since dD/dN is negative. The righthand side of Equation 14 is positive, so that downward concavity in D by A can only be achieved when the mortality function is concave up and has a more extreme slope than the stand density index line. Further, a mortality function that would combine with stand density index as a development line cannot have an inflection point ($d^2N/dA^2 = 0$), cannot have a concave down portion ($d^2N/dA^2 < 0$), cannot have a non-infinite limit as A goes to zero, must have a slope such that N increases, as A becomes small, more quickly than N increases as D becomes small and N decreases, as A becomes large, more slowly than N decreases as D becomes large.

Perhaps one or two of these conditions may be acceptable, but the combination of conditions, particularly the necessary steepness of the slope, excludes most possible mortality functions, including that represented by Equation 11. Therefore, the hypothesis of stand density index as a development line is rejected. The hypothesis of the mortality function represented by Equation 11 is still open to question.

Another approach to an SDI idea is to specify hypotheses concerning D and N with age and combine these equations to produce an N, D relation. For example, with $N = a \times \exp(b \times A)$ and $D = c \times \exp(d/A)$, where all terms are as previously defined, the resulting relation between N and D shows appropriate properties, i.e., concave down, slope increasing (negatively) with increasing D, and may be suitable as a possible line of N, D development. A further test is to develop a function between basal area and age. The basal area by age equation has a maximum, an inflection point, and shows a culmination of mean annual increment, but it also has a limit of zero as age becomes very large. The acceptability of this last property is moot.

DISCUSSION AND SUMMARY

Traditional regression-based methods of model development are not suited to examining the structure of a biological relationship; coefficients are separated from the equations representing the relationship by the statistical process, and the regression format makes biological considerations subordinate to statistical considerations.

It can be argued that the problems described above concerning regression-based models arise only because of improper application of that methodology, and that process-based models can never be successful because they lack an empirical base (Rossis, 1986). I believe, for reasons stated earlier, that even the informed and statistically correct application of regression methodology can never be more than a description of the data at hand, and it will not lead to an understanding of structural relationships. Only by attempting to discover the processes of a biological relationship can we bring the underlying structure to light.

Process models than can explicitly incorporate hypotheses about the structure of biological relationships can be used to test these hypotheses and help elucidate biological structure. These models have several other advantages over regression-based models: Data requirements are smaller, disturbances can be incorporated in a causal manner, and the model can have applicability to situations other than the one for which it was derived (Mäkelä and Hari, 1986).

Process-based structural models have been used, and compared to regression-based models, in many different fields, for example by Mäkelä and Hari (1986) in forest-stand modeling, by Sissons, Cross, and Robertson (1986) in models of biological effluent systems, and by Galea and Markatos (1987) in models of aircraft fires. Each of these authors stresses that the structural models are not yet as accurate as more traditional models, but that future development is promising.

The property-fit methodology is proposed as one means of using process models to elucidate structure. In this proposed methodology, the properties of the hypothesized processes are used to determine internal consistency of the hypotheses, to set values for coefficients not set by hypothesis, and to develop consequences than can be used for testing the hypotheses by comparison with data.

LITERATURE CITED

Clutter, J. L. 1963. Compatible growth and yield models for loblolly pine. *Forest Science* 9(3):354–371.
Dolby, G. R. 1982. The role of statistics in the methodology of the life sciences. *Biometrics* 38:1069–1083.
Galea, E. R., and N. C. Markatos. 1987. A review of mathematical modelling of aircraft cabin fires. *Applied Mathematical Modelling* 11:162–176.
Mäkelä, A., and P. Hari. 1986. Stand growth model based on carbon uptake and allocation in individual trees. *Ecological Modelling* 33:205–229.
Popper, K. R. 1968. *The Logic of Scientific Discovery.* Harper & Row, New York, NY, U.S.A. 480 pp.
Reineke, L. H. 1933. Perfecting a stand-density index for even-aged forests. *Journal of Agricultural Research* 46(7):627–638.
Rossis, G. 1986. Reductionism and related methodological problems in ecological modelling. *Ecological Modelling* 34:289–298.
Sissons, C. J., M. Cross, and S. Robertson. 1986. A new approach to the mathematical modelling of biodegradation processes. *Applied Mathematical Modelling* 10:33–40.
Zeide, B. 1987. Analysis of the 3/2 power law of self-thinning. *Forest Science* 33(2):517–537.

CHAPTER 17

ERROR PROPAGATION AND UNCERTAINTY IN PROCESS MODELING

Robert H. Gardner, Virginia H. Dale, and Robert V. O'Neill

Abstract. The principles and procedures for estimating uncertainties associated with model predictions have been extensively studied and applied to a broad spectrum of models. These studies have quantified relationships between data input, parameter estimation, and model results. Uncertainty analyses have also determined how additional data can improve the precision and reliability of predictions. The current challenge in applying error propagation techniques to forest process models lies with the development of a conceptual framework that can distinguish between (1) effects due to stochastic variables such as weather, (2) the uncertainties associated with the measurement of detailed physiological processes, and (3) the variability in results caused by local stand-level phenomena.

INTRODUCTION

Recent developments in systems analysis have revolutionized our approach to predicting the response of forest growth to environmental stress (e.g., Regional Acidification and Information Simulation [RAINS] model, Alcamo et al., 1987). Models are now routinely used for data synthesis and hypothesis testing, to explain observed phenomena, and as a basis for decision-making (Swartzman and Kaluzay, 1987). The usefulness of any given model is directly dependent on the accuracy and reliability of the predictions; but because models are imperfect abstractions of reality and because critical data are not always available, all predictions are subject to uncertainty.

The first serious look at the errors associated with ecological models was published by O'Neill in 1973. Since then a number of analytical (e.g., Clifford, 1973; Scavia, 1981) and computer (Lettenmaier and Richey, 1979; Iman and Helton, 1985) techniques have been developed to propagate errors through a model and characterize the resulting uncertainties. The large literature discusses a wide range of issues (e.g., O'Neill et al., 1980; Beck, 1981; Helton et al., 1981; Warwick and Cale, 1986) and illustrates the applicability of these methods (e.g., population dynamics: O'Neill, Gardner, Christensen et al., 1981; radioisotopes: O'Neill, Gardner, Hoffman, and Schwarz, 1981; Hoffman and Gardner, 1983; toxic substances: Bartell et al., 1984; atmospheric CO_2: Gardner and Trabalka, 1985;

Drs. Gardner, Dale, and O'Neill are employed by the Environmental Sciences Division, Oak Ridge National Laboratory, Oak Ridge, TN, 38731 U.S.A. This research was supported by the Office of Health and Environmental Research, U.S. Department of Energy, under Contract DE-AC05-84OR21400 with Martin Marietta Energy Systems, Inc., and in part by the Ecosystem Studies Program, National Science Foundation, under Interagency Agreement BSR 8516951 with the U.S. Department of Energy. Publication No. 3242. Environmental Sciences Division, Oak Ridge National Laboratory.

acid rain: Kamari et al., 1986. The results show that errors can be identified, their effect on uncertainties quantified, and the need for more and/or better data established. The purpose of this paper is to review these issues and recommend procedures for the analysis of process models concerned with estimating the effects of environmental stress on forest growth and development.

SOURCES OF ERROR

The uncertainty associated with model predictions results from all the errors involved in the process of model formation, calibration, parameter estimation, and choice of situations for which model simulations will be performed. These errors are not independent of one another, but can interact in unexpected ways to produce uncertainties in model predictions. Although there is no single best method for grouping errors for study, as evidenced by the variety of schemes which have been suggested (e.g., Burgess and Lettenmaier, 1975; O'Neill and Gardner, 1979; Hoffman and Gardner, 1983), the four categories illustrated in Figure 1 are heuristically useful. The categories show errors due to (1) inadequacies in the model; (2) the process of parameter estimation; (3) the natural variability of the processes being modeled; and (4) the context, or scenario, for the assessment question.

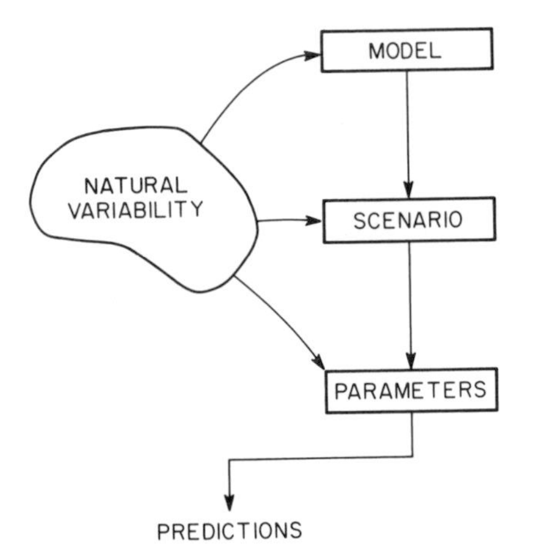

Figure 1. Schematic of sources of error in the modeling process.

Model-Related Errors

Errors due to inadequacies in the theory or the implementation of the model will always exist because of the inherent complexity of natural systems. Beck (1978, 1981, 1985) has pointed out that the exact structure of the model may be impossible to determine for many kinds of ecological systems, in part because many combinations of parameters may give acceptably good fits to data. However, errors due to inadequacies in the model may be small compared to those due to the estimation of model parameters or the variability of the natural system (O'Neill and Gardner, 1979). Thus, when assessing the effect of a particular source of error, it is helpful to express results relative to the magnitude of other sources of error.

A model can sometimes be improved by increasing the level of complexity or detail of the processes being simulated. However, theoretical results show that a complex model can produce greater prediction errors than a simpler model (Gardner et al., 1980: three articles cited), particularly if the increased complexity is based on processes that are more difficult to measure. This phenomenon has also been experienced in the verification of watershed models (Loague and Freeze, 1985).

A number of studies have addressed model simplification (O'Neill and Rust, 1979; Cale and Odell, 1979; Cale et al., 1983; Gardner et al., 1982) but have not established the level of model detail which will minimize prediction errors. The problem of determining the level of sufficient detail is particularly important when making broad-scale predictions of forest growth using models which simulate physiological changes in individual trees. The best approach, under these circumstances, is to establish confidence levels for alternative sets of models and then statistically compare simulation results (e.g., Gardner et al., 1980: three articles cited).

Parameter Errors

A number of papers have been written about the effects of parameter uncertainties on model predictions (e.g., Rose and Swartzman, 1981; Iman and Helton, 1985; Gardner, 1984). The results generally show that a few parameters dominate uncertainties, and, therefore, precise information on all parameters may not be necessary for making accurate and reliable model predictions. Methods now exist to quantify the effect of reducing the variance for any particular parameter (Iman, Conover, and Campbell, 1980; Gardner et al., 1983) or the effect of estimating the covariance among several parameters.

When parameter estimates are based on repeated measurements, a frequency distribution can be estimated that characterizes parameter uncertainties. However, replicated measurements are often unavailable, and uncertainties cannot be objectively determined. Subjective estimates which span the range of possible values (e.g., a uniform distribution, Rubinstein, 1981) have been used in these situations. The mixture of subjective approximations with actual measurements does not affect the application of methods for propagating parameter errors, but does affect the inferences which can be drawn from the results. The following two sections discuss the implications of this problem.

Natural Variability

The natural variability (Figure 1) of an empirically estimated parameter value can be defined as the variability of that value when measurements are repeated over time and/or space. However, the effects of natural variability on the accuracy and reliability of estimates are seldom known because repeated measures in time and space are rarely available. Unfortunately, the fixed scales chosen for many studies of forested systems result in natural variability dominating the uncertainty of the results. For example, the dynamics on a small forest plot are often dominated by a few large individuals. Death of a single, large individual produces a sudden change on that plot and, therefore, a large variance in the estimate of any processes. Measurements based on broader spatial scales will be less affected by individual-tree events, and processes measured at these scales will be less variable. This familiar fact makes it difficult to estimate point values but relatively easy to estimate broad-scale patterns. Because the ideal solution is likely to be prohibitively expensive, the practical solution is to avoid using measures from short time intervals or small spatial scales to extrapolate to long times or broad spatial scales.

Assessment Scenario

Systematic errors in predictions (i.e., model bias) may be important when models are used for situations that are different from those for which the model was developed. For instance, difficulties have been noted when models based on short-term observations have been used to predict the distant future (Kocher, 1982) or when models of plant uptake of radionuclides are based on greenhouse experiments and then applied to conditions with diverse soil types (Ng et al., 1982).

Validation experiments are usually performed to characterize model bias (Mankin et al., 1977), but a general knowledge of the relationship between model bias and changes in the assessment scenario is generally lacking. Because definite guidelines do not exist for the "domain of applicability" (Mankin et al., 1977) of assessment models, caution must be exercised when drawing inferences about new and different scenarios.

ERROR PROPAGATION

Concepts and Methods

The effect of parameter errors can be directly determined by analytical methods (Clifford, 1973) or numerically estimated by Monte Carlo simulation (Gardner et al., 1983; Iman, Conover, and Campbell, 1980). Although analytical expressions are difficult for complex models, they are useful for simple systems and heuristically helpful for the analysis of more complex expressions. For instance, if a model prediction, Y, is equal to the sum of its parameters, then the analytical expression for the variance of a Y is

$$S_y^2 = \Sigma_i^n(S_{Pi}^2) + 2\,\Sigma_i^n\Sigma_j^i\,(r_{ij}\,S_{Pi}\,S_{Pj}),\ i,j = 1,2,\ldots,n \quad (1)$$

where there are n parameters, p, with variance S_p^2 contributing to the variance of the prediction, S_y^2. The contribution to the variance of the prediction due to correlations among parameters is represented by multiplication of the correlation coefficient, r_{ij}, by the product of the standard deviations of the respective parameters. The usefulness of Equation 1 can be shown by a simple example. Suppose that (1) Y is the sum of three parameters, a, b, and c; (2) the parameters are normally distributed with mean values of 10.0, 1.0, and 0.1, respectively; (3) parameter variances are equal to 10% of their mean value; and (4) there are no covariances among the parameters. The expected value of Y is the sum of the means of the parameters, i.e., 11.1. The variance from Equation 1 is 1.11. Although the relative variance of the parameters is constant, parameter a contributes approximately 90% of the variance of Y. If the absolute variances of the parameters a, b, and c are set to 1.0, then the variance of Y is equal to 3.0, and all parameters contribute equally to the variance of Y. If parameters a and c are negatively correlated at −0.7, then in the first case, S_y^2 will decline by 13% to 0.97, and in the second case, S_y^2 will decline by 47% to 1.60.

An analytical method of relating parameter variance to predictions is also desirable for complex models. Sensitivity analysis has been extensively used to estimate this relationship (Brylinsky, 1972; Tomovic and Vukobratovic, 1972; Steinhorst et al., 1978; Iman, Conover, and Campbell, 1980; McRae et al., 1982). The differential sensitivity index (Tomovic, 1963), S, is the derivative of Y, the model prediction of interest, with respect to parameter p. The difficulty of obtaining the derivatives for many models, the assumption of small, instantaneous changes, and the large variance associated with most process models have motivated the development of numerical and statistical approaches to uncertainty analysis (Iman and Conover, 1979; Hoffman and Gardner, 1983; Gardner, 1984). The most widely used approach has been the Monte Carlo methods (Rubinstein, 1981) which (1) select a random set of parameter values from prespecified frequency distributions, (2) perform simulations to obtain model predictions which correspond to the random parameter values, and (3) analyze the combined set of parameters and prediction to characterize uncertainties and identify the parameters which most affect model results.

The data set produced by Monte Carlo methods allows a variety of analyses of the effect of parameter variability on model predictions (Iman and Helton, 1985; Gardner and Trabalka, 1985). Regression methods show that the slope, B, of the regression of Y on p_i is the least squares estimate of S, the classic sensitivity index. The slope, B, is an exact estimate of S if the parameter perturbations are very small (Gardner et al., 1981; Gardner, 1984). If several parameters are simultaneously and independently varied, then the multiple regression of Y on all the p_i's simultaneously estimates all sensitivities. The adequacy of this method of estimating linear relationships between model predictions and parameters can be evaluated by inspection of R^2, the ratio of regression sum of squares to total sum of squares. If R^2 is nearly 1.0, then linear methods are adequate to describe the relationship between parameters and predictions. The divergence of R^2 from 1.0 indicates that non-linear effects and interactions between parameters are important.

This multiple regression equation also provides the necessary information for a useful uncertainty index (Gardner and Trabalka, 1985). The contribution of each parameter to the regression sum of squares (i.e., the amount of the variability of Y explained by a particular parameter) divided by the total sum of squares and multiplied by 100 forms an index, U_i, representing the percent variability explained by each parameter. The values of U_i range from 0.0 to 100.0, allowing a comparison among parameters. (Note that the indices, S_i and B_i, are not normalized, being expressed in units of Y and p_i.) The regression-based indices are produced by a single computer method, and they provide a parsimonious approach for estimating model sensitivities and uncertainties. In addition, the adequacy of each index can be determined by inspection of the R^2 statistic. Recent improvements in the selection of random values for the parameters (McKay et al., 1979; Iman, Davenport, and Ziegler, 1980) plus improvements in the speed of computers have given numerical approaches a great advantage over analytical approaches.

APPLICATION TO FOREST PROCESS MODELING

The FORET (Forests of East Tennessee) model (Shugart, 1984; Solomon, 1986) of forest development has been used to estimate the influences of natural and anthropogenic disturbances on forest dynamics (Dale et al., 1986; Pastor et al., 1987; Dale and Gardner, 1987). The number and complexity of interactions within FORET make this model an interesting example in demonstrating a comprehensive approach to model uncertainties (Dale et al., 1988). FORET simulates the major processes affecting the birth, growth, and death of individual trees as a function of prevailing light, moisture, and temperature. Tree growth is based on the optimal growth pattern of each species, reduced by the limiting effects of light, temperature, moisture, and competition. Tree mortality occurs as a result of inadequate growth or as a stochastic function of the maximum age of a species. Many trees representing a variety of species can occur on a single plot, and forest characteristics can be obtained by summing the simulations of many plots.

FORET includes separate modules for tree birth, growth, and death, with the effect of each process on the major life stages treated sequentially. This modular form permits uncertainty analysis to focus separately on each module and assess parameter variability independently of stochastic elements in the model or interaction with other modules.

The growth function in FORET (Figure 2) is dependent on dbh, the diameter at breast height in cm; H, the height of the tree in m; R, a constant growth rate parameter; LA, the leaf area of the tree; and DBH_{max} and H_{max}, which are the maximum diameter and height of the species obtained directly from measures of tree growth (e.g., Botkin et al., 1972). The optimal diameter increment curve depends on the species characteristics that determine the parameter values.

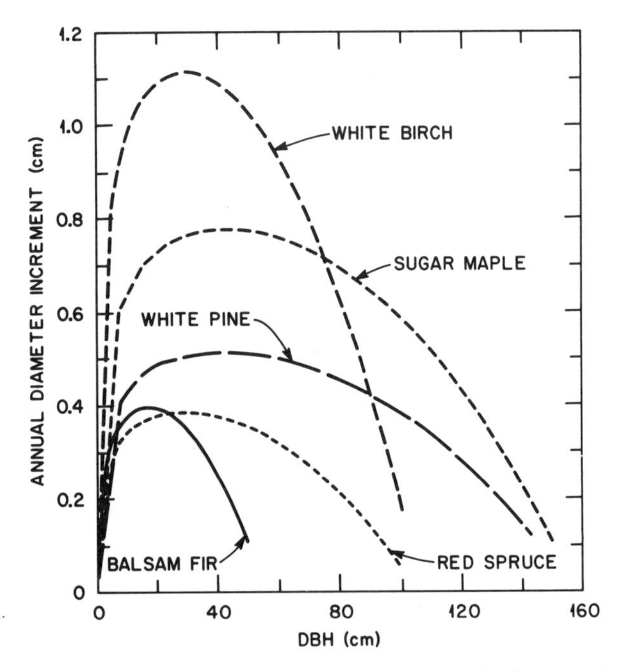

Figure 2. **Optimal annual diameter growth versus diameter for five species representing a range of the life-history characteristics of these species in the northeastern U.S.A. The species presented are white birch (*Betula papyrifera* Marsh.), sugar maple (*Acer saccharum* Marsh.), white pine (*Pinus strobus* L.), red spruce (*Picea rubens* Sarg.), and balsam fir (*Abies balsamea* L. Mill.). In the FORET model, the life-history characteristics are combined to produce different patterns of optimal annual growth as compared to the current dbh of each tree.**

Uncertainty Analysis

An uncertainty analysis was performed on FORET by simultaneously varying the parameter values by 10% of the estimated mean and using least squares regression to estimate the relationships between parameters and predictions (Dale et al., 1988). The results show that DBA_{max} and AGE_{max} contribute at least 13% and 10%, respectively, to the variability of predicted optimal growth rate, with other parameters causing less than a 4% effect. This ranking of model parameters permits a quantitative estimate of the expected improvement in predictions resulting from experiments to improve the estimates of these parameters. Because FORET has a modular structure, it is possible to examine other model components (e.g., effects of nutrients or moisture on tree growth) in a similar fashion to determine which set of parameters and processes have the greatest effect on model predictions.

Uncertainties associated with broad-scale predictions depend on how plot-specific results are extrapolated to a larger region. The spatial variability characteristic of forest inventory data is an important component of this uncertainty. Figure 3 identifies the effect of extrapolation errors by plotting contours of mean volume for typical values of volume and area in southeastern Vermont in the U.S.A. Projected volume is obtained from simulations of FORET, and measured area is from U.S. Department of Agriculture (USDA) Forest Service data. If the uncertainties associated with both the projected forest-type volume estimate and the measured forest-type area are known, then the extrapolated estimate of volume has an uncertainty which can be computed directly. Since the contours of total volume are closer together for forest types with high volumes and large area estimates, uncertainties in the higher estimates have a greater effect upon projections of total volume. For southeast Vermont, the maple (*Acer*)-beech (*Fagus*)-birch (*Betula*) type has both the highest predicted volume and the highest estimated area, indicating that the greatest potential for error lies in predicting volume of this forest type for the region. This approach identifies the forest type and area where the greatest accuracy in data collection is required.

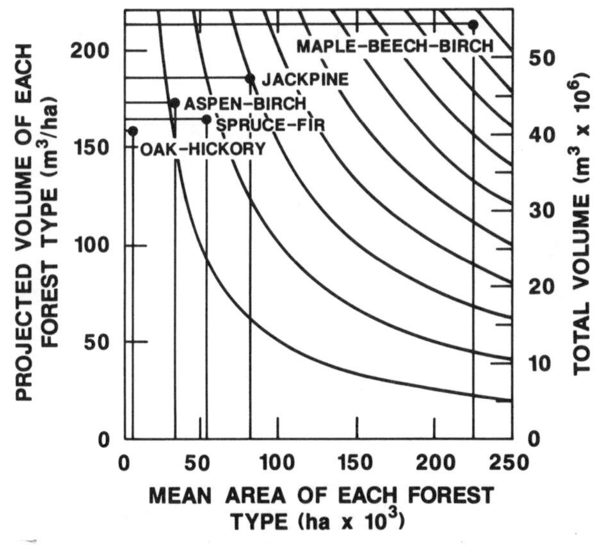

Figure 3. **Contour plot of the mean tree volume for each forest type in the southeast climatic division of Vermont, U.S.A. The five forest types considered are maple (*Acer*)-beech (*Fagus*)-birch (*Betula*), jackpine (*Pinus banksiana* Lamb.), aspen (*Populus*)-birch, spruce (*Picea*)-fir (*Abies*), and oak (*Quercus*)-hickory (*Carya*). The mean tree volume for the division is the product of the projected volume per ha and the mean area of each forest type. Greater potential for errors occurs in the top right corner of the plot because the contours are closer together.**

COMPARING PREDICTIONS AGAINST DATA

When model bias has been minimized and the uncertainties in the parameter estimates have been reduced to a level of natural variability, then the limits of predictability have been reached. This situation implies that a "stopping rule" for model development can be developed by a comparison of predictions with data. Most methods of comparing model predictions with data are based on statistical tests designed to detect significant differences between means (e.g., Reynolds, 1984). These methods are not useful when the frequency distributions of predictions and observations are different or when variability of predictions and/or data is large.

Not all differences between models and data are of equal interest or are equally critical in assessing the adequacy of the model. Key comparisons of interest are frequently the differences between the model and data in the mean, variance, and the range of predicted values. If $\overline{P}$ is the mean of the predicted values, $\overline{O}$ is the mean of the measured values, and S_o is the standard deviation of the measured values, then relative bias, rB, can be defined as $(\overline{P}-\overline{O})/S_o$. Relative bias measures model accuracy by quantifying the mean difference between the model and data in units of standard deviations of the data. If S_p^2 is the variance of the predicted effects and S_o^2 is the variance of the measured effects, then F, the ratio of the variances ($F = S_p^2/S_o^2$), indicates the relative spread of the model predictions compared to the data. For $F < 1$, the model predictions have a narrower distribution than the data, while for $F > 1$, the converse is true.

For any combination of rB and F, there exists a probability, Φ, that the distributions are the same. By assuming that the model and data are normally distributed, it is possible to express the relationships between rB, F, and Φ as a contour plot (Figure 4). If the model and data were drawn from identical distributions, then rB will equal 0.0, F will equal 1.0, and Φ will equal 1.0. The results of this procedure quantify the degree of confidence that can be placed in model predictions as a representation of the data.

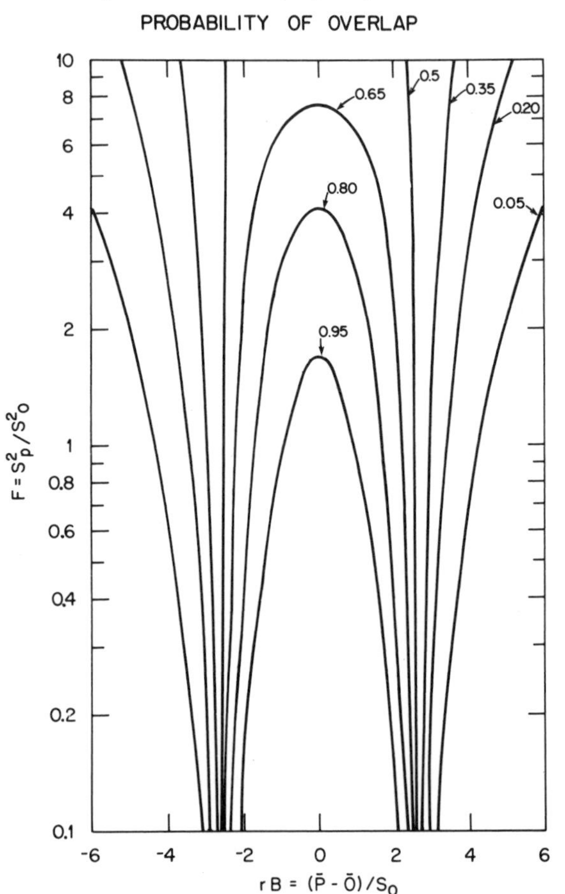

Figure 4. Contour plot of the relationship between the relative bias (rB) and the ratio of variance in model predictions (S_p) to the variance in observations (S_o) ($F = S_p^2/S_o^2$) for different degrees of overlap (Φ) between two normal distributions.

The assumption of a normal distribution is not critical to the method but is convenient because it allows calculation of Φ when only the means and variances are known. It is possible to generalize this procedure and make the results free of distributional assumptions. For instance, any percentile of the cumulative distribution can be substituted for the mean, and the results can be expressed in terms of the coincidence of two empirical frequency distributions.

The probability of overlap for normal distributions as a function of rB and F is illustrated in Figure 4. The graph can best be interpreted by an example. If rB = 3.0 (predictions overestimate observations by an average factor of 3.0 standard deviations of the observations) and F = 4.0, then the probability of overlap is 0.41. If the variance of the predictions is drastically reduced so that F = 1.0, then Figure 4 shows that the probability of overlap is only marginally increased to 0.5. However, if the relative bias is reduced to 0.0 while F remains at 4.0, then the probability of overlap, Φ, will increase to nearly 0.8. In this situation, the most efficient method of improving predictions would be to reduce the bias.

Figure 5 shows one example of the application of this method with a desirable degree of overlap defined as 0.8 (i.e., the correspondence between model predictions and data is at least 80%). Observed values are based on 1972 and 1983 continuous forest inventory (CFI) of 15 plots in Essex County, Vermont, U.S.A. (Kingsley, 1977). Predicted values are from sets of FORET simulations that were initiated with the stand structure of the 15 plots from the county measured in 1948 by the Continuous Forest Inventory of the USDA Forest Service, and monthly temperature and precipitation data from Essex County (Olson et al., 1980). When actual and predicted volume-diameter distributions were compared for each forest type, all but one value fell within the $\Phi = 0.8$ criterion. The variance of the predictions was generally less than that of the observations as indicated by most values lying within the lower left quadrant of the probability plot. Bartell et al. (1986) also found that relative bias for an aquatic model was low and that the variance of model predictions underestimated the variability obtained from direct measurements of the system. These results indicate that efforts to improve forest volume predictions should concentrate on reducing the uncertainty in actual forest measurements and/or identify those processes that operate at the spatial scales of the CFI to simulate the observed variability.

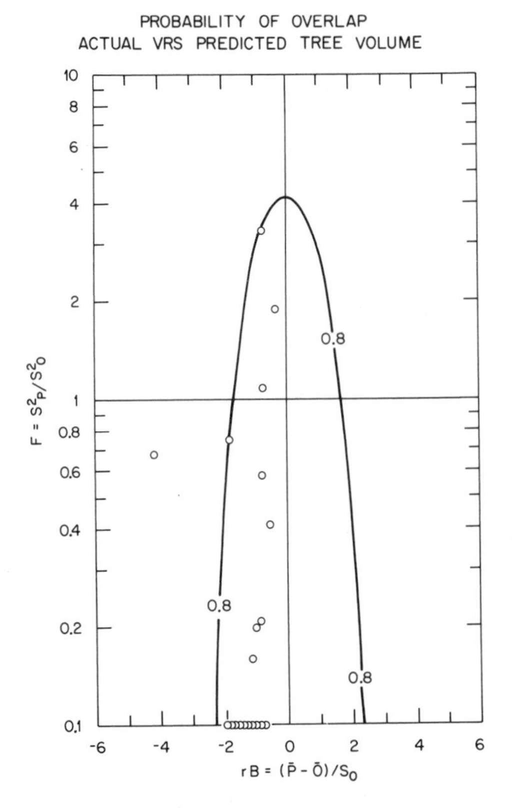

Figure 5. The points of comparison of model projections of volume by forest type and diameter class compared to USDA Forest Service data collected in 1972 and 1983 for Essex County, Vermont, U.S.A.

CONCLUSIONS

The limits of predictability of environmental models are ultimately determined by stochastic effects on the system being studied. Therefore, the objective of an uncertainty analysis is not to reduce all uncertainties to zero, but to quantify current levels of model uncertainties and indicate where new data can be used to most improve model predictions. To achieve these goals, it is necessary to consider the entire context of the assessment process (Figure 1), including the quality of existing data, errors intrinsic to the model, and the adequacy of the model and data to address the assessment question. Consideration of only a single component of error will not insure that additional data or model improvements alone will reduce prediction uncertainties.

Care must be taken in drawing conclusions from an analysis of model uncertainties because these conclusions are dependent on the quality of information used to perform the analysis. In the ideal situation, all parameters are estimated from extensive site-specific experiments. In that case, estimates of uncertainties may indicate the true limits of predictability of the model. If these "ideal" predictions correspond to the observed mean and variance of the system, then model development and parameter estimation need not continue. However, extensive information is seldom available for more than a few parameters, and others must be estimated from a literature review or expert opinion. In this "realistic" case, an analysis of uncertainties can best be used to indicate what new information is critical for model improvement.

The development and application of uncertainty analysis is still an active field of research. The current challenge for forest growth models lies with the development of a conceptual framework that can determine (1) the expected effects of broad-scale stochastic variables (e.g., weather events) and thus estimate the ultimate levels of system variability; (2) the uncertainties associated with the measurement of detailed physiological processes and thus define the level of detail needed for reliable model development; and (3) the variability in results caused by local stand-level phenomena and thus establish the uncertainties involved in the process of extrapolating from one site to another.

LITERATURE CITED

Alcamo, J., M. Amann, J.-P. Hettelingh, M. Holmberg, L. Hordijk, J. Kamari, L. Kauppi, P. Kauppi, G. Kornai, and A. Mäkelä. 1987. Acidification in Europe: a simulation model for evaluating control strategies. *AMBIO* 16:232.

Bartell, S. M., J. E. Breck, R. H. Gardner, and A. L. Brenkert. 1986. Individual parameter perturbation and error analysis of fish bioenergetics models. *Canadian Journal of Fisheries and Aquatic Science* 43:160–168.

Bartell, S. M., R. V. O'Neill, and R. H. Gardner. 1984. Modeling effects of toxicants on aquatic systems. In: Schreiber, D. L., ed. *Water for Resource Development.* American Society of Civil Engineers, New York, NY, U.S.A. Pp. 82–85.

Beck, M. B. 1978. Random signal analysis in an environmental sciences problem. *Applied Mathematical Modelling* 2:23–29.

Beck, M. B. 1981. Hard or soft environmental systems? *Ecological Modelling* 11:233–251.

Beck, M. B. 1985. Structures, failure, inference and prediction. In: Barker, H. A., and P. C. Young, eds. *Proceedings IFAC Symposium on Identification and System Parameter Estimation.* Oxford, Pergamon, England, U.K. Pp. 1443–1448.

Botkin, D. B., J. F. Janak, and J. R. Wallis. 1972. Some ecological consequences of a computer model of forest growth. *Journal of Ecology* 60:849–872.

Brylinsky, M. 1972. Steady-state sensitivity analysis of energy flow in a marine ecosystem. In: Patten, B. C., ed. *System Analysis and Simulation in Ecology,* Vol. 2. Academic Press, New York, NY, U.S.A. Pp. 81–101.

Burgess, S. B., and D. P. Lettenmaier. 1975. Probabilistic methods in stream quality management. *Water Resources Bulletin* 11:115–130.

Cale, W. G., and P. L. Odell. 1979. Concerning aggregation in ecosystem modeling. In: Halfon, E., ed. *Theoretical Systems Ecology.* Academic Press, New York, NY, U.S.A. Pp. 283–298.

Cale, W. G., R. V. O'Neill, and R. H. Gardner. 1983. Aggregation error in nonlinear ecological models. *Journal of Theoretical Biology* 100:539–550.

Clifford, A. A. 1973. *Multivariate Error Analysis.* John Wiley and Sons, New York, NY, U.S.A. 112 pp.

Dale, V. H., and R. H. Gardner. 1987. Assessing regional impacts of growth declines using a forest succession model. *Journal of Environmental Management* 24:83–93.

Dale, V. H., M. A. Hemstrom, and J. F. Franklin. 1986. Modeling the long-term effects of disturbance on forest sucessions, Olympic Peninsula, Washington, U.S.A. *Canadian Journal of Forest Research* 16:56–57.

Dale, V. H., H. I. Jager, R. H. Gardner, and A. E. Rosen. 1988. Using sensitivity and uncertainty analysis to improve prediction of broad-scale forest development. *Ecological Modelling* 42:165–178.

Gardner, R. H. 1984. A unified approach to sensitivity and uncertainty analysis. In: Hamza, M. H., ed. *Applied Simulation and Modelling.* Proceedings of the IASTED International Symposium, June 4–6, 1984, San Francisco, CA, U.S.A. ACTA Press, Anaheim, CA, U.S.A. Pp. 155–157.

Gardner, R. H., W. G. Cale, and R. V. O'Neill. 1982. Robust analysis of aggregation error. *Ecology* 63:1771–1779.

Gardner, R. H., D. D. Huff, R. V. O'Neill, J. B. Mankin, J. Carney, and J. Jones. 1980. Application of error analysis to a marsh hydrololgy model. *Water Resources Research* 16:659–664.

Gardner, R. H., R. V. O'Neill, and J. H. Carney. 1981. Spatial patterning and error propagation in a stream ecosystem model. In: *Proceedings of 1981 Summer Computer Simulation Conference,* Simulation Councils, Inc., LaJolla, CA, U.S.A. Pp. 391–395.

Gardner, R. H., J. B. Mankin, and W. R. Emanuel. 1986. A comparison of three carbon models. *Ecological Modeling* 8:313–332.

Gardner, R. H., R. V. O'Neill, J. B. Mankin, and D. Kumar. 1980. Comparative error analysis of six predator-prey models. *Ecology* 61:323–332.

Gardner, R. H., B. Rojder, and U. Bergstrom. 1983. *PRISM: A Systematic Method for Determining the Effect of Parameter Uncertainties on Model Predictions.* Technical Report, Studsvik Energiteknik AB report/NW-83/555, Nykoping, Sweden.

Gardner, R. H., and J. R. Trabalka. 1985. *Methods of Uncertainty Analysis for a Global Carbon Dioxide Model.* Technical Report DOE/OR/21400-4, U.S. Department of Energy, Washington, DC, U.S.A.

Helton, J. C., J. B. Brown, and R. L. Iman. 1981. *Risk Methodology for Geologic Disposal of Radioactive Waste: Asymptotic Properties of the Environmental Transport Model.* Technical Report SAND 79-1908, Sandia National Laboratory, Albuquerque, NM, U.S.A.

Hoffman, F. O., and R. H. Gardner. 1983. Evaluation of uncertainties in environmental radiological assessment models. In: Till, J. E., and H. R. Meyer, eds. *Radiological Assessment: A Textbook on Environmental Dose Assessment.* NUREG/CR-3332, ORNL-5968, U.S. Nuclear Regulatory Commission, Washington, DC, U.S.A. Pp. 11:1–11:55.

Iman, R. L., and W. J. Conover. 1979. The use of the rank transform in regression. *Technometrics* 21:499–590.

Iman, R. L., W. J. Conover, and J. E. Campbell. 1980. *Risk Methodology for Geologic Disposal of Radioactive Waste: Small Sample Sensitivity Analysis Techniques for Computer Models, with an Application to Risk Assessment.* Technical Report NUREG/CR-1397, U.S. Nuclear Regulatory Commission, Washington, DC, U.S.A.

Iman, R. L., J. M. Davenport, and D. K. Ziegler. 1980. *Latin Hypercube Sampling (Program User's Guide).* Technical Report SAND 79-1473, Sandia National Laboratory, Albuquerque, NM, U.S.A.

Iman, R. L., and J. C. Helton, 1985. *A Comparison of Uncertainty and Sensitivity Analysis Techniques for Computer Models.* Technical Report NUREG/CR-3904 SAND84-1461, Sandia National Laboratory, Albuquerque, NM, U.S.A.

Kamari, J., M. Posch, R. H. Gardner, and J. -P. Hettelingh. 1986. *A Model for Analyzing Lake Water Acidification on a Large Regional Scale,* Part 2. *Regional Application.* Working Paper WP-86-66, International Institute for Applied Systems Analysis, A-2361 Laxenburg, Austria.

Kingsley, N. P. 1977. *The Forest Resources of Vermont.* USDA Forest Service Resources Bulletin NE-46, Northeast Forest Experiment Station, Upper Darby, PA, U.S.A.

Kocher, D. C., ed. 1982. *Proceedings of the Symposium on Uncertainties Associated with the Regulation of the Geologic Disposal of High-Level Radioactive Waste,* March 9–13, 1981, Gatlinburg, TN, U.S.A. Conference Report NUREG/CP-0022, CONF 801372, U.S. Nuclear Regulatory Commission, Washington, DC, U.S.A.

Lettenmaier, D. P., and J. E. Richey, 1979. Use of first order analysis in estimating mass balance errors and planning sampling activities. In: Halfon, E., ed. *Theoretical Systems Ecology.* Academic Press, New York, NY, U.S.A. Pp. 79–104.

Loague, K. M., and R. A. Freeze. 1985. A comparison of rainfall-runoff modeling techniques on small upland catchments. *Water Resources Research* 21:229–248.

Mankin, J. B., R. V. O'Neill, H. H. Shugart, and B. W. Rust. 1977. The importance of validation in ecosystem analysis. In: Innis, G. S., ed. *New Directions in the Analysis of Ecosystems.* Simulations Councils Proceedings Series, Vol. 5, No. 1, pp. 63–71.

McKay, M. D., W. J. Conover, and R. J. Beckman. 1979. A comparison of three methods for selecting values of input variables in the analysis of output from a computer code. *Technometrics* 21:239–245.

McRae, G. J., J. W. Tilden, and J. H. Seinfeld. 1982. Global sensitivity analysis: a computational implementation of the Fourier amplitude sensitivity test (FAST). *Computers and Chemical Engineering* 6:15–25.

Ng, Y. C., C. S. Colsher, and S. E. Thompson. 1982. *Soil-to-plant concentration factors for use in radiological assessments.* Technical Report NUREG/CR-2975 (UCID-19463), U.S. Nuclear Regulatory Commission, Washington, DC, U.S.A.

Olson, R. J., C. J. Emerson, and M. K. Nungesser. 1980. *Geoecology: a country-level experimental data base for the conterminous United States.* Technical Report ORNL/TM-7351, Oak Ridge National Laboratory, Oak Ridge, TN, U.S.A.

O'Neill, R. V. 1973. Error analysis of ecological models. In: Nelson, D. J., ed. *Radionuclides in Ecosystems.* Conference Report CONF-710501, National Technical Information Service, Springfield, VA, U.S.A. Pp. 898–908.

O'Neill, R. V., and R. H. Gardner. 1979. Sources of uncertainty in ecological models. In: Zeigler, B. P., M. S. Elzas, G. J. Kliv, and T. I. Oren, eds. *Methodology in Systems Modelling and Simulation.* North-Holland Publishing, the Netherlands. Pp. 447–463.

O'Neill, R. V., R. H. Gardner, S. W. Christensen, W. Van Winkle, J. H. Carney, and J. B. Mankin. 1981. Effects of parameter uncertainty in density-independent and density-dependent Leslie models for fish populations. *Canadian Journal of Aquatic Sciences* 38:91–100.

O'Neill, R. V., R. H. Gardner, and J. B. Mankin. 1980. Analysis of parameter error in a nonlinear model. *Ecological Modelling* 8:297–311.

O'Neill, R. V., R. H. Gardner, F. O. Hoffman, and G. Schwarz. 1981. Effects of parameter uncertainty on estimating radiological dose in man: a Monte Carlo approach. *Health Physics* 40:760–764.

O'Neill, R. V., and B. W. Rust. 1979. Aggregation error in ecological models. *Ecological Modelling* 7:91–105.

Pastor, J., R. H. Gardner, V. H. Dale, and W. M. Post. 1987. Successional changes in nitrogen availability as a potential factor contributing to spruce declines in boreal North America. *Canadian Journal of Forest Research* 17:1394–1400.

Reynolds, M. R. 1984. Estimating the error in model predictions. *Forest Science* 30:454–469.

Rose, K. A., and G. L. Swartzman. 1981. A review of parameter sensitivity methods applicable to ecosystem models. Technical Report NUREG/CP-2016, U.S. Nuclear Regulatory Commission, Washington, DC, U.S.A.

Rubinstein, R. Y. 1981. *Simulation and the Monte Carlo Method.* John Wiley and Sons, New York, NY, U.S.A.

Scavia, D. 1981. Comparison of first-order error analysis and Monte Carlo simulation in time-dependent lake eutrophication models. *Water Resources Research* 17:1051–1059.

Shugart, H. H. 1984. *A Theory of Forest Dynamics.* Springer-Verlag, New York, NY, U.S.A. 278 pp.

Solomon, A. S. 1986. Transient response of forests to CO_2-induced climate change: simulation modeling experiments in eastern North America. *Oecologia* 68:567–579.

Steinhorst, R. K., H. W. Hunt, G. S. Innis, and K. P. Haydock. 1978. Sensitivity analysis of the ELM model. In: Innis, G. S., ed. *Grassland Simulation Model.* Springer-Verlag, New York, NY, U.S.A. Pp. 231–255.

Swartzman, G. L., and S. P. Kaluzny. 1987. *Ecological Simulation Primer.* Macmillan Publishing, New York, NY, U.S.A.

Tomovic, R. 1963. *Sensitivity Analysis of Dynamic Systems.* McGraw-Hill, New York, NY, U.S.A.

Tomovic, R., and M. Vukobratovic. 1972. *General Sensitivity Theory.* Elsevier Science Publishers, New York, NY, U.S.A.

Warwick, J. J., and W. G. Cale. 1986. Effects of parameter uncertainty in stream modeling. *Journal of Environmental Engineering* 112:489–499.

CHAPTER 18

ERROR BUDGETS: A MEANS OF ASSESSING COMPONENT VARIABILITY AND IDENTIFYING EFFICIENT WAYS TO IMPROVE MODEL PREDICTIVE ABILITY

George Gertner

Abstract. An error-propagation method was used to calculate the variance of growth and yield predictions made with a modified version of the Stand and Tree Evaluation and Modeling System (STEMS), USDA North Central Forest Experiment Station, St. Paul, MN, U.S.A. With the error-propagation method, it was possible to generate an error budget that shows the percent of variance in the overall predictions due to the individual component models of the simulation model. The primary source of variability in projected stand attributes was shown to be the individual tree mortality model. Because mortality was found to be such an important source of variability, it is suggested that the most efficient way to improve the predictive ability of the overall simulation model is to direct future research at improving the individual tree mortality models through additional data collection and model restructuring.

INTRODUCTION

Tools are being assembled that will aid in the design of new experiments and sample surveys to improve the predictive ability of forest growth simulation models. These tools will help predict forest growth response to environmental stresses. A forest growth simulation model can be considered a reservoir of information about structural and functional processes. Information in a model is usually enhanced in an iterative fashion as depicted in Figure 1. The tools being developed through this iterative process will help researchers to efficiently and systematically improve a simulation model and its components.

These tools will allow the assessment of a number of different qualitative and quantitative factors related to the simulation model and potential sample and experimental designs in the course of developing strategies for improving the model. The following factors, for example, are among those considered:

1. Ranking of the importance of individual component models in an overall simulation model.
2. Trueness of component models.

Dr. Gertner is Associate Professor, Department of Forestry, University of Illinois, Urbana, IL, 61801 U.S.A. This study was partially supported by McIntire-Stennis Cooperative Forestry Research Project MS-55-326.

3. Degree of empiricism of component models.
4. Cost of future experimentation or sampling.
5. Robustness of potential experimental or sample designs.
6. Restrictions due to biological factors, weather, measurement equipment, and so on.
7. Structural sensitivity.

A method to rank the importance of the component models of a simulation model is described. The method is applied in ranking the major component models in the Stand and Tree Evaluation Modeling System (STEMS) (Belcher et al., 1982).

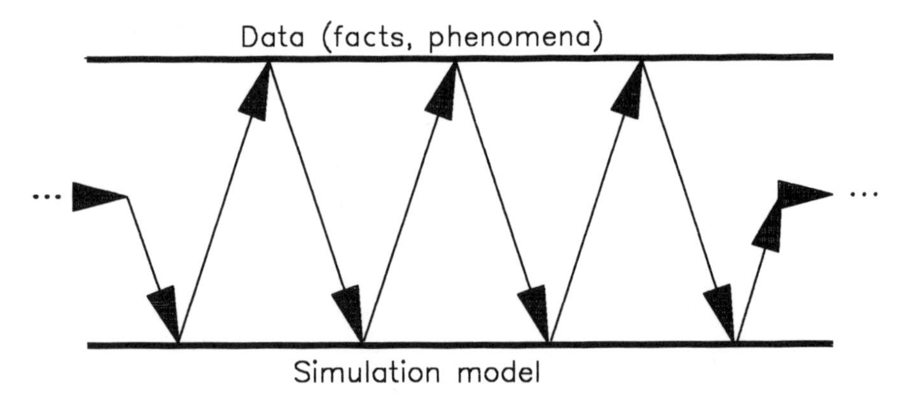

Figure 1. The iterative process of model-building.

ERROR BUDGET

STEMS is a single-tree growth projection system developed by the United States Department of Agriculture (USDA) Forest Service to evaluate a wide variety of forest types in the central U.S.A. A total of 3,000 plots with 60,000 trees were used to calibrate the central states version of the STEMS model. It has primarily been used to evaluate management and silvicultural alternatives, to update forest survey plots, and recently to estimate forest response to interacting stresses (Shifley, 1988). The component models of STEMS are not strictly empirical, but are somewhere between empirical and mechanistic. The structure of the models is based on the calibration data, as well as on the geometric and mechanistic properties that can be expected as a result of a long history of growth and yield research.

Gertner (1987 and 1988) used an error-propagation method as a computationally efficient alternative to Monte Carlo methods to obtain estimates of the precision of predictions made with a modified version of STEMS. Error propagation provides a direct method for calculating the variance of predictions. For each function in the growth model, an approximation is used to estimate the variance of prediction made with the function when there is random error in the input of the function. By approximating the variance for each function, the random errors that pass from one function to another can be estimated and accounted for. With an iterative model such as a multi-year projection system, the initial errors entered into the functions of the model might be due to errors in the state variables. After the first iteration, the errors will be due to errors in predictions from past iterations. With each additional iteration, the variance will increase as errors propagate through the system. The final variance of the prediction for the overall system will be due to the accumulation of all the errors.

Variance estimates made with the STEMS model were originally used to gauge the reliability and precision of predictions, to calculate confidence intervals, to statistically test hypotheses when experiments are performed with the model, and to weight projections used as an auxiliary source of information in combination with on-the-ground sample estimates.

Another use of this technique is to generate an error budget. An error budget shows the effects of individual errors and groups of errors on the accuracy of simulation projections. It can be considered a catalog of the contributions of the different error sources to the overall accuracy of the system.

The goal in developing the error budget is to account for all major sources of errors that can be expected with a model such as STEMS. Some types of errors that should be considered in developing the error budget for the model are the following:

1. Errors in measurements of state variables which are dependent on quality measurement equipment, time and budget constraints, and so forth.
2. Sampling errors when sampling methods are used to estimate the state variables. These errors are the result of taking only a subset of the total population when making estimates of the state variables. The size of the errors is dependent on sample size, plot size, sampling method, and so on.
3. Grouping errors which are due to grouping observations into classes (e.g., dbh classes). The grouping itself is a source of error.
4. Process errors which refer to modeling errors and stochastic errors that are not accounted for by the component models.

Once an error budget is developed for a model, the important sources of error can be ranked according to their contribution to the overall variance. On the basis of the error budget, future model refinement, experimentation, and sampling can be directed at reducing the more important sources of error.

STEMS ERROR BUDGET (PROCESS ERRORS)

Currently, an error budget is being developed for the STEMS model. Presented here is the portion of the error budget due to process error which has been found to contribute a major proportion of the variability in projected forest stand growth. The budget was developed by projecting 40 prism plots taken randomly from the oak-hickory (*Quercus-Carya*) component of Allerton Park, a mixed hardwood forest owned by the University of Illinois, Urbana, IL, U.S.A.

The variance of predictions made with the overall simulation model was partitioned according to the important component models. The four major individual tree models used in the STEMS model are (1) annual diameter increment function, (2) live crown ratio function, (3) probability of regular non-catastrophic mortality (survival) function, and (4) total stem volume function.

The annual diameter increment function consisted of two components: a potential function (POT) and a modifier function (MOD). The potential function was of the form

$$\mathrm{POT} = \mathrm{B}_1 \times \mathrm{SI}[1-\exp(-\mathrm{B}_2 \times \mathrm{CR})]\ [\exp(-\mathrm{B}_3 \times \mathrm{DBH})] \tag{1}$$

where SI is the site index (index age = 50), CR is the crown ratio class (10× (crown length/total height)), DBH is diameter at breast height, and the B's are species-specific parameters. The modifier function was of the form

$$\mathrm{MOD} = [1-\exp(\mathrm{B}_4 \times \mathrm{NT}^{\mathrm{B}_5}\mathrm{DBH}^{\mathrm{B}_6}\mathrm{BA}^{\mathrm{B}_7})] \tag{2}$$

where NT is the number of live trees per unit area and BA is live basal area per unit area. The combined diameter increment function is the product of the two components:

$$ADG = POT \times MOD \tag{3}$$

where ADG is annual diameter growth at breast height.

The crown ratio function used was of the form

$$CR = [B_8/(1+B_9 \times BA)] + B_{10}[1-\exp(-B_{11} \times DBH)] \tag{4}$$

The probability of individual tree survival was calculated with the discrete function

$$PS = \{1-[B_{12}+1/[1+\exp(B_{13}+B_{14} \times ADG+B_{15} \times DBH)]]\}^{INTERV} \tag{5}$$

where INTERV is the time interval (in years) between measurements.

The form of the individual tree volume function used to predict total cubic volume was

$$VT = B_{16} \times SI^{B_{17}}(1-\exp(B_{18} \times DBH^{B_{19}})) \tag{6}$$

The methods used for estimation and specification of process errors were as defined in Gertner (1987). A simplified flow chart of the projection system is shown in Figure 2. Sampling errors in the state variables used to initiate the system were accounted for and propagated through the models but were not partitioned. It was assumed that there were no measurement errors in the state variables.

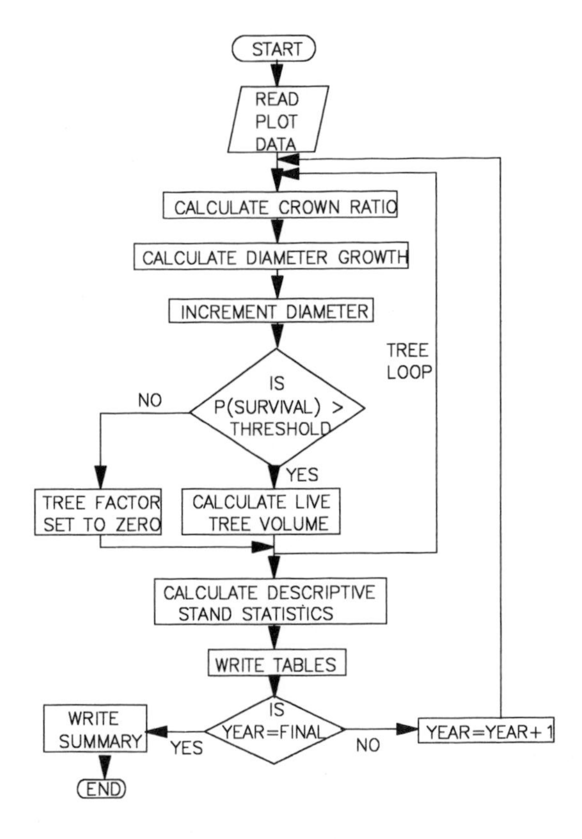

Figure 2. Flow chart of modified Stand and Tree Evaluation and Modeling System (STEMS).

Table 1 shows the percent of variability in the number of trees per hectare versus time partitioned according to the four major individual growth models. As would be expected, almost 100% of the variability in the number of trees per hectare is due to the mortality model. Other stand attributes were also greatly influenced by the mortality model. For example, Table 2 shows the partitioned variance for basal area per hectare. Although not as important in the case of stand basal area, the mortality model is still the major source of variability.

The results presented are not unique to the particular data set. Similar results were obtained with projections made for other stands and county-wide inventories from locations in both Illinois and Wisconsin in the U.S.A.

Table 1. Percent of variability for number of trees per hectare associated with component models.

Projection year	Live crown ratio	Annual diameter increment	Mortality	Total stem volume
0	0.12	0.53	99.35	0.0
10	0.11	0.43	99.46	0.0
20	0.09	0.34	99.57	0.0
30	0.08	0.28	99.64	0.0
40	0.04	0.27	99.69	0.0
50	0.02	0.25	99.73	0.0

Table 2. Percent of variability for basal area per hectare associated with component models.

Projection year	Live crown ratio	Annual diameter increment	Mortality	Total stem volume
0	2.72	29.53	67.75	0.0
10	1.92	17.57	80.51	0.0
20	1.55	11.73	86.72	0.0
30	1.23	8.72	90.05	0.0
40	1.01	6.98	92.01	0.0
50	0.92	5.69	93.39	0.0

FUTURE RESEARCH

Currently, the model used to predict the probability of annual mortality is a logistic model used in its stochastic form. It overwhelms all other component models in the system. It is fairly well agreed that the art of modeling individual tree mortality is not well developed (i.e., Buchman et al., 1983; Hamilton, 1980, 1986). With the error-propagation method, the significance of our current inability to precisely model tree mortality becomes apparent.

Based on the results presented here, it is clear that future research should be directed at improving the ability to predict non-catastrophic mortality. Even a marginal improvement in the mortality model would lead to a significant improvement in the precision of predictions made with the overall STEMS model.

A number of different approaches might be taken to improve the mortality model. The possibilities include the following:

1. The collection of new data to supplement existing data, via samples or experiments, should be considered for improving the logistic function. Currently, the data collected for the mortality model is deficient for two main reasons: The range of predictor variables is limited as a result of clustering, and previous long-term experiments were not designed for the purpose of calibrating a logistic mortality model.
2. The inputs used for the mortality model are those typically collected in mensurational surveys. These variables might be inadequate for the purpose of modeling mortality. More detailed measurements, including the measurement of physiological variables, might be considered. It is possible that these additional variables would greatly improve the precision of the logistic function.
3. Although STEMS is not a process model, consideration should be given to developing a mechanistic mortality model. If this is done, the goal should be to understand the physiological process of non-catastrophic tree mortality so that a deterministic mortality model can be developed.

LITERATURE CITED

Belcher, D., M. Holdaway, and G. Brand. 1982. A description of STEMS—the stand and tree evaluation and modeling system. General Technical Report NC-79, USDA Forest Service North Central Forest Experiment Station, St. Paul, MN, U.S.A. 18 pp.

Buchman, R., S. Pederson, and N. Walters. 1983. A tree survival model with application to species of the Great Lakes region. *Canadian Journal of Forest Research* 13:601–608.

Gertner, G. Z. 1987. Approximating precision in simulated projections: an efficient alternative to Monte Carlo methods. *Forest Science* 33:230–239.

Gertner, G. Z. 1988. Alternative methods for improving the variance approximation of single tree growth and yield projection. In: Ek, A. R., S. R. Shifley, and T. E. Burk, eds. *Forest Growth Modelling and Prediction.* General Technical Report NC-120, USDA Forest Service North Central Forest Experiment Station, St. Paul, MN, U.S.A. Pp. 739–746.

Hamilton, D. 1980. Modeling mortality: a component of growth and yield modeling. In: *Proceedings of the Workshop on Forecasting Forest Stand Dynamics,* 1980, Lakehead University, Thunder Bay, Ontario, Canada. Pp. 82–98.

Hamilton, D. 1986. A logistic model of mortality in thinned and unthinned mixed conifer stands of northern Idaho. *Forest Science* 32:989–1000.

Shifley, S. 1988. Analysis and modelling of forest growth trends along a sulfate deposition gradient in the North Central United States. In: Ek, A. R., S. R. Shifley, and T. E. Burk, eds. *Forest Growth Modelling and Prediction.* General Technical Report NC-120, USDA Forest Service North Central Forest Experiment Station, St. Paul, MN, U.S.A. Pp. 506–513.

SECTION IV

Tree and Stand Growth Modeling

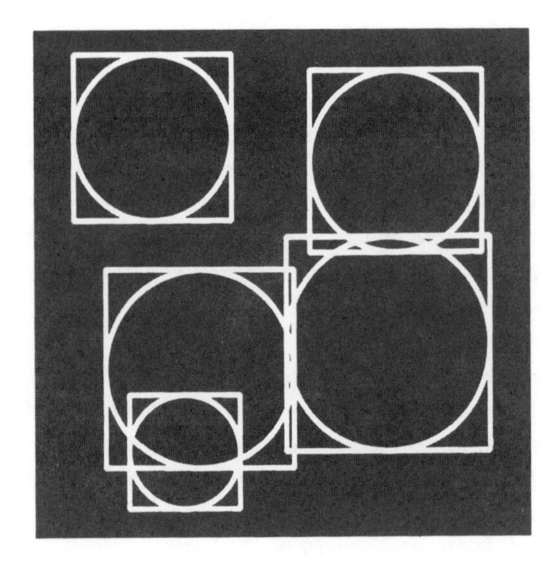

The prediction of tree and stand growth under changing environmental conditions is a broad topic. Areas of interest range from understanding and modeling basic physiological processes and mechanisms which determine and/or are related to growth, through the growth and dynamics of individual trees and forest stands, to the dynamics of ecosystems and regional patterns of growth and development. The development of modeling concepts for this diverse range of concerns and environmental conditions poses a formidable task for scientists and decision-makers.

Three major components are essential to the development of models: (1) an understanding of the process or relationships being modeled, (2) mathematical, statistical, and computational techniques and equipment capable of handling the problem, and (3) experimental or survey data. The relative importance of each of the components depends upon the purpose, the developer, users, and outputs of the desired model. This results in a broad mixture of model types, levels of detail, and ranges of applicability.

The chapters in this section represent a cross-section of current modeling efforts related to the growth and development of trees and stands. These chapters describe simple models of above- or belowground components as well as more complex individual-tree and stand-level models. In sum, they suggest that our knowledge of growth processes and our ability to model these processes are limited. We have not achieved the breadth or depth of understanding, the statistical expertise, or the necessary accumulation of data to answer the questions being posed in our time.

CHAPTER 19

GROWTH-INFLUENCING FACTORS IN DYNAMIC MODELS OF FOREST GROWTH

G. M. J. Mohren and R. Rabbinge

Abstract. The effects on forest growth of weather, soil conditions, and disturbances such as those resulting from air pollution and acidification can be analyzed with the aid of dynamic models of plant growth, based on underlying physical, chemical, and biological processes. From the photosynthesis–light response curve for individual leaves, canopy assimilation is calculated using numerical integration methods. Respiration requirements for growth and maintenance are subtracted, and net biomass growth is calculated. The results of a model for Douglas-fir (*Pseudotsuga menziesii* [Mirb.] Franco) growth based on this approach are presented. Secondary disturbances can be modeled on the basis of their direct effects on process rates, or through delayed effects resulting from changes in site conditions due to soil pollution and acidification. Comprehensive explanatory models summarize data, integrate results from different disciplines, and identify gaps in knowledge. After validation and simplification, summary models can be used for predicting system behavior under changing growing conditions, assessing exposure-effect relationships, and evaluating management alternatives and policy strategies.

INTRODUCTION

Tree growth is the outcome of a series of physical, biochemical, and physiological processes in which, driven by solar radiation, carbon dioxide (CO_2) from the air is assimilated by the foliage, and the carbohydrates produced by the photosynthetic process are converted into the structural dry matter of the living plant. From germination onwards, living biomass accumulates up to a point when production is compensated for by losses such as the dying of older tissue and subsequent litterfall or root turnover. Stem structural dry matter accumulates during stand development when living sapwood turns into dead supporting heartwood tissue.

Forest growth can be studied in many different ways. A "bottom-up" approach to forest growth starts with detailed analysis of the processes involved and aims at an accurate description of the basic elements, which are then assembled to give a description

Dr. Mohren is Forest Production Ecologist at the Research Institute for Forestry and Landscape Planning "De Dorschkamp," Bosrandweg 20, P.O. Box 23, 6700 AA Wageningen, the Netherlands. Dr. Rabbinge is Professor of Crop Ecology at the Department of Theoretical Production Ecology, Wageningen Agricultural University, Bornsesteeg 65, P.O. Box 430, 6700 AK Wageningen, the Netherlands. The authors wish to acknowledge the help of A. F. M. Olsthoorn and M. J. Kropff, who commented on an earlier draft of the present paper. An evaluation of problem areas in physiological models of tree and stand growth, together with the feasibility of using these models for the analysis of disturbances as a result of air pollution and soil acidification, was carried out as part of a CEC/COST-612 workshop held December 14–17, 1987, in Gennep, the Netherlands. Discussions and recommendations from this workshop were summarized in Mohren (1988).

of tree and stand performance. The other extreme is a "top-down" approach to primary production and growth. In this case, the analysis starts with an elementary model of the whole stand which contains as few elements as possible. This general model is subsequently extended to incorporate more growth-influencing factors as these turn out to be relevant.

Such a dynamic model, based on underlying processes, integrates knowledge of the physiological processes involved at the tree level. Modeling provides a way of scaling up from the process level to the level of the stand, and from short-term disturbances to changes in stand growth in the long run. Ecophysiological modeling is especially important in forestry, as stand development takes several decades, and long-term studies are cumbersome and expensive. Model-building and simulation are particularly useful for developing a framework for experimentation (Landsberg, 1986), and the combination of simulation models with field measurements and laboratory experiments allows rapid testing of hypotheses.

GROWTH-INFLUENCING FACTORS

In a top-down approach, a distinction is made between growth-determining factors, growth-limiting factors, and growth-reducing factors, all of which act as modifiers of primary production. Growth-determining factors set the upper limits for growth and production. The main factors determining photosynthetic production are tree physiology, temperature, and incoming radiation. Thus primary production can be calculated from the amount of intercepted radiation and from the efficiency with which the stand uses photosynthetically active radiation absorbed by the foliage (Jarvis and Leverenz, 1983; Linder, 1985).

Growth-limiting factors depend on site conditions and define the attainable production level. They include the availability of water and nutrients. Under the conditions in the Netherlands, the total potential transpiration of a closed Douglas-fir (*Pseudotsuga menziesii* [Mirb.] Franco) stand is about 400–500 mm; total nitrogen (N) and phosphorus (P) requirements can be around 100 kg/ha for N and some 10 kg/ha for P. These amounts have to be available for uptake by the roots or may be supplied in part by redistribution of the N and P already incorporated in the biomass. If the supply falls below the minimum demand, growth is retarded. Additional growth-reducing factors include a range of agents causing deviations from the attainable production level for a particular site. Examples of growth-reducing factors are incomplete canopy closure, the occurrence of air pollution, and pests and diseases. The incorporation of these factors in an ecophysiological model will be illustrated later in this paper. Various production situations may be discerned on the basis of different combinations of growth-determining, growth-limiting, and growth-reducing factors (Table 1).

Table 1. Production situations that may be discerned on the basis of growth-determining and growth-limiting factors (assuming a completely closed canopy and the absence of pests, diseases, weeds, or other growth-reducing factors).

1	Potential growth with abundant water, N, and minerals: closed canopy. Growth determined by weather and tree physiology only.
2	Growth limited by the availability of moisture in the soil during part of the growing season.
3	Growth limited by the availability of water and N and/or P during at least part of the growing season.
4	Growth limited by the availability of water, N and/or P, and other minerals during at least part of the growing season.

As indicated in production situation 1 in Table 1, potential growth is realized when water and nutrients are in ample supply, the canopy is completely closed, and disturbances are absent. In this situation, growth is solely a function of temperature and incoming radiation, as well as the physiological, geometrical, and optical characteristics of the crop that determine the photosynthetic light-use efficiency of the whole stand and the amount of incoming photosynthetically active radiation that is absorbed by the foliage. In the second production situation, available water falls short of demand during at least part of the growing season. Water deficit in the soil limits root uptake and causes the stomata to close, thereby reducing the uptake of CO_2. Shortage of N or P occurs in addition to the water shortage in production situation 3; while in production situation 4, shortage of water, N, and P may occur in combination with a shortage of other minerals such as magnesium (Mg), potassium (K), or calcium (Ca) during at least part of the growing season. Growth reductions induced by disturbances are superimposed on these production situations.

So far, models have been developed mainly for production situations 1 and 2 (e.g., De Wit et al., 1978) and to some extent for production situation 3, involving N (e.g., Van Keulen and Seligman, 1987). Some preliminary models for the effects of P on growth do exist (e.g., Mohren, 1986), but models dealing with other minerals (production situation 4) are still in the stage of infancy (see overview by Penning de Vries, 1983).

The top-down approach views the forest stand as a closed green canopy. The main state variables consist of foliage biomass (with foliage area derived from it), root biomass (usually separated into fine and coarse roots), and stem and branch biomass. Reproductive structures such as flowers and seeds usually are not taken into account in primary production models, especially when applied to coniferous forests. In typical seed-crop years, this may lead to an overestimation of bole increment, as diameter growth may be suppressed by 15–20% due to cone production (Eis et al., 1965). This effect is ignored in most general stand-growth models.

When studying the relationship between environmental conditions and stand growth using an explanatory model based on tree physiology, it is not necessary to take individual trees into account. Their contribution is encompassed through the calculation of stand properties such as horizontal canopy closure. This leads to relatively little detail in the description of stand structure in process models of tree growth. In general, it is preferable to study closed stands instead of individual trees; it is much easier to determine the resources available for a stand when a one-dimensional model is used than it is to determine the resources available to individual trees in the stand when a three-dimensional individual-tree model is used. It has been shown (Van Gerwen et al., 1987) that a stand-level model allows the calculation of individual-tree growth in the stand by taking into account competition among the trees. This will not be considered here, as incorporation of individual tree growth merely adds complexity to the model.

DYNAMIC INTERACTIONS OF CARBON, WATER, AND NUTRIENTS

Figure 1 gives an elementary representation of primary production in plants, redrawn from Mohren (1987). It is based on the production and consumption of assimilates in the plant. The diagram portrays potential production, because canopy assimilation was considered merely a function of weather conditions, i.e., solar radiation and temperature, and leaf area index (LAI). The physiological processes involved—photosynthesis, respiration, and transpiration—can be considered identical in all C_3 plant species, including most trees. Maximum values for photosynthetic rates and respiration requirements, together with the relationship of these processes to temperature, have to be

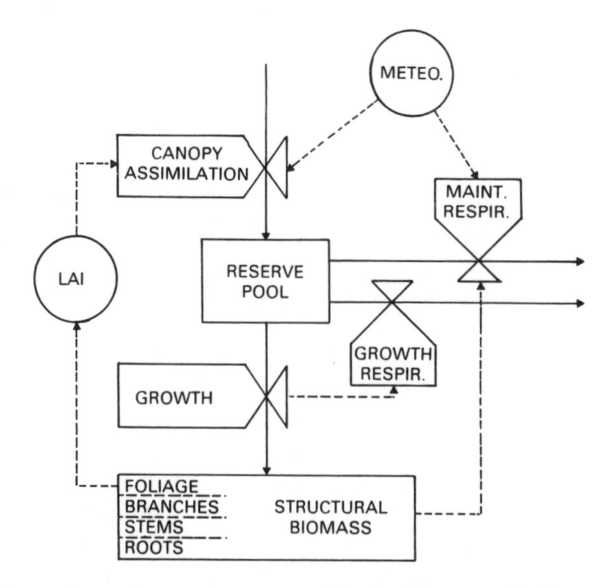

Figure 1. Diagram of an elementary summary model of plant growth. Boxes represent state variables, circles contain intermediary variables or driving functions, and valve symbols represent rate variables.

determined for a particular plant species. Based on these underlying principles, a general simulation approach has been developed at the Department of Theoretical Production Ecology of Wageningen Agricultural University (De Wit et al., 1978; Spitters et al., in press) and applied to forest stands by Mohren (1987).

A generalized growth equation for the individual biomass components in these models is

$$G_i = DWC_i \times DC_i \times (P_g - R_m) - L_i \qquad \text{(kg/ha/year)} \qquad (1)$$

where G_i is the net annual dry weight increment of biomass component i in kg/ha/year; DWC_i the corresponding dry weight conversion in kg dry weight per kg carbohydrates (CH_2O); DC_i the distribution coefficient for the carbohydrates available for growth; P_g the gross canopy assimilation in kg CH_2O/ha/year, calculated on a daily basis from light absorption and photosynthesis–light response of individual leaf layers; R_m the total maintenance respiration of the living tissue; and L_i the rate of litter loss for biomass component i in kg/ha/year.

Based on a gross CO_2 assimilation rate of 15 kg/ha/h and a light-use efficiency for C_3 species of 0.4 kg CO_2 ha/h/J/m^{2}s, or 11.2 μg CO_2/J, total gross photosynthesis of a closed forest canopy varies from some 30 CH_2O/ha/d during winter to a maximum of about 350 kg CH_2O/ha/d during June and July, using Dutch weather data for the calculations. For coniferous stands such as Douglas fir in temperate,maritime climates in western Europe, the annual total for P_g is around 60,000 kg CH_2O/ha/year. Maintenance requirements may amount to some 45% of this total, leaving 33,000 kg CH_2O for dry weight increment. The coefficient for conversion of assimilates to structural dry matter is around 0.65, giving a potential dry matter production of some 21,500 kg/ha/year. Half this is stemwood; using a value of 450 kg/m^3 for specific gravity, this gives a value for the potential annual stem volume increment of 24 m^3/ha/year. Typical values for the parameters used can be found in Mohren et al. (1984) and Mohren (1987).

In the model used here, gross canopy assimilation (P_g) is calculated from the distribution of photosynthetic active radiation over the foliage, using a gaussian technique to integrate over the day and over the layers within the canopy (Goudriaan, 1986; Spitters et al., in press). The main variables describing the forest stand are total biomass (separated into foliage, branches, stems, and roots) and the degree of horizontal canopy closure, accounting for irregular tree spacing and also for clustering of foliage within the tree crowns.

Canopy structure and light interception can be modeled with different levels of detail, ranging from a simple Lambert-Beer extinction model, based on leaf area index, to very detailed models of canopy light climate, based on individual tree and branch characteristics in combination with several levels of grouping of intercepting surfaces in the canopy (see Oker-Blom, 1986, for an overview). The type of model that is adequate depends on the degree of detail needed to analyze a particular problem. In general, a canopy that is more heterogeneous (with regard to gaps and to clustering of foliage inside the crown) requires a more detailed radiation model. The amount of photosynthetically active radiation above the canopy depends on the height of the sun, the degree cloudiness, and the radiation intensity at the outer boundary of the atmosphere. Because their extinction within the canopy differs, direct and diffuse radiation have to be treated separately, using elementary models of solar radiation (Goudriaan, 1977; Spitters, 1986).

The uptake of CO_2 is coupled to leaf transpiration, and the conversion of carbohydrates into structural dry matter is coupled to the uptake of N and mineral elements such as P, K, Ca, and Mg. This further implies that the processes of canopy assimilation and transpiration are closely coupled, with the degree of coupling determined by canopy roughness and the vapor pressure deficit of the air. When the vapor pressure deficit is low, more CO_2 can be taken up per unit of water transpired, and drought is less likely to occur. At the same time, dry matter increment and nutrient availability are linked. The ratio between the separate processes is determined by the physiological characteristics of the plant and by prevailing environmental conditions. Most nutrients are taken up by the roots and transported by the transpiration flux through the xylem tissue. Soil moisture in the root zone allows root uptake of nutrients by providing the link between the root surface and the mineral soil; when the soil is dry, nutrients cannot be taken up. Nutrient uptake may be further modified by soil conditions such as pH, the aluminum (Al) concentration and the root surface, and the presence or absence of mycorrhizae.

The nutrient content of the plant tissue is not only directly coupled to the basic requirement for incorporation in newly formed dry matter (minimum nutrient content per unit of dry matter formed), but also to the rates of the basic physiological processes. Nitrogen occurs mainly in proteins such as enzymes; process rates that depend on the amount of enzymes available are related to N content. An example of this is the high correlation between N content of leaves and the maximum net photosynthesis rate at light saturation (Van Keulen and Seligman, 1987, p. 47). Phosphorus, in the form of P_2O_4 in organic complexes such as adenosine triphosphate (ATP), is an important part of the energy metabolism of plant cells. Both N and P are, within narrow limits, stable biomass components. They are required in minimum amounts in the living tissue, and they cannot be taken up beyond a certain maximum concentration. For Douglas-fir, minimum and maximum amounts for N are 0.8 and 2.0% of dry weight respectively; and for P, minimum and maximum amounts are 0.08 and 0.30% of dry weight (Mohren et al., 1986). Other elements, such as K, are involved in transport processes within the plant and are more mobile, with high concentrations in meristematic and newly formed tissue.

At the stand level, this coupling between tissue concentration and growth is usually less clear than the interactions at the biochemical process level, because at the stand level a number of other interactions may be involved. The total amount of N and P taken up, however, is highly correlated with total net primary production, indicating that on the average, 200 kg of dry matter are formed per kg of N taken up. The corresponding figure for P is

2000 kg dry weight per kg of P taken up (Miller, 1984). At a net primary production of 20,000 kg/ha/year, this leads to estimates of N and P requirements of some 100 kg/ha/year and 10 kg/ha/year, respectively.

As a result of the interactions mentioned above, site differences in water and nutrient availability may lead to entirely different responses to fertilization or irrigation which cannot be easily separated on the basis of field data only. These interactions become even more important when additional effects of air pollution and soil acidification are investigated, as the effects of water and nutrient availability usually cannot be separated from the effects of abiotic and biotic disturbances. In addition, the effects of disturbances may differ markedly in various production situations. Under these circumstances, explanatory models in which the effects of water, nutrient, and growth disturbances on process rates are quantified separately, are very useful. Such models help to define the problem properly and may be used to formulate critical hypotheses and facilitate the design of field studies and laboratory experiments.

A MODEL OF WHOLE-STAND GROWTH

Since the mid-1960s, many physiological growth models have been developed. They range from very detailed models that explain individual physiological processes such as photosynthesis (Farquhar et al., 1980) or the functioning of roots (De Willigen and Van Noordwijk, 1987; Luxmoore and Stolzy, 1987), through single-tree models (Ågren, 1980) and models for stand growth (Mohren, 1987; Mäkelä, 1988), to models for forest micrometeorology (Goudriaan, 1977, 1979) and models for large-scale forest dynamics and succession (Shugart, 1984). Most ecophysiological simulation models, however, have been developed for agricultural crops (Penning de Vries, 1983).

An example of the outcome of a physiological stand growth model is shown in Figures 2, 3, and 4. The figures are the results of a general model based on Equation 1; a detailed description is found in Mohren (1987). The model simulates canopy assimilation, respiration, and transpiration with time-steps of one day to account for seasonal variation. Using historical weather records, stand development can be simulated for several decades to allow comparison of the simulation results with measurement series from permanent field plots. Input data for this model consist of daily weather variables such as solar radiation, air temperature, vapor pressure of the air, wind speed, and precipitation.

The soil is characterized by the water-holding capacity of the rooting zone, based on wilting point and field capacity of the soil, and by the total amounts of N and P in the rooted soil profile. The availability of N and P in the soil are based on a distinction between stable and unstable pools of these nutrients. Also taken into account is cycling of elements, through litterfall and decomposition of organic material, and rates of atmospheric deposition (see Mohren, 1986, for a full description of the nutrient cycle model).

In Figure 3, the model's overestimation at the beginning of the growing season is due to the assumption in the model that stem growth is distributed evenly over the bole, whereas in reality growth of diameter at breast height (dbh) will lag behind diameter increment higher up the trees. Total growth is somewhat overestimated by the model in this case, as can be inferred from the area underneath the curves. For this particular plot, this is due to model overestimation of horizontal canopy closure.

Figure 4 gives the simulated and measured stem volume from the same permanent plot. For the simulation runs presented in Figures 2, 3, and 4, effects of pollution or soil changes due to acidification were not taken into account, because the simulations were mainly concerned with a series of historical data.

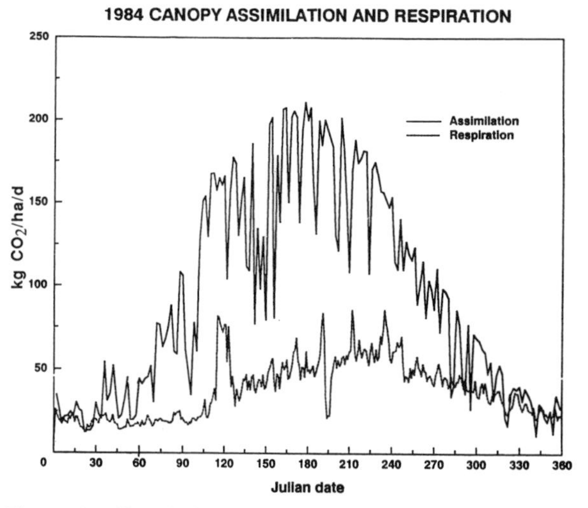

Figure 2. Simulation of daily assimilation and respiration of a Douglas-fir stand during 1984. The plot is located in the center of the Netherlands in the Veluwe area. In 1984, the trees were 65 years old, with an average height of 28 m. Stocking density was 360 trees per ha, with mean dbh of 37 cm. For the calculation, a value of 6 m^2/m^2 was assumed for leaf area index.

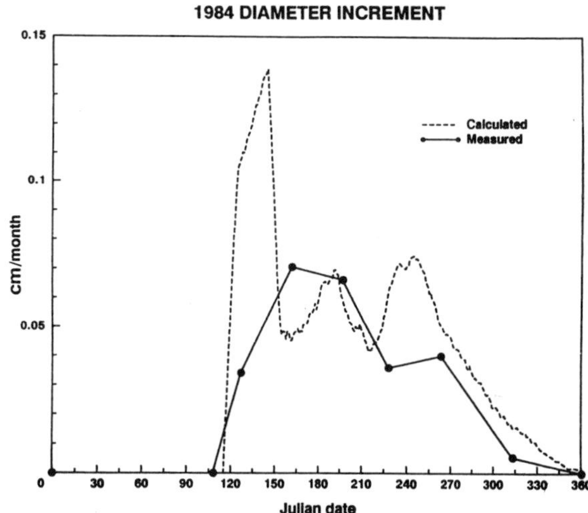

Figure 3. Simulated (daily) and measured (monthly) diameter increment during 1984 in a Douglas-fir stand. Model estimates are compared to data measured in a field plot located in the center of the Netherlands. Calculated current annual increment during 1984 equals 11.5 m^3/ha, resulting in an average diameter increment of 0.4 mm/year. Total simulated dry weight production amounted to about 13 t/ha. The overall pattern of model results agrees with experimental data.

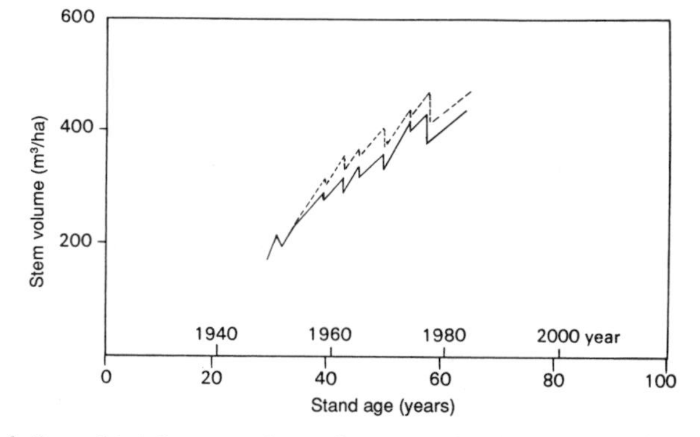

Figure 4. Simulation of total stem volume in a Douglas-fir stand. *Solid line* (—): measurement series. *Broken line* (---): simulation results. (Redrawn from Mohren, 1987.)

MODELING GROWTH DISTURBANCES

Incomplete canopy closure and the effects of growth disturbances are superimposed on and intertwined with growth limitations induced by insufficient water and nutrient availability. A general classification scheme for growth-reducing agents is given in Table 2. Disturbances can be grouped according to their effect, which may be continuous and direct (e.g., when rates of physiological processes such as photosynthesis, respiration, and aging are changed) or delayed in time (e.g., when morphological development is disrupted by means of changes in assimilate distribution or in hormonal balances). Discontinuous catastrophic disturbances occur, for instance, when windthrow or frost causes physical damage to the canopy. Examples of different effects of growth reductions caused by air pollution and soil acidification are given in Table 3. In the state-rate modeling approach used in Figures 1,2,3, and 4, these growth disturbances can be modeled by quantifying their effects on process rates, or quantifying the changes they impose on the state-variables in the model.

Table 2. Growth reductions that may be superimposed on the production situations outlined in Table 1.

- Direct effects on process rates (e.g., direct effects of air pollution on photosynthesis and transpiration).
- Changes in forcing functions or driving variables, such as effects of soil acidification on availability of nutrients for uptake, which affect soil conditions for growth.
- Delayed response to continuous disturbing agents such as changes in assimilate allocation and aging of plant tissue.
- Irregular canopy damage (injured or reduced live, functional tissue) from wind, frost, insect attack, or diseases.

Table 3. Examples of the effects on tree and stand growth of air pollution and acidification, and their classification into direct and delayed effects (after Kohlmaier et al., 1984).

Direct effects on the process level (instantaneous effects):

- Photosynthesis at light saturation, light-use efficiency, and carboxylation resistance.
- Stomatal resistance, transpiration.
- Excess respiration as a result of tissue damage.
- Assimilate translocation, increased flowering, and disturbances in morphological developments.
- Root uptake.
- Aging of foliage and roots.

Delayed effects resulting from tissue damage and changes in carbon partitioning:

- Increased aging of foliage or roots: lower leaf area index → decreased light interception; decreased root uptake capacity → water and nutrient shortage.
- Increased transpiration induced by change in stomatal resistance → increases in water deficit and effects of drought.

Delayed effects mediated by soil acidification and disturbances of soil microbial activity (plant response resulting from delayed effects of pollution on soil conditions):

- Changes in nutrient supply to the roots: leaching of nutrients as a result of increased nutrient mobility, increased supply of competing nutrients.
- Change in chemical root environment (e.g., pH) resulting in decreased root uptake of nutrients.
- Changes in biological activity in the soil: decomposition processes, mycorrhizae, etc.

The effects on various processes can be determined through detailed studies carried out in controlled laboratory environments. Changes in process rates thus determined can afterwards be incorporated into simulation models to study their consequences for growth and production of a whole forest stand. The effects on processes can be incorporated by modifying rate variables. Moreover, delayed effects on state variables can be simulated as resulting from altered process rates. Delayed effects on growth induced by changes in site conditions as a result of soil pollution and acidification can be simulated by changing driving variables in the soil.

Several models are now available for studying the effects of air pollution and acidification on the environment for plant growth, such as models of changes in soil processes resulting from proton-input, changes in acid-neutralizing capacity, and changed cation-exchange capacity as related to acidification (Van Grinsven et al., 1987; review by Reuss et al., 1986). Micro-meteorological models are available to estimate the deposition of air pollutants in terrestrial ecosystems from the concentrations in the air at some height above the canopy in combination with canopy roughness and the total amount of potentially intercepting surface. Some of these models apply to particular sites or stands only; others apply to regional scales (e.g., Alcamo et al., 1985). In combination with physiological growth models, these can be used to study the effects of air pollution and soil acidification at the level of the ecosystem.

Detailed models concerned with individual physiological processes are available for evaluating the effects of pollutants on photosynthesis (e.g., the model published by Kropff, 1987). A number of descriptive models have been published that aim to evaluate the ecosystem effects of disturbances on a larger scale, for example, with regard to succession (Harwell and Weinstein, 1983; Kercher and Axelrod, 1984; Shugart and McLaughlin, 1986). Recent reviews in this field have been published by Reuss et al. (1986), on soil acidification, and by Krupa and Kickert (1987), on the response of vegetation to pollution. Models with different levels of detail should be used to evaluate the effects on the process level or on the system level. Process models investigate biochemical aspects such as the effects on enzymatic processes or on membrane activity. In contrast, single-tree and stand-level models are used to evaluate the effects of disturbances at the process level on total production and growth. Finally, ecosystem and succession models are used to investigate the effects of changes in productivity on biomass dynamics, species composition, and ecosystem development.

PROBLEM AREAS IN ECOPHYSIOLOGICAL MODELING OF FOREST GROWTH

Assimilates produced by canopy photosynthesis are used in a variety of physiological processes in the plant. Assimilate allocation depends on the developmental stage of the plant, on growing conditions, and on supply and demand within the plant. So far, these basic processes are not well enough understood to construct comprehensive mechanistic simulation models, and most models contain empirical partitioning functions for dry matter which are based on measurements of biomass distributions in the field. This is cumbersome, because sufficient data are usually lacking, and distribution keys are either chosen to fit the field data or built around speculative elements. As disturbances and growth reductions may act upon assimilate distribution, this is an important research area. At present, the analysis of assimilate distribution based on a functional balance between biomass components in relation to environmental factors seems most likely to yield the desired results (Mäkelä, this volume; Valentine, this volume).

When modeling the effects of growth reductions as a result of air pollution, aging of photosynthetic tissue is very important. As with assimilate distribution, the process of aging is not well understood. Experimental data indicate that increased aging of assimilatory organs, leading to a lower leaf area index (in the case of evergreen conifers) or to a shorter growing period (in annual crops), results in a considerable decrease in total canopy assimilation (Kropff, 1987).

Models of dry matter accumulation can be used in combination with data on nutrient concentration in the biomass to estimate the nutrient requirements of the growing forest in a demand-oriented approach (Mohren, 1986). To analyze the effect of soil conditions on tree growth, soil nutrient availability and root uptake capacity must be known. To date, it has generally not been possible to estimate soil nutrient availability with sufficient accuracy to ascertain whether or not deficiencies and thus growth limitations will occur. In addition, root dynamics are not well understood.

CONCLUDING REMARKS

Explanatory physiological models may play an important role in the study of forest production ecology. Prospects are good for developing models based on experience from agriculture, using the large body of physiological knowledge that is currently available. A number of pilot studies have already been conducted in this area; and at this stage in the research, there is a general methodology that can be used to construct and evaluate models. As part of this approach, growth-influencing factors may be separated into growth-determining factors, which depend on climate and plant species; growth-limiting factors, which depend on the site conditions; and growth-reducing factors, encompassing all kind of disturbances. By quantifying the effects of these groups of growth-influencing factors at the level of the basic physiological, physical, and chemical processes involved in tree growth, the effects of individual disturbances can be assessed and analyzed.

With regard to pollutants and their effects, it is apparent that model-builders need more detailed physiological information on the effects of pollutants on individual process rates. To study pollution effects, research on the physiological background of these processes is needed. An overall model will be helpful in defining the problem and in constructing hypotheses that can be tested in laboratory or field experiments. Explanatory models have their main application in research. They can be used to study the relevance of individual aspects of tree growth and to identify the main growth-influencing factors for a particular site. Used in this way, the models guide experimental work and serve as a convenient means to bridge the gap between disciplines by representing theoretical understanding and present-day know-how of the functioning of trees at the whole-plant and at the stand level (Rabbinge, 1986). Thus, they may help to elucidate the ways in which growth-disturbing factors such as air pollutants affect the growth and production of forest stands under various conditions. This insight will help in designing preventive and curative measures for the detrimental effects of pollution.

LITERATURE CITED

Ågren, G. I. 1980. PT—a tree growth model. In: Persson, T., ed. *Structure and Function of Northern Coniferous Forests: An Ecosystem Study. Ecological Bulletin* (Stockholm, Sweden) 32:525–536.

Alcamo, J., L. Hordijk, J. Kämäri, P. Kauppi, M. Posch, and E. Runca. 1985. Integrated analysis of acidification in Europe. *Journal of Environmental Management* 21:47–61.

De Willigen, P., and M. van Noordwijk. 1987. Roots, plant production and nutrient use efficiency. Ph.D. Thesis, Agricultural University, Wageningen, the Netherlands. 282 pp.

De Wit, C. T. et al. 1978. *Simulation of Assimilation, Respiration, and Transpiration of Crops.* Simulation Monographs, Pudoc, Wageningen, the Netherlands. 141 pp.

Eis, S., E. H. Garman, and L. F. Ebell. 1965. Relation between cone production and diameter increment of Douglas fir (*Pseudotsuga menziesii* [Mirb.] Franco), grand fir (*Abies grandis* [Dougl.] Lindl.), and Western white pine (*Pinus monticola* Dougl.). *Canadian Journal of Botany* 43:1553–1559.

Farquhar, G. D., S. von Caemmerer, and J. A. Berry. 1980. A biochemical model of photosynthetic CO_2 assimilation in leaves of C_3 species. *Planta* 149:78–90.

Goudriaan, J. 1977. *Crop Micrometeorology: A Simulation Study.* Simulation Monographs, Pudoc, Wageningen, the Netherlands. 249 pp.

Goudriaan, J. 1979. MICROWEATHER. Simulation model applied to a forest. In: Halldin, S., ed. *Comparison of Forest Water and Energy Exchange Models.* International Society for Ecological Modelling (ISEM), Copenhagen, Denmark. Pp. 47–57.

Goudriaan, J. 1986. A simple and fast numerical method for the computation of daily totals of crop photosynthesis. *Agricultural and Forest Meteorology* 38:249–254.

Harwell, M. A., and D. A. Weinstein. 1983. Modelling the effects of air pollutants on forested ecosystems. In: Lauenroth, W. K., G. V. Skogerboe, and M. Flug, eds. *Analysis of Ecological*

Systems: State-of-the-Art in Ecological Modelling. Elsevier Science Publishing, Amsterdam, the Netherlands. Pp. 497–502.

Jarvis, P. G., and J. W. Leverenz. 1983. Productivity of temperate, deciduous and evergreen forests. In: Lange, O. L., P. S. Nobel, C. B. Osmond, and H. Ziegler, eds. *Physiological Plant Ecology,* Vol. IV. *Ecosystem Processes: Mineral Cycling, Productivity and Man's Influence.* Springer-Verlag, Berlin, F.R.G. Pp. 233–280.

Kercher, J., and M. C. Axelrod. 1984. Analysis of SILVA: a model for forecasting the effects of SO_2 pollution on growth and succession in Western coniferous forests. *Ecological Modelling* 23:165–184.

Kohlmaier, G. H., E. O. Sire, H. Broehl, W. Kilian, U. Fischbach, M. Ploechl, T. Mueller, and J. Yunsheng. 1984. Dramatic development in the dying of German spruce-fir forests: in search of possible cause-effect relationships. *Ecological Modelling* 22:45–65.

Kropff, M. J. 1987. Physiological effects of sulphur dioxide, I. The effect of SO_2 on photosynthesis and stomatal regulation of *Vicia faba,* L. *Plant, Cell and Environment* 10:753–760.

Krupa, S., and R. N. Kickert. 1987. An analysis of numerical models of air pollutant exposure and vegetation response. *Environmental Pollution* 44:127–158.

Landsberg, J. J. 1986. Experimental approaches to the study of the effects of nutrients and water on carbon assimilation by trees. *Tree Physiology* 2:427–444.

Linder, S. 1985. Potential and actual production in Australian forest stands. In: Landsberg, J. J., and W. Parsons, eds. *Research for Forest Management.* CSIRO, Melbourne, Australia. Pp. 11–35.

Luxmoore, R. J., and L. H. Stolzy. 1987. Modeling belowground processes of roots, the rhizosphere, and soil communities. In: Wisiol, K., and J. D. Hesketh, eds. *Plant Growth Models for Resource Management,* Vol. II. *Quantifying Plant Processes.* CRC Press, Boca Raton, FL, U.S.A. Pp. 129–153.

Mäkelä, A. 1988. *Models of Pine Stand Development: An Eco-Physiological Systems Analysis.* Research Notes No. 62, Department of Silviculture, University of Helsinki, Finland. 54 pp.

Miller, H. G. 1984. Dynamics of nutrient cycling in plantation ecosystems. In: Bowen, G. D., and E. K. S. Nambiar, eds. *Nutrition of Plantation Forests.* Academic Press, London, England, U.K. Pp. 53–78.

Mohren, G. M. J. 1986. Modelling nutrient dynamics in forests, and the influence of nitrogen and phosphorus on growth. In: Ågren, G. I., ed. *Predicting Consequences of Intensive Forest Harvesting on Long-Term Productivity.* Report No. 26, Department of Ecology and Environmental Research, Swedish University of Agricultural Sciences, Uppsala, Sweden. Pp. 105–116.

Mohren, G. M. J. 1987. Simulation of forest growth, applied to Douglas fir stands in the Netherlands. Ph.D. Thesis, Wageningen Agricultural University, the Netherlands. 184 pp.

Mohren, G. M. J. 1988. Report on group discussions and final recommendations. *Interrelationships between Above- and Below-Ground Influences of Air Pollutants on Forest Trees,* Proceedings, CEC/COST-612 workshop, December, 14–17, 1987, Gennep, the Netherlands. In: Bervaes, J., P. Mathy, and P. Evers, eds. CEC Air Pollution Research Report 16. Brussels, Belgium. pp. 263–270.

Mohren, G. M. J., J. Van den Burg, and F. W. Burger. 1986. Phosphorus deficiency induced by nitrogen input in Douglas fir in the Netherlands. *Plant and Soil* 95:191–200.

Mohren, G. M. J., C. P. Van Gerwen, and C. J. T. Spitters. 1984. Simulation of primary production in even-aged stands of Douglas fir. *Forest Ecology and Management* 9:27–49.

Oker-Blom, P. 1986. Photosynthetic radiation regime and canopy structure in modeled forest stands. *Acta Forestalia Fennica* 197. 44 pp.

Penning de Vries, F. W. T. 1983. Modeling of growth and production. In: Lange, O. L., P. S. Nobel, C. B. Osmond, and H. Ziegler, eds. *Encyclopedia of Plant Physiology,* New Series, Vol. 12. *Physiological Plant Ecology.* Springer-Verlag, Berlin, F.R.G. Pp. 117–150.

Rabbinge, R. 1986. The bridge function of crop ecology. *Netherlands Journal of Agricultural Science* 3:239–251.

Reuss, J. O., N. Chistophersen, and H. M. Seip. 1986. A critique of models for freshwater and soil acidification. *Water, Air, and Soil Pollution* 30:909–930.

Shugart, H. H. 1984. *A Theory of Forest Dynamics: The Ecological Implications of Forest Succession Models.* Springer-Verlag, Berlin, F.R.G. 278 pp.

Shugart, H. H., and S. B. McLaughlin. 1986. Modelling SO_2 effects on forest growth and community dynamics. In: Winner, W. E., H. A. Mooney, and R. A. Goldstein, eds. *Sulphur Dioxide Response: Plants, Plant Communities and Regulatory Agencies.* Stanford University Press, Stanford, CA, U.S.A.

Spitters, C. J. T. 1986. Separating diffuse and direct components of global radiation and its implications for modelling canopy photosynthesis, I. Components of incoming radiation. *Agricultural and Forest Meteorology* 38:225–237.

Spitters, C. J. T., H. van Keulen, and D. W. G. van Kraalingen. In press. A simple and universal crop growth simulator: SUCROS87. In: Rabbinge, R., S. R. Ward, and H. H. van Laar, eds. *Simulation in Plant Protection and Crop Management.* Simulation Monographs, Pudoc, Wageningen, the Netherlands.

Van Gerwen, C. P., C. J. T. Spitters, and G. M. J. Mohren. 1987. Simulation of competition for light in even-aged stands of Douglas fir. *Forest Ecology and Management* 18:135–152.

Van Grinsven, J. J. M., N. van Breemen, and J. Mulder. 1987. Impacts of acid atmospheric deposition on woodland soils in the Netherlands, I. Calculation of hydrological and chemical budgets. *Soil Science Society of American Journal* 51:1629–1634.

Van Keulen, H., and N. G. Seligman. 1987.*Simulation of Water Use, Nitrogen Nutrition and Growth of a Spring Wheat Crop.* Simulation Monographs, Pudoc, Wageningen, the Netherlands. 310 pp.

CHAPTER 20

LINKING MECHANISTIC MODELS OF TREE PHYSIOLOGY WITH MODELS OF FOREST DYNAMICS: PROBLEMS OF TEMPORAL SCALE

Anthony W. King, William R. Emanuel, and Robert V. O'Neill

Abstract. The individual-based forest gap models are a potential resource in the study of forest growth responses to environmental stress acting on physiological processes. This approach is currently limited by the lack of physiological detail in the gap models and by the temporal scales separating tree physiology and the annual tree growth simulated by the gap models. We describe a general procedure for integrating process-based models of tree physiology with forest gap models. The procedure involves (1) a physiologically based process model of whole-tree growth, (2) an individual-based model of forest gap dynamics, (3) derivation of a response-surface model describing the statistical relationship between annual wood production simulated by the physiological model and driving variables appropriate to the annual time scale of the gap model, and (4) modification of the gap model to include the response-surface model. The procedure is presented for consideration and testing as a promising approach to simulating the impact of tree physiology and environmental stress on forest growth and succession.

INTRODUCTION

Much of the information available on how forest systems respond to environmental stress (e.g., air pollution and climate change), especially information on mechanism, is limited to the relatively short-term behavior of physiological processes operating within individual leaves, seedlings, and, rarely, entire mature trees. Efforts to model the response of forest growth to environmental stress by using mechanistic process models of tree physiology are a response to this observation, but the models must still deal with the problem of translating physiological response across several hierarchical levels to the level of the stand or forest (Shugart et al., 1986; Urban et al., 1987).

The physiological processes affected by environmental stress and the forest dynamics of interest (e.g., growth and succession) operate at decidely different spatial and temporal scales (Figure 1). Physiological processes have rapid response times, on the order of

The authors are affiliated with the Environmental Sciences Division, Oak Ridge National Laboratory, Oak Ridge, TN, 37831 U.S.A. This research was sponsored by the National Science Foundation's Ecosystem Studies under Interagency Agreement BSR-8417923 with the U.S. Department of Energy under Contract DE-AC05-840R21400 with Martin Marietta Energy Systems, Inc. Publication No. 3251, Environmental Sciences Division, Oak Ridge National Laboratory.

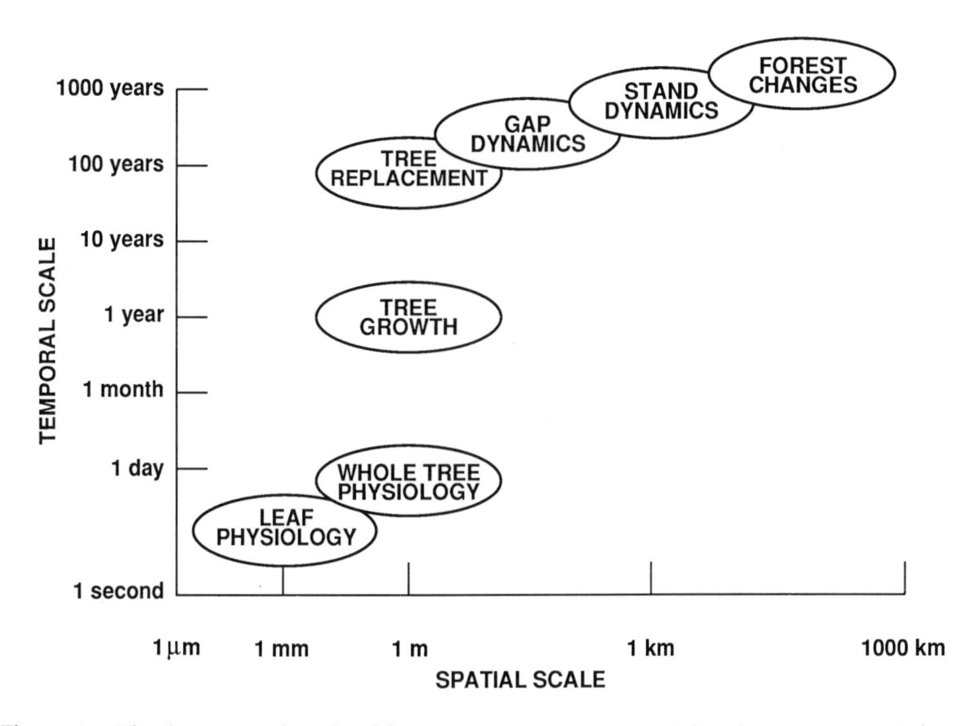

Figure 1. The time-space domain of forest system dynamics. Modified from Urban et al. (1987) and Dickinson (1988).

seconds to hours, and take place over relatively small areas (from micrometers to a very few meters for leaf and whole-tree processes, respectively). A forest operates on a temporal scale of hundreds to thousands of years (e.g., succession) and occupies hundreds or thousands of square kilometers. These widely disparate scales are linked temporally, however, by tree growth (and eventual death and replacement) and spatially and temporally by tree replacement and gap dynamics (e.g., gap-phase replacement, Shugart, 1984). The linkages between tree growth, gap dynamics, and stand dynamics (Figure 1) are described by a class of forest simulation models categorized by Shugart (1984) as gap models and exemplified by the JAnak, BOtkin, WAllis (JABOWA) (Botkin et al., 1972) and Forests of East Tennessee (FORET) (Shugart and West, 1977) models. The gap models have in turn been linked to landscape and regional forest dynamics (see review by Shugart, 1984; Solomon, 1986; Dale and Gardner, 1987; Smith and Urban, 1988).

Since tree growth is an expression of tree physiology, the gap models are a logical resource in the search for information on how physiology, altered by environmental stress, affects forest structure and function. However, the mechanistic representation of tree physiology in the gap models is very limited. Most of the gap models calculate annual diameter increment of an individual tree using a species-specific optimal growth equation that relates annual diameter increment under optimal environmental conditions to current height and diameter at breast height (dbh). Details of this formulation (and the exceptions) can be found in Shugart (1984). The gap models then reduce the optimal diameter increment according to one or more "growth multipliers" that describe the proportional reduction in optimal growth induced by suboptimal environmental conditions. Most of the differences among gap models are related to the specific consideration of growth multipliers (see Shugart et al., 1981; Shugart, 1984; Pastor and Post, 1985; Dale et al., 1985, 1986; Solomon, 1986; Smith and Huston, in press); but all consider growth limitation by shading, and most consider temperature and water limitations. Some include nutrient limitations.

An obvious solution to the problem of simulating the influence of stress-altered physiology on forest growth is the introduction of more mechanistic, physiologically based process models of tree growth as submodels in the larger structure of the gap models. There are, however, problems with this approach. The gap models perform best when predicting changes in stand composition over successional time scales of several hundred years, and the models operate on a basic time-step of one year. In contrast, mechanistic representations of tree physiology are best suited to models with hourly (or shorter) time-steps. The computational efforts associated with running an hourly model for every tree in a stand throughout an entire successional sequence are tremendous.

Forest gap models have been used to look at climate change and air pollution stress, but these studies have not explicitly considered physiological responses. Because the gap models represent tree growth as functions of annual climate, climate changes can be simulated directly (Solomon, 1986; Pastor and Post, 1988). Gap model simulation of air pollution stress is less direct. In general, the potential growth of individual trees has been reduced by stress multipliers that describe the proportional growth reduction of trees exposed to air pollution relative to nonexposed trees (West et al., 1980; Dale and Gardner, 1987). Botkin et al. (1973) used the same basic approach to increase tree growth in response to elevated carbon dioxide (simulating CO_2 fertilization). The use of proportional multipliers is both useful and consistent with the formulation of the gap models, and it avoids the aforementioned problems of including a detailed physiological tree growth submodel in the gap model structure. However, the stress multipliers do not incorporate the explicit physiological response, the mechanistic cause-and-effect information, that could be useful in predicting forest response to specific stresses.

The need to better incorporate whole-tree physiology in the gap models is widely recognized. For example, Smith and Huston (in press) deal with this issue in detail in their ongoing study of the relationship between physiological constraints on resource utilization and patterns of succession, distribution, and community structure (Huston and Smith, 1987). They derive "functional growth responses" for 15 functional types (analogous to resource utilization guilds) from multidimensional response surfaces describing the simultaneous light and water limitation of net carbon gain. Luxmoore et al. (this volume) use the annual output from a physiological model of tree growth to calculate a CO_2 fertilization factor. Dale et al. (1986) included a physiologically based moisture index in their Computer Linked Integrated Model for Assessing Community Structure (CLIMACS) model, and this work is continuing.

In this paper we propose an alternative to the stress multipliers. We submit that the scales separating tree physiology and forest growth can be spanned by explicitly linking process models of tree physiology with forest gap models. In the conceptual approach we present for consideration, this linkage is accomplished by rescaling a tree physiology model so that it is more compatible with the temporal scale of the gap model. The temporal rescaling is based on the statistical derivation of predictable relationships between the temporally integrated behavior of the physiological model and the driving variables appropriate to the annual time scale of the gap model. These rescaled relationships are then used to modify the gap model.

We present our approach as a general procedure for using models to translate information across scales (to scale up; see Shugart et al., 1986). This problem is nicely exemplified by the incorporation of physiology into gap models. We refer to our approach as a temporal translation or rescaling of a fine-scale model, in this case a physiologically based tree model. We describe the proposed procedure as a general methodology with several basic components.

A PROCEDURE FOR LINKING TREE PHYSIOLOGY AND GAP DYNAMICS

The Physiological Response Model

The physiological model should be a mechanistic or process model describing a tree's physiological response to the pollutant(s) or other environmental stress(es) of interest. The model describes the observed physiological response that is to be translated into forest growth. The model must

1. describe the available physiological response data at the appropriate time scale (e.g., hourly, perhaps daily);
2. describe whole-tree physiology (even if the stress response data is limited to a single component or process);
3. simulate the physiological response over the entire lifetime of a tree (i.e., include age-dependent effects); and
4. simulate annual wood production (AWP) as part of tree growth.

The existence of mechanistic, physiologically based models of whole-tree growth is debated. Gap modelers feel that such models either do not exist or cannot be applied beyond time scales of 1–5 years (Shugart, 1984; Shugart et al., 1986; M. Huston, pers. comm; D. Urban, pers. comm.). On the other hand, Reynolds et al. (1980) presented process models of creosotebush (*Larrea tridentata* Cov.) and loblolly pine (*Pinus taeda* L.) growth, and Luxmoore et al. (this volume) and Isebrands et al. (this volume) present whole-tree process models for the investigation of forest growth. Landsberg (1986, p. 430) described the conceptual base of a "process-based model of the growth of a tree." The proceedings edited by Luxmoore et al. (1986) and the present volume are evidence of the growing desire to use plant process models in the study of forest growth, presumably with the assumption that these models exist or can be developed.

Some of the debate is over semantics (e.g., the meaning of mechanistic or process model) and the level of physiology required for qualification as a physiologically based model. However, while we are inclined to disagree with the extreme position of the gap modelers, we do not contest their specific arguments, especially the question of whether physiologically based models can be applied over the time scales necessary to address forest growth. Notably, discussions by process modelers often mention the challenges, particularly the data requirements, of the approach. Our point is simple: If process models of tree growth are going to be used in the study of forest growth, they must be integrated over the long time scales of forest growth, and their performance at these larger scales must be assessed. The approach we propose here is a potential tool for addressing these issues. We accept, for our present purpose, the position of the process modelers that physiologically based models either exist or can and should be developed, and we assume the existence of a physiologically based model of tree growth that meets our requirements. The qualifications we set forth above may be viewed either as screening criteria for existing models or as requirements for future model development.

The Individual-Based Forest Dynamics Model

The model of forest dynamics or growth must (1) be an individual-tree-based model, and (2) simulate annual tree growth in terms of volume or diameter increment. Forest gap models satisfy these requirements.

Gap models calculate the woody biomass of individual trees as a function of dbh. AWP is the net increase in woody biomass calculated before and after the tree's dbh is incremented with the realized diameter increment obtained from the growth equation (e.g., Pastor and Post, 1985). The relationship between AWP and annual diameter increment provides an obvious and explicit linkage between the physiologically based tree growth model and the forest gap model.

Annual Integrated Physiological Response As a Function of Driving Variables in the Forest Model

Temporal integration of the fine-scale physiological model provides a process-based prediction of AWP as it is affected by the environmental stress. The physiological model is used to simulate AWP independently of the gap model. Ideally, the physiological model should be executed across a factorial design of important environmental conditions (see Downing et al. (1985) for a discussion of screening for "influential" inputs). Each element in the factorial is an annual time series of input to the physiological model, with the temporal resolution appropriate for the model (e.g., hourly). The factorial should encompass the appropriate range of exposures to the agent of environmental stress being investigated (e.g., air pollution) and the range of natural conditions (e.g., drought stress, light availability) that trees on the site in question might encounter. This factorial should be replicated for all species under study with significant interspecific differences in the modeled physiology. This may require the use of several species-specific calibrations of the model. If age-dependent response is important, tree age should also be included in the factorial. Cumulative effects of the stress might be addressed by executing the model over several years of simulated growth.

For example, a frequency distribution describing the variability in annual time series of model input (e.g., weather) appropriate to a particular site (from observation or generated by a stochastic weather model) might be combined with high, intermediate, and low ambient concentrations of an atmospheric pollutant and the range of incident solar radiation experienced by trees growing at different heights in the canopy. These input data are propagated through the physiological model by repeated simulation across the factorial. Monte Carlo simulation may be appropriate when the variability across time series input is described by frequency distributions (e.g., the distribution of annual moisture regimes). Note that the annual time series (the elements of the factorial) used as model input involve continuous variables; the factorial presentation of the individual time series is a necessary compromise between the need to represent an appropriate range of environmental variability and the demands of computation, data management, and analysis.

The AWP values from the factorial execution of the physiological model and their associated inputs are used to define new relationships among the integrated annual output of the physiological model (i.e., AWP) and environmental driving variables with the annual resolution of the gap model. First, the high-resolution time series used as input to the physiological model are converted to integrated annual variables. Where possible, the high-resolution (e.g., hourly or daily) input should be converted to the annual variables used in the gap model. For example, daily temperatures and soil water potentials can be converted to annual growing degree days and annual drought days, respectively. Incident solar radiation can be converted to percent of full annual sunlight. If there is no corresponding variable in the gap model (e.g., ambient concentration of the air pollutant), other conversions can be made. For example, hourly or daily concentrations of an air pollutant used as input to the physiological model can be converted to mean (or maximum) annual concentration.

The next step, and the heart of the temporal rescaling, is the derivation of a response-surface model (see Downing et al., 1985) describing AWP simulated by the physiological model as a function of the converted annual driving variables. The response-surface model can be produced by a least-squares fit between AWP and the annual driving variables. For example, a significant and acceptable proportion of the variance in AWP might be explained by a linear combination of the annual driving variables, or AWP might be better predicted by a non-linear model. The form of the function may not be critical, but the behavior of the surface (e.g., the variance explained and the bias in the surface) is critical. Downing et al. (1985) discuss "adequate," "acceptable," and "representative" response-surface models. The requirements for this application are (1) that a useful and predictable relationship be found between AWP and the annual driving variables, and (2) that the influence of percent of available sunlight be included in the relationship. In this discussion the term "useful" means that the predictive variables can be easily calculated (e.g., the driving variables in the gap models). We use the term "predictable" to mean that the response-surface model explains an acceptable level of the variation in AWP (e.g., an r^2 of 0.80) and that the surface exhibits minimal (or acceptable) bias. If the response-surface modeling does not define a statistically significant relationship between annual available sunlight and AWP simulated by the physiologically based model (we doubt the likelihood of such a result), linkage with gap models in which light interception and shading are critical components is inappropriate, and an alternative approach to rescaling tree physiology is required.

The transformation of time series of hourly or daily input data to annual values is straightforward. Design of an appropriate factorial simulation procedure requires careful consideration, and the factorial simulation will be computationally intensive. Nevertheless, with time and effort, a sample of simulated AWP values can be acquired. However, the question of whether annual integrated inputs can capture the aspects of high-resolution input that are important to physiological processes in such a way that annual inputs can be used to predict annual integrated physiological response with acceptable precision can only be determined through the application and testing of the technique we propose here.

The Revised Model of Forest Growth Response

The final step in our proposed approach to linking a fine-scale physiologically based model of tree growth and a coarse-scale forest gap model involves modification of the gap model. The derived annual response-surface model is added as a new subroutine that calculates AWP as a function of annual-scale environmental variables, and the gap model's growth subroutine is modified to calculate diameter increment as a function of AWP (the reverse of the traditional gap model operation of AWP as a function of change in dbh). In an iterative loop across all trees in the gap, the revised growth subroutine calculates the light available to each tree (as a function of overshading canopy) and passes that information to the new subroutine that calculates AWP. The growth subroutine might also have to be modified to calculate the influence of canopy structure on other environmental variables (e.g., interception of atmospheric pollutants) that are also passed to the AWP subroutine. The calculated AWP is passed back to the growth subroutine, the wood mass is converted to a diameter increment (i.e., a tree ring), and the tree grows accordingly. (Note that the diameter increment can simply be back-calculated by using the gap model's equation for woody biomass as a function of dbh, but more sophisticated functions could be substituted.)

The proposed gap model revision does not calculate an optimal diameter increment and does not calculate the growth multipliers that have generally been used to reduce this optimal increment under suboptimal conditions. In other attributes, the revised gap model can be considered a typical gap model.

DISCUSSION

We have proposed a general procedure for using models to translate information across temporal scales. In the specific context of forest growth, we have described a procedure for linking mechanistic models of whole-tree physiology with forest gap models. The difference in temporal scales between physiology and gap dynamics makes this a nontrivial task (Shugart et al., 1986). However, our proposed method explicitly integrates across the intervening scales. The approach does not avoid the many simulations with the fine-scale, high-resolution model needed to integrate the small-scale behaviors, but the approach does avoid the use of a fine-scale, high-resolution physiology model as a subroutine in an annual-scale gap model of forest succession. We believe that in the long run our approach will require fewer simulations with the fine-scale model, and that our explorations of annual integrated physiology as a function of annual environmental variables and the large-scale expression of small-scale phenomena are additional benefits not provided by the alternative.

Our general procedure complements other efforts to link tree physiology and gap dynamics. The explicit temporal integration of a whole-tree physiologically based simulation model is also used by Luxmoore et al. (this volume), but the methods differ in the incorporation of those results in the gap model. The response-surface modeling approach to the derivation of an annual response function could be used to describe the response surfaces in the gap model used by Smith and Huston (in press) in their study of functional types and adaptive strategies. The use of a physiological model is a surrogate for the whole-plant factorial experiments they discuss.

We do not suggest that implementation of our proposed technique will be simple or that gap modelers have been remiss in not pursuing a comparable approach. Fine-scale physiological process models of tree growth may indeed prove to be incapable of simulating environmental stress over the long time scales of tree growth. Design of the proposed factorial simulation will require careful consideration. The factorial simulation will be computationally intensive, and, while response-surface methodologies are well defined, application of the techniques are not trivial, and useful positive results are not guaranteed (although negative results may be revealing). These are issues that research must address. We believe our proposed approach for translating information across scales is theoretically sound, that it is a viable alternative to the use of stress multipliers or fine-scale, high-resolution subroutines, and that, indeed, difficulties in the approach reflect the difficulties inherent in using small-scale information and models to address large-scale phenomena. If physiological process models are going to be used to investigate the influence of environmental stress on forest growth, the large differences in temporal scale must be dealt with. Our proposed procedure is one approach to that problem. Realization of the promise in the approach must be determined through application and testing of the procedure.

LITERATURE CITED

Botkin, D. B., J. F. Janak, and J. R. Wallis. 1972. Some ecological consequences of a computer model of forest growth. *Journal of Ecology* 60:849–873.

Botkin, D. B., J. F. Janak, and J. R. Wallis. 1973. Estimating the effects of carbon fertilization on forest composition by ecosystem simulation. In: Woodwell, G. M., and E. V. Pecan, eds. *Carbon and the Biosphere.* National Technical Information Service, U.S. Department of Commerce, Springfield, Virginia, U.S.A. Pp. 328–342.

Dale, V. H., T. W. Doyle, and H. H. Shugart. 1985. A comparison of tree growth models. *Ecological Modelling* 29:145–169.

Dale, V. H., and R. H. Gardner. 1987. Assessing regional impacts of growth declines using a forest succession model. *Journal of Environmental Management* 24:83–93.
Dale, V. H., M. Hemstrom, and J. Franklin. 1986. Modeling the long-term effects of disturbances on forest succession, Olympic Peninsula, Washington. *Canadian Journal of Forest Research* 16:56–67.
Dickinson, R. 1988. Atmospheric systems and global change. In: Rosswall, T., R. G. Woodmansee, and P. G. Risser, eds. *Scales and Global Change: Spatial and Temporal Variability in Biospheric and Geospheric Processes (SCOPE 35).* John Wiley and Sons, Chichester, England, U.K. Pp. 57–80.
Downing, D. J., R. H. Gardner, and F. O. Hoffman. 1985. An examination of response-surface methodologies for uncertainty analysis in assessment models. *Technometrics* 27:151–163.
Huston, M., and T. Smith. 1987. Plant succession: life history and competition. *The American Naturalist* 130:168–198.
Isebrands, J. G., H. M. Rauscher, T. R. Crow, and D. I. Dickmann. 1989. Whole-tree growth process models based on structural-functional relationships. In: Dixon, R. K., R. S. Meldahl, G. A. Ruark, and W. G. Warren, eds. *Process Modeling of Forest Growth Responses to Environmental Stress.* Timber Press, Portland, OR, U.S.A.
Landsberg, J. J. 1986. Experimental approaches to the study of the effects of nutrients and water on carbon assimilation by trees. *Tree Physiology* 2:427–444.
Luxmoore, R. J., J. J. Landsberg, and M. R. Kaufman, eds. 1986. *Coupling of Carbon, Water and Nutrient Interactions in Woody Plant Soil Systems: Proceedings of a Symposium of the International Union of Forestry Research Organizations.* Heron Publishing, Victoria, Canada. 467 pp.
Luxmoore, R. J., M. L. Tharp, and D. C. West. 1989. Simulating the physiological basis of tree-ring responses to environmental changes. In: Dixon, R. K., R. S. Meldahl, G. A. Ruark, and W. G. Warren, eds. *Process Modeling of Forest Growth Responses to Environmental Stress.* Timber Press, Portland, OR, U.S.A.
Pastor, J., and W. M. Post. 1985. *Development of a Linked Forest Productivity–Soil Process Model.* ORNL/TM-9519, Oak Ridge National Laboratory, Oak Ridge, TN, U.S.A. 155 pp.
Pastor, J., and W. M. Post. 1988. Response of northern forests to CO_2-induced climate change. *Nature* 334:55–58.
Reynolds, J. F., B. R. Strain, G. L. Cunningham, and K. R. Knoerr. 1980. Predicting primary productivity for forest and desert ecosystem models. In: Hesketh, J. D., and J. W. Jones, eds. *Predicting Photosynthesis for Ecosystem Models.* CRC Press, Inc., Boca Raton, FL, U.S.A. Pp. 169–207.
Shugart, H. H. 1984. *A Theory of Forest Dynamics: The Ecological Implications of Forest Succession Models.* Springer-Verlag, New York, NY, U.S.A.
Shugart, H. H., M. Y. Antonovsky, P. G. Jarvis, and A. P. Sandford. 1986. CO_2, climatic change and forest ecosystems. In: Bolin, B., B. R. Doos, J. Jager, and R. A. Warrick, eds. *The Greenhouse Effect, Climatic Change, and Ecosystems (SCOPE 29).* John Wiley and Sons, New York, NY, U.S.A. Pp. 475–521.
Shugart, H. H., and D. C. West. 1977. Development of an Appalachian deciduous forest succession model and its application to assessment of the impact of the chestnut blight. *Journal of Environmental Management* 5:161–179.
Shugart, H. H., D. C. West, and W. R. Emanuel. 1981. Patterns and dynamics of forests: an application of simulation models. In: West, D. C., H. H. Shugart, and D. B. Botkin, eds. *Forest Succession: Concepts and Application.* Springer-Verlag, New York, NY, U.S.A. Pp. 74–98.
Smith, T., and M. Huston. In press. Plant functional types: linking physiology and vegetation dynamics. *Vegetatio.*
Smith, T. M., and D. L. Urban. 1988. Scale and resolution of forest structural pattern. *Vegetatio* 74:143–150.
Solomon, A. M. 1986. Transient response of forests to CO_2-induced climate change: Simulation modeling experiments in eastern North America. *Oecologia* 68:567–579.
Urban, D. L., R. V. O'Neill, and H. H. Shugart. 1987. Landscape ecology. *Bioscience* 37:119–127.
West, D. C., S. B. McLaughlin, and H. H. Shugart. 1980. Simulated forest response to chronic air pollution stress. *Journal of Environmental Quality* 9:43–49.

CHAPTER 21

DEVELOPING A PROCESS-BASED GROWTH MODEL FOR SITKA SPRUCE

Anthony R. Ludlow, Timothy J. Randle, and Jennifer C. Grace

Abstract. In developing a single-tree, process-based growth model for Sitka spruce (*Picea sitchensis*), we have extended McMurtrie and Wolf's (1983) stand model in a number of ways. First, following the work of Valentine (1985) and Mäkelä (1986), who combined the pipe theory with a carbon balance model, we have assumed that new foliage must be accompanied by new sapwood, and that this determines the way carbon is allocated between foliage and sapwood. Second, the model predicts the height growth for each year, and this is made equal to the depth of new foliage. That, in turn, depends on the quantity of new foliage and the area over which it is spread. Third, we use a modified version of the Lambert-Beer law in which photosynthesis rate increases with the surface area of the exposed crown, rather than the projected area. Fourth, competition between neighboring trees tends to reduce the exposed surface area of the crown, which is calculated from the distance and relative height of neighbors. For this calculation, we assume that the exposed surface of spruce crowns is conical and that the crown angle remains constant throughout life.

The model is being tested with a mixture of methods. The relationship between photosynthesis and crown surface area was tested by regression analysis, which showed that exposed crown surface area accounted for 83% of the variance in timber-volume growth rate in Sitka spruce. Adding the area available to each tree explained only 0.4% more of the variance. The assumption was also tested by simulations with a more detailed light interception model used to estimate the photosynthesis of Monterey pine (*Pinus radiata* D. Don) in New Zealand. Linear functions of crown depth or crown surface area gave equally good approximations to the photosynthesis rate calculated from the more detailed model (Grace, Jarvis, and Norman, 1987; Grace, Rook, and Lane, 1987; Grace, 1988; Grace, this volume). Finally, the model's predictions conform in a qualitative way with the common forestry experience that height growth is fairly independent of stocking density (Figure 4b) but much more related to site quality (Figure 5a) and to tree morphology. In addition, since height growth and basal-area growth in the model are both proportional to the quantity of new foliage, the cumulative volume of an unthinned stand, as well as its top height, depends on how much foliage there has ever been. Hence, the cumulative volume is strongly correlated with top height (Figure 5b), and the relationship is little influenced by site quality. The model therefore conforms with the fundamental assumption on which the United Kingdom yield tables were constructed (Edwards and Christie, 1981).

INTRODUCTION

The research described here is aimed at building a process-based growth model for Sitka spruce in the United Kingdom. In the short term, the model is to be used to predict growth and yield for management purposes, especially where it is necessary to predict responses to novel treatments. In the longer term, the model will be extended to handle

Dr. Ludlow and Mr. Randle are at the Biometric Modelling Section, Forestry Commission, Forest Research Station, Alice Holt Lodge, Farnham, Surrey, GU10 4LH, U.K. Dr. Grace is at the Forest Research Institute, Ministry of Forestry, Rotorua, New Zealand.

the likely responses to stress. For both these purposes, we are forced to extrapolate, which is made a little less dangerous if the model embodies the processes occurring in real trees.

It is never easy, though, to find an appropriate level of detail. To decide this, we have used a mixture of methods: testing approximations by regression analysis, comparing simple models with more complex ones, and fitting the full model to data. But always we insist that parameters used in the model should represent biological quantities that can be measured. In that way, we test both the model's goodness-of-fit and the sense or absurdity of the parameter values needed to make it fit.

At the outset, the project involved A. Ludlow (A.L.) and T. Randle (T.R.), who developed the single-tree model outlined below. Parts of this model parallel closely the one developed by Grace (1980): Both models handle competition among trees by calculating the effect of neighbors on crown size, and both make photosynthesis rate a function of crown surface area. In the present study, A.L. and T.R. tested the second assumption by regression analysis, while J. Grace tested it against results from her more detailed light-interception model (Grace, Jarvis, and Norman, 1987; Grace, Rook, and Laver, 1987; Grace, 1988; Grace, this volume).

DEVELOPMENT OF A SINGLE-TREE MODEL

McMurtrie and Wolf (1983) described a simple whole-stand model which makes a natural starting point for further developments. Photosynthesis depended on the leaf area index, which in turn depended on the foliage dry weight. But there were diminishing returns as new leaves shaded the old, and they used the Lambert-Beer law to calculate the change in photosynthesis rate with leaf area. The carbon assimilated was then allocated to each of three compartments: foliage, fine roots, and all the woody tissue in between. The proportion allocated to each compartment was constant through the stand's life and was set by the three parameters η_f, η_r, and η_s. Losses through litterfall and foliage respiration were proportional to foliage dry-weight, and corresponding equations were used for losses in fine roots and woody tissues.

Using this simple model, McMurtrie and Wolf showed that increasing the proportion allocated to the fine roots, η_r, reduced woody production enormously without an appreciable change in photosynthesis. Keyes and Grier (1981) had found just these differences when they compared Douglas-fir stands on nutrient-rich and nutrient-poor sites.

The clarity of McMurtrie and Wolf's model exposes the issues which need to be faced and helps one to see ways in which it could be improved. Perhaps the most important insight is that η_f and η_s should not be constant since, according to the pipe theory (Shinozaki et al., 1964a, b), more sapwood is required to serve each leaf as the tree gets taller. Valentine (1985, 1987, 1988) and Mäkelä (1986; 1988a, b) have, independently, published models incorporating this idea.

We also combined the pipe theory with McMurtrie and Wolf's model, with one slight difference. Mäkelä (1986) calculates the allocation to the stem so that total sapwood area is proportional to total foliage dry weight, while we make the area of new sapwood proportional to the dry weight of new foliage. Mäkelä's assumption suggests a feedback loop increasing sapwood growth when some variable such as leaf turgor is low. Our assumption implies that the constant foliage/sapwood ratio occurs because the development processes forming new foliage and new sapwood are correlated. In addition, we need to assume that foliage mortality and sapwood mortality are correlated.

WHAT DETERMINES HEIGHT GROWTH?

If the pattern of carbon allocation changes with tree height, as assumed in Valentine's and Mäkelä's models, and if the tree's running costs also change with height, it is clear that we need to understand height growth better. That, in turn, raises two separate questions: What determines the height growth of a tree? And how does the tree benefit from height growth? No model of tree growth is complete unless it answers the first question, and we cannot calculate both the costs and the rewards of height growth unless we answer the second.

To model height growth, we have assumed that height increases because new foliage is usually put out above existing foliage, that it is put out in the most efficient way, and that the most efficient way does not change as the tree grows. Hence, we assume that a new layer of foliage is added over the existing foliage each year, and that the new layer has the same depth over the whole crown. We assume, too, that the depth depends on the weight of new foliage and the area over which it is spread. These assumptions give a simple equation for height growth:

$$H' = \frac{\nu_f N_f}{C_p} \tag{1}$$

where H' is height growth rate (m/year), N_f is the rate of new foliage growth (kg/year), ν_f is a constant giving the volume of new space occupied by each kg of new foliage (m^3/kg), and C_p is the projected area of the crown (m^2) at any instant. In a closed canopy, C_p will change very slowly as natural mortality increases the space per tree.

A key assumption behind Equation 1 is that height growth is a consequence of foliage growth. This view receives some support from development studies. For example, Pollard and Logan (1977), working on spruce (*Picea*) species, write that in most northern coniferous species, the annual complement of new foliage is predetermined as primodia formed in the previous year, and that extension of branches and height growth are similarly determined. Lanner (1976), working on pine (*Pinus*) species, points out that shoot growth depends to a large extent on primordium numbers laid down, so that environmental factors influence year-to-year extension more by modifying primordium numbers than by modifying internode elongation between those primordia. The results of Bollmann et al. (1986) are consistent with this view; they say that "the numbers of primordia laid down in buds determine foliage, branch and flower production and hence strongly affect crown structure, allocation and productivity."

Combined with the rest of the model, Equation 1 predicts that diameter growth should respond strongly to thinning, while height growth does not; this result is illustrated in Figures 4a and 4b. It happens because large crowns in the model assimilate more carbon than small ones and so generate more new foliage. But the greater quantity of foliage is spread over a larger area, and the two effects of crown size cancel out. Hence, height growth changes little when crown area is increased by thinning, but diameter growth is strongly affected.

Height growth in the model is affected by differences in rate of foliage production, so it is sensitive to the parameters which depend on site quality (Figure 5a). Moreover, height growth depends on tree morphology, which is reflected in the parameter ν_f. Hence, the model is consistent with the common forestry experience that height growth is fairly independent of stocking density but is much more related to site quality and tree morphology.

HOW DOES THE TREE BENEFIT FROM HEIGHT GROWTH?

To understand how the individual tree benefits from height growth, we need to understand how a tree's height affects its interception of light. There are, broadly, two approaches. We may ask: How much of the sky can each leaf see? Or we may ask: How much of the tree can see the sky? Using the first approach, we would calculate the rate of photosynthesis in each leaf or class of leaves and sum this to find the total crown photosynthesis. This makes heavy demands on computing time, and we use the second approach as an approximation to the first. We have tested the approximation by regression analysis of growth data and by comparing our approach with a detailed light interception model based on the first approach (see the discussion below).

We assume that the tree will arrange its foliage to make the best use of the space available but that, within this space, new leaves will shade the old. Hence, there is a limit to the amount of effective foliage. There comes a point when adding further foliage increases respiration costs more than it increases photosynthesis. If the tree has enough foliage to reach that point, then photosynthesis will be closely related to the space available. Thus, if we can calculate the space available, we can estimate the photosynthesis rate for that tree.

Photosynthesis will fall short of this rate where foliage amount is limited by nutrient shortage, say, but for the moment we consider the case where it is not. For the moment, then, we concentrate on how the photosynthesis rate of a tree with abundant foliage can be estimated from the space available.

The effectiveness of foliage is clearly limited by shading by neighbors and by self-shading from other leaves on the tree. But these types of shading act in different ways. Figure 1 shows how the crowns of two typical Sitka spruce trees changed over 19 years. In Tree 115, from an unthinned plot, the crown size remained much the same as the tree grew taller. Its lower branches were killed by the shade from close neighbors. Tree 20, from a heavily thinned plot, grew a large crown because lower branches were not killed and lateral growth was uninterrupted by neighbors.

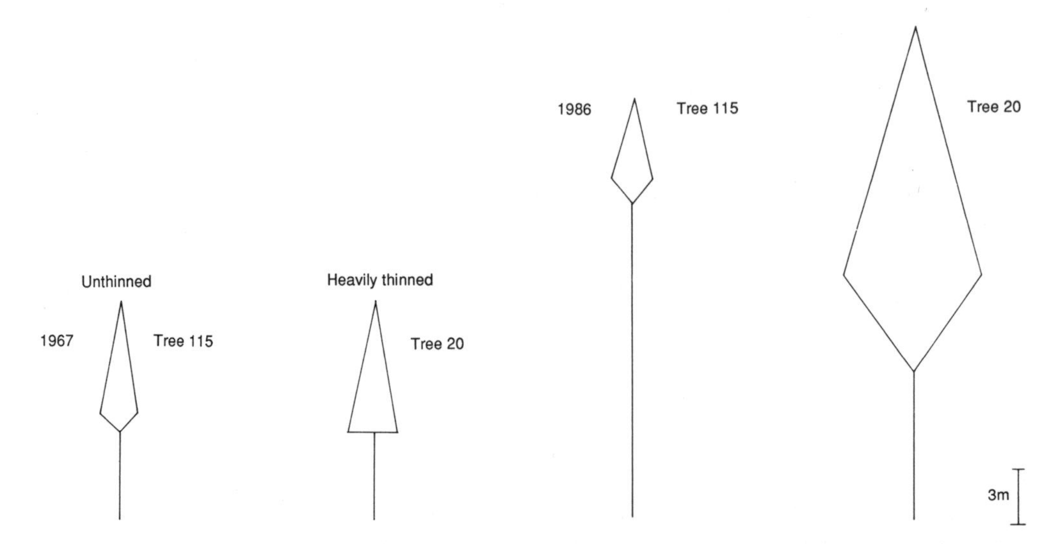

Figure 1. Changes in crown size of two individual Sitka spruce trees over 19 years. Tree 115 was in an unthinned plot with 3,500 stems/ha. Tree 20 was in a plot thinned at 1.5 times the recommended management rate. Schematic diagrams show tree height, height of the lowest live whorl, height of the lowest live branch, and crown diameter, assuming that the lowest live whorl is the widest part of the conical crown.

To model these responses in Sitka spruce, we have followed Grace (1980) with minor differences. We have assumed that each crown is a cone with a characteristic angle. This condition would arise if lateral growth were proportional to height growth, until the branches of neighboring trees touched. Then, we assume, the branches below the point of contact are killed by shading, so that crown size is determined by height growth, crown angle, and the distance and growth rate of neighbors. These assumptions allow us to calculate the crown's exposed surface area and the way this area depends on competition from other trees.

We have also followed Grace (1980) in assuming that photosynthesis is related to crown surface area; but we have used a different function based on a modified version of McMurtrie and Wolf's (1983), approach. They assumed that gross photosynthesis of the whole stand would follow the Lambert-Beer law, so that

$$P_s = P_0(1 - e^{-CL^*}) \tag{2}$$

where P_s is the rate of photosynthesis (kg C/m^2/year) averaged over the whole stand, P_0 is the rate that would be achieved if all light were intercepted (kg C/m^2/year), C is the canopy extinction coefficient, and L^* is the leaf area index, or the average number of foliage layers between the sky and the ground.

To calculate the photosynthesis for an individual tree in a closed canopy, we simply note that $P = C_pP_s$, where P is the photosynthesis rate of the individual tree (kg C/tree/year) and C_p the projected area of its crown (m^2). This assumes that the projected areas of neighboring crowns do not overlap and that the depth of foliage is the same over the whole crown. Hence, we multiply the righthand side of Equation 2 by C_p to obtain the photosynthesis rate of a single tree.

$$P = C_pP_0(1 - e^{-CL^*}) \tag{3}$$

Strictly, Equations 2 and 3 apply only to canopies which are horizontally uniform, but it is worth asking what errors are made if we extend Equation 3 to an open canopy of well-separated crowns. To do this, we interpret the leaf area indes, L^*, as the number of layers of foliage between the sky and the ground, averaged over the projected area of the tree's crown instead of being averaged over the whole stand. Hence, $L^* = A_l/C_p$ where A_l is the projected leaf area (m^2). Substituting this in Equation 3 gives

$$P = C_pP_0(1 - e^{-CA_l/C_p}) \tag{4}$$

which allows automatically for gaps between crowns because the photosynthesis of each tree depends on the ground it covers. Calculating P for all trees in a stand and then summing will give an increasing rate of photosynthesis as the canopy closes.

However, Equation 4 could only be correct if all light fell vertically downwards, so the errors from this equation depend partly on the proportion of light from lower angles. These errors come from two sources: Neighboring crowns will shade each other from side light, and the tree's own foliage will cause self-shading. For the moment we ignore the effect of neighboring trees, apart from their effects in restricting each other's crown size, which is already calculated in the model; we concentrate on self-shading.

According to Equation 4, increasing height would have no effect on P, but increasing height reduces self-shading. It ought to affect photosynthesis rate because it reduces the number of layers of foliage between the innermost and outermost leaves. Since this number of layers is given by A_l/C_s, where C_s is the crown surface area, we can modify Equation 4 by replacing C_p with crown surface area, giving

$$P = C_sP_0(1 - e^{-CA_l/C_s}) \tag{5}$$

Note that P_0 has a slightly different definition in Equations 4 and 5. In the former, it is the maximum photosynthesis rate per square meter of projected crown area that would be achieved if all light above the canopy were intercepted. In the latter, it is the maximum photosynthesis rate per square meter of crown surface that would be achieved if all light were absorbed.

Like Equation 4, Equation 5 allows automatically for gaps in the canopy, and elsewhere in the model we take account of the way trees restrict each other's crown size. But Equation 5 does not allow for the shading effect of neighboring crowns, so it is an approximation which needs to be tested before being built into the full model. In the following section, we test these ideas.

EMPIRICAL STUDIES OF GROWTH RATE AND CROWN SIZE

While there are no data relating the photosynthesis rate of the whole tree to its crown size, it is possible to test the assumptions behind Equation 5 by comparing volume increment with various measures of crown size. For trees with abundant foliage, Equation 5 leads us to expect photosynthesis to be proportional to crown surface area, while volume increment should reflect both photosynthesis rate and the proportion allocated to stems. Trees with sparse crowns should fall short of this maximal growth rate. Hence, we should expect much of the variation in volume increment to be explained by variation in crown surface area, some by changes in allocation coefficients as the tree gets taller, and some by the foliage density.

Allocation coefficients and crown density have not been measured systematically; but Hamilton (1969) measured volume increments fortnightly for 60 trees in a 23-year-old stand of Sitka spruce and found that crown surface area accounted for 88% of the variation in increment. It is unlikely that allocation coefficients varied much in these trees of equal age and similar height. Crown volume, dbh, projected crown area, and tree height accounted for 80%, 79%, 71%, and 58% of the variation, respectively.

Grace (1980), noting that the most efficient new foliage lay in the crown's outer layers, also concluded that photosynthesis should be proportional to crown surface area, all other things being equal. She described regressions of height and diameter increment, measured over one or more years, in Sitka spruce stands in the United Kingdom. She found that crown surface area accounted for 64% of the variation in height and 74% of the variation in diameter (Grace, unpublished data).

We examined the influence of crown size further, using data from a thinning experiment at Dyfi, North Wales, U.K. (52°41′N, 3°45′W), already described by Rennolls (1985). We limited our analysis to 149 trees whose crowns and timber volumes had been measured at least twice. This allowed us to calculate 299 volume growth rates, although the measurement interval varied from 2–16 years, and there were between one and six measurement intervals per tree (mean = 2).

For each interval and each tree, we calculated the exposed surface area of the crown, crown depth, crown volume, and projected crown area at the start and end of the interval. These values were then used to calculate the interval mean. The calculations assumed that the crown formed a cone, and only the upper slanting surface of the cone was included in calculating the surface area.

Ideally, we would like to explain photosynthesis rate in terms of crown size at any instant. But instead of photosynthesis rate at any instant, we had volume growth rate averaged over several years. Moreover, the crown variables changed through the measurement interval, dramatically in the case of the heavily thinned treatments. We therefore regressed the volume growth rate during the measurement interval against the interval mean of each crown variable (Figure 2). As a check, we also regressed the volume

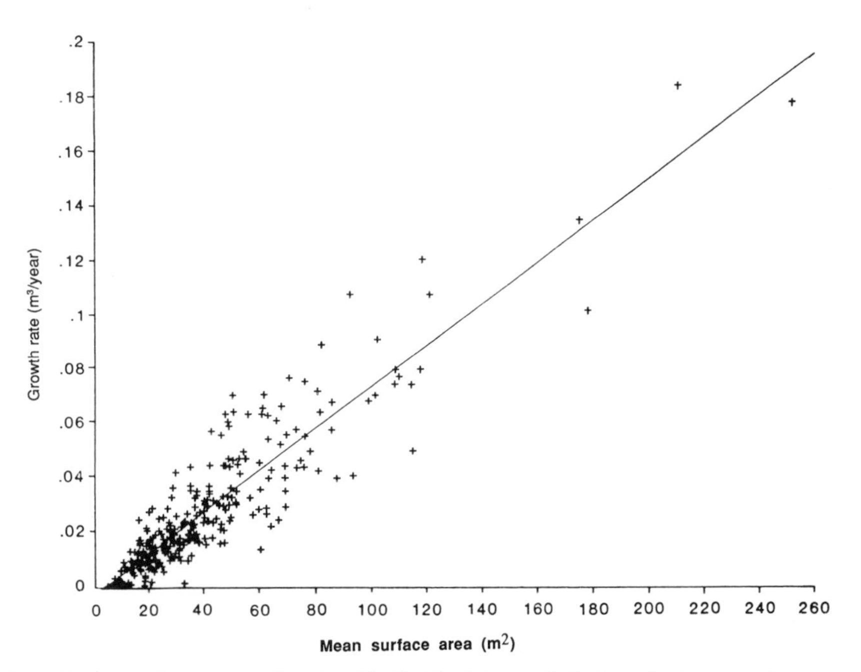

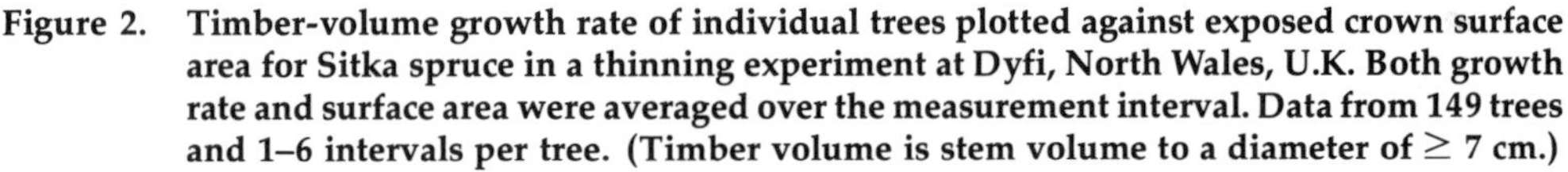

Figure 2. Timber-volume growth rate of individual trees plotted against exposed crown surface area for Sitka spruce in a thinning experiment at Dyfi, North Wales, U.K. Both growth rate and surface area were averaged over the measurement interval. Data from 149 trees and 1–6 intervals per tree. (Timber volume is stem volume to a diameter of $\geq$ 7 cm.)

growth rate against the values at the start of the interval. The results were qualitatively the same.

Figure 2 shows, and the analysis confirmed, that there is a strong relationship between volume growth rate, measured over any interval, and the mean crown surface area averaged over the same interval. Some 83% of the variance was explained, and crown surface area was the best predictor.

A method previously used to calculate competition among trees is to calculate the area potentially available (APA) to each tree (Nance et al., 1987). This is a polygonal area based on tree size and was calculated using software by Nance et al. On its own, it explained 70% of the variance in volume increment; when added to crown surface area, it explained only 0.4% extra. It is not surprising that crown surface area should be a better predictor than APA, because APA measures the way area available changes after thinning, while crown surface area measures the extent to which the tree actually uses the extra space available. Knowing the foliage density should improve the prediction still further.

A key question here is: How large are the errors from using Equation 5? In particular, how much do we lose by ignoring the way neighboring crowns shade each other from side light? If this factor were a large source of error, we should expect that adding relative height of the tree ($H/\overline{H}$) to a regression equation including crown surface area would improve the prediction substantially. In fact, it added only 0.5% to the amount of variance explained.

As a further check, a thinning experiment on Norway spruce (*Picea abies*) at Bowmont, U.K. (55°33'N, 2°25'W), was analyzed in a similar way. While the Dyfi trees were measured from the ages of 18–38 years and were growing vigorously (yield class 18–22 m^3/ha/year), the trees at Bowmont were older (46–72 years), in a cooler site, and described as "on their last legs when clear-felled" (Fletcher, pers. comm.). The average yield class was 8–10 m^3/ha/year.

Figure 3 shows volume growth rate plotted against crown surface area for this site with, superimposed, the regression line fitted to the Dyfi data. There is evidence of a spine of points lying close to the superimposed line, while a number of trees are not achieving the growth rates their crown surface area would suggest. This is consistent with Equation 5, which predicts that, for a given crown size, there is a maximum effective foliage weight. Trees that reach this foliage weight will have photosynthesis rates proportional to crown surface area. Those which fall short are unable to make full use of the space available. Unfortunately, there are no data available on foliage weight at this site to test the interpretation.

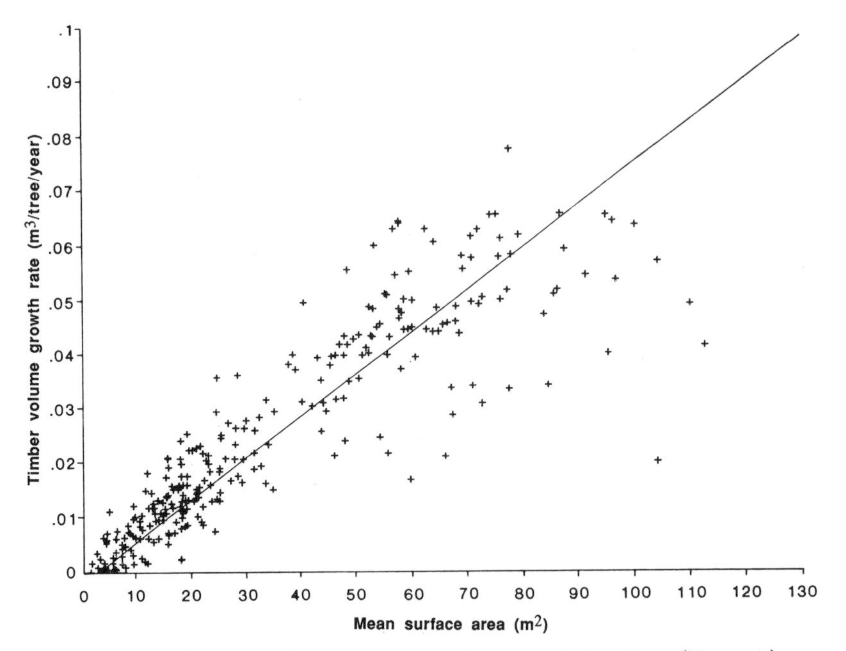

Figure 3. Regression line from Sitka spruce at Dyfi, North Wales, U.K. (Figure 2), superimposed on data for Norway spruce (*Picea abies*) at Bowmont, U.K. Timber-volume growth rate for individual trees and crown surface area calculated as in Figure 2.

MODELING STUDIES OF PHOTOSYNTHESIS AND CROWN SIZE

An alternative test of Equation 5 is to calculate how much of the sky each leaf can see and to check whether that more detailed approach gives a similar answer when the approach is scaled up to a yearly basis and simplified. To do this, J. Grace has calculated the way photosynthesis changes with crown surface area, using a more complex model of light interception.

Grace and colleagues (Grace, Jarvis, and Norman, 1987; Grace, Rook, and Lane, 1987; Grace, 1988; Grace, this volume) developed a model for estimating photosynthesis for individual trees within a stand. For this study, the model was used to estimate annual net photosynthesis for the innermost 18 trees within a 328 m^2 plot containing 71 radiata pine near Rotorua, New Zealand (38°17'S, 176°7'E).

All 71 trees were measured in March, 1984, when approximately 5 years old, and their location was calculated. Measurements on each tree included dbh (1.3 m), height of base of green crown, diameter 10 cm below green crown base, and horizontal crown radius in four perpendicular directions. The heights of half the trees were measured, and the heights of the other half were estimated from a height-diameter regression. In November, 1984, the leaf surface area was measured for 10 trees, and a linear regression of leaf surface area versus basal area at the base of the green crown was calculated. This was used to estimate what the leaf area of all 71 trees had been in March, 1984.

Annual net photosynthesis for each tree was estimated, using the light interception and photosynthesis models (Grace, Jarvis, and Norman, 1987; Grace, Rook, and Lane, 1987), and then regressed against crown depth, crown volume, and exposed surface area of the crown (defined as the surface area above the widest part of the crown, which was assumed to be a complete ellipsoid).

Net photosynthesis was linearly related to crown surface area ($r = 0.75$), the best regression being

$$P_n = 0.376C_s \tag{6}$$

where P_n is the net annual photosynthesis (kg C/tree/year) and C_s the exposed surface area of the crown (m^2).

Quadratic terms were needed to relate photosynthesis to crown depth and crown volume (unless an outlying data point was removed). Nevertheless, the correlation between net photosynthesis and crown depth was the same as that between net photosynthesis and crown surface area, and both were barely higher than the correlation between net photosynthesis and crown volume ($r = 0.74$).

Hence, crown surface area was less a clear winner than at Dyfi or Bowmont. Furthermore, the residuals from Equation 6 were linearly related to relative height, so that a better regression was

$$P_n = 0.377C_s\frac{H}{\overline{H}} \tag{7}$$

where H and $\overline{H}$ are the tree height and mean height respectively. Equation 7 explained 67% of the variation as opposed to 57% with just crown surface area.

So the results in this section only partly confirm the conclusions from Dyfi and Bowmont. Although crown surface area is as good as the other measures of crown size, it is far less a clear winner. And relative height, which is important here, was unimportant at Dyfi.

There are several differences between the three sites, quite apart from the difference in tree species. The plots at Dyfi and Bowmont were part of a thinning experiment with a very large variation in crown size, while the single stand near Rotorua had a much smaller range. The light regime, too, was quite different; the heavier cloud at Dyfi and Bowmont would reduce direct radiation and favor diffuse.

Hence, we should use Equation 5 with care. The approximation may be poor when direct radiation is high. One obvious point is that competing trees not only restrict the size of each other's crowns, they also shade each other's crowns. Only the first effect is included in Equation 5, which should clearly be extended to take account of the second effect. Nevertheless, the simulations described below show that the full model, even in its preliminary form, can be made to fit the data surprisingly well. In a climate where much of the light is diffuse, then, the errors that come from using Equation 5 are not large. It is clear that approximate models can be useful (see also Grace, this volume), and more checks of this type are urgently needed.

SIMULATIONS WITH THE FULL MODEL

The main aim of this research is to develop a full process model, and simulations using the current version have been used to check progress at every stage. In the most recent checks, we have matched the model to the Forestry Commission yield table curves for unthinned Sitka spruce planted at 1.7 m spacing and growing at yield class 24 m^3/ha/year (Edwards and Christie, 1981). Mortality was not simulated. Instead we used a look-up table with the number of trees per ha at any given instant. The number of trees was then used to calculate the area available to the simulated tree and its crown size. The look-up table was taken directly from the published yield tables. Entry to the table was by top height, not age, because we could then use the same look-up table for all yield classes. Since the model calculates the growth of the mean tree, some errors are involved in equating mean height in the model to top height in the tables. On the other hand, these should be seen as preliminary checks on a developing model. The yield tables themselves may be wrong in detail, and the model must be validated against sample plot data. However, extensive use of the tables has failed to reveal any bias, and they cover a wide range of spacings and yield classes.

Growth was simulated from planting so that the same initial conditions could be used for each yield class. Hence, the tree was grown from a height of 0.15 m, and, at a spacing of 1.7 m, it was unrestricted by competition for the first few years.

Following Mäkelä (1988a, b), we used a Monte Carlo approach to test the model. Sensible ranges for each parameter were selected from the literature; for each simulation run, the parameter values were selected at random from these ranges. A run was deemed to have predicted diameter correctly if the predicted value was within 10% of the yield-table mean diameter at all times between the ages of 15 and 78 years. Similarly strict criteria were required for height and stand volume. The parameters used in each of the first 100 successful runs were written to a file and analyzed in various ways before adjusting the ranges (still within limits derived from the literature) to improve the success rate. This iterative cycle was repeated a number of times.

Having matched the model to the growth of unthinned Sitka spruce at yield class 24 m^3/ha/year, we then changed the look-up table (giving the number of trees per ha) from unthinned to that for a line-thinning treatment of normal intensity for that yield class. The results are shown in Figure 4. The parameters of the model were not changed, and the results in both Figures 4 and 5 were obtained with the same set of parameter values selected by the Monte Carlo method.

For the unthinned simulation, the diameter, height, and stand volume were bound to be within 10% of the yield table curve. As Figure 4a shows, the effect of thinning on diameter is overestimated by the model; as Figure 4b shows, the effect of stand volume is slightly underestimated; and as Figure 4c shows, there is almost no effect of thinning on height, either in the model or in the yield tables.

Figure 5 shows the results of simulations with different values of P_0 (see Equation 5). The simulation with $P_0 = 1.5$, strictly $P_0 = 1.495$, is identical to the unthinned run shown in Figure 4. In other words, the look-up table gave the number of trees per ha in an unthinned stand, and the parameters were identical to those used for Figure 4.

P_0 is a site-specific parameter, the maximum photosynthesis rate that could be achieved if all light were intercepted; so Figure 5 shows the effect on growth of differing levels of income for the tree. Conforming with common forestry experience, height growth is very sensitive to site quality, while, in an unthinned stand, the stand volume for a given mean height is not. Indeed, the United Kingdom yield tables (Edwards and Christie, 1981), along with others, are based on the empirical observation that trees behave qualitatively as shown in Figure 5.

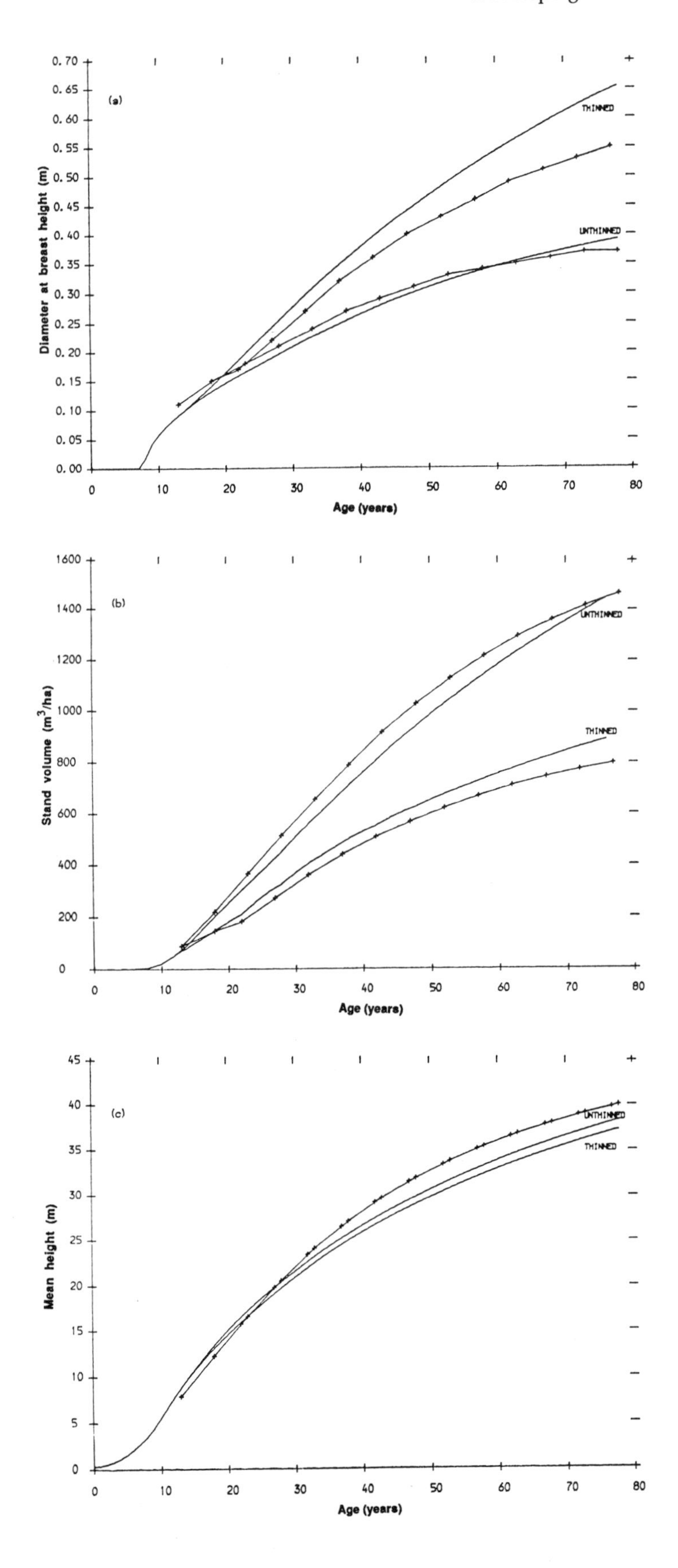

Figure 4.
Response of model, calibrated for an unthinned stand, to line-thinning at recommended management rate. Smooth curves show model predictions for yield class 24 m^3/ha/year; marked curves show predictions from yield tables (Edwards and Christie, 1981) for (*a*) mean diameter at breast height, (*b*) stand volume/ha, and (*c*) mean height of stand from the model but top height from the yield tables. Only the number of trees/ha was changed; parameters of the model are the same for both simulations.

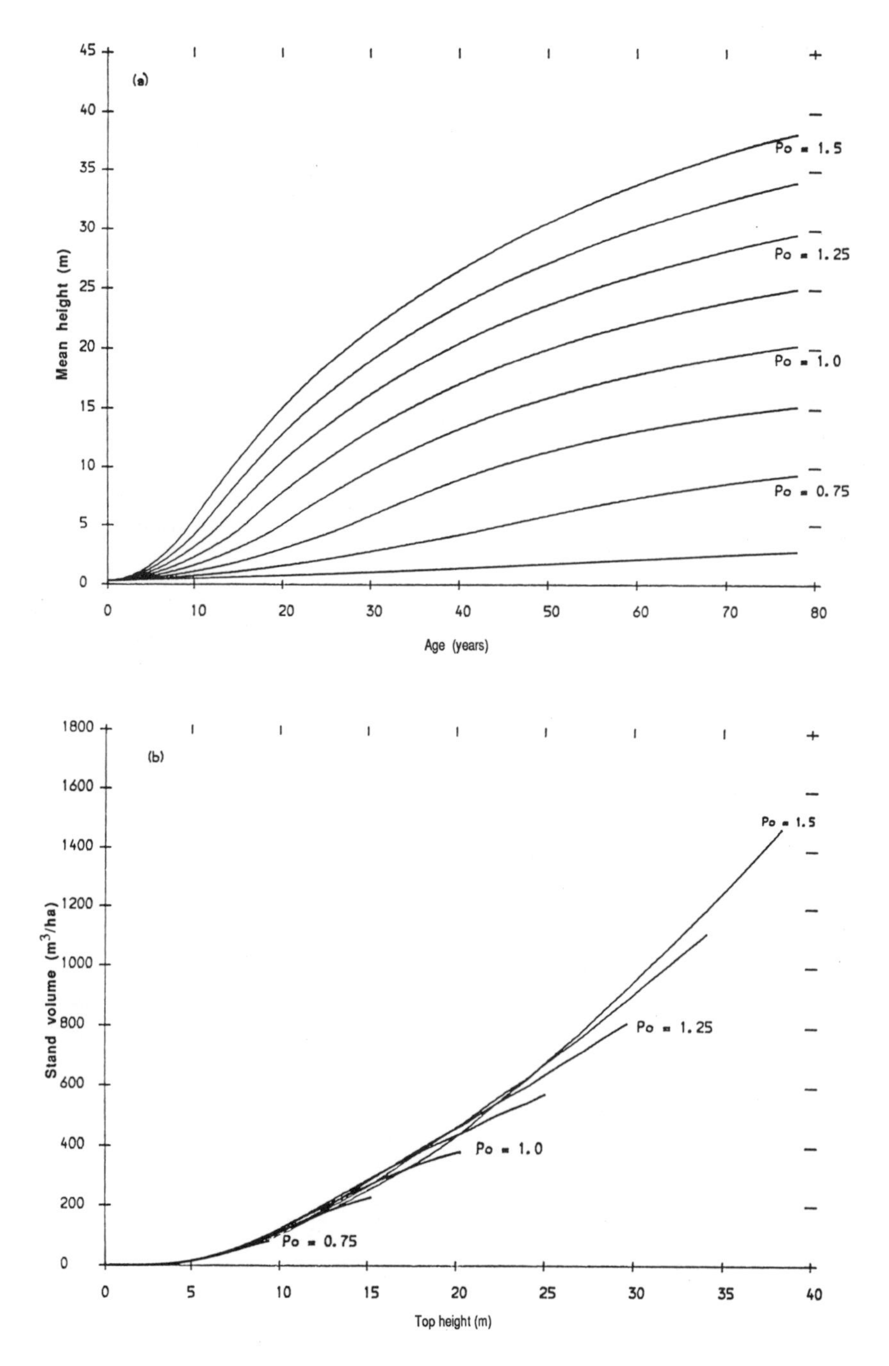

Figure 5. Response of model to changing the site parameter, P_0, in Equation 4. P_0 sets the maximum photosynthesis rate kg C/m²/year that would occur if all light were intercepted. (*a*) Mean height against age, and (*b*) stand volume against mean height. Apart from P_0, the parameters are the same as in Figure 4, and the unthinned curve in Figure 4c is identical to that in Figure 5b, with $P_0 = 1.5$.

DISCUSSION

The model we are developing has, as its first aim, an increased understanding of growth and yield in Sitka spruce. This is necessary to help us predict growth in conditions for which there are not data. In part, these new conditions arise because of changes in management practice, in part because of the effects of atmospheric pollution. In either case, it is not enough to replicate the existing yield tables; they are not necessarily correct. Also, we need to gain insight into why they worked and where they are likely to break down.

In that context, it is interesting to compare the yield tables with the process model. Both describe the relationship between top height and volume (Figure 5b) or the response to thinning. But the yield tables contain no further information and can only be tested by testing their goodness-of-fit. The process model, on the other hand, can be rejected if it fits the data badly, if the parameter values needed to fit the data are absurd, if any of the intermediate variables reach silly values during a simulation, or if the assumptions underlying the model can be shown to be false. Whether or not the process model survives these tests, it contains more information than the yield tables because it offers both a description of the data and an explanation of why volume and top height should be related in the way that they are. Although the explanation may turn out to be wrong, in advancing it we raise specific questions that are not raised by the yield tables. For example, is new sapwood area always proportional to new foliage dry weight? If not, then what form would we expect the volume top-height curve to take? Will it be such a simple and consistent curve?

Another important difference is that, once the assumptions of the process model are written down, the relationships between all its variables are fixed. The volume top-height relationship is fixed because of two assumptions: first, that new sapwood area is proportional to new foliage, and, second, that height increment is proportional to new foliage and inversely proportional to the projected crown area. Since both height and diameter growth depend on new foliage growth, it is inevitable that stand volume growth is a function of total foliage growth. In other words, the cumulative volume of an unthinned stand at any instant depends on how much foliage there has ever been. This determines the height growth, which in turn determines the number of pipes that have been produced to support that foliage. Moreover, the cumulative volume depends very little on how long the stand took to produce that foliage; thus two stands with the same top height will tend to have the same cumulative volume even when they differ in yield class and have taken very different times to reach that height. These relationships may be explained by other models and in other ways, but the point here is that the process model inevitably links them. By contrast, the yield tables have independent functions for height and diameter, and these have been fitted separately to the data.

Third, it is worth asking if the process model improves on the regression analysis described above. We showed that crown surface area was a good predictor of volume increment, at least in the diffuse light of the United Kingdom; but is that enough? The answer must be no, because we should always have to repeat the regression analysis for new sites or treatments, and we cannot wait for the data to grow. Nevertheless, the regression analysis provides a rapid test for part of the model. By noting that relative height is important in New Zealand but not at Dyfi in Wales, we have identified one set of conditions in which our approach to modeling light competition breaks down. But our study has also suggested what should be done, to produce a better model showing when Equation 5 is adequate and when it must be improved.

LITERATURE CITED

Bollmann, M. P., G. B. Sweet, D. A. Rook, and E. A. Halligan. 1986. The influence of temperature, nutrient status, and short drought on seasonal initiation of primordia and shoot elongation in *Pinus radiata. Canadian Journal of Forest Research* 16(5):1019–1029.

Edwards, P. N., and J. M. Christie. 1981. *Yield Models for Forest Management.* Booklet 48, Forestry Commission, Farnham, Surrey, U.K.

Grace, J. C. 1980. Computer modelling of individual tree growth. Ph.D. Thesis, Oxford University, Oxford, England, U.K.

Grace, J. C. 1988. Effect of foliage distribution within tree crowns on intercepted radiant energy and photosynthesis. In: Werger, M. J. A., P. J. M. van der Aart, H. J. During, and J. T. A. Verkoeven, eds. *Plant Form and Vegetation Structure: Adaptation, Plasticity and Relation to Herbivory.* S.P.B. Academic Publishing, the Hague, the Netherlands.

Grace, J. C. 1989. Modeling the interception of solar radiant energy and net photosynthesis. In: Dixon. R. K., R. S. Meldahl, G. A. Ruark, and W. G. Warren, eds. *Process Modeling of Forest Growth Responses to Environmental Stress.* Timber Press, Portland, OR, U.S.A.

Grace, J. C., P. G. Jarvis, and J. M. Norman. 1987. Modelling the interception of solar radiant energy in intensively managed stands. *New Zealand Journal of Forestry Science.* 17:193–209.

Grace, J. C., D. A. Rook, and P. M. Lane. 1987. Modelling canopy photosynthesis in *Pinus radiata* stands. *New Zealand Journal of Forestry Science* 17:210–228.

Hamilton, G. J. 1969. The dependence of volume increment of individual trees on dominance, crown dimensions and competion. *Forestry* 42(2):133–144.

Keyes, M. R., and C. C. Grier. 1981. Above- and below-ground net production in 40-year-old Douglas-fir stands on low and high productivity sites. *Canadian Journal of Forest Research* 11:599–605.

Lanner, R. M. 1976. Patterns of shoot development in *Pinus* and their relationship to growth potential. In: Cannell, M. G. R., and F. T. Last, eds. *Tree Physiology and Yield Improvement.* Academic Press, London, England, U.K. Pp. 223–243.

Mäkelä, A. 1986. Implications of the pipe-model theory on dry matter partitioning and height growth in trees. *Journal of Theoretical Biology* 123:103–120.

Mäkelä, A. 1988a. Parameter estimation and testing of a process-based stand growth model using Monte Carlo techniques. In: Ek, A. R., S. R. Shifley, and T. E. Burk, eds. *Forest Growth Modelling and Prediction.* General Technical Report NC-120, USDA Forest Service North Central Forest Experiment Station, St. Paul, MN, U.S.A. Pp. 315–322.

Mäkelä, A. 1988b. Performance analysis of a process-based stand growth model using Monte Carlo techniques. *Scandinavian Journal of Forest Research* 3:315–331.

McMurtrie, R., and L. Wolf. 1983. Above- and below-ground growth of forest stands: a carbon budget model. *Annals of Botany* 52:437–448.

Nance, W. L., J. E. Grissom, and W. R. Smith. 1988. A new competition index based on weight and constrained area potentially available. In: Ek, A. R., S. R. Shifley, and T. E. Burk, eds. *Forest Growth Modelling and Prediction.* General Technical Report NC-120, USDA Forest Service North Central Forest Experiment Station, St. Paul, MN, U.S.A. Pp. 134–142.

Pollard, D. F. W., and K. T. Logan. 1977. The effects of light intensity, photoperiod, soil moisture potential, and temperature on bud morphogenesis in *Picea* species. *Canadian Journal of Forest Research* 7:415–421.

Rennolls, K. 1985. The Dyfi thinning experiment on Sitka spruce. In: Gallagher, G., ed. *The Influence of Spacing and Selectivity in Thinning and Stand Development, Operations and Economy.* Proceedings, IUFRO Project Group P.4.02.02. Pp. 108–113.

Shinozaki, K., K. Yoda, K. Hozumi, and T. Kira. 1964a. A quantitative analysis of plant form: the pipe-model theory. I. Basic analyses. *Japanese Journal of Ecology* 14(3):97–105.

Shinozaki, K., K. Yoda, K. Hozumi, and T. Kira. 1964b. A quantitative analysis of plant form: the pipe-model theory. II. Further evidence of the theory and its application in forest ecology. *Japanese Journal of Ecology* 14(4):133–139.

Valentine, H. T. 1985. Tree-growth models: derivations employing the pipe-model theory. *Journal of Theoretical Biology* 117:579–585.

Valentine, H. T. 1987. *A Carbon-Balance Model of a Self-Thinning Stand with the Pipe-Model Theory.* WP-87-56, IIASA, Vienna, Austria.

Valentine, H. T. 1988. A growth model of a self-thinning stand based on carbon balance and the pipe-model theory. In: Ek, A. R., S. R. Shifley, and T. E. Burk, eds. *Forest Growth Modelling and Prediction.* General Technical Report NC-120, USDA Forest Service North Central Forest Experiment Station, St. Paul, MN, U.S.A. Pp. 353–360.

CHAPTER 22

A SIMPLIFIED CARBON PARTITIONING MODEL FOR SCOTS PINE TO ADDRESS THE EFFECTS OF ALTERED NEEDLE LONGEVITY AND NUTRIENT UPTAKE ON STAND DEVELOPMENT

Eero Nikinmaa and Pertti Hari

Abstract. A carbon-budget–based forest-stand development model is presented for Scots pine (*Pinus sylvestris* L.). Tree needles and the structure supporting them in branches, stem, and roots are considered to form functional units. Needle growth within the crown is assumed to be such that carbohydrates required for the formation of the structure of the functional unit are a constant fraction of expected annual photosynthesis of the needles at any height in the crown. The model is used to examine the effects of altered environmental conditions on forest growth and development. Both decreased nutrient uptake efficiency and increased needle mortality caused substantial changes in stand development. Increased needle mortality appeared to be more serious. However, the model does not account for possible tissue nutrient imbalances.

INTRODUCTION

The carbon budget has often been the foundation for process-based stand development models (e.g., Hari et al., 1982; McMurtrie and Wolf, 1983; Rook, 1985; Mäkelä and Hari, 1986; Mohren, 1987). In all these models, carbohydrate allocation for growth of different tree parts is a crucial step. However, physiological factors controlling allocation are still rather poorly understood (Hari et al., 1982, 1985; Waring, 1983; Rook, 1985).

Our approach to understanding carbon allocation and its dynamics is to examine the implications of the formation of the allometric structure of trees expressed by the pipe-model theory of Shinozaki et al. (1964) (Hari et al., 1985; Valentine, 1985; Mäkelä, 1986). Past research has indicated that the conversion of water-conducting sapwood into non-conducting heartwood (Mäkelä, 1988) and the vertical distribution of needle biomass in the crown are critical factors in the dynamics of carbon allocation.

We present a model in which the formation of new needles at different heights is related to their expected photosynthetic production balanced by the carbohydrate consumption required to maintain the allometric structure of the rest of the tree. The model was used to examine the effects of enviromental changes on Scots pine stand

Mr. Nikinmaa and Dr. Hari are at the University of Helsinki, Department of Silviculture, Unioninkatu 40 B, 00170 Helsinki, Finland.

growth and development. Effects were assumed to (1) directly affect the average life span of needles, or (2) indirectly affect the functional efficiency of roots.

THE STAND DEVELOPMENT MODEL

General Structure

The general structure of the model is presented in Figure 1. It follows the same principles presented by Mäkelä and Hari (1986) except for the allocation submodel. The time-step of the model is one year. A stand is defined as an aggregation of different size-classes of trees. Each size-class consists of a number of trees, which are presented by a mean tree in the simulations. The environment above the stand is described by the annual photosynthetic capacity of foliage in unshaded conditions, which was estimated from 15 years of field measurement (Korpilahti, 1988). Soil environment is described indirectly with a variable describing the mean annual average concentration of all nutrients in living tissue times average photosynthetic efficiency of needles divided by nutrient uptake efficiency of roots of all nutrients. This parameter is derived from Brouwer's (1962) functional balance principle.

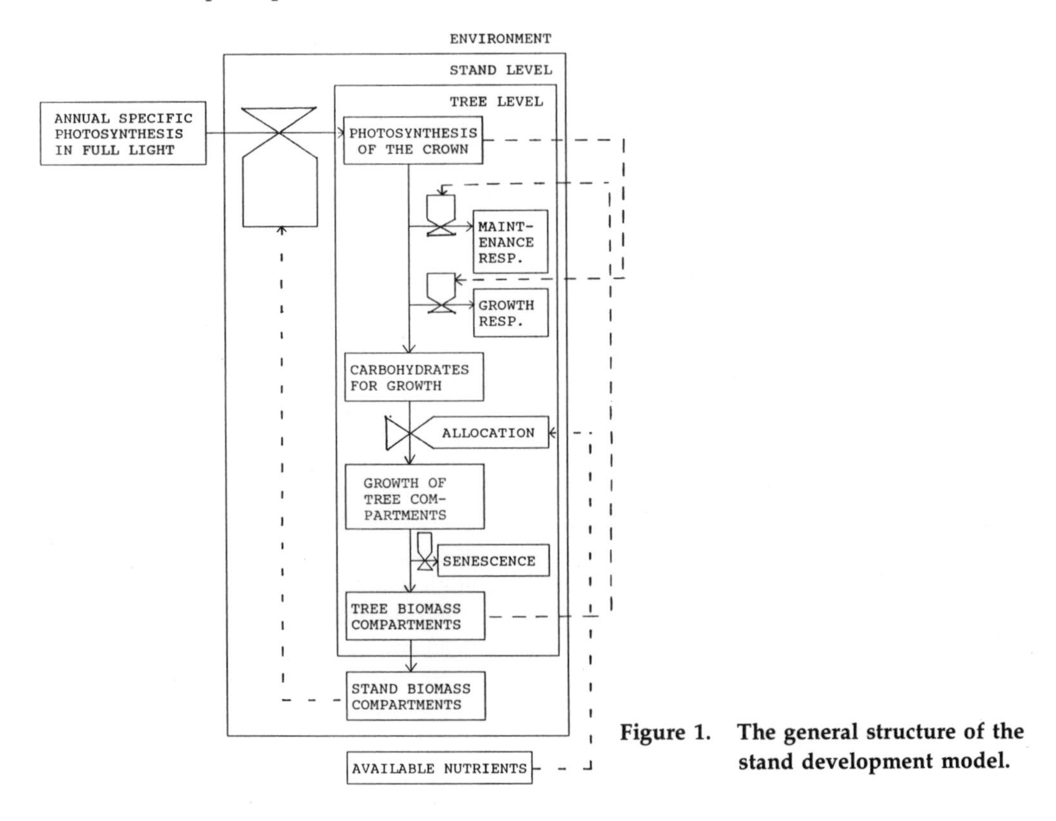

Figure 1. The general structure of the stand development model.

Stands are considered even-aged with an initial height difference of 10 cm between the size-classes, the largest size-class having an initial height of 150 cm. Stand-level interactions among trees are assumed to result from shading. The degree of shading at any height is calculated as a function of total needle biomass above that height. The needle biomass distribution within the canopy is assumed to be horizontally homogeneous. Total photosynthesis of a tree is computed by integrating photosynthate production at each height (unshaded annual photosynthesis × shading × needle biomass) over the length of the crown.

The photosynthetic production of a single tree is assumed to be completely consumed in respiration and growth. No carbohydrate storage is considered. Maintenance respiration is assumed to be a function of the biomass of different tree parts. Growth respiration is assumed to be proportional to photosynthesis.

Senescence of branches and roots is assumed to be proportional to their biomass. Needles are assumed to function for three years at equal efficiency (Linder and Troeng, 1980) after which they die if no accelerated needle mortality occurs. The transformation of water-conducting sapwood into non-conducting wood is considered to be related to crown dynamics. According to the experimental results of Kaipiainen and Hari (1985), the number of water-conducting tree rings is equal to the number of living whorls of branches, at least in young trees. In the model, the proportion of water-conducting tissue which becomes non-functioning as the needles below the lowest, live remaining branch die, is removed from from the total water-conducting capacity of the stem.

The death of an individual tree is assumed to be dependent on the growth of its needle biomass. When the net growth of the needle biomass is negative, the probability of a tree dying in that particular size-class increases.

THE ALLOCATION OF CARBOHYDRATES

General Principles

It is assumed that the growth and subsequent death of needles imply formation and release, respectively, of transport and support structure in branches, stem, and coarse roots, and growth and death of fine roots in proportional ratios (Valentine, 1985; Mäkelä, 1986). A balanced formation of new needles in the tree crown is assumed to occur when the gain in photosynthesis by the new needles (unshaded potential photosynthesis × shading × needle mass) per consumed carbon on the structure (needles + corresponding supporting structure in other tree parts) is equal at each height.

The Allocation Structure

The principle described above can be formulated as follows. If we call E(i) a new needle area formed per amount of carbohydrates required to build necessary structure in needles, branches, stem, and roots, and if P(i) is the (photosynthetic production per unit needle area at height i, then a balanced formation of needles takes place when

$$E(i)\ P(i) = K \tag{1}$$

$i = 0, \ldots, h$, where h = the top of the crown and 0 = the lowest living branch, and where K is constant.

P(i) can be written as (Mäkelä and Hari, 1986)

$$P(i) = P_0\ PLR(i) \tag{2}$$

where P_0 = potential yearly photosynthesis in full light (kg C/year/m^2 of needles) and PLR(i) = photosynthetic light ratio at depth i in the canopy. The latter is determined as a function of stand needle mass above point i (Mäkelä and Hari, 1986).

E(i) can be written as

$$E(i) = (a(i)\ m_1(i))/(m_1(i)+m_2(i)+m_3(i)+m_4(i)) \tag{3}$$

where a(i) = needle area / needle mass (m^2/kg), $m_1(i)$ = new needle biomass (kg dw), and

$m_2(i) - m_4(i)$ = new biomass required to support needle mass m_1, in branches, stem, and roots (kg dw), respectively.

Assuming the unit pipe model (Shinozaki et al., 1964), $m_2 - m_4$ can be written as a function of m_1. We used the following relations (for each height):

$$m_2 = (1\ m_1 + 1_b\ m_1 - \text{REL})\ g \tag{4}$$

in which

$$1 = (b\ m_1 + c)\ /\ (6!/(6-A)! \times 1_b/d) \text{ if } A < 6 \tag{5}$$

$$1 = (b\ m_1 + c)\ /\ (6! \times 1_b/d) \text{ if } A \geq 6 \tag{6}$$

where 1 = average length of new shoots (cm), 1_b = average length of branch (cm), A = age of the whorl (years), REL = released transport structure in branches by dying needles (kg dw × cm), and g = conversion factor (kg wood/cm/kg needles), b and c are empirical parameters relating the length of the shoot and its needle biomass (b = 5121.0, c = 1.627, when 1 is in cm and m_1 in kg). The latter term in parenthesis in Equations 5 and 6 divides the needle biomass at height i between different shoots, its minimum value being 3. The value for d is 200.

$$m_3 = (h\ m_1 - \text{REL}_2)\ g_2 \tag{7}$$

where h = height (cm), REL_2 = released transport structure in stem by dying needles, and g_2 = conversion factor (unit same as above).

For roots, assuming Brouwer's functional balance principle (1962) and the unit pipe model, we can write

$$m_4 = (\text{par1}\ (m_1 - \text{nm}) + \text{par2 mr}) + 1_r g_3\ \text{par1}\ (\text{ml} - \text{nm}) \tag{8}$$

where par1 = nutrient concentration in living tissue × specific photosynthetic production of needles per year per specific uptake of all nutrients by fine foots per year (kg n/kg C) (kg C/kg dw needles)/(kg n/kg dw roots), par2 = root biomass turnover percentage, mr = total mass of roots for height i (kg dw), nm = dying needle biomass (kg dw), 1_r = average length of roots (cm), and g_3 = conversion factor (kg dw wood/cm/kg dw fine roots).

As mentioned earlier, P(i) is calculated as a function of stand needle biomass above the point i. Assuming E(i) P(i) = K, where K is constant, m_1(i) can be solved. However, an additional assumption on the maximal value of E(i) has to be made to allow for self-pruning of trees. This means assuming a maximal proportion of expected carbohydrate production which can be used on needle growth.

The possible indirect and direct effects of environmental changes on forest growth and development were examined using the simulation model. The indirect effects were simulated by changing the value of the parameter describing the ratio between nutrient demand and nutrient uptake efficiency (par1 in Equation 8).

The direct effects on the needles were simulated by simultaneously increasing the mortality of both second- and third-year needles up to 90%.

RESULTS AND DISCUSSION

The vertical needle biomass distribution of dominant trees moved slightly when the ratio between nutrient demand and uptake efficiency (par1) was changed from 0.1 to 0.4 (Figure 2). Small values of par1 correspond to higher uptake efficiencies of roots. It was assumed that the specific photosynthesis of the needles or the average nutrient concentrations in the living tissue would not change significantly at the annual time-step with the increasing nutrient uptake efficiency, but that the effects would show up in the structural

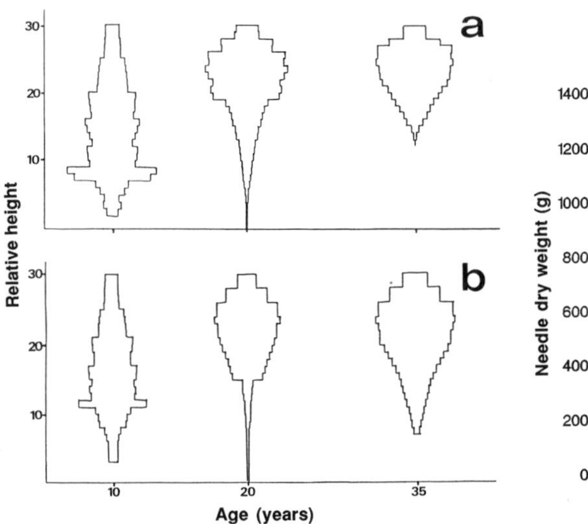

Figure 2. The simulated development of crown needle biomass distribution in the largest size-class when the ratio between nutrient demand and uptake efficiency (par1) is (*a*) 0.1 and (*b*) 0.4. (Biomass per unit height is directly proportional to the width of the crowns.)

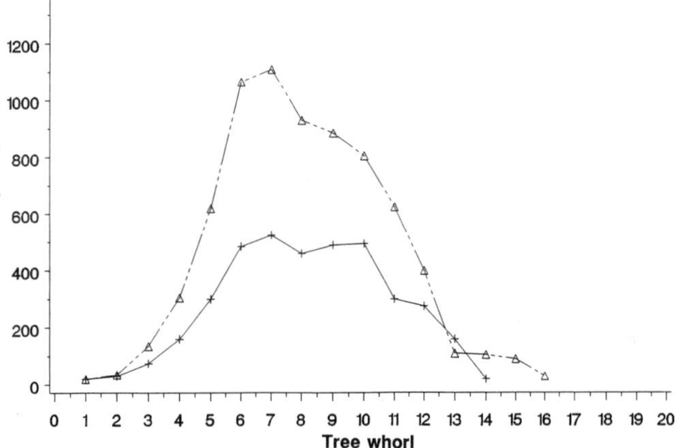

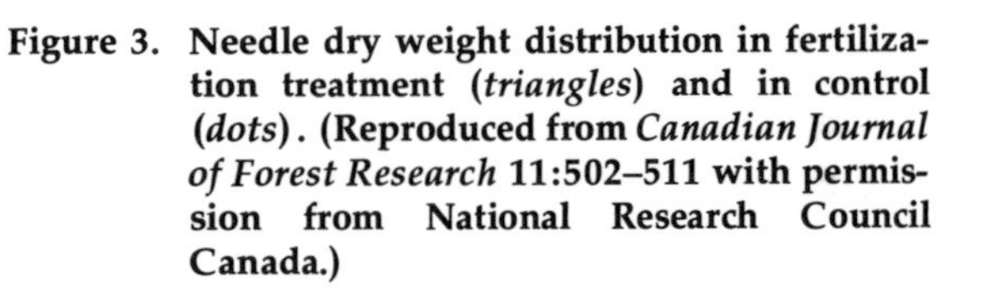
Figure 3. Needle dry weight distribution in fertilization treatment (*triangles*) and in control (*dots*). (Reproduced from *Canadian Journal of Forest Research* 11:502–511 with permission from National Research Council Canada.)

acclimation. Assuming this, the results are in agreement with the results obtained for Douglas fir (*Pseudotsuga menziesii* [Mirb.] Franco) by Brix (1981). He found that in fertilized trees, the needle biomass distribution was concentrated more in the top part of the crown than in control trees (Figure 3). However, direct comparisons between the simulation results and the results by Brix (1981) cannot be made. His experiments were fertilization experiments, whereas in the simulations it was assumed that the different nutrient uptake efficiencies had affected stands throughout their development.

The total needle biomass of the mean tree is relatively small when the parameter value is smaller. Smaller values of par1 allow suppressed trees to survive a longer time, thus also the stem density is higher than could be expected from the observation that the stem density on better sites generally decreases more rapidly than on poorer sites (e.g., Koivisto, 1959).

The development of needle, branch, stem, and root biomass with varying the par1 is presented in Figure 4. The overall shape of the development curves obtained at one parameter value closely resembles those observed by Albrektson (1980). As expected, an increase in the par1 parameter value decreased needle biomass and increased root biomass. At later phases of stand development, the decreased allocation to needles resulted in smaller root biomass production at extreme parameter values. It is noteworthy that very small parameter changes significantly affected stem biomass development.

Changing the mortality of old needles clearly affected the development of all biomass compartments (Figure 5). At 30% needle mortality, the biomass levels were only about half the levels when no accelerated mortality was assumed. At later simulation ages, the highest needle mortality rate caused the whole stand to die.

Results presented to this point reflect the effects of parameter values which were kept constant through the whole simulation. Effects of parameter values that change during stand development can be evaluated by observing the values across the three-dimensional

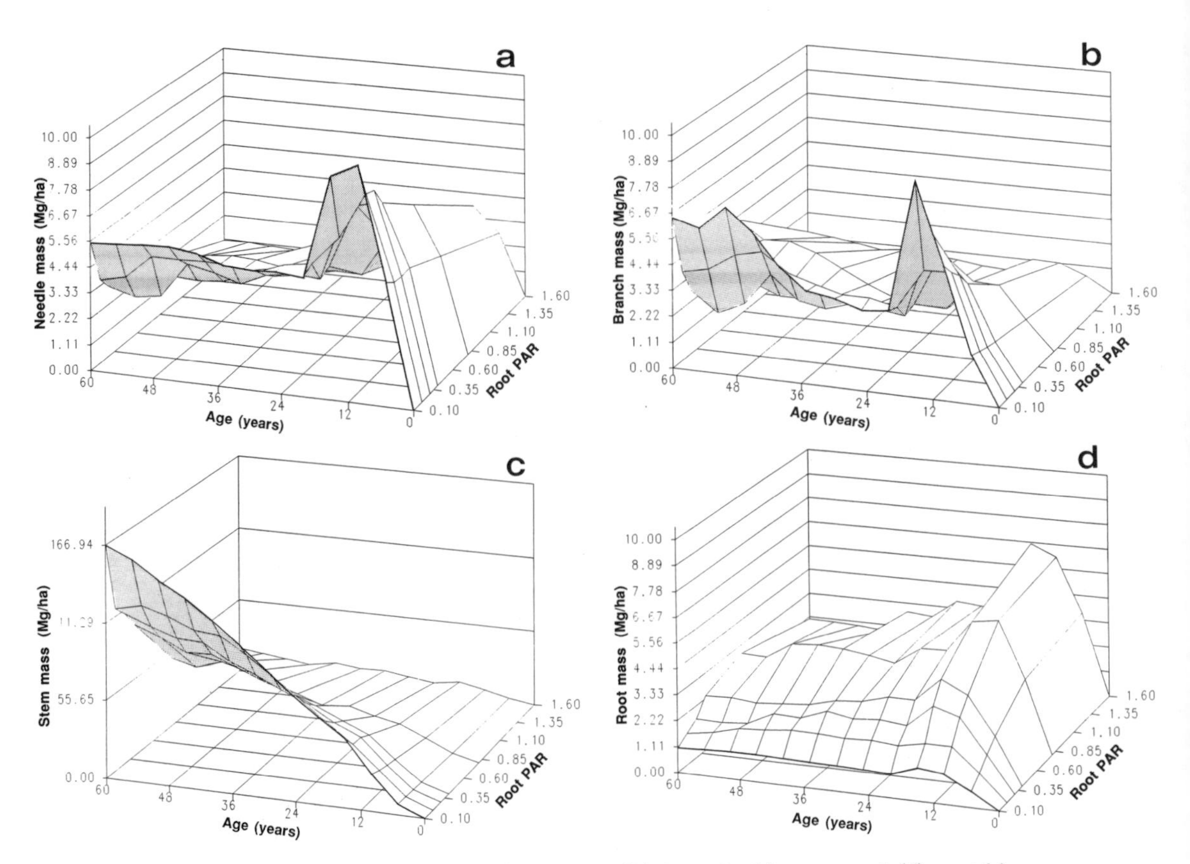

Figure 4. Simulated development of (*a*) needle, (*b*) branch, (*c*) stem, and (*d*) root biomasses when changing the ratio between nutrient demand and uptake efficiencies.

planes in Figures 4 and 5. In these cases, changes can be more significant, because the already obtained size of trees can be too large to adjust to the new worse conditions (Mäkelä, 1986).

From the results presented here, it appears that factors which directly affect needle mortality are more harmful to trees than the indirect effects of decreased nutrient availability. It must be recognized, however, that because all nutrients were dealt with as a group, the effects of nutrient imbalances could not be examined.

Our model uses as one of its basic assumptions the allometric structure of trees. Although this is the strong feature of the model in simulating the natural development of stands, it is also one of the major drawbacks of the model in predicting the effects of environmental changes. Environmental stresses can affect trees in ways that disturb the observed regular structure (Waring, 1987). Therefore, in order to construct more reliable models, more knowledge of the functional bases involved in the formation of structure is necessary. In this respect, the work by Zimmermann (1983) and Tyree et al. (1987) is very promising.

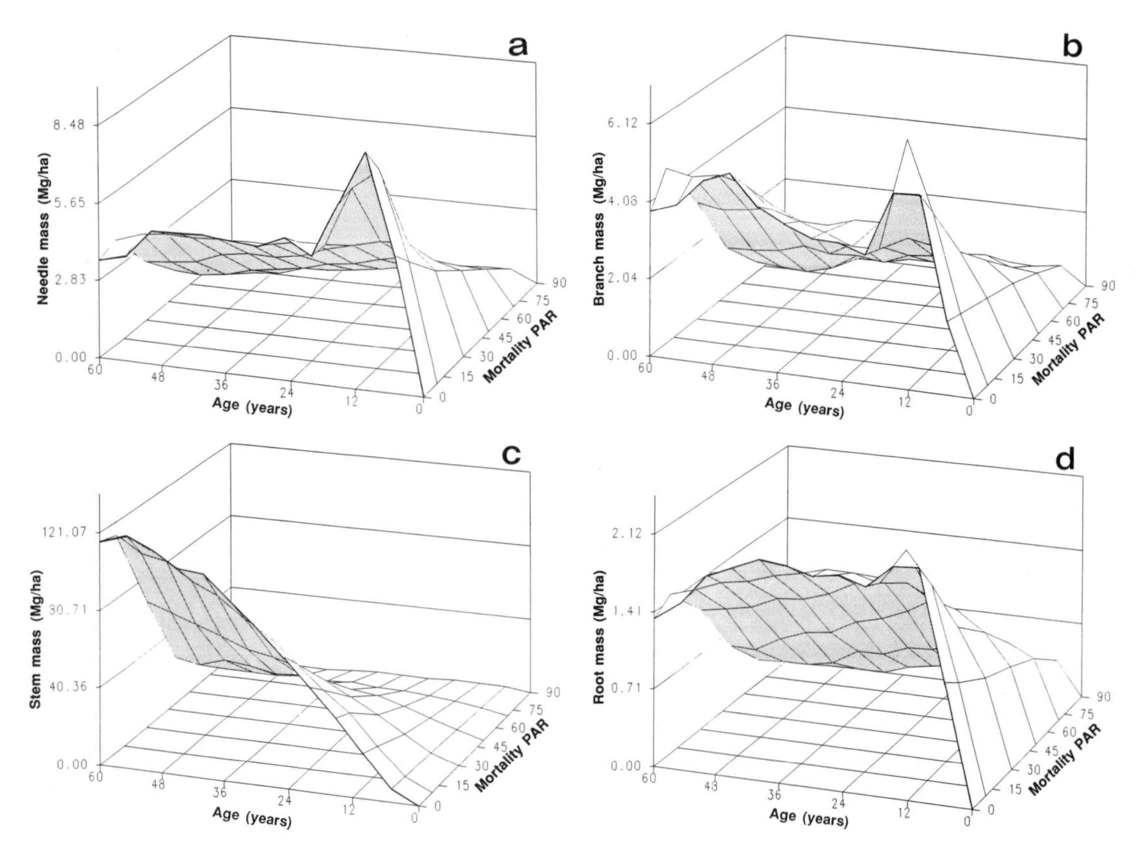

Figure 5. Simulated development of (*a*) needle, (*b*) branch, (*c*) stem, and (*d*) root biomasses when changing the needle mortality parameter.

LITERATURE CITED

Albrektston, A. 1980. Total tree production as compared to conventional forestry production. In: Persson, T., ed. *Structure and Function of Northern Coniferous Forests: An Ecosystem Study. Ecological Bulletin* (Stockholm, Sweden) 32:315–328.

Brix, H. 1981. Effects of thinning and nitrogen fertilization on branch and foliage production in Douglas-fir. *Canadian Journal of Forest Research* 11(3):502–511.

Brouwer, R. 1962. Distribution of dry matter in the plant. *The Netherlands Journal of Agricultural Sciences* 10(5):361–376.

Hari, P., L. Kaipiainen, E. Korpilahti, A. Mäkelä, T. Nilsson, P. Oker-Blom, J. Ross, and R. Salminen. 1985. Structure, radiation and photosynthetic production in coniferous stands. Research Note 54, Department of Silviculture, University of Helsinki, Helsinki, Finland. 233 pp.

Hari, P., S. Kellomäki, A. Mäkelä, P. Ilonen, M. Kanninen, E. Korpilahti, and M. Nygren. 1982. Dynamics of early development of tree stand. *Acta Forestalia Fennica* 177. 39 pp.

Kaipiainen, L., and P. Hari. 1985. Consistencies in the structure of Scots pine. In: Tigersted, P. M. A., P. Puttonen, and V. Koski, eds. *Crop Physiology of Forest Trees.* Helsinki University Press, Helsinki, Finland. Pp. 32–37.

Koivisto, P. 1959. Growth and yield tables (in Finnish). *Communicationes Instituti Forestalia Fenniae* 51(8).

Korpilahti, E. 1988. Photosynthetic production of Scots pine in the natural environment. *Acta Forestalia Fennica* 202. 71 pp.

Linder, S., and E. Troeng. 1980. Photosynthesis and transpiration of 20-year-old Scots pine. In: Persson, T., ed. *Structure and Function of Northern Coniferous Forests: An Ecosystem Study. Ecological Bulletin* (Stockholm, Sweden) 32:165–182.

Mäkelä, A. 1986. Implications of the pipe-model theory on dry matter partitioning and height

growth in trees. *Journal of Theoretical Biology* 123:103–120.
Mäkelä, A. 1988. Models of pine stand development: an ecophysiological systems analysis. Research Note 62, Department of Silviculture, University of Helsinki, Helsinki, Finland. 267 pp.
Mäkelä, A., and P. Hari. 1986. Stand growth model based on carbon uptake and allocation in individual trees. *Ecological Modelling* 33:205–229.
McMurtrie, R., and L. Wolf. 1983. Above- and below-ground growth of forest stands: a carbon budget model. *Annals of Botany* 52:437–448.
Mohren, G. M. J. 1987. *Simulation of Forest Growth, Applied to Douglas Fir Stands in the Netherlands.* Pudoc, Wageningen, the Netherlands. 184 pp.
Rook, D. 1985. Physiological constraints on yield. In: Tigersted, P. M. A., P. Puttonen, and V. Koski, eds. *Crop Physiology of Forest Trees.* Helsinki University Press, Helsinki, Finland. Pp. 1–19.
Shinozaki, K., K. Yoda, K. Hozumi, and T. Kira. 1964. A quantitative analysis of plant form: the pipe-model theory. I. Basic analyses. *Japanese Journal of Ecology* 14(3):97–105.
Tyree, M. T., L. B. Flanagan, and N. Adamson. 1987. Response of trees to drought. In: Hutchinson, T. C., and K. M. Meema, eds. *Effects of Atmospheric Pollutants on Forests, Wetlands and Agricultural Ecosystems.* NATO ASI Series, Vol. G16. Springer-Verlag, Berlin/Heidelberg, F.R.G. Pp. 201–216.
Valentine, H. T. 1985. Tree-growth models: derivations employing the pipe-model theory. *Journal of Theoretical Biology* 117:579–584.
Waring, R. H. 1983. Estimating forest growth and efficiency in relation to canopy leaf area. In: Macfadyen, A., and E. D. Ford, eds. *Advances in Ecological Research,* Vol. 13. Academic Press, London, England, U.K. Pp. 327–354.
Waring, R. H. 1987. Characteristics of trees predisposed to die. *BioScience* 37:569–574.
Zimmermann, M. H. 1983. *Xylem Structure and the Ascent of Sap.* Springer-Verlag, Berlin/Heidelberg, F.R.G. 143 pp.

CHAPTER 23

AN EVALUATION OF COMPETITION MODELS FOR INVESTIGATING TREE AND STAND GROWTH PROCESSES

Thomas W. Doyle

Abstract. Process modeling of forest growth has provided a significant adjunct to field studies by enabling us to test ecological constructs of tree growth and stand-level stresses of both natural and anthropogenic origin. Competition is a major component of these models, though there is little uniformity in the conceptual design and mathematical detail that is employed. The behavior and performance of various competition models were evaluated from both simulation trials and comparative tests using actual field data. Understanding the spatial and temporal aspects of forest competition can only foster our ability to model and asses more accurately the cumulative stresses and interactions of environmental influences.

INTRODUCTION

Process models of forest growth provide a significant adjunct to experimental studies by enabling us to test ecological constructs of tree and stand response to environmental stress. The cause of these stresses may be either natural or anthropogenic. The term stress denotes an inhibitive growth factor or effect, such that tree and forest growth is less than a given potential in the absence of stress. Researchers are attempting to identify the pathways and mechanisms of various stress agents on the cell, organelle, organ, organism, and community levels of tree and forest structure and function. The diversity of stress agents and effects on any of these organizational levels of varied resolution and scale has thwarted any unified approach to growth-stress analysis and modeling. This study focuses on modeling efforts for one such natural yet major stress agent, competition.

Competition may be defined as the degree to which proximal trees modify each other's environment, such that they alter the distribution and availability of resources deemed essential for growth and survival. This dynamic process includes biotic and abiotic interactions as well as aboveground and belowground determinants. Consequently, competition presents a difficult stress component to quantify because it defies direct measurement. In mesic forest communities, competition accounts for most of the observed variation in forest structure and individual tree growth (Doyle, 1983a). Subsequently, whole-tree and stand development under natural conditions comprises a spatiotemporal response to an ever-changing competitive environment. Because of the com-

Dr. Doyle is a Research Scientist with the MAXIMA Corportion under contract to Environmental Sciences Division, Oak Ridge National Laboratory, Martin Marietta Energy Systems, Inc., Oak Ridge, TN, 37831 U.S.A.

plexities of studying competition in the field and laboratory, it remains a largely undefined process influencing growth and mortality.

Competition may also play an important role in influencing the degree of tree and species response in the presence of other stresses. Models and analyses of forest and tree response to anthropogenic stress cannot ignore the possible confounding and/or compounding effects under varying competitive conditions (Doyle, 1983b). Modeling efforts to date are based on experimental results and assumptions that have largely gone unvalidated with respect to system behavior under combinative stresses. The question of how competition and other stress agents behave combinatively has only recently been explored (Blum et al., 1983; Zangerl and Bazzaz, 1984).

In greenhouse and outplanting studies, competition is often evaluated by growth performance in relation to mean spacing and density (Harper, 1961). Relating the within-stand variation of growth response to a mean stand condition, however, fails to adequately explain the biology and distribution of stress on an individual basis. This is particularly true in natural stands where the proximity, size, and species of competing individuals may vary greatly from the mean stand condition and composition. This fact has warranted the development of spatially explicit and size-weighted expressions known as competition indices. These models provide a logical construct for assessing the stress effect on an individual tree basis by considering size-ratio and distance relationships with surrounding competitors.

Competition measures represent a key component of many individual tree and stand growth models (Dale et al., 1985). These submodels, or indices, as they are often called, attempt to predict a tree's physiological vigor in reference to its ability to exploit the available resources relative to its neighbors. While competition does not in itself constitute a physiological process, it does determine the biotic and abiotic conditions that affect tree structure and function.

Many studies have investigated the performance of competition models for given sets of data but have failed to identify model characteristics and behavior that account for their success. In this study, three major categories of model types are reviewed and evaluated: (1) influence-zone overlap indices, (2) size-weighted ratios, and (3) growing-area polygons. These models similarly consider the size, number, and distance of neighboring trees but vary greatly in mathematical design and detail. All require distance measures to approximate available growing area and are, therefore, considered distance-dependent models.

Because these models also utilize size differences to weight competitor influence, this approach relies on the competitive effect rather than the causal condition of explicit spatial and temporal characteristics of species response and stand development. As a result, there is a degree of circularity in using the product of the parameter one wants to predict in the modeling process.

For this reason, this chapter considers only the spatial aspects and behavior of some noted competition models of the types already described. A Spatially Explicit Landscape Vegetation Analysis (SELVA) was used to test these models under a range of spacing regimes. This exercise was designed as an initial step toward understanding the spatial component and contribution of these models to efforts to mimic the process of competitive stress and interaction in forests.

METHODS

An integrated tree and stand model (SELVA) was used to investigate the behavior and performance of noted distance-dependent competition models under different spacing regimes. SELVA simulates both plantation and natural stands under uniform-to-random spatial distribution input. In order to consider the spatial contribution of competition indices, SELVA operated under a plantation mode assuming equal growing units or space for each tree. This application allowed the basic assumption that equal growing space renders the same competitive stress level, thus equal growth. Assuming equal growth and size from initial establishment, size-ratio effects of these competition models are nullified to no effect, thus allowing a true evaluation of the spatial attributes of the same models with increasing development and spacing. Growth was apportioned equally in incremental steps as a percentage of the spacing trial. SELVA simulated a square planting design of 4 (1.2), 6 (1.8), 8 (2.4), 10 (3.0), 12 (3.7), and 16 (4.9) ft (m) distances between rows.

Competition models of various design and complexity were evaluated for their predictive attributes with respect to assessing growing space under different planting densities. These included distance-dependent measures categorized as influence-zone overlap indices, distance-weighted size ratios, and growing-space polygons.

The influence-zone concept is based on an assumed circular zone around every tree, roughly equivalent to the crown area of an open-grown tree, wherein direct competition occurs (Staebler, 1951). This zone is thought to be related to the expected growing space given full crown and root development (Opie, 1968). The extent to which this area overlaps the influence zones of neighboring trees represents a measure of the subject tree's optimum functional environment. Zonal overlap indices vary with the type of overlap expression and weighting of tree size. Staebler's (1951) model calculates the linear overlap of intersecting crowns of subject tree and competitors. Newnham (1964) expanded on this approach by considering instead the sum angular overlap of competitors with the subject tree. Gerrard's (1969) Competition Quotient and Bella's (1971) Competitive Influence Overlap (CIO) models are similar in that they relativize the degree of areal overlap of intersecting neighbors based on subject tree size.

Distance-weighted size ratios specify the competitive effect as the sum of size ratios multiplied by the distance of selected competitors from the study tree. Hegyi's (1974) index, the first and most noted index of this design, is a function of the sum of the competitor–study-tree size ratios and the inverse distance between them. A modified Hegyi index tested by Daniels (1976) differs from the above only in that the model terms are squared. Monserud and Ek (1977) introduced a hybrid of Hegyi's model and the influence-zone overlap type by adding an exponential term onto the distance measure that accounts for linear overlap of maximum crown widths of subject tree and competitor.

Another class of competition model approximates actual or available growing space using elaborate geometrical constructions, appropriately termed polygons and polargrams. Growing area is represented by a polygon or polargram with non-overlapping crown boundaries defined by the proximity and size of neighboring trees. Brown (1965) was the first to propose such a model wherein crown boundaries were defined by bisectors fixed at half the distance between competing trees. Moore et al. (1973) modified Brown's design by weighting the placement of crown expanse on relative size of the subject tree and its competitors. Alemdag (1978) formulated a competitive stress index based on the average sector size per competitor using a polar approach. Doyle (1983a) calculated relative crown area from a polargram formed by the summed area of sections between subject tree and competitors as compared to maximum expected crown area under open-grown conditions. These assume that only one competitor is eligible to affect the limits of crown expanse in any one direction. The degree to which any one competitor confines crown development is determined by the diameter ratio of the subject tree and competitor or the maximum expected crown size, whichever is least.

Along with the many types of competition models, there are also different methods and ideas for determining which neighbors are competitors and to what degree. Because of the plantation spacing design used in this study, this task is made easier by selecting only those immediate trees within crown contact of the subject tree. Assessments of model behavior were curtailed at the point where the crown radius of the subject tree overlapped its secondary competitors.

Results were analyzed graphically to assess the model's systematic behavior and biological consistency with increasing size and crowding for each spacing regime. Where possible, these results were simplified based on the degree of overlap relative to the spacing distance. Data from a slash pine (*Pinus elliottii* Engelm.) spacing trial experiment (Harms and Collins, 1965) were calculated and graphed similarly to the competition model output to serve as a benchmark for validation purposes.

RESULTS

These simulation trials revealed five distinct behavioral types for all models tested. Figure 1 represents the simplest type, which shows an inverse relationship to spacing distance. This response was characteristic of the size-weighted distance ratios devised by Hegyi (1974) and Daniels (1976). These functions simply state that the greater the spacing distance between trees, the less the competitive stress that the trees will encounter. With these models, no distance is so great that there is no competition, although the measured effect becomes fairly negligible after a moderate distance. Squaring the distance term suggests that little if any competition is likely beyond 8 ft (2.4 m), which would seem to invalidate this approach for forest systems after only a few years following establishment. One other apparent limitation with these models involves the static relationship with distance such that the developmental phase of the tree and stand does not alter the competitive effect through time.

Figure 2 illustrates the spatial effect for a host of influence-zone overlap indices (Arney, 1973; Bella, 1971; Gerrard, 1969; Monserud and Ek, 1974; Newnham, 1964; Staebler, 1951). This finding suggests that while these indices measure the degree of overlap in different units, their behavior is the same due to a proportionality factor. These results differ from Model I types in that trees can be sufficiently distant so as to have no effect on one another. Type II models also differ from Type I results in that competition is a function of the degree of development for a given spacing distance rather than a function of the spacing distance alone. Model II behavior logically increases in a systematic manner with increasing stand closure.

The hybrid model of Monserud and Ek (1977) comprises a mix of the above-mentioned size-weighted and influence-zone indices. This model therefore represents a special case as shown in Figure 3. It behaves more similarly to the Type II models because its distance function is an expression of crown overlap rather than just a spacing factor. The behavior of this model suggests that the rate of competitive stress increases more rapidly with partial overlap than with full overlap. It also reflects a sensitivity to this anomaly when overlap is initiated with the primary diagonal competitors.

Alemdag's (1978) Proportionally Available Growing Space Area of Sectors (PAGAS) model offers yet another individual case similar to the aforementioned Type III model that demonstrates a somewhat chaotic response with overlap from the primary diagonal competitors (Figure 4). In this situation, the reversal in the rate and degree of competitive stress with greater crowding indicates an errant model design brought about by expressing the competitive condition on the basis of number of competitors divided by their crown sector sum. And because this approach is fixed on non-overlapping crown sector area, it is possible to show no change in competitive stress while increasing the crowding factor. The Area Potentially Available (APA) models of Brown (1965) and

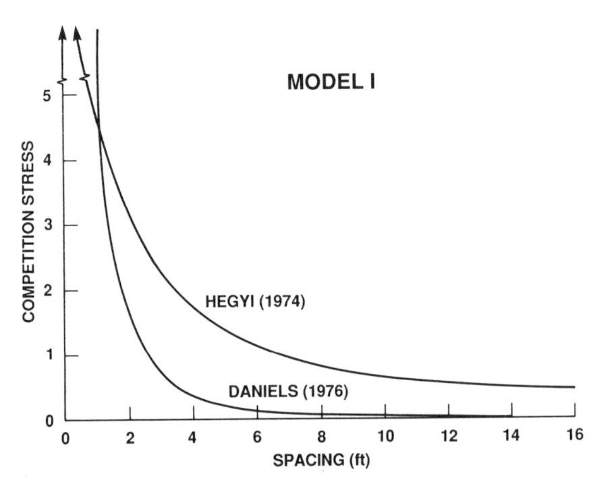

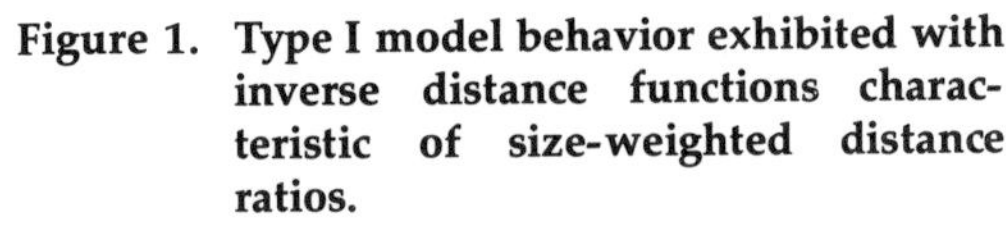
Figure 1. Type I model behavior exhibited with inverse distance functions characteristic of size-weighted distance ratios.

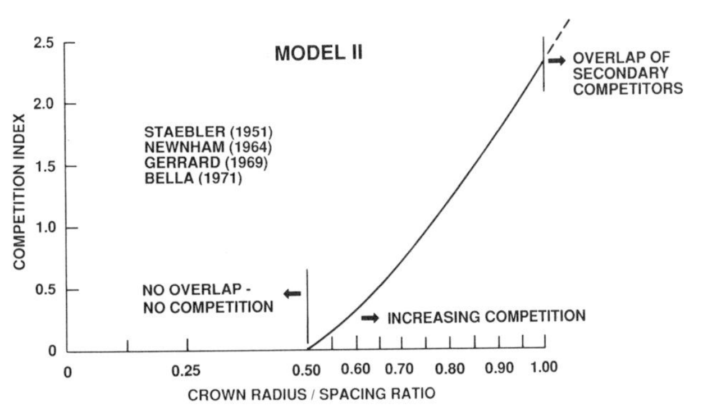

Figure 2. Type II model behavior characteristic of influence-zone overlap indices. The scale of competitive stress varies with each model tested, yet the functional response is identical.

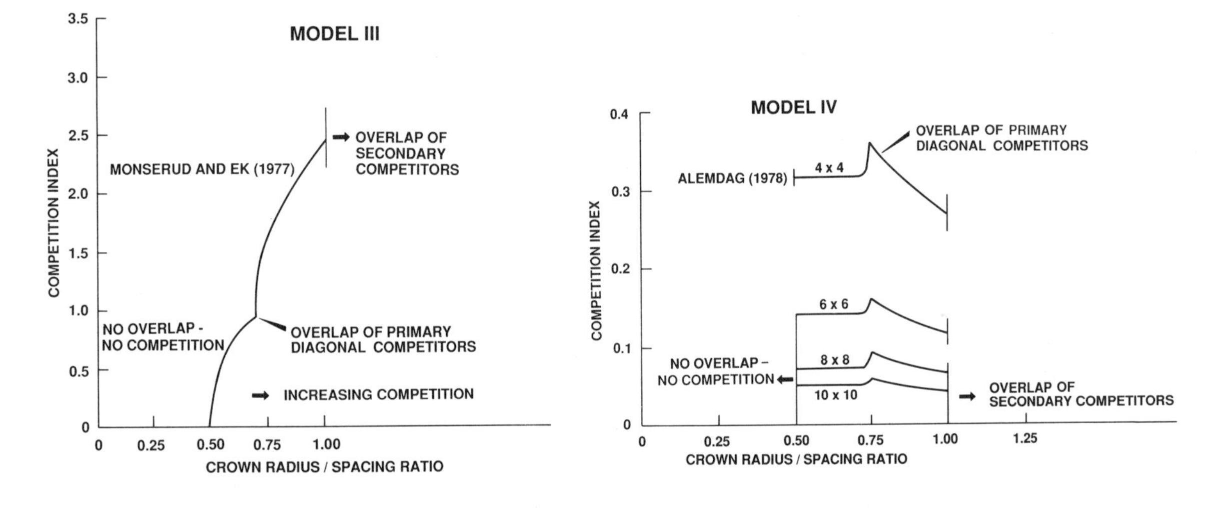

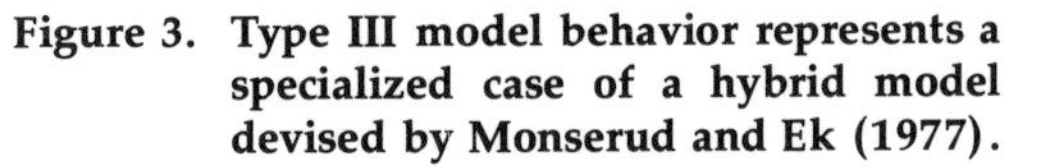
Figure 3. Type III model behavior represents a specialized case of a hybrid model devised by Monserud and Ek (1977).

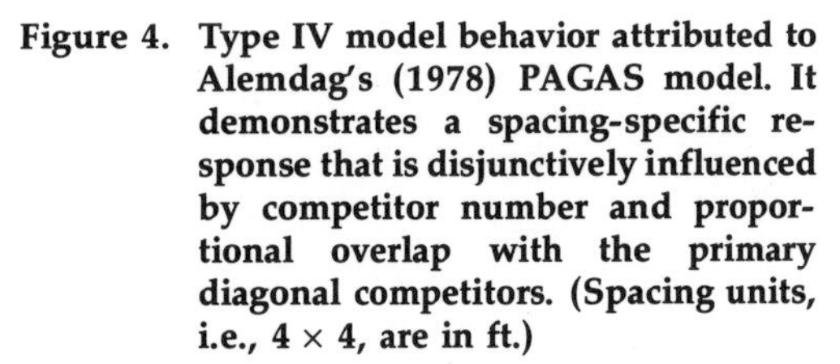
Figure 4. Type IV model behavior attributed to Alemdag's (1978) PAGAS model. It demonstrates a spacing-specific response that is disjunctively influenced by competitor number and proportional overlap with the primary diagonal competitors. (Spacing units, i.e., 4 × 4, are in ft.)

Moore et al. (1973) follow this same pattern in that only one estimate for crown area is possible under the restraints outlined for this study, regardless of stand spacing and development.

Doyle's (1983a) relative crown area polargram behaved most similarly to the Type II model with no competition prior to crown overlap under presupposed open-grown conditions (Figure 5). Unlike other polygon models, this model considers the ratio of realized crown area to the given potential for the same tree under optimal conditions. This approach relativizes the index in such a way that as the size and density of trees for a cer-

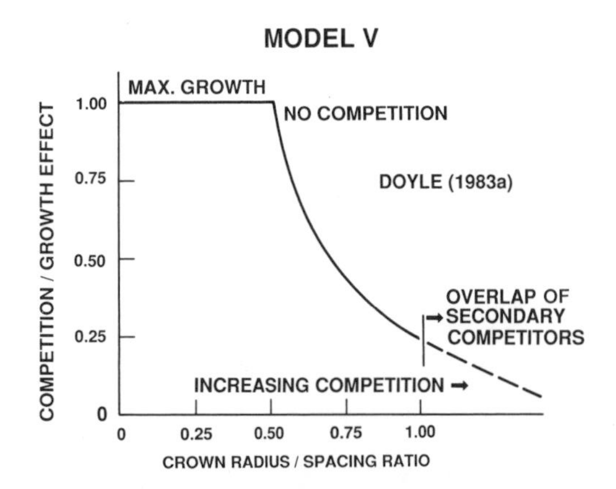

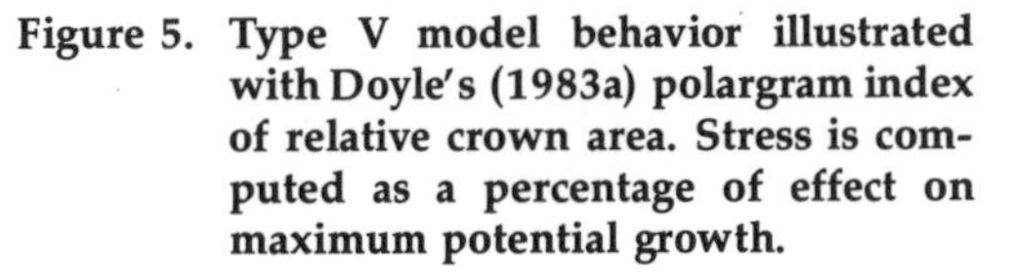
Figure 5. Type V model behavior illustrated with Doyle's (1983a) polargram index of relative crown area. Stress is computed as a percentage of effect on maximum potential growth.

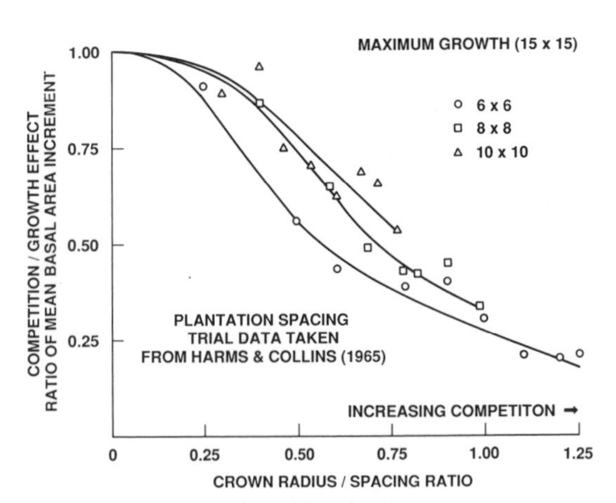

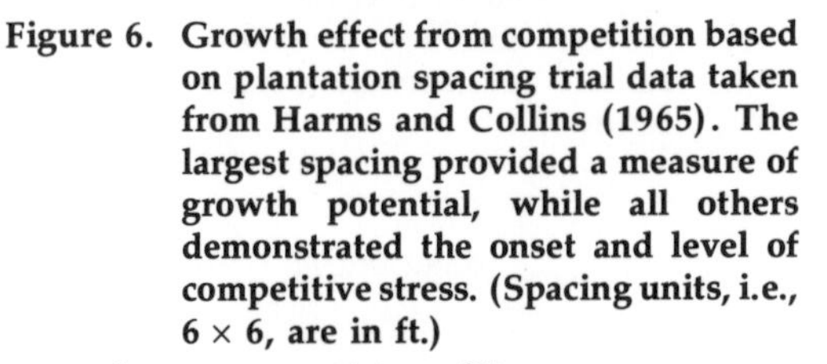
Figure 6. Growth effect from competition based on plantation spacing trial data taken from Harms and Collins (1965). The largest spacing provided a measure of growth potential, while all others demonstrated the onset and level of competitive stress. (Spacing units, i.e., 6 × 6, are in ft.)

tain spacing situation increase, so also does stress from competition. Given as a percentage of maximum growth potential, this index decreases growth in proportion to the increase in competition.

Figure 6 shows data from a spacing trial study of slash pine (Harms and Collins, 1965) presented in a form comparable to the behavioral trends just described. As might be expected, there is little or no competition in early years followed by a systematic and consistent increase until self-thinning begins to occur, verifying model behavior as illustrated with the Type II (Figure 2) and Type V (Figure 5) examples. The pattern appears to replicate itself across all spacing though some response lag may exist with greater spacings.

CONCLUSION

This study demonstrates the need for and feasibility of rigorous testing of competition models and their components to determine whether their performance and success can be validated. It also identifies the importance of considering the biological basis for a given construction and application of process models of this sort. As scientists investigate the more complex questions of combinative stresses and the role competition might play in abating or intensifying the impact of other environmental stresses, modeling exercises as demonstrated herein may prove a valuable tool for examining the pervasive nature of competitive interactions.

LITERATURE CITED

Alemdag, I. S. 1978. Evaluation of some competition indices for the prediction of diameter increment in planted white spruce. Information Report FMR-X-108, Canadian Department of the Environment, Forest Management Institute, Ottawa, Ont., Canada. 39 pp.

Bella, I. E. 1971. A new competition model for individual trees. *Forest Science* 17:364–372.

Blum, U., A. S. Heagle, J. C. Burns, and R. A. Linthurst. 1983. The effects of ozone on fescue-clover forage: regrowth, yield, and quality. *Environmental and Experimental Botany* 23:121–132.

Brown, G. S. 1964. Point density in stems per acre. Forest Research Notes No. 38, Forest Service, Forest Research Institute, Rotorua, New Zealand. 11 pp.

Dale, V. H., T. W. Doyle, and H. H. Shugart. 1985. A comparison of tree growth models. *Ecological Modelling* 29:145–169.

Daniels, R. F. 1976. Simple competition indices and their correlation with annual loblolly pine tree growth. *Forest Science* 22:454–456.

Doyle, T. W. 1983a. Competition and growth relationships in a mixed-aged, mixed-species forest community. Ph.D. Thesis, University of Tennessee, Knoxville, TN, U.S.A. 85 pp.

Doyle, T. W. 1983b. Investigating the role of intertree and interspecies competition in forests subject to air pollution exposure. In: Davis, D., ed. *Air Pollution and Forest Productivity.* Symposium sponsored by the Isaac Walton League, October 4–5, 1983, Washington, DC, U.S.A. Pp. 95–104.

Gerrard, D. J. 1969. Competition quotient: a new measure of the competition affecting individual forest trees. Research Bulletin No. 20, Michigan State University, East Lansing, MI, U.S.A. 32 pp.

Harms, W. R., and A. R. Collins. 1965. Spacing and twelve-year growth of slash pine. *Journal of Forestry* 63:909–912.

Harper, J. L. 1961. Approaches to the study of plant competition. *Symposium Society for Experimental Biology* 15:1–39.

Hegyi, F. 1974. A simulation model for managing jack-pine stands. In: Fries, J., ed. *Growth Models for Tree and Stand Simulation.* Royal College of Forestry, Stockholm, Sweden. Pp. 74–90.

Monserud, R. A., and A. R. Ek. 1977. Prediction of understory tree height growth in northern hardwood stands. *Forest Science* 23:391–400.

Moore, J. A., C. A. Budelsky, and R. C. Schlesinger. 1973. A new index representing individual tree competitive status. *Canadian Journal of Forest Research* 3:495–500.

Newnham, R. M. 1964. The development of a stand model for Douglas-fir. Ph.D. Thesis, Faculty of Forestry, University of British Columbia, Vancouver, B.C., Canada. 201 pp.

Opie, J. E. 1968. Predictability of individual tree growth using various definitions of competing basal area. *Forest Science* 14:314–323.

Staebler, G. R. 1951. *Growth and Spacing in an Even-Aged Stand of Douglas-Fir.* M.F. Thesis, University of Michigan, Ann Arbor, MI, U.S.A. 46 pp.

Zangerl, A. R., and F. A. Bazzaz. 1984. The response of plants to elevated CO_2. II. Competitive interactions among annual plants under varying light and nutrients. *Oecologia* 62:412–417.

CHAPTER 24

A CONCEPTUAL MODEL OF FOREST GROWTH EMPHASIZING STAND LEAF AREA

James M. Vose and Wayne T. Swank

Abstract. A conceptual model of forest stand growth based on radiation interception, conversion efficiency, respiration costs, and carbon allocation patterns is developed. A critical model component is leaf area index (LAI) which influences the amount of radiation interception, an important determinant of forest growth. Changes in resource availability (water, nutrients) influence the amount and vertical distribution of stand radiation interception. Conversion efficiency is modeled as a function of environmental conditions (photon flux density, water, nutrients, temperature). Little is known about the effects of resource supply on respiration, an important process influencing the availability of fixed carbon for stemwood growth. Carbon allocation is also partially regulated by stand environmental conditions (e.g., more carbon is allocated below-ground on dry and infertile sites). The proposed modeling approach has practical utility in that it can be easily understood, parameterized, and tested.

INTRODUCTION

Predicting changes in forest growth in response to altered environmental conditions requires process-based models. Top-down versus bottom-up process models are alternative approaches for simulating forest stand growth. Top-down models typically begin at the stand level and use empirical representations of physiological processes. Bottom-up models begin at the leaf level (or lower) and contain physiological functions scaled up to the stand. For example, Reynolds et al. (1980) used detailed physiological measurements to simulate seasonal net canopy photosynthesis in loblolly pine (*Pinus taeda* L.). At the other extreme, Vose and Allen (1988) and Teskey et al. (1987) demonstrated linear relationships between loblolly productivity and leaf area index (LAI) or foliar biomass, indicating that a single factor such as light interception may be an adequate predictor of growth in single-species stands. Extremes of both approaches have limited practical utility in predicting growth response to stress: Detailed bottom-up models are too data-intensive, and simplified top-down models are too empirical. Thus a merged approach is needed to develop useful predictive models of forest growth response to environmental stress. In this paper we propose such a conceptual framework for modeling growth responses to altered environmental conditions in pine plantations.

Drs. Vose and Swank are Research Ecologist and Principal Plant Ecologist, respectively, at the USDA Forest Service, Coweeta Hydrologic Lab, 999 Coweeta Lab Road, Otto, NC, 28763, U.S.A. The authors appreciate the helpful reviews provided by J. B. Waide, P. M. Dougherty, and two anonymous reviewers.

THE MODEL

Similarly to the approach discussed in Landsberg (1986), growth of forest stands can be modeled as a function of four factors: (1) radiation interception (Q); (2) conversion efficiency (net carbon fixation/unit radiation interception) (E); (3) growth and maintenance respiration (R); and (4) carbon allocation to stemwood growth (A_{stem}). Functionally, these factors can be expressed as follows:

$$\text{stemwood production (kg/ha/t)} = \sum_{i=1}^{t} (Q_i \times E_i - \sum_{c=1}^{n} R_{i,c}) \times A_{stem} \qquad (1)$$

where the subscript $_c$ in $R_{i,c}$ is respiration rate for specific plant components (leaves, stem, roots, etc.). As shown, the model predicts total carbon gain, a proportion of which is then allocated to stemwood by the A_{stem} coefficient. The environment regulates productivity through the effects of resource availability and stress (e.g., nutrients, water, temperature, air pollution) on the four key factors. Among recent efforts to parameterize models similar to ours (e.g., Mohren et al., 1984; McMurtrie, 1985), only McMurtrie (1985) directly addressed the effects of altered environmental conditions on physiological processes.

Recent theoretical treatments of forest productivity (Jarvis and Leverenz, 1983) have emphasized that radiation interception (primarily a function of LAI) is the major determinant of growth. Existing data (Monteith, 1977; Linder, 1985) support this contention. Therefore, our approach concentrates on stand LAI development (annual and seasonal), canopy LAI distribution, and the effects of altered environmental conditions on LAI. We also discuss the importance of conversion efficiency, respiration, and carbon allocation in determining forest growth. Radiation interception and conversion efficiency are treated at two levels of resolution: (1) whole stand, and (2) within the canopy. The information discussed here provides the basis for the functional relationships upon which the proposed modeling approach depends.

LAI AND RADIATION INTERCEPTION

Whole Stand

The proposed modeling approach requires an understanding of changes in LAI with stand age. Data on leaf area development in pine plantations, particularly data from repeatedly measured stands, are limited. Repeated measurements from an eastern white pine (*Pinus strobus* L.) plantation in North Carolina, U.S.A. (Figure 1), showed that projected LAI reached a stable maximum value of about 5.5 by age 15 (Swank and Schreuder, 1973). Gholz and Fisher (1982) examined a chronosequence of slash pine (*Pinus elliotii* Engelm.) stands in Florida and found that LAI increased rapidly from 0–10 years, reached a maximum of 2.0 between 10–20 years, and then declined to about 1.5 in a 30-year-old stand.

Differences among species in stand LAI development are due to the physiological characteristics of each species and to environmental conditions that regulate leaf area. Fertilization studies have demonstrated that the maximum potential leaf area is limited by nutrient supply (Vose and Allen, 1988). Fertilization may also increase the rate at which this maximum is reached (Miller, 1984). The implications of altered nutrient relations on radiation interception are shown in Figure 2, where the highest nitrogen fertilization rate resulted in an increased interception of 65 MJ/m^2 above controls during August.

Stand leaf area is also regulated by soil moisture, as demonstrated by strong relation-

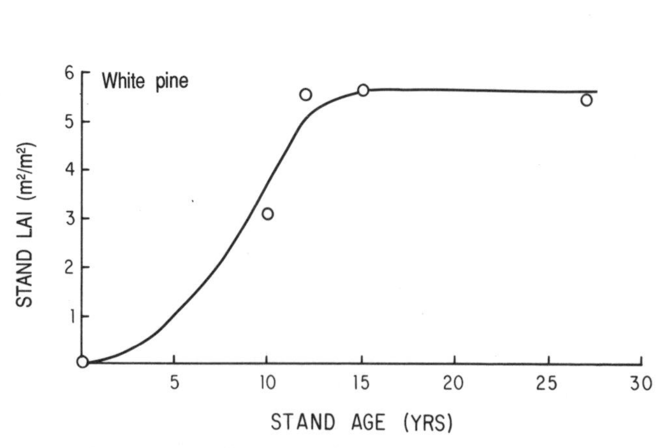

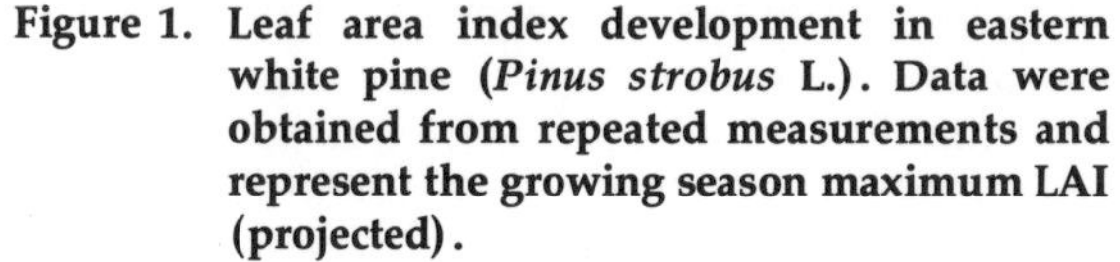
Figure 1. Leaf area index development in eastern white pine (*Pinus strobus* L.). Data were obtained from repeated measurements and represent the growing season maximum LAI (projected).

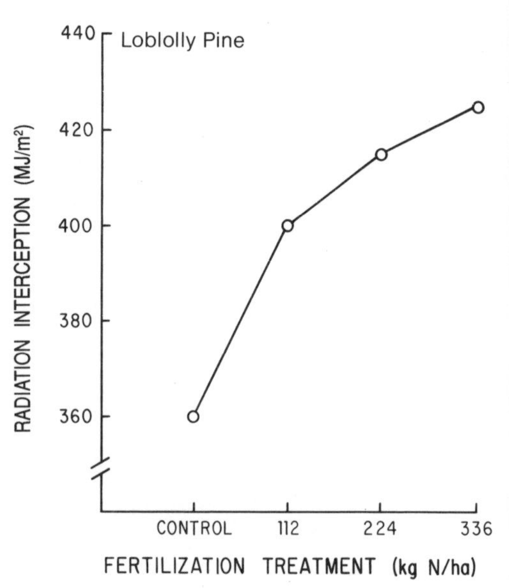

Figure 2. Effects of increased nitrogen supply on stand radiation interception in loblolly pine (*Pinus taeda* L.). Radiation interception data were estimated from stand LAI using the Lambert-Beers equation and an extinction coefficient of 0.46.

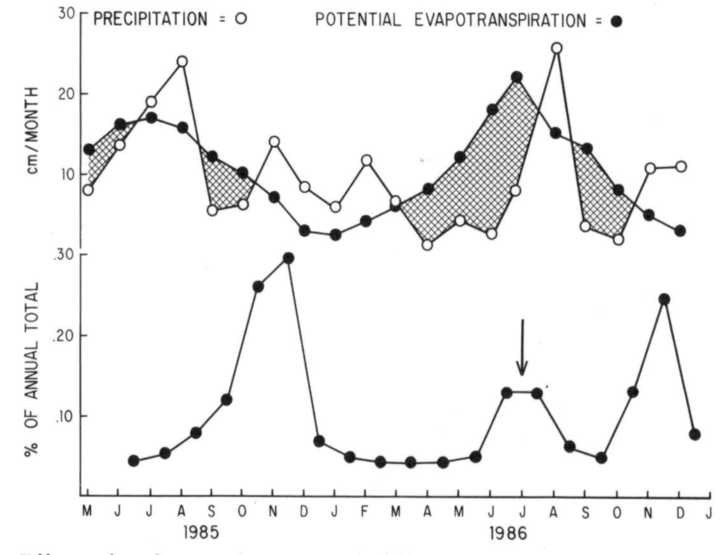

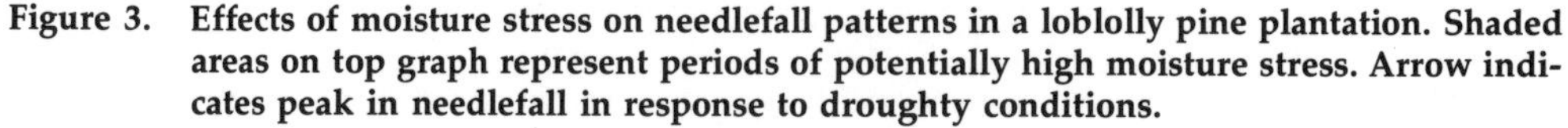
Figure 3. Effects of moisture stress on needlefall patterns in a loblolly pine plantation. Shaded areas on top graph represent periods of potentially high moisture stress. Arrow indicates peak in needlefall in response to droughty conditions.

ships between LAI and site water balance for many western conifers (Grier and Running, 1977). No studies have directly examined soil moisture effects on LAI in southern pine stands. However, in loblolly pine stands, needlefall patterns indicate that moisture stress reduces leaf area by accelerating needlefall (Figure 3). Additionally, drought limits needle elongation (Linder et al., 1987) and probably limits the maximum leaf area a stand is able to maintain. Temperature may also influence leaf area (Gholz, 1986). Respiration rates depend on temperature, and needle respiration rates are higher than those for other tissues (Kinerson, 1975).

For fully stocked stands, it should be possible to model the relationships between stand age and LAI. However, model parameters are dependent upon site resources. One approach is to model the pattern of stand LAI development under optimal site conditions, and to make adjustments based on deviations of actual site conditions (e.g., based on soil type, site water balance, nitrogen availability indices) from the optimal. Region-wide fertilizer field trials provide an excellent data base for this approach.

Seasonal Patterns Within the Stand

In many pine species, stand LAI changes seasonally. Figure 4 summarizes data on seasonal solar radiation input and patterns of leaf area development (based on Kinerson et al., 1974) in a 13-year-old loblolly plantation and a 27-year-old white pine plantation. The fraction of solar radiation intercepted (at solar noon) was estimated using the Lambert-Beers equation and light extinction coefficients of 0.46 for loblolly (Sinclair and Knoerr, 1982) and 0.45 for white pine (Jarvis and Leverenz, 1983). The amount of solar radiation intercepted by the canopy varies seasonally in response to radiation input, development

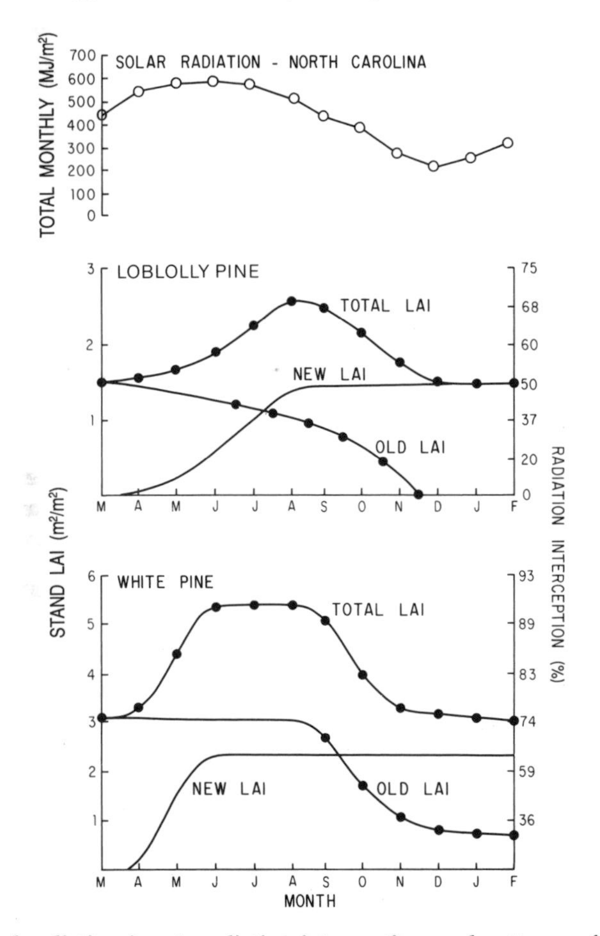

Figure 4. Seasonal radiation input, radiation interception, and patterns of stand LAI development in loblolly and eastern white pine plantations. Changes in old foliage were determined from litterfall. Changes in new foliage were estimated from actual measures of total new LAI and typical flushing patterns. Total LAI = old LAI + new LAI. X-axis for the top graph is the same as for the bottom graphs.

of the current year's foliage, and loss of older foliage. Additionally, environmental conditions such as drought alter needlefall and LAI development patterns. Thus, process models with a resolution of less than one year must account for the dynamic nature of radiation interception.

Within the Canopy

At a finer resolution, radiation interception should be modeled by canopy depth because photosynthetic rates vary by canopy position (Higginbotham, 1974). Vertical distribution of leaf area changes with stand development because as the canopy closes, shading in the lower canopy causes branch mortality and restricts leaf growth. For example, the vertical leaf area distribution with stand development for a white pine plantation (Schreuder and Swank, 1974) showed a shift from a skewed to a normal LAI distribution as the canopy closed (Figure 5). Vertical leaf area distribution in both loblolly and white pine have been successfully modeled with the Weibull function (Vose, 1988; Schreuder and Swank, 1974). These studies have shown that the Weibull function can fit many of the potential foliage distribution patterns in forest canopies (i.e., negative skew, normal, positive skew).

In closed canopies with high leaf areas, increased nutrient supply may alter leaf area distribution and thus alter patterns of radiation interception (Ford, 1984) and photosynthate production (Linder and Axelsson, 1982). Similarly, stresses such as reduced nutrient supply, drought, or ozone damage may reduce canopy LAI sufficiently to stimulate branch retention and leaf growth in the lower crown. Figure 6 shows the LAI distribution (modeled with the Weibull) within the canopy and the amount of radiation intercepted in each 1 m layer for a 13-year-old loblolly pine stand during August. Based on crown development patterns, the theoretical amount of radiation intercepted by each foliage age class is indicated. Separation by foliage age class is important because conversion efficiency varies with foliage age (Higginbotham, 1974); however, phenological data on leaf area development are limited.

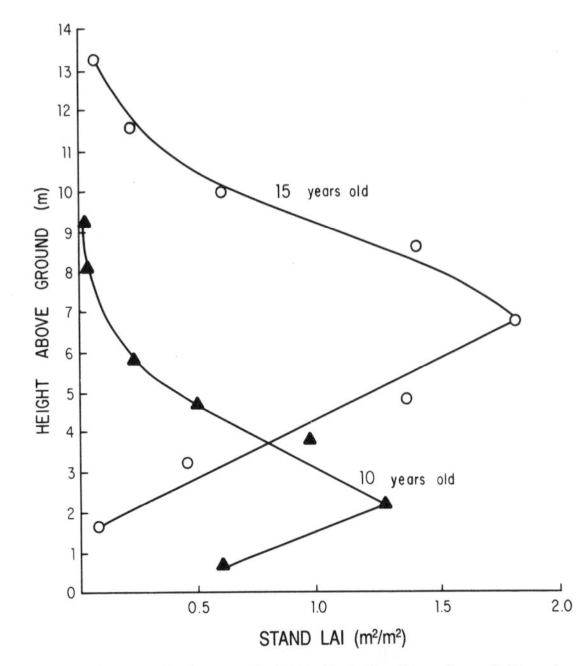

Figure 5. Changes in vertical stand LAI distribution in white pine with stand age.

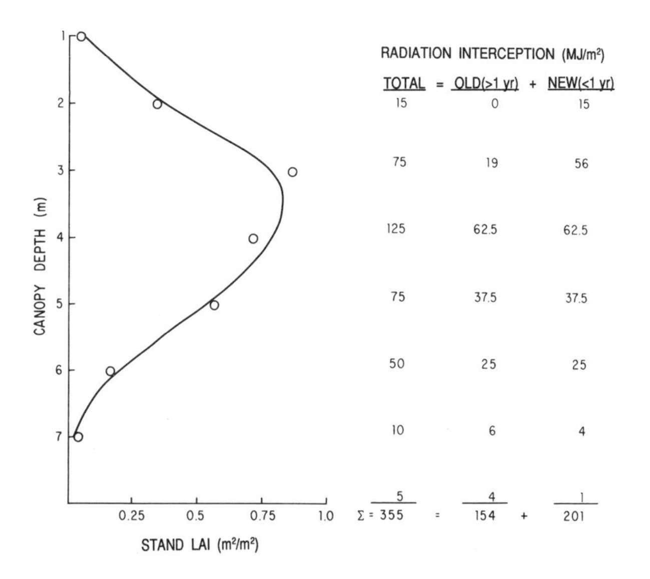

Figure 6. Vertical stand LAI distribution and radiation interception (during August) in loblolly pine. LAI distribution curve was fitted with a Weibull function.

CONVERSION EFFICIENCY

At the stand level, Jarvis and Leverenz (1983) reported an average annual conversion efficiency (E) value of 1.42 g/MJ for a Scots pine (*Pinus sylvestris* L.) stand; the highest seasonal value occurred in August (E = 3.14 g/MJ). Seasonal variations in E reflect the importance of environmental conditions (i.e., light, temperature, water) and canopy characteristics in determining photosynthetic rate. As an example of nutrient and water limitations on E, an average annual value of E = 1.80 g/MJ was found on irrigated and fertilized plots from the same Scots pine study. Manogaran (1973) demonstrated a relationship between several measures of evapotranspiration and growth in loblolly pine, suggesting that evapotranspiration models may have some utility in predicting stand E. Growth efficiency (stemwood growth/unit LAI) has often been used as a measure of stand E (Brix, 1983) in fertilization studies. However, it is difficult to attribute changes in growth efficiency directly to changes in E because growth efficiency is also a function of carbon allocation patterns. Comeau and Kimmins (1986) used Ågren's (1983) concept of canopy nitrogen use efficiency (ANPP/kg canopy N) as an integrator of nitrogen availability and photosynthetic rate in lodgepole pine (*Pinus contorta* Dougl.). Similar approaches may permit incorporation of environmental conditions into canopy conversion efficiency models.

More detailed conversion efficiency models must account for the effects of canopy position, needle age, and environmental conditions. Linder and Axelsson (1982) found that photosynthate production in Scots pine varied by canopy position due both to patterns of vertical LAI distribution and to photosynthetic rate. Photosynthetic rate is a function of photon flux density in each whorl position, and photon flux density in each whorl may vary with environmental conditions. For example, Linder and Axelsson (1982) found that irrigation and fertilization increased photosynthetic rates and leaf production in the upper canopy. In the lower canopy, however, the same treatments increased shading and reduced photosynthetic rates. The net effect of irrigation and fertilization on canopy photosynthesis was only a 20% increase. The model of Reynolds et al. (1980) showed that newly developing loblolly pine needles were sinks for fixed carbon until May for the first flush and July for the second flush, and that the major source of photosynthate in the course of the year was 2-year-old foliage. Thus, detailed canopy conversion models should account for differences in foliage age and radiation interception within the canopy. An approach to modeling canopy dynamics is to combine the data provided in Figure 1 (total LAI), Figure 4 (seasonal LAI patterns), and Figure 6 (vertical radiation interception) to estimate radiation interception by foliage age class, month, and canopy position. Combined with a model of Q (e.g., Q = f[photon flux density, environment, foliage age class]), canopy carbon gain can be simulated (Figure 7).

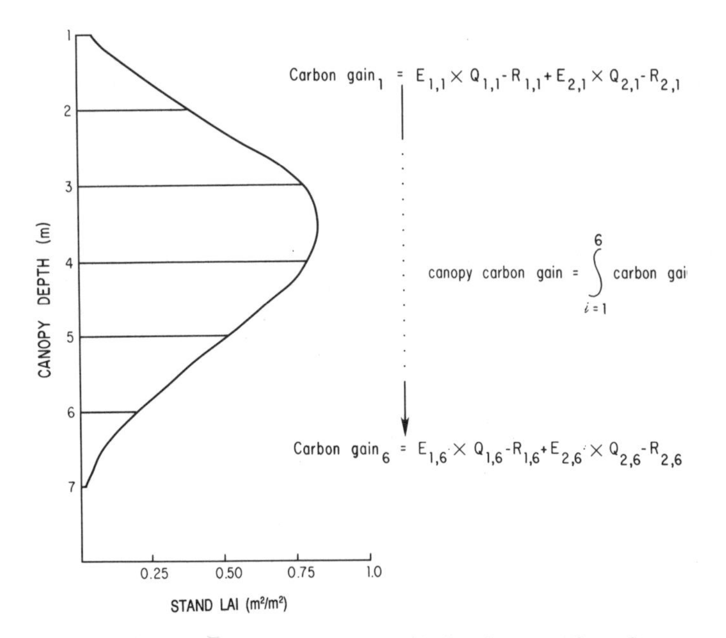

Figure 7. Canopy carbon gain model, where E = radiation interception, Q = conversion efficiency, and R = respiration. Parameter subscripts represent foliage age class and canopy position, respectively.

RESPIRATION

Growth and maintenance respiration are major processes utilizing fixed carbon. Total respiratory carbon losses depend upon environmental conditions and upon the surface area, physiological activity, and growth of living tissue. Maintenance respiration rate is exponentially related to temperature, while growth respiration rate is less sensitive to environmental conditions (Linder and Rook, 1984). Kinerson (1975) estimated that about 60% of gross primary production in a loblolly pine plantation in North Carolina, U.S.A., was utilized in respiration. Highest respiration rates were observed in summer. Leaf tissue had the highest respiration rates (12.5 g CO_2/m^2 ground area/hr), followed by stems (0.95 g CO_2/m^2 ground area/hr), branches (0.2 g CO_2/m^2 ground area/hr), and roots (0.04 g CO_2/m^2 ground area/hr).

Respiration for each tissue type can be estimated by determining tissue surface area and calculating a temperature-dependent respiration rate per unit surface area. As a rough approximation of respiratory losses, Mohren et al. (1984) used constant annual maintenance respiration coefficients for needles, branches, stemwood, and root biomass and a single constant growth respiration coefficient for all plant components. In contrast,

Reynolds et al. (1980) used temperature-dependent respiration rate functions which accounded for shifts in Q_{10} before and after initiation of spring cambial activity.

More research is needed on the effects of other environmental parameters on respiration. For example, Lavigne (1987) found that respiration rates in balsam fir (*Abies balsamea* [L.] Mill.) depended upon stem water content and supplies of reducing sugars. Linder and Rook (1984) reported that nitrogen fertilization increases respiration in many species, and that there is a linear relationship between tissue nitrogen concentration and respiration. Stresses such as ozone damage may also increase respiration rates (Barnes, 1972).

CARBON ALLOCATION

Carbon allocation is strongly regulated by environmental variables (Linder and Rook, 1984; Cannell, 1985). The proportion of net primary productivity allocated belowground is highest on dry or infertile sites. Because stemwood growth has lower priority for fixed carbon than root growth (Waring and Schlesinger, 1984), less carbon is allocated to stemwood under dry and infertile conditions.

Data on belowground carbon allocation for southeastern pines are limited. In a 14-year-old loblolly plantation, carbon allocation ratios were 0.40 for stemwood (and bark), 0.05 for branches, 0.20 for foliage, and 0.35 for roots (Kinerson et al., 1977; Harris et al., 1977). In a 12-year-old plantation, allocation ratios were 0.64 for stemwood (and bark), 0.08 for branches, 0.14 for foliage, and 0.12 for roots (Nemeth, 1973). Differences between studies may be related to site quality and sampling methodology. Increased nutrient and water availability decreases the proportion of net primary production allocated belowground and increases the proportion allocated to foliage and stemwood (Linder and Rook,1984). Carbon allocation patterns also vary with stand age.

Approaches to modeling carbon allocation range from detailed sink-source models (Reynolds et al., 1980) to constant allocation ratios that allocate fixed carbon to various tissues (Mohren et al., 1984). It is clear from experiments and observations that fixed allocation ratios are inappropriate for modeling forest productivity under variable site or environmental conditions. We advocate allocation ratios for top-down approaches, but such ratios must account for changes in site water and nutrient status (cf. McMurtrie, 1985).

CONCLUSIONS

A conceptual approach for modeling forest growth is proposed. In contrast to traditional growth and yield models, this approach is based on functional relationships that regulate productivity. A simple, physiologically based model will have immediate practical utility because (1) it can be parameterized and tested; (2) it can be understood by the user (e.g., a forest manager); and (3) it can be adapted and modified for specific conditions. Leaf area is a critical component of the proposed approach because significant variation in forest growth can be explained by this single parameter. In addition to influencing radiation interception, LAI is a valuable parameter because of its interactions with respiration, photosynthesis, and carbon allocation patterns. These processes and interactions must be considered in physiologically based process models. Unfortunately, limited information is available on stand-level measures of respiration, photosynthesis, and carbon allocation. Including these parameters in stand-level models will require several assumptions, generalizations, and scaling-up procedures.

Models that do not consider the effects of "site quality" on physiological processes will have limited practical utility in predicting forest growth response to altered environmental conditions. Traditional approaches, such as the use of a site index, are too general and lack a physiological base. As discussed in this paper, knowledge gained from fertilization and irrigation studies provides insight and data for incorporating environmental effects into process models.

Most growth and yield models cannot accommodate the impacts of air pollution, disease, and insect damage. The present model can incorporate the effects of mortality and foliage loss through reductions in LAI. Air pollution, disease, and insect damage may also be simulated via the impacts on carbon allocation, photosynthesis, and respiration.

LITERATURE CITED

Ågren, G. I. 1983. Nitrogen productivity of some conifers. *Canadian Journal of Forest Research* 13:494–500.

Barnes, R. L. 1972. Effects of chronic exposure of air pollution on photosynthesis and respiration of pines. *Environmental Pollution* 3:133–138.

Brix, H. 1983. Effects of thinning and nitrogen fertilization on growth of Douglas-fir: relative contribution of foliage quantity and efficiency. *Canadian Journal of Forest Research* 13:167–175.

Cannell, M. G. R. 1985. Dry matter partitioning in tree crops. In: Cannell, M. G. R., and J. E. Jackson, eds. *Attributes of Trees As Crop Plants.* Institute of Terrestrial Ecology, Huntingdon, England, U.K. Pp. 160–193.

Comeau, P. G., and J. P. Kimmins. 1986. The relationship between net primary production and foliage nitrogen content, and its application in the modeling of forest ecosystems: a study in lodgepole pine (*Pinus contorta*). In: Fujimori, T., and D. Whitehead, eds. *Crown and Canopy Structure in Relation to Productivity.* Forestry and Forest Products Research Institute, Ibaraki, Japan. Pp. 202–223.

Ford, E. D. 1984. The dynamics of plantation growth. In: Bowen, G. D., and E. K. S. Nambiar, eds. *Nutrition of Plantation Forests.* Academic Press, New York, NY, U.S.A. Pp. 17–52.

Gholz, H. L. 1986. Canopy development and dynamics in relation to primary productivity. In: Fujimori, T., and D. Whitehead, eds. *Crown and Canopy Dynamics in Relation to Productivity.* Forestry and Forest Products Research Institute, Ibaraki, Japan. Pp. 224–242.

Gholz, H. L., and R. F. Fisher. 1982. Organic matter production and distribution in slash pine (*Pinus elliotii*) plantations. *Ecology* 63:1827–1839.

Grier, C. C., and S. W. Running. 1977. Leaf area of mature northwestern coniferous forests: relation to site water balance. *Ecology* 58:893–899.

Harris, W. F., R. S. Kinerson, and N. T. Edwards. 1977. Comparison of belowground biomass of natural deciduous forest and loblolly pine plantations. *Pedobiologia* 17:369–381.

Higginbotham, K. O. 1974. The influence of canopy position and the age of leaf tissue on growth and photosynthesis in loblolly pine. Ph.D. Thesis, Duke University, Durham, NC, U.S.A.

Jarvis, P. G., and J. W. Leverenz. 1983. Productivity of temperate, deciduous and evergreen forests. In: Lange, O. L., P. S. Nobel, C. B. Osmond, and H. Ziegler, eds. *Physiological Plant Ecology,* Vol. IV, *Encyclopedia of Plant Physiology,* New Series 12D. Springer-Verlag, New York, NY, U.S.A. Pp. 233–280.

Kinerson, R. S. 1975. Relationship between surface area and respiration in loblolly pine. *Journal of Applied Ecology* 12:965–971.

Kinerson, R. S., K. O. Higginbotham, and R. C. Chapman. 1974. The dynamics of foliage distribution within a forest canopy. *Journal of Applied Ecology* 11:347–353.

Kinerson, R. S., C. W. Ralston, and C. G. Wells. 1977. Carbon cycling in a loblolly pine plantation. *Oecologia* 29:1–10.

Landsberg, J. J. 1986. *Physiological Ecology of Forest Production.* Academic Press, New York, NY, U.S.A. 198 pp.

Lavigne, M. B. 1987. Differences in stem respiration responses to temperature between balsam fir trees in thinned and unthinned stands. *Tree Physiology* 3:225–233.

Linder, S. 1985. Potential and actual production in Australian forest stands. In: Landsberg, J. J., and W. Parsons, eds. *Research for Forest Management.* CSIRO, Melbourne, Australia. Pp. 11–35.

Linder, S., and B. Axelsson. 1982. Changes in carbon uptake and allocation patterns as a result of irrigation in a young *Pinus sylvestris* stand. In: Waring, R. H., ed. *Carbon Uptake and Allocation in*

Subalpine Ecosystems As a Key to Management. Forest Research Laboratory, Oregon State University, Corvallis, OR, U.S.A. Pp. 38–44.

Linder, S., M. L. Benson, B. J. Meyers, and R. J. Raison. 1987. Canopy dynamics and growth of *Pinus radiata.* I. Effects of irrigation and fertilization during a drought. *Canadian Journal of Forest Research* 17:1157–1165.

Linder, S., and D. A. Rook. 1984. Effects of mineral nutrition on carbon dioxide exchange and partitioning in trees. In: Bowen, G. D., and E. K. S. Nambiar, eds. *Nutrition of Plantation Forests.* Academic Press, New York, NY, U.S.A. Pp. 211–236.

Manogaran, C. 1973. Economic feasibility of irrigating southern pines. *Water Resources Research* 9:1485–1495.

McMurtrie, R. E. 1985. Forest productivity in relation to carbon partitioning and nutrient cycling: a mathematical model. In: Cannell, M. G. R., and J. R. Jackson, eds. *Attributes of Trees As Crop Plants.* Institute of Terrestrial Ecology, Huntingdon, England, U.K. Pp. 194–207.

Miller, H. G. 1984. Dynamics of nutrient cycling in plantation ecosystems. In: Bowen, G. D., and E. K. S. Nambiar, eds. *Nutrition of Plantation Forests.* Academic Press, New York, NY, U.S.A. Pp. 53–78.

Mohren, G. M. G., C. P. Van Gerwen, and C. J. T. Spitters. 1984. Simulation of primary production in even-aged stands of Douglas-fir. *Forest Ecology and Management* 9:27–49.

Monteith, J. L. 1977. Climate and efficiency of crop production in Britain. *Philosophical Transactions Royal Society,* London, U.K. Series B, 281:277–294.

Nemeth, J. C. 1973. Dry matter production in young loblolly (*Pinus taeda* L.) and slash pine (*Pinus elliotii* Engelm.) plantations. *Ecological Monographs* 43:21–41.

Reynolds, J. F., B. R. Strain, G. L. Cunningham, and K. R. Knoerr. 1980. Predicting primary productivity for forest and desert ecosystem models. In: Hesketh, J. D., and J. W. Jones, eds. *Predicting Photosynthesis for Ecosystem Models.* CRC Press, Boca Raton, FL, U.S.A. Pp. 170–207.

Schreuder, H. T., and W. T. Swank. 1974. Coniferous stands characterized with the Weibull distribution. *Canadian Journal of Forest Research* 4:518–523.

Sinclair, T. R., and K. R. Knoerr. 1982. Distribution of photosynthetically active radiation in the canopy of a loblolly pine plantation. *Journal of Applied Ecology* 19:183–191.

Swank, W. T., and H. T. Schreuder. 1973. Temporal changes in biomass, surface area, and net production for a *Pinus strobus* L. forest. In: *IUFRO Biomass Studies.* University of Maine, Orono, ME, U.S.A. Pp. 173–182.

Teskey, R. O., B. C. Bongarten, B. M. Cregg, P. M. Dougherty, and T. C. Hennessey. 1987. Physiology and genetics of tree growth response to moisture and temperature stress: an examination of the characteristics of loblolly pine (*Pinus taeda* L.). *Tree Physiology* 3:41–61.

Vose, J. M. 1988. Patterns of leaf area distribution within stands of nitrogen-and phosphorus-fertilized loblolly pine trees. *Forest Science* 34:564–573.

Vose, J. M., and H. L. Allen. 1988. Leaf area, stemwood growth, and nutrition relationships in loblolly pine. *Forest Science* 34:547–563.

Waring, R. H., and W. H. Schlesinger. 1985. *Forest Ecosystems: Concepts and Management.* Academic Press, New York, NY, U.S.A.

CHAPTER 25

AN ECOLOGICAL GROWTH MODEL FOR FOUR NORTHERN HARDWOOD SPECIES IN UPPER MICHIGAN

David D. Reed, Michael J. Holmes, Elizabeth A. Jones,
Hal O. Liechty, and Glenn D. Mroz

Abstract. An individual tree diameter growth model is presented for four northern hardwood species growing in Upper Michigan. The model consists of four separate components representing physiological potential growth, inter-tree competition, site effects (including chemical, physical, and climatic properties), and seasonal growth pattern. All relationships in the model have been verified on intensively measured field sites. Field measurements will continue for several more years and allow model refinement and verification under varied climatic conditions.

INTRODUCTION

The objective of this study is to develop a model which can be used to quantify differences in productivity of mixed-species northern hardwood forests due to natural factors: site physical and chemical properties, climate, and competition. This study is a part of a larger project attempting to determine the effects of an imposed factor, in this case electromagnetic fields, against a background of natural variability in climate and other factors (Mroz et al., 1987). Ultimately, the emphasis is on the response of the natural ecosystem; this includes the need to quantify the effects of natural factors on individual trees in order to detect changes in system productivity which may be due to imposed environmental stresses.

Productivity is defined as the annual increase in overstory biomass. There is a strong relationship between a tree's diameter at breast height (dbh) and total tree biomass. Furthermore, cambial activity is strongly related to climatic variation, competition from neighboring trees, and site physical and chemical properties. While monitoring of actual biomass production over time is difficult in field situations, it is relatively easy to accurately and precisely measure cambial development. For these reasons, diameter increment was chosen as the response variable representing biomass productivity.

Several existing models attempt to describe annual diameter growth as a function of

Dr. Reed and Mr. Holmes are at the School of Forestry and Wood Products, Ms. Jones is at the Department of Mathematical Sciences, Mr. Liechty and Dr. Mroz are at the School of Forestry and Wood Products, Michigan Technological University, Houghton, MI, 49931 U.S.A. This study was funded by a grant from the U.S. Navy Space and Naval Warfare Systems Command through a subcontract to the Illinois Institute of Technology Research Institute under Contract E06595-88-C-001.

tree and stand characteristics while accounting for the effects of site physical, chemical, and climatic properties. Diameter growth functions in the JAnak, BOtkin, WAllis (JABOWA) (Botkin et al., 1972) and Forests of East Tennessee (FORET) (Shugart and West, 1977) models and models of the type described by Reed (1980) are examples. Several of these models were tested in Upper Michigan, U.S.A., by Fuller et al. (1987) and found to perform poorly when compared to actual field measurements.

One reason for this poor performance may be that an annual time-step might be inadequate when attempting to quantify the effects of environmental stress on productivity. Charles-Edwards et al. (1986) indicate the amount of time required for individual plant growth processes to recover from a perturbation in nutrient status of the rooting environment to be on the order of 10^5 seconds (a few days), and the recovery time of a natural system to be on the order of 10^9 seconds (many years). It is illogical to use a time-step which is longer than the recovery time of the system of interest, whether that system is an individual plant or a plant community. It is also counter-productive to use a time-step that is many orders of magnitude less than the recovery time of the system of interest. In this study, the objective was to quantify diameter growth of individual trees while accounting for the effects of neighboring trees as well as soil and climatic conditions in a biologically rational fashion. Since this involves individual-plant as well as system responses to the environment, a time-step of one week was utilized in developing a diameter growth model of the type described by Reed (1980).

METHODS

This study was undertaken with the objective of determining the effects of an imposed environmental factor on a forest ecosystem. For this reason, intensively measured field plots were established. Data from these plots are used to verify assumptions concerning plant level processes incorporated into the growth model. The field measurements are also used to calibrate the model for the various species in the study area. The mathematical forms representing the relationships between tree, stand, and environmental factors and diameter growth are consistent with the results of previous field studies and controlled experiments. These relationships were verified by the field data prior to incorporating them into the diameter growth model.

Site Description

Two study sites are located in the central Upper Peninsula of Michigan in the U.S.A. Site one is in Iron County (46°10'N, 88°30'W) and site two is in Marquette County (46°20'N, 88°10'W). Both sites have relatively undisturbed second-growth northern hardwood vegetation consisting principally of red maple (*Acer rubrum* L.) and northern red oak (*Quercus rubra* L.), with minor components of quaking aspen (*Populus tremuloides* Michx.), bigtooth aspen (*Populus grandidentata* Michx.), and paper birch (*Betula papyrifera* Marsh.). The sites are characterized as the *Acer-Quercus-Vaccinium* habitat type (Coffman et al., 1983).

Field Measurements

Measurement of radial increment was accomplished using a band dendrometer as described by Cattelino et al. (1986). The dendrometer bands were read weekly to the nearest 0.0254 cm of circumference. Readings began in early April and continued through the fall when over 50% of leaf fall had taken place. There were 274 trees banded on site one and 197 trees banded on site two prior to the 1985 growing season. Weekly measurements have been made over the 1985, 1986, and 1987 growing seasons. Locations of the individual trees are mapped on a Cartesian coordinate system with a 0.1 m resolution.

The second category of field measurements includes climate and soil properties which may affect plant growth processes. Each of the study sites is equipped with a Handar 540A data collection platform located in a cleared area adjacent to the sites. The main data collection platform contains sensors measuring precipitation, air temperature, relative humidity, solar radiation, soil temperature, and soil moisture. Each of three 30 m $\times$ 35 m subplots at each site contains sensors measuring air temperature, soil temperature at 5- and 10-cm depths, and soil moisture at 5- and 10-cm depths. Data are retrieved eight times daily via NOAA satellite transmissions. Sensors are queried every 30 minutes and computed into 3-hour mean values by the platform microprocessor. Precipitation data are logged once every 3 hours. The daily climatological and soil data are summarized into weekly averages to coincide with the dendrometer band readings for analysis. Soils at each plot are sampled monthly for determination of nutrient levels.

GROWTH MODEL FORMULATION

The basic diameter growth model formulation follows the conceptual pattern described by Botkin et al. (1972) and Reed (1980). In the model, the amount of growth during a given week, d_t, is represented as a function of tree, stand, climate, and site physical and chemical factors. These factors are incorporated in four model components: (1) annual physiological potential growth (PPG), (2) the reduction of annual potential growth due to inter-tree competition (IC), (3) the reduction of annual potential growth due to site physical, chemical, and annual climatic properties (SPCC), and (4) the seasonal growth pattern and further reduction of annual potential growth due to weekly climatic factors (SGP_t). Each of the last three components is expressed as a proportion of the physiological potential growth, so the weekly diameter growth (d_t) is expressed as the product of the four components:

$$\begin{aligned} d_t = {} & \text{(physiological potential diameter growth)} \\ & \text{(effect of inter-tree competition)} \\ & \text{(effect of site physical, chemical, and climatic factors)} \\ & \text{(seasonal growth pattern)} \end{aligned} \tag{1}$$

Physiological Potential Growth

A tree's physiological potential growth is defined as the growth which the tree could achieve, given its species and size, in the absence of any resource limitations. The model form utilized in this study was first proposed by Botkin et al. (1972). In the derivation of this model, they assumed various proportional relationships between total tree biomass increment, leaf area, leaf biomass, and the relative height (H) and diameter (D) of the tree compared to the maximum height (H_{max}) and diameter (D_{max}) recorded for the species. Utilizing these assumptions, they derived the following model form for annual physiologically potential diameter growth (PPG):

$$PPG = \frac{G\ D\ (1 - D\ H\ /\ D_{max}\ H_{max})}{(274 + 3\ b_2\ D - 4\ b_3\ D^2)} \tag{2}$$

They imposed constraints on the coeffiecients so that $H = H_{max}$ and $dH/dD = 0$ when $D = D_{max}$. The species-specific growth constant, G, is a function of the proportional relationships assumed during model derivation. When tested on the data from the study sites using the values for G, b_2, and b_3 given by Botkin et al. (1972), the model performed poorly and the coefficients required re-estimation. As discussed by Botkin et al. (1972), this may be due to H_{max} and D_{max} being site-specific within a species rather than constant for all sites. Also, the proportionality constants involved in the model derivation may be a function of local site, stand, and genetic factors affecting tree form.

Inter-tree Competition

In most JABOWA-type models, competition effects of neighboring trees have been expressed by limiting resource availability, usually by reducing the amount of light intercepted by a tree's crown. Other forest growth and yield models have used simple indices to indicate the relative resource requirements of an individual and its neighbors, often including distances between competing individuals in the index. In most comparisons, the use of simple indices has performed as well as attempting to define the mechanistic effects of competition on a tree and its physiological requirements.

In this study, indices of competition between a tree and its neighbors were used to represent competitive stress. Due to the variety of species and their different resource requirements, there was no common index which performed well for all species. Indices representing crown overlap performed better for the more intolerant species, while indices incorporating sizes of the subject tree and its neighbors performed better for the more tolerant species. Whatever the species or index, little diameter growth reduction was expected at low levels of competition; as competition increased, growth was expected to decrease until there was very little growth and the tree was near mortality. In mixed northern hardwood stands, very few trees are subject to low levels of competition. Most individuals will be under some level of competitive stress, with some individuals experiencing extreme stress. In the growth model, the proportion of annual potential growth which can be expected is expressed as a simple negative exponential function of the competition index (CI) best suited for a species:

$$IC = e^{-a\,CI} \tag{3}$$

Site Physical, Chemical, and Climatic Conditions

Incorporating site physical, chemical, and climatic factors in the model accounted for a large portion of the variation in diameter growth rates between study sites and among study years for each tree species. The factors explaining this variation differ by species but include soil (5 cm) or air temperature growing degree days (4.4°C basis), soil water-holding capacity (cm/cm), and potassium concentration (ppm) in the upper 15 cm of soil during the growing season. Temperature was strongly related to growth rate for more shade-intolerant species. Nutrient availability increased in importance for more shade-tolerant species.

For each of these factors, there is expected to be a range of values where a species responds positively to increased amounts of the factor, a range of values where the factor is adequate for the species and there is little response to increases or decreases, and a range of values where the species responds negatively to increased amounts. To date, all species in the study have a significant negative response to increased growing degree days, with this response being most prominent in aspen ($r = -0.86$) and paper birch ($r = -0.69$). Increased soil water-holding capacity is beneficial to red maple ($r = 0.66$) and detrimental to aspen ($r = -0.38$). Increased potassium availability has a more prominent effect on the more tolerant species of red maple ($r = 0.48$) and northern red oak ($r = 0.26$). Tree productivity appeared to respond linearly to changes in the levels of these factors over the range of conditions on the study sites.

These factors are accounted for in the model by a linear function which is constrained to produce the proportion of annual potential growth which might be expected:

$$SPCC = \frac{(D + b_1 X_1 + b_2 X_2 + b_3 X_3)}{D} \tag{4}$$

The particular factors (X_k) included in the model and the associated constants (b_k) are species-specific. Currently efforts are under way to further identify the breakpoints in the relationships between levels of these factors and species productivity. Once these are identified, this model component will be modified to cover a wider range of conditions than those previously observed on the study sites.

Seasonal Growth Pattern

Cumulative air temperature growing degree days to time t appears to be the most important variable affecting the seasonal growth pattern for all tree species on the study sites. The proportion of annual growth expected in a given week is estimated using a difference form of a modified Chapman-Richards growth function and the cumulative air temperature growing degree days at the beginning and end of the week. This involves the implicit assumption that each species will respond to temperature up to a certain point and that further additions of degree days will not lead to increased growth. The coefficients quantifying this response are assumed to be genetically controlled and related to climatic conditions in a local area.

Increased air temperature may lead to decreased levels of soil moisture due to increases in evaporation and plant respiration. The expected growth, given the cumulative air temperature degree days, may not be achieved if moisture is limiting. In the model, average soil water tension (−MPa) at a depth of 5 cm is used to indicated the moisture stress. Moisture is not assumed to be limiting if tension levels are below .101 −MPa (1 atm). Plant response is assumed to be a simple negative exponential function of increasing average soil water tension. The coefficient of this function can be considered as an index of drought tolerance for the species. Like the coefficients quantifying the response to temperature, this coefficient may also be under genetic control and related to climatic conditions in a local area.

The model combines the effects of cumulative air temperature degree days at the beginning (ATD_{t1}) and end (ATD_{t2}) of week t and average soil water tension at 5 cm in week t (SWT_t):

$$SGP_t = (\exp[-(ATD_{t1}/b)^c] - \exp[-(ATD_{t2}/b)^c])(\exp[-a(SWT_t - .101)]) \tag{5}$$

SUMMARY

A diameter growth model using a weekly time-step was developed for selected northern hardwood species growing in mixed stands in Upper Michigan. Prior to inclusion in the model, expected ecological relationships had to be verified on intensively measured field sites. This was judged to be absolutely necessary in order to identify changes in forest productivity due to changes in natural environmental factors. Only after we undertand the effects of natural factors can the effects of imposed factors, such as air pollution or electromagnetic fields, on the productivity of natural systems be confirmed in the field.

LITERATURE CITED

Botkin, D. B., J. F. Janak, and J. R. Wallis. 1972. Some ecological consequences of a computer model of forest growth. *Journal of Ecology* 60:849–873.

Cattelino, P. J., C. A. Becker, and L. G. Fuller. 1986. Construction and installation of homemade dendrometer bands. *Northern Journal of Applied Forestry* 3:73–75.

Charles-Edwards, D. A., D. Doley, and G. M. Rimmington. 1986. *Modelling Plant Growth and Development.* Academic Press, Sydney, Australia.

Coffman, M. S., E. Alyanak, J. Kotar, and J. E. Ferris. 1983. *Field Guide, Habitat Classification System for the Upper Peninsula of Michigan and Northeastern Wisconsin.* CROFS, Department of Forestry, Michigan Technological University, Houghton, MI, U.S.A.

Fuller, L. G., D. D. Reed, and M. J. Holmes, 1987. Modeling northern hardwood diameter growth using weekly climatic factors in northern Michigan. In: Ek, A. R., S. R. Shifley, and T. E. Burk, eds. *Forest Growth Modelling and Prediction,* Vol. 1. General Technical Report NC-120, USDA Forest Service North Central Forest Experiment Station, St. Paul, MN, U.S.A. Pp. 467–474.

Mroz, G. D., K. T. Becker, R. H. Brooks, J. N. Bruhn, P. J. Cattelino, M. R. Gale, M. J. Holmes, M. F. Jurgensen, H. O. Liechty, J. A. Moore, D. D. Reed, E. A. Jones, D. L. Richter, and Y. F. Zhang. 1987. Annual Report of the Herbaceous Plant Cover and Tree Studies. In: *Compilation of 1987 Annual Reports of the Navy ELF Communications System Ecological Monitoring Program,* Vol. 1. Technical Report EO6595-2, Illinois Institute of Technology Research Institute, Chicago, IL, U.S.A. Pp. 1–316.

Reed, K. L. 1980. An ecological approach to modeling growth of forest trees. *Forest Science* 26:35–50.

Shugart, H. H., and D. C. West. 1977. Development of an Appalachian deciduous forest succession model and its application to assessment of the impact of the chestnut blight. *Journal of Environmental Management* 5:161–179.

CHAPTER 26

THE STATIC GEOMETRIC MODELING OF THREE-DIMENSIONAL CROWN COMPETITION

W. Rick Smith

Abstract. The geometric modeling of a tree's crown defines its geometric attributes. However, modeling becomes complex when inter-tree competition alters a crown's geometric attributes from the open-grown state. Several methods to calculate the geometric attributes of a tree's crown are presented to facilitate a diversity of modeling approaches for between-tree and within-tree processes. To reduce the computational complexity of this task, efficient algorithms for modeling the crown of a tree are presented and the biological adequacy and applications of this model in forestry are discussed. The geometric model of crown competition offers one method of bridging process and stand-level models to examine the effects of atmospheric pollution on forests.

INTRODUCTION

The resources a tree depends upon for sustenance and growth, whether minerals, water, light, or air, occupy space. Inter-tree competition for resources may be viewed in its most basic form as a competition for space. Thus, the resolution of a crown model is limited by its ability to define the space occupied by the crown of a tree.

The geometric modeling of the crown is a complex and difficult task. The primary aim is to define the space a tree's crown occupies to facilitate the modeling of the biological, chemical, and physical processes that define a tree's growth. Geometric models may be classified with respect to time (static or dynamic) and space (two- or three-dimensional). This paper is concerned with the development of a three-dimensional static crown competition model, though the methods discussed are also applicable to the development of two-dimensional static geometric models of crown competition.

The modeling of a static three-dimensional crown is developed in such a way as to be biologically reasonable and computationally efficient. The biological assumptions of this model and its applications are discussed. The computational complexity of the geometric modeling of tree crowns has limited its use in forestry research. Algorithms to efficiently compute the crown of a tree based on its position with respect to its neighbors are used to make three-dimensional modeling of the crown a practical tool for forest scientists.

LITERATURE REVIEW

Two-dimensional static geometric models are generally concerned with resources available and competitive status, and may be viewed as implicit models of the crown. Brown (1965) used the Voronoi polygon to estimate the area potentially available (APA)

Mr. Smith is at the U.S. Forest Service Southern Forest Experiment Station, Institute for Quantitative Studies, 701 Loyola Avenue, New Orleans, LA, 70113 U.S.A.

to a tree. The Voronoi polygon results from points in a plane that spread isotropically (equally in all directions) at an equal rate until all the area in the plane is covered. If crowns spread horizontally at an equal rate in an even-aged plantation, at crown closure the projected crowns (PC's) would be similar to Voronoi polygons. Moore et al. (1973) modified APA by weighting the midpoints of the distance between competitors by basal area used to create the polygons. The weighting of midpoints may be viewed as an empirical method to adjust for unequal crown growth rates. Bella (1971) developed a competition index known as competition influence-zone overlap (CIO). A disc was assigned to each tree based on the relationship between an open-grown tree's crown radius and diameter at breast height (dbh). The index was calculated from the ratio of the area of overlap of competitors' discs to the area of the disc of the subject tree and their relative dbh's. The ratio of area of overlap is an empirical adjustment of the open-grown crown for competition. These two approaches, using polygons and using discs, were combined to create a static geometric two-dimensional model of the projected crown, adjusted for the size of competitors (Nance et al., 1988; Smith, 1987) to adjust for weaknesses in both types of indices.

Three-dimensional crown models use geometric attributes such as surface area and volume, attempting to model the crown explicitly. Hatch et al. (1975) developed a competition index based on the surface area of a crown exposed to direct sunlight. The exposed surface area was then adjusted for height to the base of the live crown of the subject tree and the relative basal area of the competitors. Mitchell (1969, 1975) used a dynamic three-dimensional crown competition model for a distance-dependent individual tree model of white spruce (*Picea glauca* [Moench] Voss) (TASS I) and Douglas fir (*Pseudotsuga menziesii* [Mirb.] Franco) (TASS II). A branch-length function (BLF) which decreased with base-of-branch distance from the top of the crown was used to model the crown. The resulting shape of the crown appeared parabolic. Mitchell conceptualized the crown to be a series of shells, each shell containing a year's branch growth and the needles produced in that year. The shape of the crown was maintained by keeping track of the PC on the ground (x-y plane). Goudie (1980) adapted this model to lodgepole pine (*Pinus contorta* var. *latifolia* [Engelm. ex S. Wats.] stat. nov.). He included height as a variable in the BLF and assumed that the photosynthetic efficiency of foliage throughout the crown was equal.

Other geometric models that have been used to simulate the crown are the cone and the ellipsoid. The cone has been used for research in forest actinometry (Oker-Blom, 1987) and in a primary productivity model of Douglas fir (Van Gerwen et al., 1987). An ellipsoid with four shells was used to calculate radiant energy interception for an aboveground productivity model of radiata pine (*Pinus radiata* D. Don) (Grace et al., 1987).

THE STATIC MODEL

The development of a static model can be considered a four-stage process: (1) identification of crown competitors, (2) identification of trees on the border of the plot, (3) partitioning of crown intersections, and (4) calculation of the geometric attributes of the crown.

Assumptions, Strategy, and Data

A tree's crown is assumed to be isotropic, which results in the crown being symmetric about the bole (z-axis). The vertical projection of an open-grown crown onto the ground is a disc. The quality of this assumption may be a function of geographic latitude and the sun angle for the population being modeled. The crowns of adjoining trees are assumed to be non-interlocking, and the space assigned to each tree is unique. This seems biologically reasonable, based on reduced light interception and wind action. Finally, the PC areas do not overlap representing total shade intolerance. This may be a reasonable approximation in even-aged monocultures but would certainly be questionable in more diverse stand structures.

The calculation of the intersection of objects in three-dimensional space is complex and computationally prohibitive when large numbers of trees are simulated. Mitchell (1975) solved this problem with a BLF that used the distance from the acropetal terminus to the point of a branch's attachment on the bole as the argument. This allowed him to map from three-dimensional space to two-dimensional space. The same strategy is used in this paper to allow the calculation of the intersection of the tree crowns in two-dimensional space.

The required tree-level variables for this model are the x-y coordinates, total height, height to the base of the live crown (HBLC), and the BLF with the distance from the top of the tree as one of the predictor variables.

Identifying Crown Competitors

Initially, all pairs of competing crowns must be identified. Using the projected disc to simplify the task, the question of interest becomes which pairs of discs intersect. Let n denote the number of trees being studied throughout this paper. Up to $n(n - 1)/2$ pairwise tests must be made if all possible pairs are to be tested for intersection. This brute force approach would result in an algorithm of $\theta(n^2 + k)$ operating time and $O(n)$ storage, where k is the number of pairwise intersections reported. This would incur a high computing cost.* Algorithms from the field of computational geometry, a branch of computer science, can be used to increase the computational speed and decrease the memory requirements of software used to solve the problems encountered in calculating the geometric attributes of a tree's crown. All pairs of intersecting discs can be calculated in optimal $O(n \log n + k)$ time requiring $O(n)$ storage (Edelsbrunner, 1982). This is accomplished by transforming the problem into one of detecting all pairs of intersecting isothetic (parallel to the major axes) rectangles by circumscribing discs with rectangles (Figure 1). Due to this transformation from discs to rectangles, a post-processing step must be added requiring $\theta(k)$ time to prune false positives (where rectangles intersect but not the discs) from the list of intersections. The same approach may be applied to other shapes that can be approximated by an isothetic rectangle.

Identifying Trees on the Border of a Plot

Another consideration in the modeling of the crown is the identification of trees on the border of a plot due to the edge effect. This influence may be due to unaccounted-for crown competition from trees that have not been stem-mapped. Also, it may be necessary to flag trees on the border of the plot that are partially open-grown because only stand-grown trees are of interest. If we view each tree's location as a point in the x-y plane, the points comprising the convex hull are the trees on the border of the plot (Figure 2). The convex hull of a set of points in a plane can be computed in $O(nh)$ time and $O(n)$ storage (Jarvis, 1973), where h is the number of points on the convex hull.

*The notation used follows that of Knuth (1976) for the analysis of algorithms. The notation "$O(f(n))$" means of the same or lesser order of $f(n)$, and "$\theta(f(n))$" means in the order of $f(n)$. Order is used in the mathematical denotation, e.g., $y = ax^2$ is a second-order equation. When referring to the speed of the algorithm, $f(n)$ is the number of primary operations an algorithm must perform based on the number, n, of datum input. This notation with respect to memory indicates the amount required.

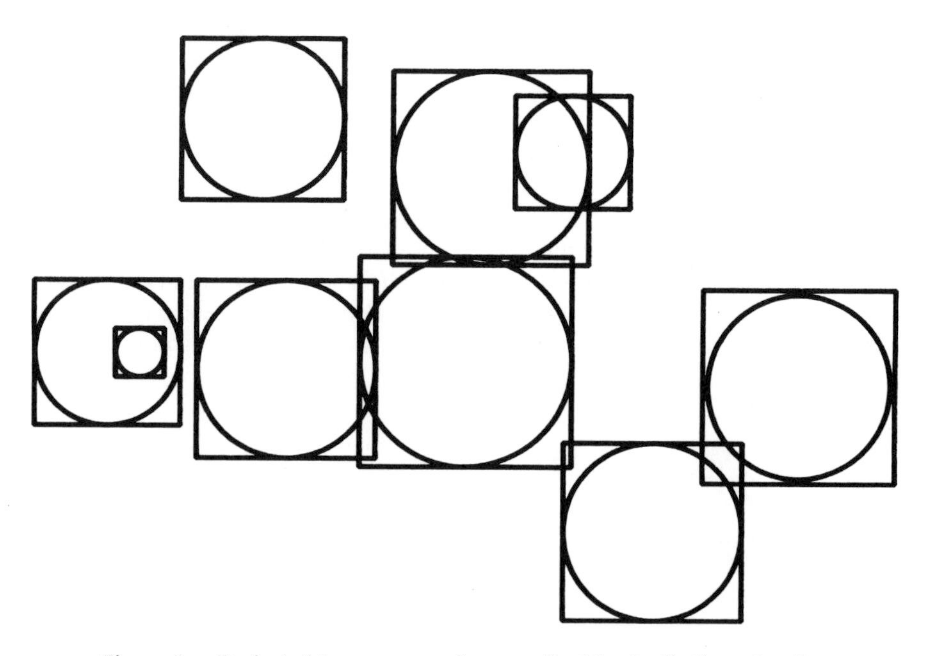

Figure 1. Projected tree crowns circumscribed by isothetic rectangles.

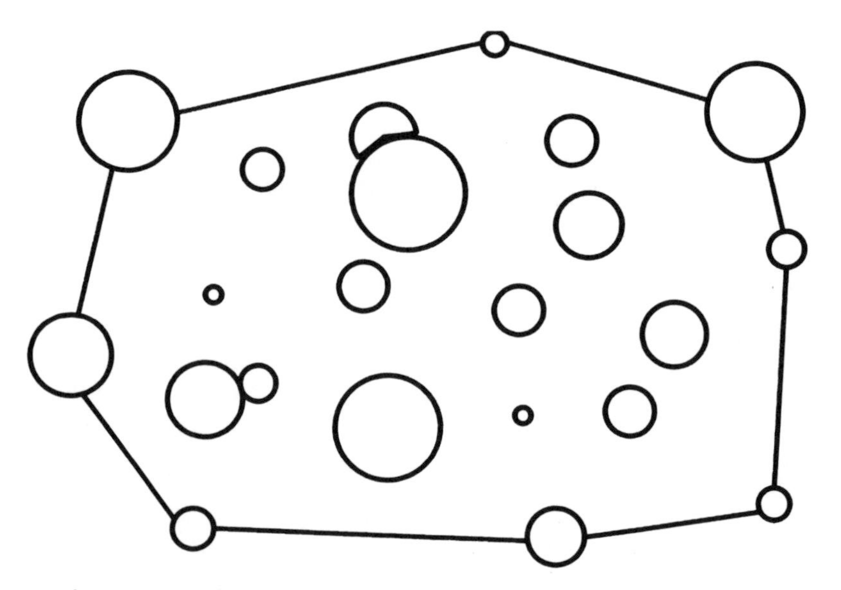

Figure 2. The PC's joined by edges define the convex hull (trees on the border of the plot).

The Partitioning of Crown Intersections

In the preceding two stages, only pairwise intersection and border trees have been identified. Next, the area of the disc overlaps must be assigned to each tree. Three categories of intersection must be considered: (1) equal HBLC and intersecting crowns, (2) unequal HBLC and intersecting crowns, and (3) unequal HBLC and crowns that do not intersect (Figure 3). In the first category, it is necessary to partition the overlaps of the discs used in identifying competitors. Here the convention is to use the common chord of the two intersecting discs to define the boundary. This is a very simple method of partitioning the overlap which lends itself well to computing the PC of the tree based on its competitors. (For a more thorough discussion of the partitioning of the overlap of discs, see Gates et al., 1979.)

In the case of unequal HBLC, it is necessary to determine whether the crowns truly intersect or merely overlap. To do this, the disc at the HBLC of the tree with the higher HBLC must be calculated on the tree with the lower HBLC. This new disc and the disc of the tree with the higher HBLC define the common chord which is a portion of the boundary between the two crowns. The remainder of the boundary would be defined by the projection of the disc of the tree with higher HBLC on the disc with the lower HBLC. Finally, if the crowns do not intersect but do overlap, the area of overlap would be assigned completely to the tree with the highest HBLC.

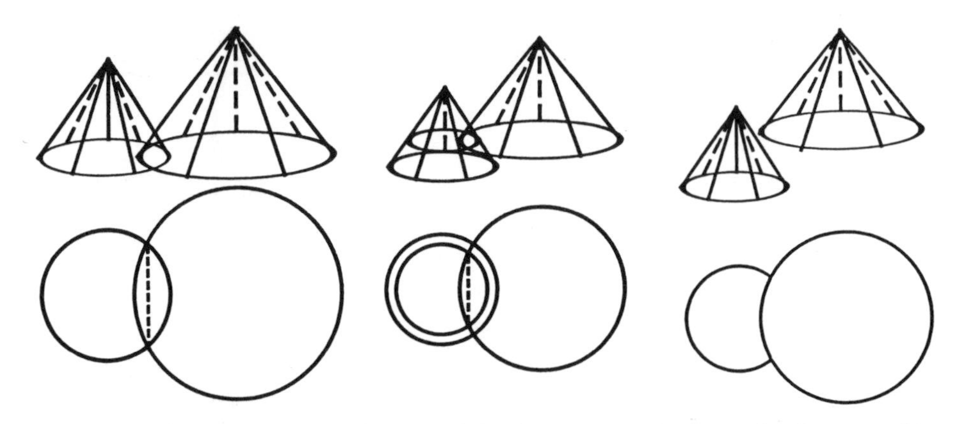

Figure 3. **(*Left*) Equal height to the base of the live crown and crowns that intersect. (*Center*) Unequal height to the base of the live crown and crowns that intersect. (*Right*) Unequal height to the base of the live crown and crowns that do not intersect.**

Geometric Attributes

The geometric attributes of the crown can now be calculated. The attributes of primary interest are the perimeter and area of the PC and the exposed surface area (ESA) of the crown, crown volume (CV), and foliar volume (FV), i.e., the portion of the CV where foliage is present. The exact area of the PC can be calculated by a plane-sweep algorithm used for shading in computer graphics (Lee, 1981). This algorithm works by drawing a horizontal line at m points and then calculating the area within the slab defined by adjoining lines (Figure 4). A contour is either an arc or a chord based on the assumptions. The time required for this algorithm is $O(k \log k + m)$, and it uses $O(k)$ storage, where k is the number of minima, maxima, or vertices of a PC. From this worst-case analysis, it is clear that the resources required to calculate the area of the PC are a direct function of the amount of competition being experienced within the simulated stand.

Another method of calculating the PC area is to approximate it by means of a polygon (Figure 4). The quality of the approximation is a function of the number (m) of line segments (edges) used to define the contours of the PC that are arcs. This algorithm requires $O(km \log m + km + e)$ time and $O(km + e)$ storage, where k is the number of arcs in the PC and e is the number of chords in the PC. The PC area can then be calculated by taking the determinant of the counterclockwise-ordered vertices and dividing by two. There is a trade-off between computational efficiency and precision based on the number of edges used to approximate the PC. This approach is reasonable, given that the PC of an open-

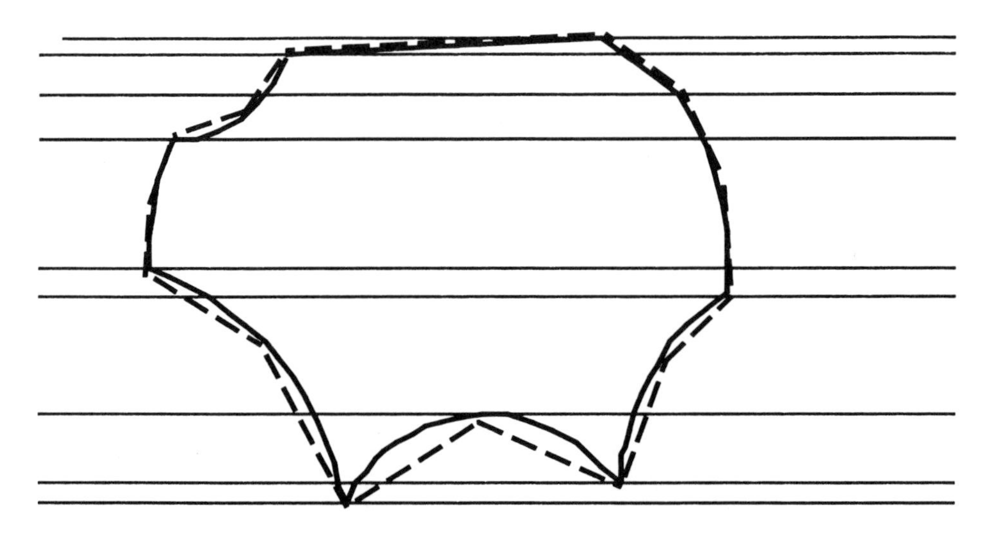

Figure 4. Calculation of PC area by slabs and polygons. The horizontal lines define the slabs. The dashed lines define the polygon used to approximate the solid-lined PC.

grown tree as a disc is also an approximation.

The ESA of the crown can be calculated by taking the inverse function of the BLF and integrating over x and y.

$$\mathrm{ESA} = \iint \sqrt{\left[\frac{\partial x}{\partial z}\right]^2 + \left[\frac{\partial y}{\partial z}\right]^2 + 1}\ \partial A \tag{1}$$

The PC area must be partitioned to be fed into the equation to calculate the ESA. This can be difficult with the curved contours of the PC. The poly gonal approach would lend itself much more readily to establishing bounds since all contours are edges.

FV and CV are the attributes of greatest interest to most forest scientists. Mitchell (1975) shows the calculation of FV and CV. Using his assumptions of foliar distribution, the FV can be calculated by $FV_i = HG_i \cdot PC_i$, where i is the year and HG is the height growth. The summation of the FV_i for all the years of foliage gives the total FV.

To calculate the approximate CV, Mitchell (1975) integrated over the BLF. A difficulty of this approach is determining the bounds of integration. The equating of the PC with a disc allowed the average base-of-live-crown branch length and average HBLC to be solved for. The BLF can then be integrated between zero and average HBLC. The same approach can be applied to the calculation of ESA.

The above methods do not give exact information on the quantity of crown area (CA) at a given height, but only the CV or FV per tree, similar to a competition index. For precise modeling of the allocation of wood volume growth where wood area growth along the bole is a function of the foliage above (Pressler's law) or the within-crown light regime, it would be advantageous to be able to calculate the CA at any given height. The CA at any height can be calculated by use of a space-sweep algorithm. This algorithm slices the crown with a plane to calculate the exact foliar area at that height. The integration of these slices will result in the CV using $O((k \log k) + k(s-1))$ time and $O(k + s)$ storage (Figure 5) (Lee, D. T., 1988, pers. comm.), where k is the number of contours of the PC and s is the number of slices. This crown must be sliced only up to the height at which it is symmetric; at this point it can be integrated based on the BLF.

Figure 5. Horizontal planes slicing paraboloid crown.

DISCUSSION

The crown model as presented may be used as an index for competitive status using only the geometric attributes of the crown as an approximation of the foliar component of the crown. This empirical application has use in experimental designs where the individual tree is of interest and where it is necessary to account for the competitive effects of neighbors. In the same vein, the geometric attributes of the crown can be used in a distance-dependent individual-tree model. In this application, the geometric attributes would again be used as an index of competitive status, unless within-crown components such as foliage and branches have been attached to the spatial model of the crown. The model is capable of distributing wood growth along the bole by using one of many empirical relationships that can be derived from the geometric attributes.

The static nature of the model is a limitation which also requires that the geometric attributes of the crown be used as an empirical index in a distance-dependent individual-tree model. The shape of the crown must be recreated (static) for every growth interval rather than being updated (dynamic) as the crown grows and encounters other crowns. This leads to a misrepresentation of the shape of the projected crown as a disc. The primary factor causing the misrepresentation of the crown is the need to input the HBLC into the model for every iteration. To input the average HBLC, it must be solved for by equating the PC area calculated from a previous iteration with a disc. The use of such an approach in even-age plantations, where crown competition is relatively uniform on all sides, may not lead to large prediction errors. However, the performance of this index in less uniform situations is dubious.

Beyond the empirical applications, the crown competition model is a developmental step towards building a bridge between models of physiological processes such as photosynthate production and carbon allocation and models of the stand where the effects of adjoining trees on a subject tree's processes must be accounted for. A foliage model is one within-tree process model where a geometric crown competition model has application. A foliage model may have inputs such as nutrients, light, carbon dioxide, and water.

In order to integrate a foliage process model to a tree model and a tree model to a stand model, two problems must be overcome: (1) the foliage of the model must be attached to the tree in space so that it may interact with other within-tree processes such as cambial respiration and carbon allocation which will vary by location in a tree, and (2) inputs to the foliar model such as light and carbon dioxide must be adjusted based on the influence of neighboring trees. The use of a crown model makes it possible to attach the foliage to the crown in a spatially correct manner, and the crown model can provide the information necessary for adjusting inputs to the foliar model based on a tree's position with respect to its neighbors.

The study of the effects of atmospheric depositants (e.g., ozone) on forest growth has been limited to controlled environments where only branches or seedling may be evaluated for changes in growth. The information derived from these experiments provides the data necessary for physiological process meodels that incorporate the effects of atmospheric depositants. However, the question of interest is: What is the effect of these pollutants on the mature forest? The information from dose-response experiments cannot be directly extrapolated to the mature forest due to the effects of age, stand dynamics, and climate. The geometric model of crown competition presented offers one method of bridging process and stand-level models to examine the effects of atmospheric depositants on the mature forest.

CONCLUSIONS

The geometric modeling of crown competition has been presented to establish its practicality in forestry research. How these methods may be applied are a direct function of the assumptions of the modeler and the processes that are to be modeled. The materials presented establish the theory of static geometric models of crown competition and lay the groundwork for more advanced geometric models of crown competition. The geometric model of crown competition presented offers one method of bridging process and stand-level models to examine the effects of atmospheric depositants on the mature forest.

LITERATURE CITED

Bella, I. E. 1971. A new competition model for individual trees. *Forest Science* 17:364–372.

Brown, G. S. 1965. Point density in stems per acre. Forest Research Notes 38, Forest Research Institute, New Zealand Forest Service, Rotorua, New Zealand. 11 pp.

Edelsbrunner, H. 1982. A new approach to rectangle intersections, Part II. *International Journal of Computer Mathematics* 13:221–229.

Gates, D. J., A. J. O'Connor, and M. Westcott. 1979. Partitioning the union of discs in plant models. *Proceedings, Royal Society of London,* Series A 367:59–79.

Goudie, J. W. 1980. Yield tables for managed stands of lodgepole pine in northern Idaho and southeastern British Columbia. Master's Thesis, University of Idaho, Moscow, ID, U.S.A. 110 pp.

Grace, J. C., P. G. Jarvis, and J. M. Norman. 1987. Modelling the interception of solar radiant energy in intensively managed forests. *New Zealand Journal of Forestry Science* 17:193–209.

Hatch, C. R., D. J. Gerrard, and J. C. Tappeiner, II. 1975. Exposed crown surface area: a mathematical index of individual tree growth potential. *Canadian Journal of Forest Research* 5(2):224–228.

Jarvis, R. A. 1973. On the identification of the convex hull of a finite set of points in the plane. *Information Processing Letters* 2:18–21.

Knuth, D. E. 1976. Big omicron and big omega and big theta. *SIGACT News* 8(2):18–24.

Lee, D. T. 1981. Shading of regions on vector display devices. *Association for Computing Machinery* 15(3):37–43.

Mitchell, K. J. 1969. Simulation of the growth of even-aged stands of white spruce. Bulletin No. 75, Yale University School of Forestry, New Haven, CT, U.S.A. 48 pp.

Mitchell, K. J. 1975. Dynamics and simulated yield of Douglas-fir. *Forest Science Monograph* 17. 39 pp.
Moore, J. A., C. A. Budelsky, and R. C. Schlesinger. 1973. A new index representing individual tree competitive status. *Canadian Journal of Forest Research* 3(4):495–500.
Nance, W. L., J. E. Grissom, and W. R. Smith. 1988. A new competition index based on weighted and constrained area potentially available. In: Ek, A. R., S. R. Shifley, and T. E. Burk, eds. *Forest Growth Modelling and Prediction.* General Technical Report NC-120, USDA Forest Service North Central Forest Experiment Station, St. Paul, MN, U.S.A. Pp. 134–142.
Oker-Blom, P. 1986. Photosynthetic radiation regime and canopy structure in modeled forest stands. *Acta Forestalia Fennica* 197. 44 pp.
Smith, W. R. 1987. Area potentially available to a tree: a research tool. In: McKinley, C. R., W. J. Lowe, and J. P. van Buitjenen, eds. *Proceedings of the 19th Southern Forest Tree Improvement Conference.* Publication No. 41, Southern Forest Tree Improvement Conference, College Station, TX, U.S.A. Pp. 22–29.
Van Gerwen, C. P., C. J. T. Spitters, and G. M. J. Mohren. 1987. Simulation of competition for light in even-aged stands of Douglas fir. *Forest Ecology and Management* 18:135–152.

CHAPTER 27

DEVELOPMENT OF A SOIL- AND ROOT-BASED PRODUCTIVITY INDEX MODEL FOR TREMBLING ASPEN

Margaret R. Gale and David F. Grigal

Abstract. A non-regression, multiplicative model was recently developed that integrates the genetic potential of belowground root characteristics with soil/site characteristics. The Productivity Index (PI) model is a process-related model using empirically and theoretically based functional relationships between mineral soil properties and belowground root morphology. The PI model was modified and tested for sites of trembling aspen (*Populus tremuloides* Michx.) in Wisconsin with and without the occurrence of fire. Results indicate that the PI model can be used to predict site quality and aboveground production on sites where fires have not occurred. Significant lower site index values and poorer relationships with PI were observed for sites where fires have occurred, indicating the need to include other soil characteristics, such as nitrogen or organic matter content, in order to explain these differences.

INTRODUCTION

Regression techniques have been used extensively in the past to identify soil properties that explain significant differences in site quality for trembling aspen (Fralish and Loucks, 1975; Stoeckeler, 1960). Complex interactions among soil properties and their resulting effect on aboveground production have been ignored (Stone, 1978). The emphasis of most research concerning site quality has been on the indirect effects of soil properties on aboveground production instead of on the direct effects of these properties on belowground production. A soil productivity index (PI) model, recently developed for forested sites, integrates soil and site characteristics and their relationship to root growth and to the vertical root distribution of the tree (Gale and Grigal, 1988). The model is process-related in that it attempts to describe the effect of limiting soil properties on root growth and thus on the proportional root pattern with depth. The PI model has proved to be a good predictor of site quality and stand production for trembling aspen (*Populus tremuloides* Michx.) in areas where fire has not occurred (Gale and Grigal, 1988). The objective of this study was to test whether the PI model accurately predicted site index and stand production on burned trembling aspen areas and to compare these predictions to estimates for unburned areas.

Dr. Gale is Assistant Professor at the School of Forestry and Wood Products, Michigan Technological University, Houghton, MI, 49931 U.S.A. Dr. Grigal is Professor in the Departments of Soil Science and Forest Resources, University of Minnesota, St. Paul, MN, 55108 U.S.A.

ASSUMPTIONS

The PI model integrates the response of vertical root distribution to soil and site properties. The following assumptions were made in order to test the PI model:

1. Root responses to soil/site conditions are genetically controlled.
2. Cumulative vertical root distributions (Y) with depth (d) can be described using a non-linear equation of the form $Y = 1 - \beta^d$ (Gale and Grigal, 1987).
3. Under optimum soil and site conditions a species will consistently produce a specific vertical root distribution.
4. Deviations from the optimum vertical root distribution affect aboveground production.
5. Soil and site characteristics multiplicatively modify vertical root distributions.
6. Potential available water, pH, percent of slope, precipitation, evaporation rates, and depth to water table affect belowground production and distribution.

DATA BASE

The data used to test the PI model were taken from work by Stoeckeler (1960) on 26 stands of trembling aspen in northern Wisconsin. Stands were even-aged with less than 25% mixture of other species. Stands varied in age from 15 to 51 years and were selected from a complete range of soil textures: sands, loamy sands, sandy loams, silt loams, and clays. Sites were also selected to include soils of glacial and lacustrine origin as well as lime-rich and lime-poor areas. Sites with shallow water tables were included in the analysis. Eleven of the 26 stands had some history of fire occurrence since the stand was established. Site index was estimated using curves developed by Kittredge and Gevorkiantz (1929) on measurements of height and age from dominant or codominant trees. Texture and pH were measured for each horizon using methods described by Gale and Grigal (1988).

THE PRODUCTIVITY INDEX (PI) MODEL

The form of the PI model used in the analysis was

$$PI = \sum_{i=1}^{r} ([AVW \times pH]^{1/2} \times WF)_i \times [CL \times SL \times WBL]^{1/3} \quad (1)$$

where AVW is the horizon sufficiency rating for potential available water, pH is the horizon sufficiency rating for pH, CL is the site sufficiency rating for climate, SL is the site sufficiency rating for percent of slope, and WBL is the site sufficiency rating for depth to water table. The ½ and ⅓ powers in the PI model are associated with the geometric means of soil and site sufficiency rating, respectively. Sufficiency ratings range in value from 0.0 to 1.0 (Figure 1). A sufficiency rating of 1.0 represents the optimum level of a soil/site property for root growth, while a rating of 0.0 represents an absolutely limiting level of a soil/site property. The number of horizons within maximum rooting depth is represented by r. The weighting factor (WF) for each horizon is the optimum vertical proportion of roots in each horizon and is based on an equation developed by Gale and Grigal (1987) for intolerant species,

$$Y = 1 - 0.98^d \quad (2)$$

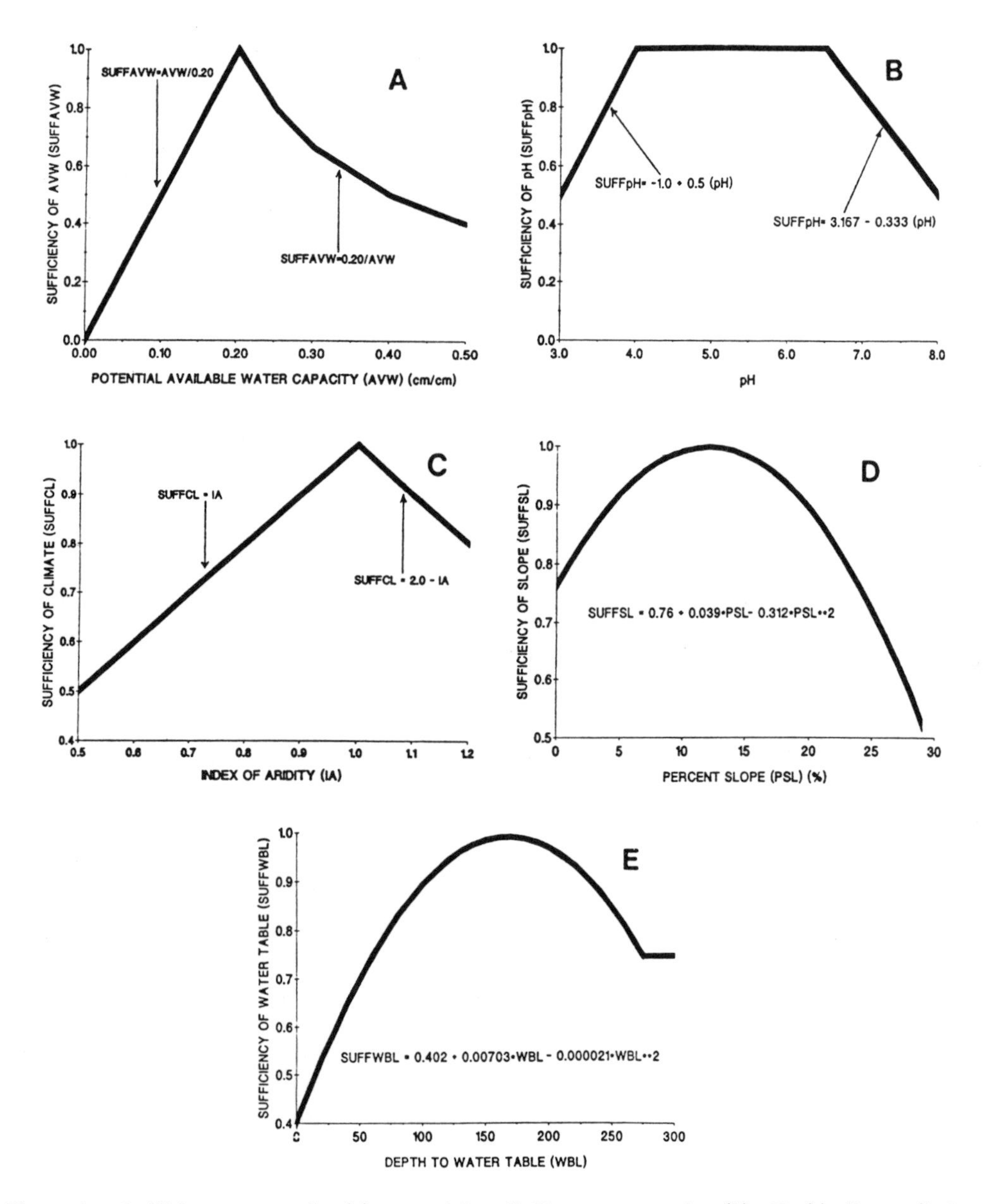

Figure 1. Sufficiency curves for (*a*) potential available water capacity, (*b*) pH, (*c*) climate (index of aridity), (*d*) percent of slope, and (*e*) depth to water table (adapted from Gale and Grigal, 1988).

where Y is the cumulative proportion of roots to a depth d in cm. The coefficient 0.98 was used to describe the median coefficient for intolerant species (Gale and Grigal, 1987) and assumed to be the optimum vertical root distribution for trembling aspen. Precise information was not available on optimum vertical root distribution for trembling aspen. The weighting factor is calculated as the difference in the cumulative proportion of roots at the upper and lower horizon boundaries. If all soil sufficiency ratings for each horizon are 1.0 and all site sufficiency ratings are 1.0, PI will equal 1.0.

SOIL SUFFICIENCY CURVES

Simple equations were used to describe the relationships from low AVW values to optimum AVW (i.e., 0.20 cm/cm) and from optimum AVW to 0.50 cm/cm (Gale and Grigal, 1988) (Figure 1a). Sufficiency curves for pH (SUFFpH) were developed using optimum pH values cited by Spurway (1941) and Stone et al. (1962). Since pH's lower than 3.0 and higher than 8.0 are uncommon in forested soils, the sufficiency rating for pH was set to 0.50 at these values, and a linear relationship was assumed between these and optimum values (Gale and Grigal, 1988) (Figure 1b).

SITE SUFFICIENCY CURVES

Site sufficiency curves for climate were developed using an index of aridity (IA) proposed by Vysockiy (1905; cited by Czarnowski, 1964). The index was calculated by dividing precipitation by total evaporation for homoclimatic regions as tabulated by Rauscher (1984). Climatic sufficiency for root growth was considered to be the actual index up to a value of 1.0. IA's above 1.0 were considered to be less than optimum (Gale and Grigal, 1988) (Figure 1c). Due to the lack of information on the effect of slope on vertical root distributions for trembling aspen, the sufficiency curve for percent of slope was described using an equation developed by Gale and Grigal (1988) for white spruce *Picea glauca* (Moench) Voss (Figure 1d). We believe this curve to be applicable for trembling aspen sites. The sufficiency curve for depth to water table was developed from work by Stoeckeler (1960) on trembling aspen in Wisconsin and Minnesota, U.S.A. (Gale and Grigal, 1988) (Figure 1e).

RESULTS AND DISCUSSION

A significant linear correlation existed between site index and PI on unburned sites ($r = 0.90$, $p < 0.01$) and on burned sites ($r = 0.54$, $p = 0.05$) (Figure 2). The assumption that burned sites have the same slope and intercept as unburned sites, but with lower PI values, was tested using analysis of covariance. The slopes and intercepts for the equations describing the two site conditions were found to be significantly different ($p < 0.01$). The difference in slopes was attributed to a low value observation (PI = 0.58, site index = 51) on the burned site (Figure 2). This value was determined to be a significant influential case. When it was dropped from the analysis, slopes were determined to be non-significant ($p = 0.05$). Intercepts, however, were still determined to be significantly different ($p < 0.01$), indicating that the PI model does not describe site quality of aspen on burned sites as accurately as it describes site quality on unburned sites.

Similar results were found when comparing Schumacher-type equations for volume production and mean annual increment between burned and unburned sites (Table 1). When added with age, the productivity index explained 29% of the variation in volume on unburned sites, while it explained 54% of the variation in mean annual increment on unburned sites (Table 1). PI did not add any additional explained variation to volume nor to mean annual increment predictions on burned sites when age was added. These results indicate that the PI model is inadequate for predicting site quality on burned sites. Although age is important in predicting volume and mean annual increment, age was not significantly related to volume and mean annual increment on burned sites ($p > 0.5$) nor to mean annual increment on unburned sites ($p > 0.5$). Coefficients for the PI variable in the equations for burned sites were significantly different from the PI coefficients for unburned sites ($p = 0.05$).

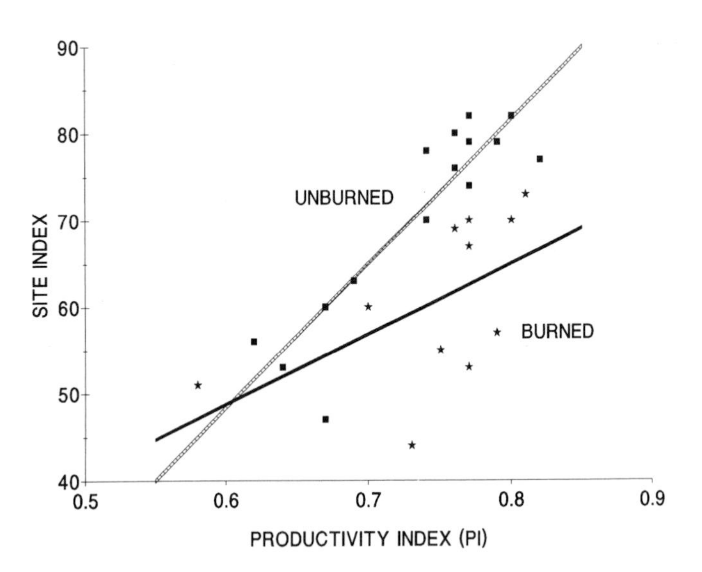

Figure 2. Scatterplot of site index versus productivity index (PI) on unburned (■) versus burned (★) sites.

Table 1. Volume (ft^3/ac) and mean annual increment ($ft^3/ac/year$) equations for unburned and burned trembling aspen sites.

VOLUME

Equation	R^2	Partial R^2	$S_{y.xx}$
Unburned (11 sites)			
LNVOL = 4.58 − 25.6 (1/AGE) + 5.00 (PI)	0.83**	0.29**	0.20
Burned (15 sites)			
LNVOL = 7.55 − 13.7 (1/AGE) + 0.07 (PI)	0.23^{NS}	0.00^{NS}	0.37

MEAN ANNUAL INCREMENT

Equation	R^2	Partial R^2	$S_{y.xx}$
Unburned (11 sites)			
LNMAI = 0.12 + 8.2 (1/AGE) + 4.90 (PI)	0.69**	0.54**	0.20
Burned (15 sites)			
LNMAI = 3.50 + 11.9 (1/AGE) + 0.27 (PI)	0.25^{NS}	0.00^{NS}	0.32

**, significant at the 0.01 p level ($p < 0.01$)
NS, not significant at the 0.10 p level ($p > 0.10$)

The decrease in site quality of trembling aspen stands due to the occurrence of fire has been documented by Stoeckeler (1960). These changes have been attributed to losses of soil organic matter and nitrogen and to bark damage related to fire intensity (Stoeckeler, 1948; Rouse, 1986). The PI model does not include these soil/site characteristics, as data concerning soil organic matter and nitrogen content and fire intensity were not available. Inclusion of soil sufficiency curves for organic matter and nitrogen content may improve the overall relationship between PI and site index, volume, and mean annual increment on burned as well as unburned sites. The addition of information concerning stand age at the time of the fire and fire intensity may also help explain potential bark damage and improve the applicability of the PI model for burned sites.

CONCLUSIONS

Because the system that we are attempting to model is very complex and dynamic, inclusion of soil/site characteristics and their effect on a tree's optimum vertical root distribution may aid in our understanding of the system. Regression techniques can provide needed information on the response of roots to one or two soil/site factors if all other factors are considered optimum. The PI model, however, uses multiple interactions among soil and site factors in relation to a tree's vertical root distribution. Sufficiency curves are used in the PI model to describe a tree's root response to individual soil/site conditions using standardized curves, interrelating root growth responses to aboveground productivity responses. When sufficiency curves are multiplied together, as in the PI model, interactions among soil and site characteristics are included. Because roots and shoots interact, incorporating root growth responses with soil/site properties into process models can add to our knowledge and understanding of the dynamics of aboveground productivity.

The main assumptions of the PI model are (1) root responses to soil/site conditions are genetically controlled, (2) under optimum soil and site conditions a species will consistently produce a specific vertical root pattern, and (3) aboveground production is related to the response of a tree's vertical root pattern to soil/site characteristics. If these assumptions are invalid, additional refinement of the PI model will not increase the accuracy of site quality predictions. Additional studies are needed to determine the validity of these assumptions.

Although the sample size for this analysis was small, these stands of trembling aspen represented a wide range of site qualities and soil characteristics. More research is needed, however, to determine whether the PI model may be applied to other geographic areas and whether it can adequately describe the range of soil/site characteristics on which trembling aspen grows. It is possible that optimum vertical root distributions for trembling aspen vary due to the high genetic variability of the species. Additional work is needed to determine the optimum vertical root distribution for trembling aspen and whether there are significant genetic differences in optimum vertical rooting patterns for trees of this species.

LITERATURE CITED

Czarnowski, M. S. 1964. *Productive Capacity of Locality As a Function of Soil and Climate with Particular Reference to Forest Land.* Biological Sciences Series, Louisiana State University Press, Baton Rouge, LA, U.S.A. 174 pp.

Fralish, J. S., and O. L. Loucks. 1974. Site quality evaluation models for aspen (*Populus tremuloides* Michx.) in Wisconsin. *Canadian Journal of Forest Research* 5:523–528.

Gale, M. R., and D. F. Grigal. 1987. Vertical root distributions of northern tree species in relation to successional status. *Canadian Journal of Forest Research* 17:829–834.

Gale, M. R., and D. F. Grigal. 1988. Performance of a soil productivity index model used to predict site quality and stand production. In: Ek, A. R., S. R. Shifley, and T. E. Burk, eds. *Forest Growth Modelling and Prediction,* Vol. 1. General Technical Report NC-120, USDA Forest Service North Central Forest Experiment Station, St. Paul, MN, U.S.A. Pp. 403–410.

Kittredge, J., Jr., and S. R. Gevorkiantz. 1929. *Forest Possibilities of Aspen Lands in the Lake States.* Technical Bulletin 60, University of Minnesota, St. Paul, MN, U.S.A. 84 pp.

Rauscher, H. M. 1984. *Homogeneous Macroclimatic Zones of the Lake States.* Research Paper NC-240, USDA Forest Service North Central Forest Experiment Station, St. Paul, MN, U.S.A. 39 pp.

Rouse, C. 1986. *Fire Effects in Northeastern Forests: Aspen.* General Technical Report NC-102, USDA Forest Service North Central Forest Experiment Station, St. Paul, MN, U.S.A. 8 pp.

Spurway, C. W. 1941. *Soil Reaction (pH) Preferences of Plants.* Special Bulletin 306, Michigan State University, East Lansing, MI, U.S.A. 36 pp.

Stoeckeler, J. H. 1948. The growth of quaking aspen as affected by soil properties and fire. *Journal of Forestry* 46:727–737.

Stoeckeler, J. H. 1960. *Soil Factors Affecting the Growth of Quaking Aspen Forests in the United States.* Technical Bulletin 233, University of Minnesota Agricultural Station, St. Paul, MN, U.S.A. 48 pp.

Stone, E. L. 1978. A critique of soil moisture–site productivity relationships. In: Balmer, W. E., ed. *Soil Moisture–Site Productivity Symposium Proceedings.* USDA Forest Service, Southeast State and Private Forestry, Atlanta, GA, U.S.A. Pp. 377–387.

Stone, E. L., R. Feuer, and H. M. Wilson. 1962. *Judging Land for Forest Plantations in New York.* Cornell Extension Bulletin Number 1075, Cornell University, Ithaca, NY, U.S.A. 16 pp.

SECTION V

Modeling Responses to Environmental Stress

The preceding sections represented the current state of our knowledge about tree metabolism and growth, tree structure and function, model structure and evaluation, and tree growth models. Earlier chapters characterized genetic, physiologic, and ecologic relationships and processes in terms of observation, experimentation, mathematics, and/or hypotheses; together they form a basis from which further models can be developed. Current models and concepts were developed to advance our understanding of important relationships and to estimate responses to given environmental conditions or physiological inputs.

Section V represents a sample of recent efforts to integrate our current knowledge into models which will allow us to predict or evaluate past, current, and potential responses to environmental stresses. Some of the models focus on specific processes of tree growth, while others characterize the development of forest stands. Alternative modeling approaches, such as tree-ring analysis, are also described. The variety of information presented illustrates the complexity of modeling the myriad interacting environmental stresses on forests today.

Process modeling of the response of forest and tree growth to environmental stresses is a developing scientific field. As our base of knowledge and experience grows, many of the current models are evolving rapidly. Model complexity, as well as model reliability, will undoubtedly continue to increase in the future.

CHAPTER 28

EVALUATING EFFECTS OF POLLUTANTS ON INTEGRATED TREE PROCESSES: A MODEL OF CARBON, WATER, AND NUTRIENT BALANCES

David A. Weinstein and Ron Beloin

Abstract. Hypotheses concerning the causes of forest decline can be related to the physiological tree process they affect. However, most hypotheses ultimately involve the creation of imbalances in the plant carbon, water, and nutrient acquisition and storage systems. A model that follows the interactions between pollutants and these systems is described. Using this model to compare scenarios of the response of trees exposed to ozone with the response of unexposed trees, the effects of ozone upon the ability of the tree to build large stores of carbon during the period immediately preceding budbreak is identified as a potential mechanism leading to long-term tree degradation.

INTRODUCTION

The forests of the northeastern United States may be experiencing a growth rate alteration as a result of exposure to atmospheric pollutants, particularly ozone, sulfur dioxides, nitrous oxides, and acid precipitation (Johnson and Siccama, 1982; McLaughlin, 1985). Analysis of the recent growth patterns of trees in New York, U.S.A., shows growth rate decreases during a period when pollutant loadings to forests have been increasing (Puckett, 1982). The levels of pollutants are sufficient to be causing these growth declines. For example, average ozone exposure of 60 ppbv across the state of New York would be expected to cause 10–20% growth reductions in sugar maple seedlings (Reich and Amundson, 1985).

Evaluating the causes of forest decline has been a particularly difficult problem. There are many different potential causes for the observed decline symptoms, each of which can logically be related to their development. Two problems confound clear resolution of cause and effect. First, trees are integrated systems. As a consequence, many different

Dr. Weinstein and Mr. Beloin are Staff Scientist and Research Support Specialist, respectively, at the Ecosystems Research Center, Corson Hall, Cornell University, Ithaca, NY, U.S.A. This publication is ERC-177 of the Ecosystems Research Center (ERC), Cornell University, and was supported in part by the U.S. Environmental Protection Agency Cooperative Agreement Number CR812685-02. Additional funding was provided by Cornell University and by the Electric Power Research Institute, Inc., Agreement RP2799-1. The work and conclusions published herein represent the views of the authors and do not necessarily represent the opinions, policies, or recommendations of the Environmental Protection Agency.

stresses may manifest in similar symptoms. For example, low water availability, low light availability, or impairment of carbon fixation by air pollutants all result in decreases in the pool of available carbon a tree may use for maintenance and growth. Lower growth rates may be attributable to any of these initiating causes. Second, the chain of responses a tree will exhibit may be identical regardless of cause if the organism is responding to low carbon reserves rather than to the peculiarities of the form of stress.

HYPOTHESES BASED ON PROCESS DISRUPTION

The variety of causal agents with the potential to initiate a decline sequence are identified in Figure 1. Several of these involve processes that would be expected to occur without any anthropogenic interaction. Examples of these include forest successional processes, changes in climate moisture and temperature regimes, wind damage, and pest attacks. Most, however, are associated with human-induced processes. Increased nitrogen and hydrogen ion input has been documented for forested ecosystems throughout the northeastern U.S.A., for example. This increased fertilization from atmospherically derived nitrogen could create water or nutrient imbalances within plants, or could delay the onset of hardening in leaves, making them susceptible to frost damage. Acid deposition could directly injure foliage, reducing effective light-capturing leaf area or damaging water regulation. Alternatively, acidity could increase the leaching of cations from leaves, creating foliar nutrient deficiencies. Increased levels of acidity entering the soil system could result in the leaching of essential cations away from the rooted zone, decreasing plant nutrient uptake, or could increase the mobilization of heavy metals in the root zone, causing root damage. Ozone is also present in high concentrations during periods of the growing season in this region and could cause damage to the photosynthetic apparatus. Finally, any of these effects can weaken trees and make them more susceptible to insect or pathogen invasion.

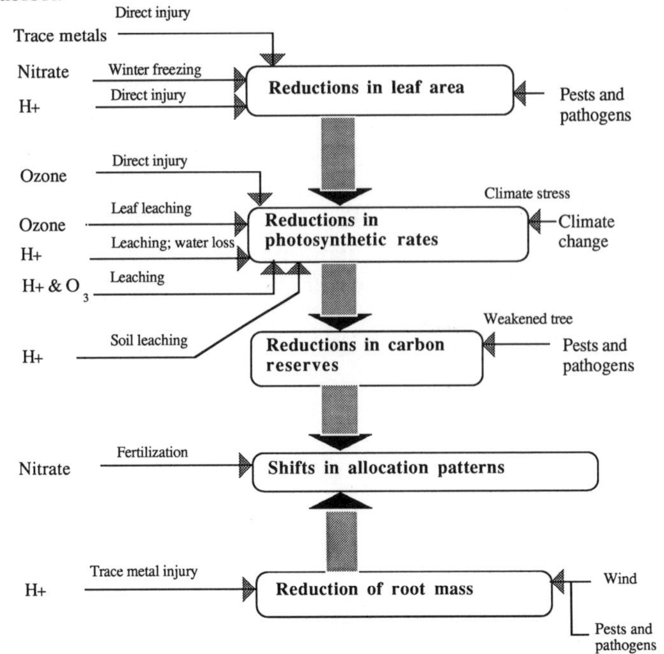

Figure 1. Hypotheses of the causes of forest decline. Anthropogenic causes are listed together with the pathway of their mode of action on the left side of the diagram. Natural causes are listed on the right. Processes within the affected tree are described within boxes in the center of the diagram.

Many of these hypotheses have similarities in the patterns of plant response that they cause. In Figure 1, hypotheses are grouped by the processes affected. For example, it is suggested that leaf area reductions may result from direct injury by hydrogen ions, but may also result from needle loss associated with nitrogen fertilization. Alterations in assimilation rate may occur as a result of climate stress, direct ozone injury, water stress caused by cuticle damage or root damage, or lowered nutrient status resulting from root damage or cation loss from either leaves or soil.

Both leaf area loss and reductions in photosynthetic rate may interact to reduce the total amounts of carbon fixed. Reductions in carbon fixation lead to decreases in carbon reserves. This reduction affects the ability of a plant to perform operations required in phenological development. The result may be a decrease in the amount allocated to leaf tissue construction, decreasing the maximum potential carbon which can be fixed by the tree. Alternatively, less carbon may be allocated to root mass, affecting the potential rate of resource acquisition and, thus, photosynthetic rate.

MITIGATION SYSTEMS IN TREES

Trees function in an environment that periodically poses stressful conditions. As a consequence, trees contain response systems that are capable of mitigating the detrimental effects of a stress. Understanding the causes of tree decline will depend on understanding the limits of this mitigation system. In the next section, we will examine an example of trees' pathways of response to air pollutants and identify critical points where the potential for mitigation of these effects exists.

Figure 2 shows a schematic of the pathways of the hypothesized effects of ozone exposure on a tree. The initial injury involves a suppression of the rate of photosynthesis, through either permanent or temporary disruption of the carboxylation sites (Pell, 1987). Unless respiration rates are also reduced, the reduced rate of photosynthesis will inevitably lead to a decrease in the quantity of carbon available for basic plant maintenance and allocation to growth. Little evidence is presently available suggesting declines in respiration.

During periods of severe or extended stress, photosynthetic fixation rates may decrease to such low levels that the quantities of new carbon available are insufficient to meet all respiration and growth needs. Logically, in order to meet these needs, the plant must draw upon carbon reserves accumulated during periods of more rapid fixation. Further, plants tend to respond to ozone exposure by adding more leaf tissue, thus mitigating the ozone damage by increasing the quantity of new leaf surface. This is accomplished by preferentially increasing the proportion of carbon being allocated to leaf and branch growth (Oshima et al., 1979). If carbon reserves are sufficiently plentiful, enough new leaf area may be added through this process to recover to pre-stress photosynthetic fixation rates.

Mitigation can never be achieved without a cost, however. The cost may be decreased allocation of carbon to root system production and maintenance (Oshima et al., 1979). In soils with large amounts of available water and nutrients, the resulting loss of root biomass may be inconsequential to plant performance. Even in these cases, however, it will render the plant more susceptible to limitations caused by periodically occurring fluctuations in the availability of water and nutrient resources. When there is insufficient root surface area to permit water and/or nutrient uptake in the quantities required to meet the needs of new tissue production, carbon fixation will decline. Consequently, the mitigation of stress from one source may indirectly increase the plant's susceptibility to stress from another source.

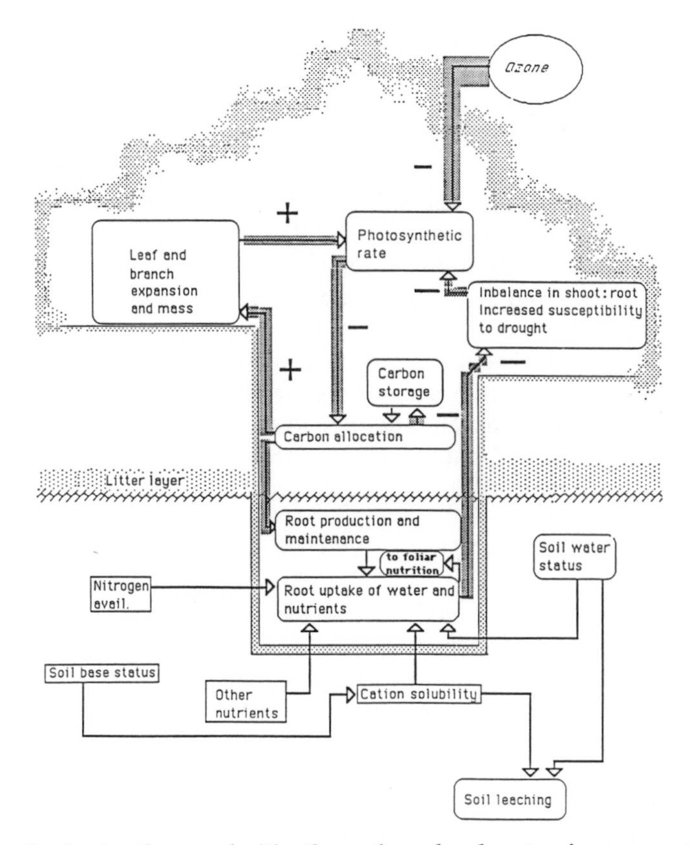

Figure 2. Hypothesized pathways of mitigating action taken by a tree in response to ozone injury to its photosynthetic mechanism. Highlighted pathways are the principal ones changed. The direction of change is indicated by a plus or minus sign.

Trees rely heavily upon the reserve pool of carbon (Holl, 1985; Loach and Little, 1973). Under conditions of continuing stress, it is reasonable to assume that the pool may eventually be depleted. It can be hypothesized that under these circumstances, there will be insufficient carbon from the sum of new fixation and reserve carbon to fully satisfy the requirements for leaf and branch expansion. Thus, insufficient reserves of carbon could potentially cause prematurely shortened branches, dense leaf clusters, and decreased leaf area. These conditions increase self-shading within the plant canopy and reduce the amount of photosynthate that can be fixed when conditions improve.

Consequently, we hypothesize that a tree's capability to withstand stress is tightly correlated with the amount of carbon reserves within the plant. It is reasonable to assume that the response systems within trees depend on mobilizing carbon to rebuild injured tissue or to increase the resource-collecting area. In plants with small carbon reserves, changes in carbon allocation can occur; but decreases in rates of carbon allocation to certain tissues in preference to others are likely to increase the sensitivity of the plant to other stresses by creating imbalances in tissue mass. As a consequence, it is important to understand the state of the carbon balance in the trees.

MODELING THE CARBON RESERVE SYSTEM

The tree system must be modeled as an integrated entity if we are to understand the linkage between the initial cause of stress, the demands upon the carbon reserves, and the eventual response of the tree. We have constructed a computer model, Response of Plants to Interacting Stresses (ROPIS), to simulate the response of red spruce (*Picea rubra Sarg.*) trees to ozone and acidic precipitation stress in conjunction with a field experiment in which 15-year-old trees are being exposed to combinations of these pollutants. Through the model, the flux of carbon, water, and nutrients through the plant-soil system can be monitored (Laurence et al., 1988). The model is being used to evaluate the long-term effects of pollutants upon resource availability, the ability of plants to utilize resources to fix carbon, the allocation patterns used by the plant to maintain carbon fixation, and the ability of the plant to maintain and repair pollution damage.

The model calculates the photosynthesis of an entire red spruce tree each hour as a function of the availability of light, water, and nutrients and in response to ambient environmental conditions. On a daily basis, the model estimates the quantity of essential nutrients and water available in each of three soil horizons, and the amount of these resources taken up by the tree. The chemical interactions of leaves with precipitation passing through the canopy and with ozone exposure are simulated. The interactions of the soil matrix and solution with percolation water and the effects of these interactions upon root growth and functioning are simulated. Due to space limitations, we will describe in detail only the carbon allocation methodology. A complete description of the other aspects of the model is forthcoming.

Model Description

The allocation process is perhaps the least understood of all major plant functions (Ågren, 1981). Knowledge concerning carbon partitioning in conifers was derived from the mature-tree budgets of Keyes and Grier (1981) and Cannel (1985) and from studies with seedlings performed by Cannell and Willett (1976), Drew (1982), and Reid et al. (1983). The approach used here represents a hybrid between the fixed ratio of allocation to needles, branches, and roots used by Mohren et al. (1984) and a method in which flow to a component is controlled by the magnitude of the sink (Dixon et al., 1978). In the present formulation, allocation tends to follow a fixed ratio pattern, but it deviates from this pattern when essential resources become limiting. In this respect, the method is similar to the control of partitioning by water deficit, as proposed in several theoretical models (Borchert, 1973; Schulze et al., 1983), and control by the ratio of carbon to nitrogen (Reynolds and Thornley, 1982). By combining various aspects of these ideas, the model can consider the influences of sources, sinks, and flux resistances, all of which may be important in determining allocation (Gifford and Evans, 1981).

The model calculates the allocation of newly synthesized carbohydrate and sugars remobilized from starch reserves to various tree compartments. Dynamics of pools are calculated for each of four needle age classes, branches, stem, coarse roots, and fine roots. Allocation to coarse- and fine-root masses in each of three soil horizons is considered. The carbohydrate within each compartment is divided into pools of sugar (for the purposes of the model, containing all non-structural carbohydrate substances other than starch), starch, structural compounds, and wood (where appropriate).

The pattern of allocation varies with the period of the growing season (Figure 3) and is drawn from data being collected in a field experiment with 15-year-old red spruce saplings (Laurence et al., 1988). The timing of transition from one growth period into the next is determined by the accumulation of growing degree hours, an indicator of the duration of conditions permitting growth. Cool weather will delay the phenological development. Dormancy exists from late October or early November until early March. The only operation that occurs during dormancy is a minor amount of respiration. The pre-growth period runs from March through mid-April. During this period, any photosynthate fixed is stored in branches, stem, and/or coarse roots. Needle starch is used for needle demand and never exported. Root growth and bud formation (growth period 1 in Figure 3) occurs between

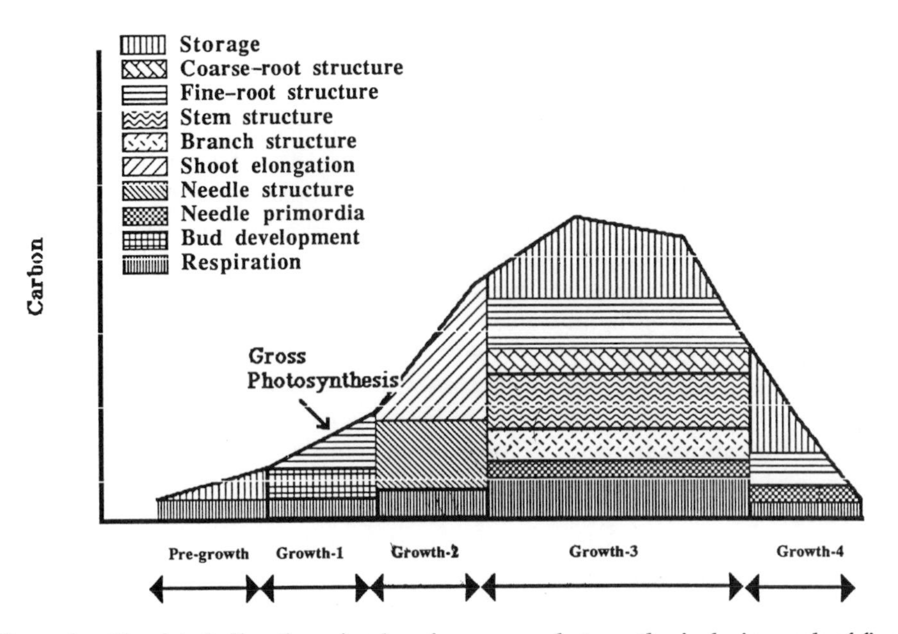

Figure 3. **Simulated allocation of carbon from gross photosynthesis during each of five growth periods. Note that respiration always receives the first priority. Within each growth period, indicated on the x axis, priority of allocation is given first to the component closest to the x axis, then to the next closest, and so on, until all available carbon is utilized. The approximate growth periods include:**

P-G: March to mid-April
G-1: mid-April to mid-May
G-2: mid-May to July
G-3: July to late September
G-4: late September to early November

mid-April and mid-May. Growth (swelling) of this year's buds occurs provided there is adequate photosynthate and nutrient material. Growth of fine roots also occurs, using photosynthate not used in bud swelling. At this point, any excess photosynthate is stored as starch in coarse roots.

Needle elongation (growth period 2 in Figure 3) begins in mid-May and lasts until early July. When this growth is initiated, the carbon stored in the bud compartment is transferred to a new needle compartment. Needles are approximately 25% of their final size at this point; they grow to their full length following an exponentially decreasing growth-rate curve. Carbon for this growth comes from current photosynthate, and starch is drawn first from branches, then from stem, then from coarse roots. During this period, root growth is suspended. Such suspensions in root growth in the middle of the growing season have been observed by Harris et al. (1973). Shoot elongation then begins, using carbon in excess of that needed for needle expansion.

Allocation to all other structural compartments, primordia formation, wood formation, and starch buildup (growth period 3 in Figure 3) occurs from July to late September. It is initiated when starch reserves fall to some minimum or when branch expansion is completed. During this period, carbon is preferentially allocated to leaf primordia formation, creating the primordia for next year's leaves at a rate governed by the daily temperature and the rate of nutrient flux. The remaining carbon is allocated, in order of priority, to branch structure, stem structure, coarse-root structure, and starch formation. Presently, these priorities are hypothetical, based in part upon patterns reported by Waring and Schlesinger (1985). Needle starch is restored to maximum using photosynthate generated from within the same leaf class only. Starch is stored in branches, stem, and coarse roots based upon greatest demand (the pool with the greatest difference between maximum potential storage and present level of storage). Finally, excess carbon not used in the above distributions is allocated to belowground starch reserves and to fine-root growth. This allocation is larger if nutrient limitation exists. Excess carbon not needed for fine-root growth is converted to root starch. Structural carbon is converted to wood in stem, branches, and coarse roots during this period.

Needle loss, root growth, and primordia development (growth period 4 in Figure 3) occur between late September and late October or early November. If photosynthate continues to be in excess of respiration needs, primordia formation will continue during this period, as will fine-root growth and coarse-root starch storage. Allocation to branch and stem structural growth is stopped, as is conversion of structural carbon to wood.

Certain activities can occur only during certain time periods. For example, in growth period 2, carbon is used first to satisfy respiration demands, second for needle elongation, and finally for shoot elongation. After respiration, both the remaining processes occur at a rate determined by the availability of carbon, water, and nutrient resources and the demand for the services provided by that plant part. Root growth does not occur during this period. Even during an appropriate time period, tissue growth can be prematurely terminated by poor environmental conditions or by insufficient carbon storage. Initially, carbon will be delivered to the highest priority item at the fastest rate possible; then the remaining carbon will be delivered to lower priority sites at the fastest rate possible.

Figure 4 illustrates the way in which growth of multiple components can proceed concurrently and the way in which priorities can shift, using the allocation to branches and

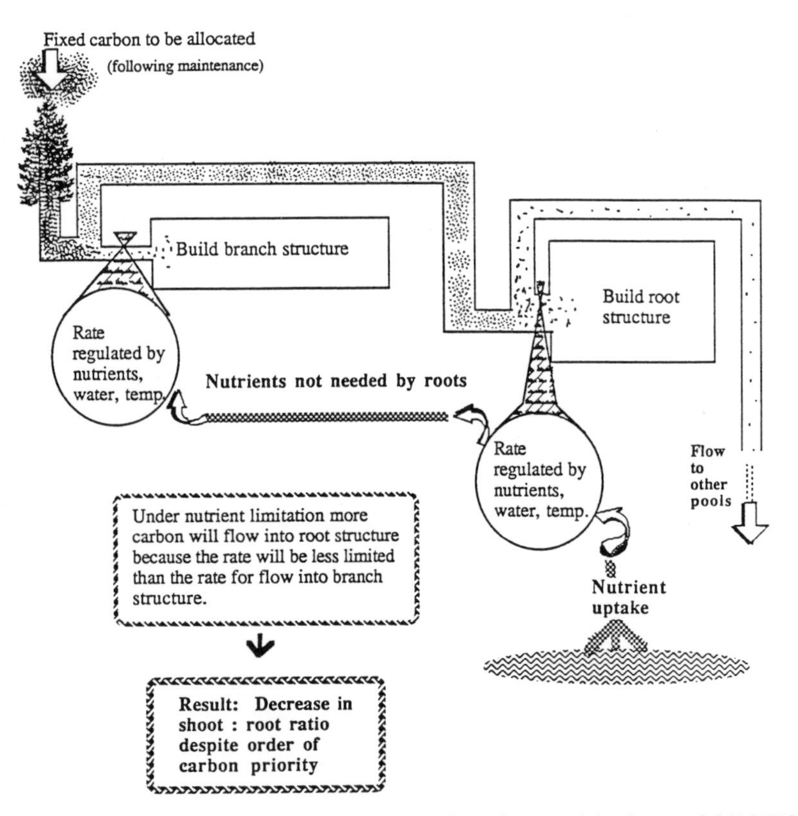

Figure 4. Schematic example of the mechanism of allocation used in the model ROPIS. Allocation to branch and root structure during growth period 3 is shown.

to roots as an example. During growth period 3, carbon traveling from the leaves is first used to build branch structure. The rate at which carbon can be utilized by branches is schematically depicted here as the width of the entrance into the branch box, a width that is determined by environmental conditions. Next, carbon in excess of what can be used by branches is utilized to build root structure. Typically, sufficient carbon is fixed by leaves to satisfy the growth needs of both components. However, when nutrients are in short supply, the rate at which branch structure can be built is limited. Roots, on the other hand, can utilize nutrients as rapidly as they can be absorbed. The proportion of absorbed nutrients used by roots is limited by the demand to make new root tissue. Consequently, under nutrient-limited conditions, the rate of root growth is less limited then the rate of branch growth. Although branches continue to have the first opportunity to utilize available carbon, their rate of utilization is extremely restricted. Most carbon will flow past the branches and on to the roots. The rate at which carbon is received and utilized in root growth will exceed the rate with which it goes to branch growth, causing the shoot:root ratio to decrease. Alternatively, if nutrients are in abundant supply, branch growth demand for carbon will be satisfied first, causing the shoot:root ratio to increase.

The regression equations from the tree measurement studies of ROPIS are being used for maximum rate parameters. Where these numbers have yet to be generated, the regression equations of Czapowskyj et al. (1985) were used. Currently, field data for root mass in two horizons are used. Stem mass was estimated from stem volumes, derived from diameter measurements. Branch mass was estimated from needle mass, using Czapowskyj et al. (1985). For all structural tissue, respiration is assumed to be 1% of the dry tissue weight lost per day. This assumption results in up to 40% of fixed carbon going to respiration in mid-summer, which may be high. The rates of use of assimilates by the root system were estimated from Bowen (1985), Persson (1983), and Keyes and Grier (1981). Of the list of parameters required for operating the model, most are easily measured (and are being monitored in our field experiment) or are obtainable from the literature, including rates of photosynthesis, rates of tissue growth, and periods of phenological development. The exceptions to this rule are the estimates of respiration for most tissues, which are not fully understood at present and which can greatly influence the estimates of carbon balance.

SIMULATIONS OF CARBON POOLS

Figure 5 shows the results from a simulation of carbon allocation to the growth of the principal structural components of a single young red spruce tree during an entire growing season. For this simulation, meteorological data for 1987 were used, and tree component sizes were initialized for the start of the year using data collected from an average-sized tree approximately 15 years old. Structural growth appeared first in the manufacture of new needles. A significant amount of carbon utilized for building structural material was allocated to new needle production. Increases in branch structure and root structure followed the termination of needle creation. Minimal expansion of the mass of coarse roots and stem occurred throughout the latter part of the season.

The expansion of new needles and branches occurred at a critical time for the plant. The nature of this critical period was demonstrated by the dynamics of the carbon storage pool in the most critical aboveground carbon reserve component, the branches (Figure 6). In the curve representing the behavior of branch starch in the tree unexposed to pollution, maintenance costs drove pools of starch gradually downward during the period before photosynthesis became very active. Photosynthate fixed during the early spring was stored in branch tissue as starch, causing a rapid rise in the starch levels of these tissues. Beginning at approximately day 100, fine-root growth began, utilizing newly fixed

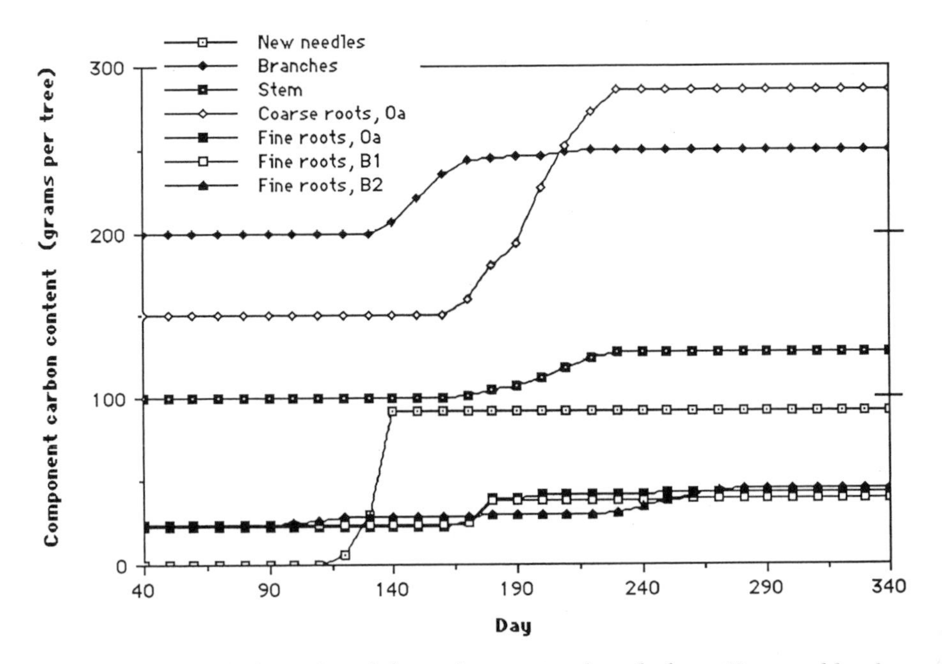

Figure 5. Simulated dynamics of the major structural pools for a 15-year-old red spruce tree during an entire growing season. Note that earliest, most rapid, and largest changes occur in the growth of coarse roots and the flushing of new needles.

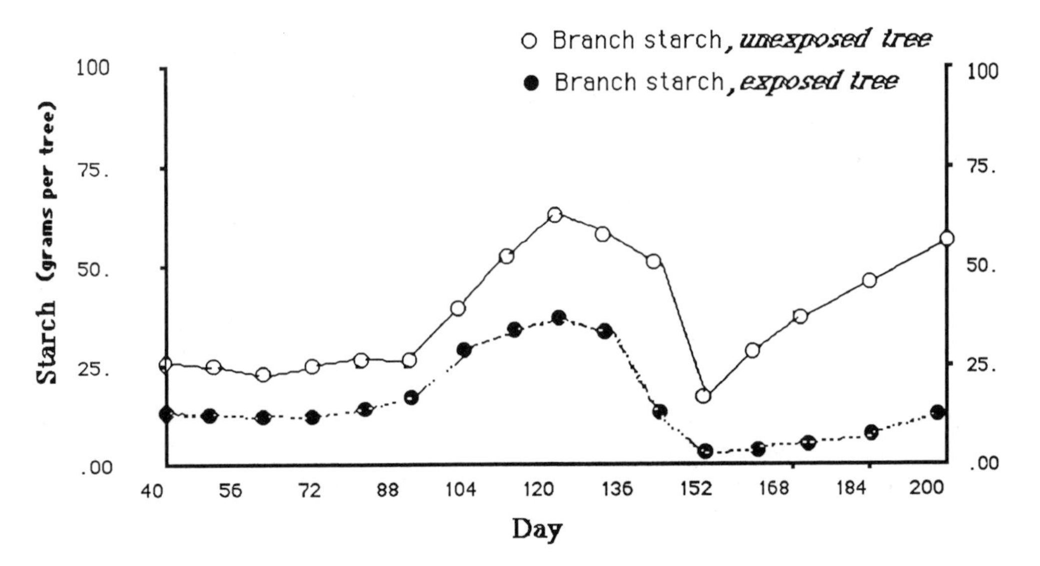

Figure 6. Simulated dynamics of branch starch pool from day 40 to day 200. The top curve shows behavior when no pollution is included in the model. The bottom curve shows behavior when an exposure to 80 ppbv ozone occurs from day 72 to day 86.

photosynthate as well as starch reserves in coarse-root tissue.

Initiation of growth of new needles occurred at approximately day 140, and the rapid growth of these leaves quickly depleted the reserves of starch. Shoot expansion further depleted carbon reserves. In order to achieve full expansion of needles and shoots, it was critical that sufficient starch reserves were available immediately preceding this growth period. The utilization of carbon reserves for needle and branch expansion reduced branch starch from 60 g per tree to a low of 18 g, a drop of 42 g. Once needle expansion was completed, however, starch reserves built back up again. During this mid- to late-summer growth period, photosynthate fixation was ample to foster sizable growth in fine roots, small additional increments in branch structure, and significant increases in stem and coarse roots in addition to providing excess to be stored as starch.

An identical simulation was conducted assuming that the tree was exposed to 80 ppbv ozone during the two-week period immediately preceding budbreak and leaf expansion. The ozone exposure was assumed to reduce photosynthesis by 20%. In this simulation, branch starch in the exposed tree never accumulated to the levels of the unexposed tree. In fact, branch starch levels at the time of budbreak were approximately half those for the unexposed tree. There were insufficient quantities of carbon stored in branches to meet the needs of needle and branch expansion. Starch in the branches was reduced to minimal levels. As a result, needles did not completely expand, and new shoots were shorter than normal. This incomplete expansion left the tree with shorter, denser needle clusters, reducing the exposure of each needle to full sunlight and making it more difficult for the tree to maintain a high rate of carbon fixation during the growing season.

CONCLUSIONS

The ultimate effects of gaseous pollutants upon trees are likely to be mediated through alterations in the carbon reserve and distribution system. The decreases in the amount of carbon fixed are likely to be small, but they can accumulate into significant reductions over long periods. The model described here permits us to evaluate the degree of alteration of carbon balance to be expected under conditions of elevated ozone exposure. Further, the ramifications of short-term events upon long-term plant processes are extremely difficult to understand in an experimental setting without the aid of a model of carbon balance. Through continued use of this model to complement our experimental exposures, we will have a much more powerful tool with which to evaluate the complex interactions of plant resource balances and pollution exposures.

The model serves as conceptual framework in which the implications of each hypothesized response within the plant's entire physiological system, as well as the plant's ecological capabilities, can be evaluated. It provides a means of addressing the plant as a balanced system in which all gains and losses of basic resources (carbon, water, and nutrients) for all plant components must be in line with one another. In this sense, the model can be used as a convenient tool for bookkeeping mass balances of plant component pools and material exchanges. Through these mass balance calculations, the additive effects of multiple injuries to a tree can be evaluated. The model will allow us to project the implications of the behavior observed during the experiment beyond the sets of conditions explicitly tested, thus permitting evaluation of the potential for gradual repair of tissues damaged by pollution.

LITERATURE CITED

Ågren, G. I. 1981. Problems involved in modelling tree growth. In: Linder, S., ed. *Understanding and Predicting Tree Growth. Studia Forestalia Suecica* 160:7–18.

Borchert, R. 1973. Simulation of rhythmic tree growth under constant conditions. *Physiologia Plantarum* 29:173–180.

Bowen, G. D. 1985. Roots as a component of tree productivity. In: Cannell, M. G. R., and J. E. Jackson, eds. *Attributes of Trees As Crop Plants.* Institute of Terrestrial Ecology, Huntingdon, England, U.K.

Cannell, M. G. R. 1985. Dry matter partitioning in tree crops. In: Cannell, M. G. R., and J. E. Jackson, eds. *Attributes of Trees As Crop Plants,* Institute of Terrestrial Ecology, Huntingdon, England, U.K. Pp. 160–193.

Cannell, M. G. R., and S. C. Willett. 1976. Shoot growth phenology, dry matter distribution and root:shoot ratios of provenances of *Populus trichocarpa, Picea sitchensis,* and *Pinus contorta* growing in Scotland. *Silvae Genetica* 25:49–59.

Czapowskyj, M. M., D. J. Robison, R. D. Briggs, and E. H. White. 1985. *Component Biomass Equations for Black Spruce in Maine.* Research Paper NE-564, USDA Forest Service Northeastern Forest Experiment Station, Broomall, PA, U.S.A.

Dixon, K. R., R. J. Luxmoore, and C. L. Begovich. 1978. CERES: model of forest stand biomass dynamics for predicting trace contaminant, nutrient, and water effects, Vol. I. Model description. *Ecological Modelling* 5:17–38.

Drew, A. P. 1982. Shoot-root plasticity and episodic growth in red pine seedlings. *Annals of Botany* 49:347–357.

Gifford, R. M., and L. T. Evans. 1981. Photosynthesis, carbon partitioning, and yield. *Annual Review of Plant Physiology* 32:485–509.

Harris, W. F., R. S. Kinerson, and N. T. Edwards. 1973. A comparison of belowground biomass in deciduous forests and loblolly pine ecosystems. In: Marshall, N. K., ed. *The Belowground Ecosystem.* Range Publication No. 26. Colorado State University, Ft. Collins, CO, U.S.A.

Holl, W. 1985. Seasonal fluctuation of reserve materials in the trunkwood of spruce (*Picea abies* [L.] Karst.). *Journal of Plant Physiology* 117:335–362.

Johnson, A. H., and T. G. Siccama. 1982. Acid deposition and forest decline. *Evironmental Science and Technology* 17:294–305.

Keyes, M. R., and C. C. Grier. 1981. Above- and below-ground net production in 40-year-old Douglas-fir stands on low and high productivity sites. *Canadian Journal of Forest Research* 11:599–605.

Laurence, J. A., R. J. Kohut, R. G. Amundson, T. J. Fahey, and D. A. Weinstein. 1988. *Response of Plants to Interacting Stresses.* Annual Report to the Electric Power Research Institute, RP2799-1. Boyce Thompson Institute, Ithaca, NY, U.S.A.

Loach, K., and C. H. Little. 1973. Production, storage, and use of photosynthate during shoot elongation in balsam fir (*Abies balsamea*). *Canadian Journal of Botany* 51:1161–1168.

McLaughlin, S. B. 1985. Effects of air pollution on forests: a critical review. *Journal of the Air Pollution Control Association* 35(5):512–534.

Mohren, G. M. J., C. P. Van Gerwen, and C. J. T. Spitters. 1984. Simulation of primary production in even-aged stands of Douglas fir. *Forest Ecology and Management* 9:27–49.

Oshima, R. J., P. K. Braegelmann, R. B. Flagler, and R. R. Teso. 1979. The effects of ozone on the growth, yield, and partitioning of dry matter in cotton. *Journal of Environmental Quality* 8:474–479.

Pell, E. J. 1987. Ozone toxicity: is there more than one mechanism of action? In: Hutchinson, T. C., and Meema, K. M., eds. *Effects of Atmospheric Pollutants on Forests, Wetlands, and Agricultural Ecosystems. NATO* ASI *Series,* Vol. G15. Pp. 229–240.

Persson, H. A. 1983. The distribution and productivity of fine roots in boreal forests. *Plant and Soil* 71:87–101.

Puckett, L. J. 1982. Acid rain, air pollution, and tree growth in southeastern New York. *Journal of Environmental Quality* 11:376–380.

Reich, P. B., and R. G. Amundson. 1985. Ambient levels of ozone reduce new photosynthesis in tree and crop species. *Science* 230:566–570.

Reid, C. P. P., F. A. Kidd, and S. A. Ekwebelam. 1983. Nitrogen nutrition, photosynthesis and carbon allocation in ectomycorrhizal pine. *Plant and Soil* 71:415–432.

Reynolds, J. F., and J. H. M. Thornley. 1982. A shoot:root partitioning model. *Annals of Botany* 49:585–597.

Schulze, E.-D., K. Schilling, and S. Nagarajah. 1983. Carbohydrate partitioning in relation to whole plant production and water use on *Vigna unguilata* (L.) Walp. *Oecologia* 58:169–177.

Waring, R. H., W. H. Schlesinger. 1985. *Forest Ecosystems: Concepts and Managements.* Academic Press, New York, NY, U.S.A. 350 pp.

CHAPTER 29

MODELING THE EFFECTS OF POLLUTANTS ON THE PROCESSES OF TREE GROWTH

E. David Ford and A. Ross Kiester

Abstract. Policy-relevant models of the effects of pollutants on tree and forest growth need to achieve two goals. First, they must encapsulate our understanding of the processes by which trees grow and by which pollutants act. Second, they must provide the ability to make plausible projections of the response of trees and forests to alternative pollution scenarios in order to assess the potential effects of regulatory action. Here we present the background considerations and an account of work in progress for a modeling system of tree growth designed to meet these needs. The modeling system consists of a series of related models, each of which can be studied alone or in concert as they form a whole tree. These individual models are characterized by consideration of both physiology and morphology. Taken together as a Simple Whole Tree (SWT) model, they can account for synergies and compensations at the level of the whole tree as it responds to pollutant effects. This approach allows the relative evaluation of different hypothesized mechanisms of action related to pollution.

INTRODUCTION

We describe scientific issues relevant to modeling the effects of pollutants on tree and forest growth at the mechanistic level and how they can be resolved within the context of a large national research program. We use models as vehicles for achieving an integration of knowledge and for making predictions about likely affects under a range of different pollution scenarios. Results must be synthesized from a variety of experiments and investigations. How effective such a synthesis can be depends not only on the extent of our understanding of pollutant influence on component tree growth processes, but particularly upon our quantitative understanding of tree growth as an integrated process. As we research pollutant effects, we increasingly ask new questions about tree physiology and forest processes, and it is these questions that dominate our uncertainty about pollutant influences.

Dr. Ford is Director at the Center for Quantitative Science, University of Washington, Seattle, WA, 98195 U.S.A. Dr. Kiester is Project Leader of the Synthesis and Integration Project, Forest Response Program, Environmental Protection Agency, Corvallis, OR, 97333 U.S.A. The research is sponsored by the Synthesis and Integration Project of the Joint U.S. Environmental Protection Agency USDA Forest Service Response Program, a part of the National Acid Precipitation Assessment Program. The research reported in this chapter has been funded (in part) by the U.S. Environmental Protection Agency under the cooperative agreement CR814640 to the University of Washington. It has been subjected to Agency review and approved for publication. The authors are grateful to Ms. Anne Avery, Ms. Susan Bassow, Mr. Rupert Ford, and Ms. Jackie Haskins for permission to refer to as yet unpublished work.

THE PURPOSE AND CHALLENGE OF MODELING TREE RESPONSE TO POLLUTANTS

The National Acid Precipitation Assessment Program (NAPAP, 1988) seeks to evaluate the complex of evidence describing the effect of acid precipitation on trees and forests and also to provide answers to the multiple questions that policy-makers ask. These include considerations of different levels of acid precipitation, both more and less than currently measured; changes in the balance of chemical elements contributing to the deposition; and the effects of regional variation in control or mitigation.

The Forest Response Program (FRP) is the component program of NAPAP currently investigating forests (Schroeder and Kiester, 1989). Much of its work is to survey the potential damage. A parallel program of experimental work and analysis of forest processes is investigating possible mechanisms by which acid precipitation may affect tree and forest growth. That work is organized to investigate particular theories of potential influence—for instance, that acid precipitation causes changes in soil properties that influence root growth or function, or that acid precipitation increases nutrient loss from foliage.

The Need for Process-Based Models

These requirements—prediction for policy purposes and synthesizing results of very different types of investigation—dictate two features of the modeling activity being developed by the Synthesis and Integration Team of the FRP. First, the requirement for prediction beyond current conditions demands that models simulate mechanisms of interaction between pollutant influence and growth. We cannot depend solely upon statistical models produced from the results of field investigations made under current conditions because such investigations do not allow us to predict what may happen under different deposition patterns. Nor can we depend solely upon experiments conducted with seedlings and necessarily with restricted sets of treatments; these cannot supply sufficient information about the response of mature trees and forests. In our view, the most important requirement is that model structure should be functionally correct in describing the mechanisms of response, particularly for mature trees and forests. Scientifically, this is the most demanding aspect of our modeling task. Theories of pollution influence and of the processes that control the growth of large trees are developing continuously, and this has important implications for the way in which we develop and use our models.

Second, models must be constructed to assess the relative importance of different mechanisms, singly and in combination, perhaps with different emphasis in different regions and under different conditions of pollutant load. Models are quantitative representations of theory, and in different geographical regions, theories of pollution influence with different emphases are being developed. This means that no single, global model will be adequate for the purpose of the FRP. We can expect to construct and use different models, not just one model with different parameters.

Synergistic Effects and Compensating Mechanisms in the Response to Pollutants

To extrapolate from experiments under controlled conditions, we must be particularly careful to estimate the effects of synergisms and compensating mechanisms. Synergisms occur where the effect of acidic precipitation on two or more processes is greater than can be estimated from studies of the individual processes themselves. Compensating processes may reduce the impact of damage that may be estimated from experiments on single processes. However, compensating and synergistic mechanisms are not immediately apparent; thus we must develop theories about them that can be incorporated into a more complete theory of pollution influence. Typically, synergisms and compensations operate through mechanisms at higher levels of organization than can usually be investigated through direct experiment, as, for example, through the organizational structure of the whole tree or the forest stand.

It is conceivable, for instance, that a synergistic effect could occur through the effects of acid precipitation on soil and foliar processes—that, taken together, a small increase in foliar leaching and a small increase in soil leaching could, over time, have a marked effect upon growth. Hari et al. (1987) proposed a complex sequence for the influence of acid precipitation on Finnish forests where there is first an increased nutrient mobilization in the soil and a concomitant increase in tree growth, but where, under sustained acid precipitation, nutrient leaching reduces nutrient levels in the whole ecosystem in such a way as to produce a rapid and dramatic decrease in growth. To examine a theory of synergistic effects between root-soil and foliar responses, we may need to simulate the rates at which increased nutrient uptake or mobilization within the tree can compensate for nutrient loss from foliage.

A possible compensating mechanism in the response of forest stands to acid precipitation might involve the interaction between a genetically based differential in the response of individual trees to acidic deposition and the competition process that also results in differential growth rates. A theory could be proposed that while acidic deposition might reduce the growth of some individuals within a stand, this would in effect mitigate the competition process; as a result, the reduction in growth for the whole stand would be less than might have been estimated through direct experimentation related to the primary effect of acid precipitation. To assess this, it would be important to simulate the spatial and temporal characteristics of differential growth rates between trees as due to the two processes, that is, genetically determined sensitivity to acidic precipitation, and competition (Webb and Burkhart, 1988).

Implications for Model Development, Use, and Structure

We emphasize that process-based models should be treated as developing theory. We suggest that they should be modular in construction, that the modules should be documented as components of a theory, and that the modules should be replaceable. In particular, because investigation of compensating and synergistic effects seems likely to require the use of models, we need to pay attention to tree and forest properties of a higher order than we might expect to investigate experimentally. Consequently, in our simulation approach, we are developing a structure that focuses on how component physiological processes are connected in time and space.

This leads us to formulate models of tree and forest growth that use tree morphology and phenology as concepts for organizing physiological processes. This is a new development in whole-tree models that challenges some of the concepts most frequently used in the past. (See Bassow et al. (this volume) for an examination of two such concepts, i.e., allocation and maximum foliage amount.)

DEVELOPMENT OF A THEORY FOR TREE AND FOREST GROWTH

Our starting point for a theory of individual tree growth is similar to that for indeterminate plant growth described by Thornley (1972) (Figure 1). He described growth as the result of feedback control between root growth, with its function of nutrient uptake, and foliage growth, with its function of photosynthetic production. He modeled growth as a balance between the rates of nitrogen (N) uptake and transfer to foliage, and production of

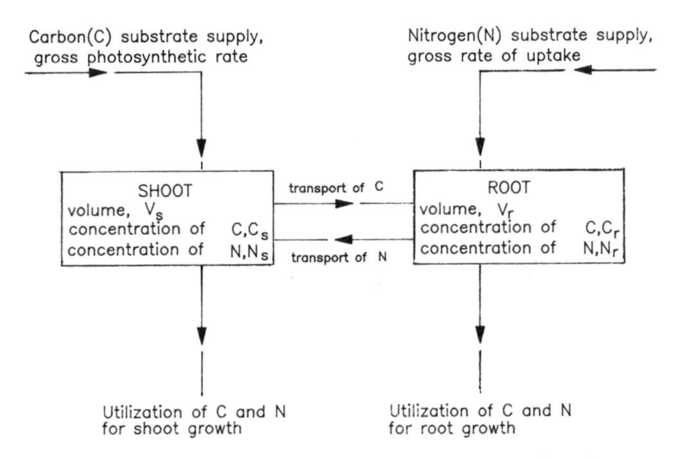

Figure 1. Schematic representation of the components in a model for plant growth (Thornley, 1972). Growth rate is the result of a dynamic interaction between photosynthesis and nitrogen uptake.

photosynthate (C) and transfer to roots. The rates of uptake were determined by the respective sizes of roots, and foliage, and a controlling point in this model was the maintenance of constant internal soluble carbon and nitrogen concentrations.

The value of Thornley's approach is that the equations for growth, particularly the relative growth rates of roots and foliage, are fully specified in terms of the growth processes themselves. The model does not require terms for the relative allocation of either photosynthate or nitrogen to different plant parts, terms which can only be obtained through empirical studies of weight increment itself. Thus the model can be used as a predictive tool, the value of which rests upon the theory that internal concentrations of nitrogen and soluble carbon remain constant as growth takes place, and upon the accuracy with which it represents the processes of photosynthesis, nutrient uptake, translocation, and conversion of nitrogen and carbon into dry weight.

One of the six suggested mechanisms of action of acidic precipitation on tree growth is an influence on allocation (Schroeder and Kiester, 1989). While this may be measured in dose-response experiments with seedlings, accurate, direct measurement for large trees and forests, particularly under different conditions of acid precipitation, is a task that is beyond our capabilities and that is not being attempted within the FRP (Ford and Bassow, in press). In our assessments, we will be dependent upon model-based predictions. Clearly, Thornley's (1972) approach has great utility for this purpose if it can be modified for application to tree growth. In particular, we propose that the following must be considered.

1. In trees, unlike the indeterminate plants which Thornley (1972) modeled, the onset, cessation, and some aspects of the rate of meristem growth are controlled through phenological processes.
2. The structure of the tree may determine the rates of growth processes (see the discussion below). Tree structure determines rates of transport of water, nutrients, and photosynthate between roots and foliage, and also the amounts of those quantities that may be stored and available for the growth process (McMurtrie and Wolf, 1983).
3. We cannot assume, as did Thornley (1972), that relative growth of roots and foliage is set by a mechanism within the plant that maintains a balance in internal nutrient and carbohydrate concentration. Ingestad (1982) has proposed a theory, based on experiments with tree seedlings, that growth is a function of internal nutrient

concentrations and has shown that such concentrations may change as growth proceeds. Further, we know that substantial storage and re-translocation of both carbohydrate and nutrients occur within perennials and that we must account for these processes in a model-based approach.

The first two propositions above require that we introduce concepts of form and timing into our theory. The third recognizes that the experimental evidence for a mechanism that maintains constant internal concentrations does not extend in its simplest form to tree species, and that within trees there are substantial patterns of storage and re-distribution of both carbohydrate and nutrients, patterns which also require that we understand the spatial distribution of different tissues.

The Mechanism of Tree Growth

The theory of tree growth that we are using (Fig. 2) is developed from the general theory described by Thornley (1972). This represents a Simple Whole Tree (SWT) model where each compartment has a structure which can be expressed in terms of mass and morphology and which is the result of carbon gain. The structure of each compartment influences the processes of carbon, water, and nutrient uptake and/or movement and so defines the functional properties that determine the amount of further growth that will be made. The interactions among the three compartments are determined by the rates and absolute mass of transfer among them of carbon, water, and nutrients.

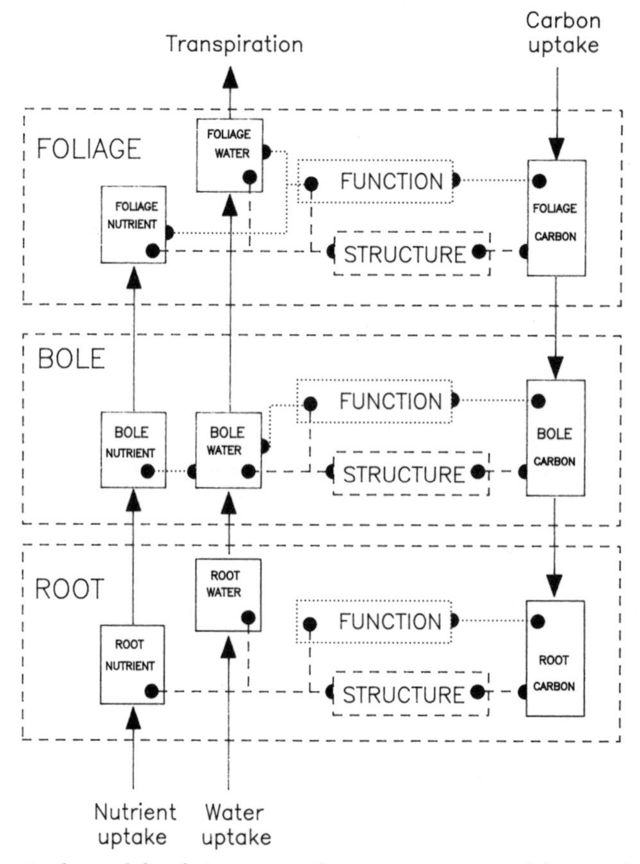

Figure 2. Conceptual model of tree growth processes comprising carbon gain through photosynthesis and nutrient uptake. The rates of both these processes and that of water and nutrient uptake and translocation are influenced by the morphology of the tree.

This is not an appropriate venue for a detailed review of the literature that has been required to formulate propositions for the three compartments. What follows is a summary of work in progress for two aspects of SWT: the distribution of foliage on a branched structure, and the translocation of photosynthate. These examples are selected to describe the aspects of SWT that differ from other modeling approaches to whole-tree growth.

The Distribution of Foliage on a Branched Structure

Our starting point is the model BRANCH (Ford and Ford, 1988) for branch growth (Figure 3), a model that simulates the following interactions:

1. *Phenology*, which is determined by a set of hypotheses defining the timing of different growth processes and in which shoot extension plays a major role.
2. *Morphology*, which is determined by hypothesized processes defining the numbers of branch segments produced from each bud, their orientation, and their absolute and relative lengths.
3. *Foliage development and function*, a set of hypotheses defining the production of foliage on branchlets, its rate of production of biomass and how this changes with time, and the rate of foliage mortality.
4. *Branchwood thickening*. Hypotheses are proposed that define the amount of wood increment at different positions along a branch. The first algorithm used has been to calculate branchwood thickening in relation to the structural requirements to maintain a particular curvilinear relationship of branch posture (McMahon and Kronauer, 1976).

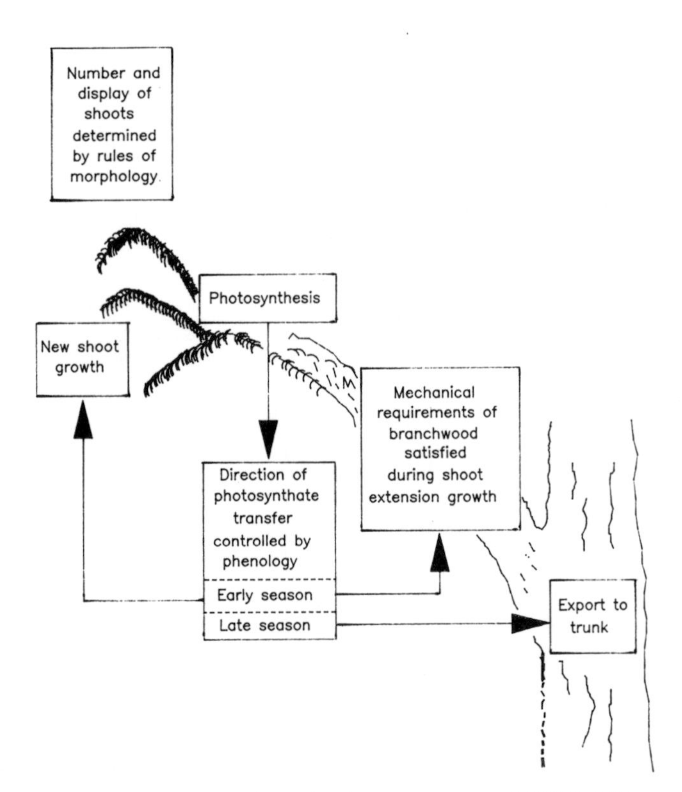

Figure 3. Schematic representation of the basic components of BRANCH, a model for branch growth and the export of photosynthate from the branch to the trunk. The model runs for successive annual cycles.

Export to the main stem is the residual production after shoot extension, foliage growth, and branch thickening have been satisfied under the various hypotheses. Simulation is at the level of the whole branch; the carbon balances of individual branch segments are not considered separately.

Two versions of BRANCH exist. BRANCH 1 (Figure 4) requires a statistical description of the morphological development of the branch that is species-specific. BRANCH 2 (Figure 5) requires that the branching at each growing tip of the branch as it develops follows rules specified by the environment of the tip. The two models are currently being used to explore the structure of the model itself, and particularly to investigate various alternative hypotheses for such functions as the control of wood increment along the branch as it grows.

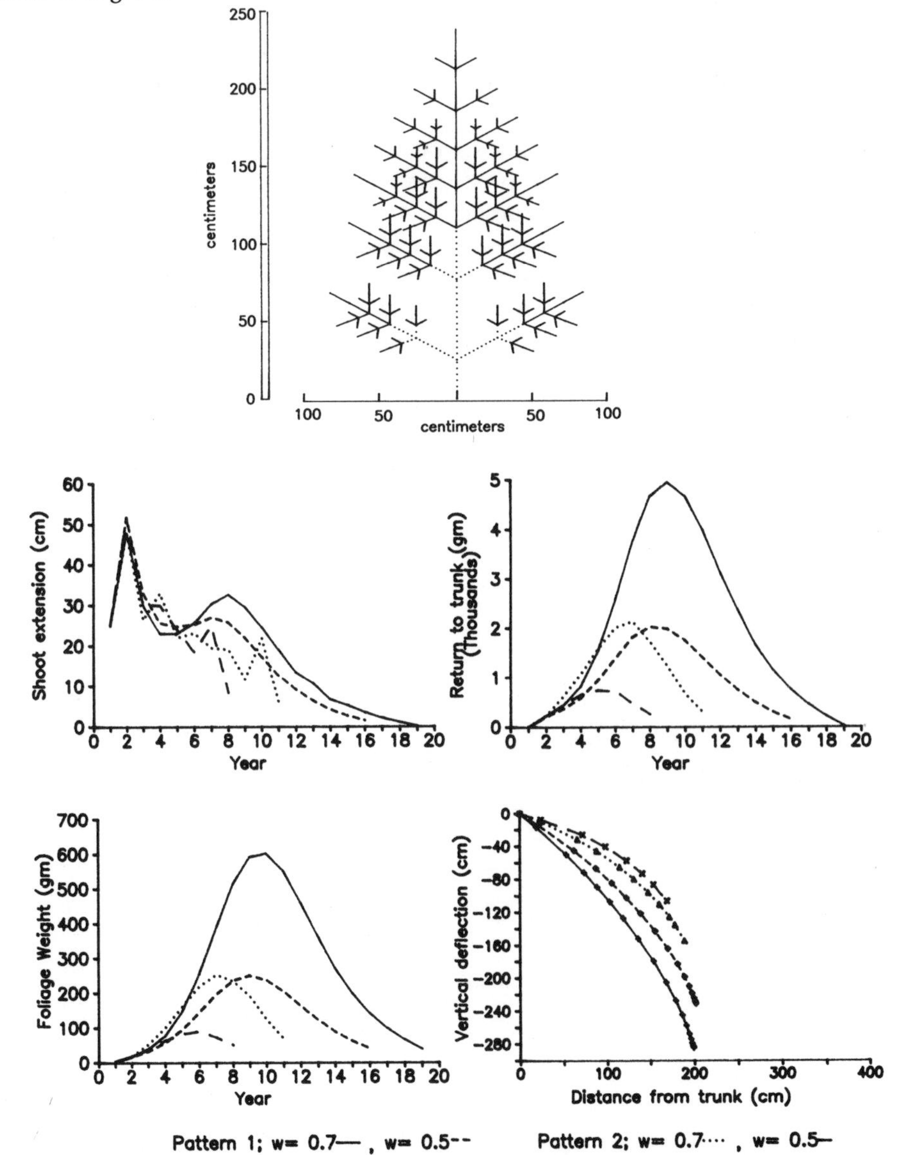

Figure 4. Simulations with BRANCH 1. (*Above*) A two-dimensional representation of branch morphology (Pattern 1, w = 0.7; see text). (*Below*) Outputs from four simulations that illustrate the differences in shoot extension, export to stem, foliage produced, and branch angle—all influenced by differences in branch morphology.

From a simulation with BRANCH 1, the branch drawn in Figure 4 has a structure similar to that typically found in spruce, referred to as Pattern 1, and is drawn after 8 years of simulated growth. A ratio of the lengths of side branches to apices of 0.7 was used in that simulation; the outputs from the simulation, in terms of annual leading shoot extension, foliage weight on the branch, and material exported to the trunk, are drawn in solid lines. The lower righthand graph depicts the angular distribution of the branch at its death. The branch is assumed to die when it makes no more extension of its main shoot axis, and this occurs when all the photosynthate produced is required to maintain the branch structure. When the side branch ratio is reduced to 0.5, and still considering Pattern 1 (dashed line), then, although the extension of the main axis in not reduced, the growth of side branches is reduced, as are the production of foliage and the amount of material exported to the trunk.

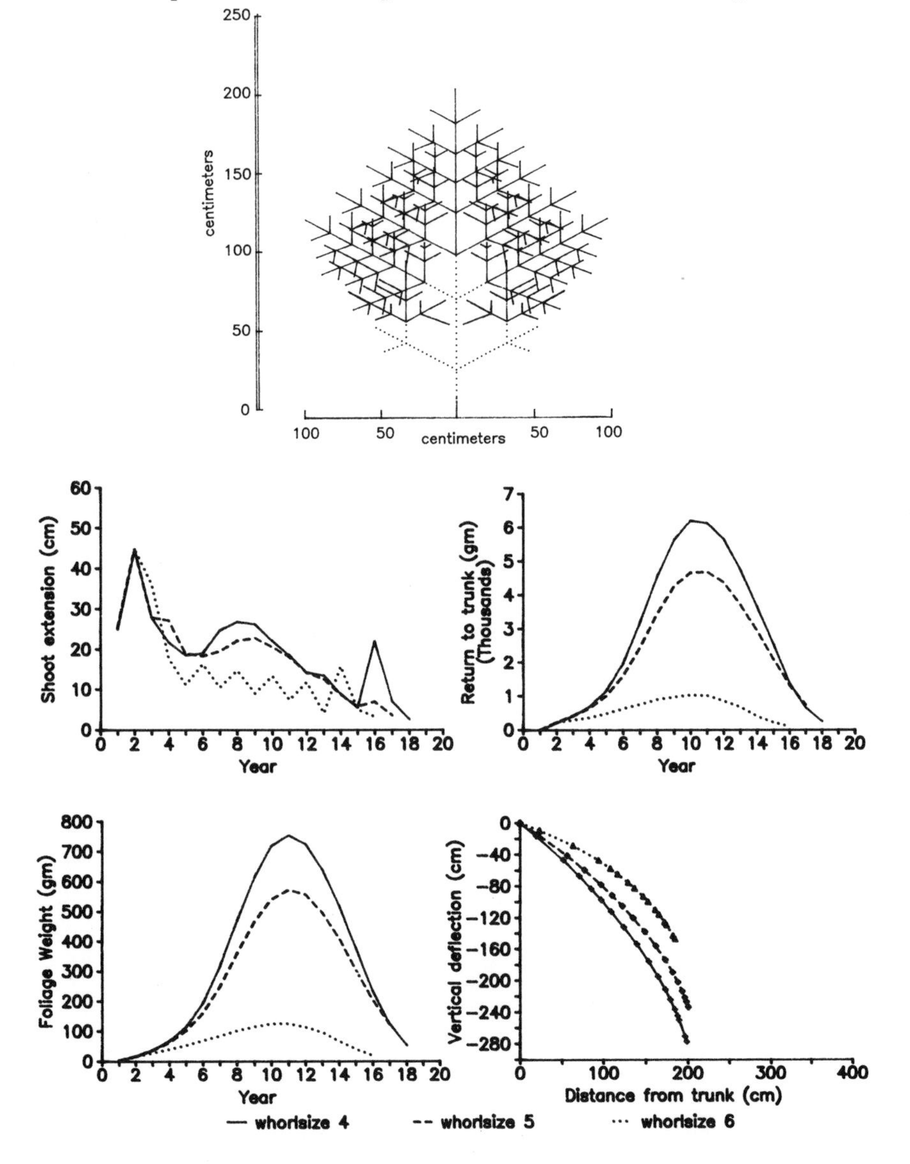

Figure 5. **Simulations with BRANCH 2. (*Above*) A two-dimensional representation of branch morphology, whorlsize 5 (see text). (*Below*) Outputs from four simulations that illustrate the differences in shoot extension, export to stem, foliage produced, and branch angle—all influenced by differences in the number of branches in a whorl.**

Simulations using Pattern 2 (Figure 4) describe a distinct relationship between the extension length of a segment and the number of branchlets arising from it (>40 cm, 7 branchlets; >30, 5; >20, 3; >10, 1; <10, 0). This turns out to be a more stringent limitation of branchlet production and results in earlier accumulation of foliage weight and export of material to the main trunk. A ratio of side branch to main branch length of 0.7 for Pattern 2 is approximately equivalent in total amounts of these quantities to that of a ratio of 0.5 for Pattern 1.

In Figure 5, using BRANCH 2, we explore the consequences of varying the numbers of branches in a whorl. This is possible with this model because the interaction between branchlets is simulated as a form of lateral competition. In the figure, it can be seen that the greatest return to stem is attained from a branch in a whorl of 4. When the cumulative export to the trunk over the full life of the branches and for the whole whorl was considered, then a branch number per whorl of 4 was greatest of those shown and also exceeded that for 3 branches per whorl.

Simulating the Transport of Water and Photosynthate

Two types of observations suggest that water and carbohydrate transport rates may vary according to the size of the tree. First is the observation that in older trees, the duration of periods of low shoot water potential is extended (Waring et al., 1979) over that typically found in younger forest trees (e.g., Milne et al., 1983). This may have implications not only for direct effects of low water potential on such processes as stomatal conductance or cell expansion, but also for nutrient flux to the foliage and carbohydrate flux to the roots. Second is the observation that much of the annual root increment in trees may be constructed from stored carbohydrate reserves (e.g., Deans and Ford, 1986). As trees age, pre–growth period starch concentrations in roots appear to decrease. We need to explore the extent to which this may be a response to an increased transport pathway, a decrease in photosynthetic production relative to requirements for growth and respiration, or some other mechanisms.

The basic theory that we are using to model translocation is Münch pressure flow (e.g., Tyree et al., 1974; Smith et al., 1980). In this theory, water potential gradients have a controlling influence upon translocation (Figure 6) and must be modeled directly.

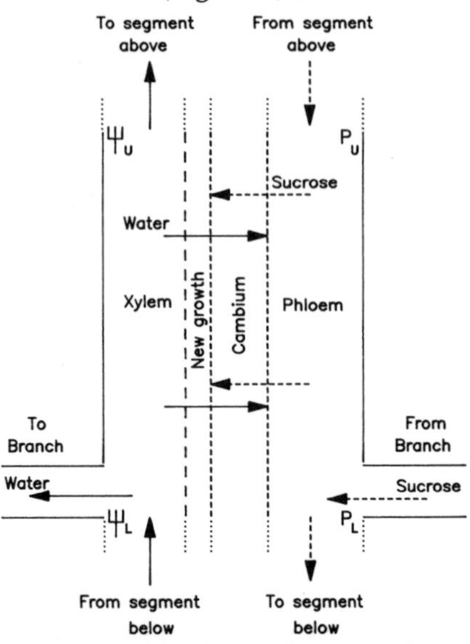

Figure 6. A schematic representation of the flows of water and translocated photosynthate in the region of phloem, cambium, and xylem.

We are developing a water transport model, HYDRA. To date, construction of a process-based model of tree water relations that can predict gradients of water potentials within the tree has been hampered by (1) lack of inclusion of spatial detail in whole-tree morphology and (2) lack of ability to predict resistance to water flow from tissue characteristics. With HYDRA, we are developing in both these directions by exploring the implications of wood structure on both macroscopic and microscopic scales. Haskins and Ford (this volume) examine relevant cellular and whole-tree models and use simulation to explore possible interaction of flow through an individual tracheid and flow through tissues with a variety of tracheid size distributions. Important problems are to achieve a definition of the capacitance of trunks, and how this may vary both with wood structure and amount and in relation to actual water potential gradients.

We are also developing a non-steady state model of Münch pressure flow, TRANS. As a first step, phloem sieve tubes are considered to be long, tubular columns of sieve cells, separated by sieve plates. The sieve plates act as a resistance to flow down the sieve tube, causing successive pressure drops. Figure 7 illustrates the sieve tube and the variables incorporated in TRANS (based on Tyree et al., 1974; and Smith et al., 1980). The variables of the model are given in Table 1. The effects of all the sieve plates are averaged together in the calculation of sieve tube hydraulic conductivity, denoted γ.

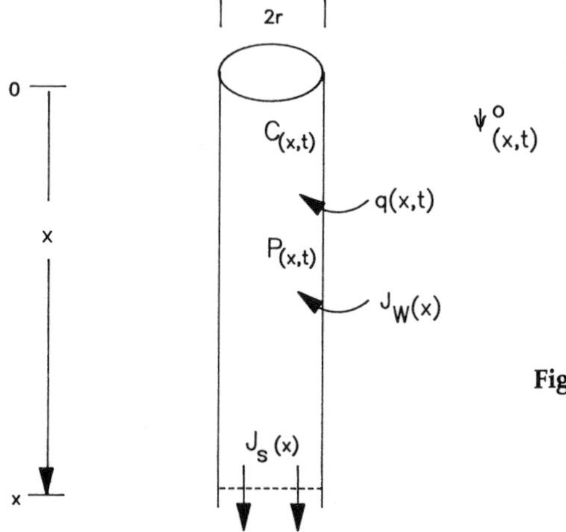

Figure 7. A schematic representation of a cylindrical phloem sieve tube showing flows of material and concentration and pressure variables. See Table 1 for named variables.

Table 1. List of symbols used in the model TRANS (after Smith et al., 1980).

Symbol	Definition	Units
T	Temperature	K
t	Time	s
R	Gas law constant	cm^3 bar/mol K
$\overline{V}$	Partial molar volume of sucrose	cm^3/mol
L	Length of sieve tube	cm
$r(x)$	Radius of sieve tube at x	cm
$A(x)$	Cross-sectional area at x	cm^2
L_p	Hydraulic conductivity of peripheral membrane at x	cm/bars
$\gamma(x)$	Conductivity coefficient at x	cm^2/bars
$\psi_0(x)$	Water potential of apoplast at x	bars
$\alpha_u(x)$	Unloading coefficient	cm^3/s cm
$q(x)$	Loading or unloading rate at x	mols/s cm
$J_s(x)$	Velocity of solution at x	cm/s
$J_w(x)$	Volume flux of water through peripheral membrane at x	cm^3/s cm^2
$P(x)$	Pressure at x	bars
$C(x)$	Concentration at x	mols/cm^3

Flow of sucrose solution down the sieve tube, caused by the mechanism of Münch pressure flow, can then be calculated by the following two partial differential equations for conservation of water volume and sucrose (Smith et al., 1980).

Conservation of water volume is represented by the equation

$$\frac{\partial}{\partial t}[1-\overline{V}C(x,t)]A\Delta x = J_s(x_1)[1-\overline{V}C(x_1)]\ A+2\pi r\Delta x J_w(x)-J_s(x_2)[1-\overline{V}C(x_2)]A \qquad (1)$$

with the passive water influx defined as

$$J_w(x) = L_p[\psi_0(x)-P(x)+C(x)RT] \qquad (2)$$

and where the velocity or volume flux at x is

$$J_s(x) = -\gamma\frac{\partial P}{\partial x}(x) \qquad (3)$$

Equation 1 for conservation of water volume can be rewritten as

$$\frac{\partial}{\partial t}[1 - \overline{V}C(x,t)] + \frac{\partial}{\partial x}\left[-\gamma\frac{\partial P}{\partial x}(1-\overline{V}C(x,t))\right] = \frac{2L_p}{r}[\psi_0(x,t)-P(x,t)+C(x,t)RT] \qquad (4)$$

Conservation of sucrose is represented by the equation

$$\frac{\partial C}{\partial t}(x,t)A\Delta x = q(x)\Delta x+AJ_s(x_1)C(x_1)-AJ_s(x_2)C(x_2) + \frac{\delta}{\Delta x}[C(x_1)-C(x_2)] \qquad (5)$$

Equation 5 can be rewritten as

$$\frac{\partial C}{\partial t}(x,t) + \frac{\partial}{\partial x}\left[-\gamma\frac{\partial P}{\partial x}C(x,t) + \delta\frac{\partial C}{\partial x}\right] = \frac{q(x)}{A} \qquad (6)$$

where δ is the diffusion coefficient and q(x) is the loading and unloading function.

These two conservation equations can be solved to describe the sieve tube flow as it may vary with time. Under certain conditions, however, mathematical instabilities may develop. To date, only transport in short plants has been studied adequately—either empirically or mathematically. We are using this model to explore the variability that may arise in long-distance transport as would be required in trees, and the effects of varying water potential gradients, with both diurnal and seasonal time scales. Tyree et al. (1974) found the need for a long, slow sink along the path region in order for the Münch pressure flow theory to account for long-distance transport in trees.

The function for loading and unloading of photosynthate (q(x)) ultimately links the production by the foliage to the diameter growth of the bole. This crucial linkage will be explored in a time-dependent model in relation to prediction of both growth patterns along the bole and predicted transport rates. These may vary with different tree morphologies and phenologies, and these variations will be considered as different species are simulated.

CONSTRUCTION OF A SIMULATOR OF TREE GROWTH

Biological Considerations

The Simple Whole Tree (SWT) model comprises a series of component models or "kernels," each of which can be justified as an independent model based on its own theory. We envisage the development of SWT taking place in stages, where the overall structure remains as described in Figure 2, but the precise formulation becomes more integrated, particularly in terms of morphology and the seasonal control of growth patterns. The requirement is to simulate the features that control allocation with increasing effectiveness.

An important challenge in developing SWT will be simulating the vertical pattern of wood increment as a function of differences in sucrose gradient, water potential, and a vertical gradient of growth potential. The sink strength of roots will be simulated at various levels to examine the effect this may have, both on root growth itself and upon the growth of the rest of the tree.

Construction of SWT at this level will permit assessment of information on the effects of pollutants on tree net photosynthetic rates and growth. It will be particularly valuable for SWT to compare the effects on pollutant influence of different assumptions about how carbon is distributed to different meristems (i.e., components of the BRANCH model and the distribution functions within the stem).

Mathematical and Engineering Considerations

The mathematics underlying much of SWT may be expressed as partial differential equations. These result naturally from considerations of flows and diffusion between segments of a compartment and between compartments. They are often of the reaction-diffusion type and are relatively complicated. In some cases, the underlying finite difference equation is more realistic because the steps of the finite differencing correspond to units of morphology. This is the case, for example, with internodal segments of a branch or the individual cells of a phloem tube. In these cases, the partial differential equation is an approximation to the finite difference equation, and it may make more sense to simulate the finite difference equation directly. In other cases, the partial differential equation is more realistic, and it is the finite difference equation that is the approximation. This would be true, for example, for diffusion within a single cell. Thus the partial differential equations that arise in the context of this model are of two types depending on the assumed underlying morphology. The distinction between these two types makes a difference in the strategy adopted for simulation. In the cases where the morphology is continuous, the partial differential equation should be solved with whatever method is most efficient. This may result in the use of the finite difference equation. In the discrete case, it is usually best to simulate that process directly. However, it may be that solving some variant of the partial differential equation which approximates the finite difference equation is more efficient.

The model is constructed to test hypotheses both about the normal mechanisms of tree growth and about the impact of pollutants on these mechanisms. Therefore, it is important that the model be engineered in such a way as to allow easy modification of these hypotheses. This consideration leads us to adopt the style of programming known as "object oriented" (Meyer, 1988). The necessity of using a modern production language that is highly portable has dictated that we use the C programming language. However, our goal is to write in the style of the C++ language, which is an object-oriented extension of C (Stroustrup, 1986). When C++ becomes a solid production language available on many operating systems, we will switch to that language.

DISCUSSION

Modeling Under Mechanistic Uncertainty

It is worth re-emphasizing that the modeling undertaken here is uncertain both in terms of the details of proposed mechanisms and in terms of the functional forms of parametric values used to model them. This reflects the current state of our knowledge about the general mechanisms of tree physiology and growth. It also reflects the fact that any investigation of the effects of pollutants must consider alternative hypothesized mechanisms as does the Forest Response Program. However, this must be carried out in an explicit way.

The Role of Functional Morphology

Finally we wish also to emphasize the role of morphology and phenology in the construction of these models. Morphology provides constraints on the modeling process, for the shape of the tree must be predicted as well as the mass. Although this may seem an added burden for the modeler, it actually gives an added, and very different, way of assessing the effectiveness of a model. The strategy of mixing morphology and physiology into the unified study of "functional morphology" has accounted for many recent successes in animal physiology, and we expect that it may do the same for tree physiology.

LITERATURE CITED

Bassow, S. S., E. D. Ford, and A. R. Kiester. 1989. A critique of carbon-based tree growth models. In: Dixon, R. K., R. S. Meldahl, G. A. Ruark, and W. G. Warren, eds. *Process Modeling of Forest Growth Responses to Environmental Stress.* Timber Press, Portland, OR, U.S.A.

Deans, J. D., and E. D. Ford. 1986. Seasonal patterns of radial root growth and starch dynamics in plantation grown Sitka spruce trees of different ages. *Tree Physiology* 1:241–251.

Ford, E. D., and S. L. Bassow. In press. Modeling the dependence of forest growth on environmental influences. In: Pereira, J. S., ed. *Biomass Production by Fast-Growing Trees.* NATO Advanced Research Workshop, Lisbon, Portugal.

Ford, R., and E. D. Ford. 1988. A simulator for branch growth in the Pinacea: structure and basic equations. Center for Quantitative Science, University of Washington, Seattle, WA, U.S.A.

Hari, P., T. Raunemaa, and A. Hautojarvi. 1987. The effects on forest growth of air pollution from energy production. *Atmospheric Environment* 20:129–137.

Haskins, J. L., and E. D. Ford. 1989. Flow through conifer xylem: modeling in the gap between spatial scales. In: Dixon, R. K., R. S. Medahl, G. A. Ruark, and W. G. Warren, eds. *Process Modeling of Forest Growth Responses to Environmental Stress.* Timber Press, Portland, OR, U.S.A.

Ingestad, T. 1982. Relative addition rate and external concentration: driving variables used in plant nutrition research. *Plant, Cell and Environment* 5:443–453.

McMahon, T. A., and R. E. Kronauer. 1976. Tree structures: deducing the principle of mechanical design. *Journal of Theoretical Biology* 59:443–466.

McMurtrie, R., and L. Wolf. 1983. Above- and below-ground growth of forest stands: a carbon budget model. *Annals of Botany* 52:437–448.

Meyer, B. 1988. *Object-Oriented Software Construction.* Prentice Hall International, Hertfordshire, England, U.K.

Milne, R., E. D. Ford, and J. D. Deans. 1983. Time lags in the water relations of Sitka spruce. *Forest Ecology and Management* 5:1–25.

NAPAP. 1988. *Plan and Schedule for NAPAP Assessment Reports. 1989–1990. State of Science, State of Technology, Integrated Assessment.* National Acid Precipitation Assessment Program, Washington, DC, U.S.A.

Schroeder, P., and A. R. Kiester. 1989. The Forest Response Program: national research on forest decline and air pollution. *Journal of Forestry* 87:27–32.

Smith, K. C., C. E. Magnuson, J. D. Geoschel, and D. W. DeMichele. 1980. A time-dependent mathematical expression of the Münch hypothesis of phloem transport. *Journal of Theoretical Biology* 86:493–505.

Stroustrup, B. 1986. *The C++ Programming Language.* Addison-Wesley Publishing Company, Reading, MA, U.S.A.

Thornley, J. H. M. 1972. A balanced quantitative model for root:shoot ratios in vegetative plants. *Annals of Botany* 36:431–441.

Tyree, M. T., A. L. Christy, and J. M. Ferrier. 1974. A simpler iterative steady state solution of Münch pressure flow systems applied to long and short translocation paths. *Plant Physiology* 54:589–600.

Waring, R. H., D. Whitehead, and P. G. Jarvis. 1979. The contribution of stored water to transpiration in Scots pine. *Plant, Cell and Environment* 2:309–317.

Webb, C. D., and H. E. Burkhart. 1988. Modelling the effects of air pollution on forest productivity: sensitivity analysis using PTAEDA. In: Ek, A. R., S. R. Shifley, and T. E. Burk, eds. *Forest Growth Modelling and Prediction.* General Technical Report NC-120, USDA Forest Service North Central Forest Experiment Station, St. Paul, MN, U.S.A. Pp. 530–537.

CHAPTER 30

MODELING TREE RESPONSES TO INTERACTING STRESSES

Carl W. Chen and Luis E. Gomez

Abstract. A physiologically based tree model was developed to simulate tree growth according to the tree's genetic blueprint, expressed by a series of rate equations which describe the physiological processes of light interception, photosynthesis, respiration, evapotranspiration, nutrient uptake, water uptake, and the growth and mortality of plant parts. The effects of environmental stresses were included through equations which modify the physiological rates. Tests of the model suggest that crowding produces a smaller tree, drought reduces tree growth, and elevated ozone reduces tree growth and needle retention. The combination of drought and high ozone and the combination of high ozone and high shading produce the smallest tree at the end of a 10-year simulation. It was concluded that the model results were reasonable. The model could be used as a starting point and further refined.

INTRODUCTION

A tree living in nature is constantly subjected to multiple stresses, including nutrient deficiency, water deficit (drought), and air pollution (ozone and acid deposition). The stress factors change dynamically and interactively. In any given day, one factor can enhance tree growth while another factor retards it. At one time, a chemical factor can be low and harmless or even beneficial to tree growth. At another time, the same factor can become high and toxic to the tree (Chen and Goldstein, 1985). Various combinations of the environmental stresses occurring in real time control the vigor of an individual tree and the health of a forest (Hakkarinen and Allan, 1986; Hakkarinen, 1987; EPRI, 1985).

To evaluate the impact of environmental stresses on plants, a project was initiated to develop a physiologically based tree model that integrates the effects of air pollution

Dr. Chen and Mr. Gomez are Principal Engineer and Environmental Engineer, respectively, at Systech Engineering, Inc., 3744 Mt. Diablo Boulevard, Suite 101, Lafayette, CA, 94549 U.S.A.
This work was funded by the National Council of the Paper Industry for Air and Stream Improvement, Inc. (NCASI), and the Southern California Edison Company (SCE). Workshop participants included Dr. Andrzej Bytnerowicz of the University of California (Riverside), Dr. Michael Coffman of Champion International Corp., Dr. Philip Dougherty of the University of Georgia, Dr. Paul Dunn of the USDA Forest Service, Dr. Mark Fenn of the USDA Forest Service, Dr. David Gates of the University of Michigan, Dr. Cheryl Gay of the U.S. Environmental Protection Agency, Dr. Robert Goldstein of the Electric Power Research Institute, Dr. George Ice of NCASI, Mr. Jack Kawashima of SCE, Dr. Alan Lucier of NCASI, Dr. Paul Miller of the USDA Forest Service, Dr. Craig Murray of SCE, Dr. David Olszyk of the University of California (Riverside), Dr. David Peterson of the USDA Forest Service, Dr. Lou Pitelka of the Electric Power Research Institute, Dr. Gregory Ruark of the USDA Forest Service, Dr. Patrick Temple of the University of California (Riverside), Dr. Roger Timmis of Weyerhaeuser Company, Dr. Charles Webb of NCASI, and Dr. Stan Zarnoch of the USDA Forest Service, U.S.A.

(ozone and acid deposition) and climate (drought, flood, and frost) (Chen et al., 1987). Concurrent with the model development, data for model calibration are being collected in a number of research projects, including the air Quality/Forest Health Program of the National Council of the Paper Industry for Air and Stream Improvement (NCASI) and the Response of Plant to Interacting Stresses (ROPIS) study of the Electric Power Research Institute (EPRI).

To facilitate communication between the modelers and plant scientists, a number of workshops were held. During the workshops, the plant scientists postulated hypotheses to test with a preliminary version of the tree model. The results were evaluated to determine whether the tree model can accurately simulate a tree (Chen and Gomez, 1988). The purpose of this paper is to present the simulated tree responses to the interacting stresses.

THE PLANT-GROWTH–STRESS MODEL

The model was conceptualized after reviewing the literature with respect to 4 crop models and 13 forest-management models, several textbooks on plant physiology, and numerous papers on stress physiology related to drought, flood, air pollution, aluminum toxicity, and nutrient deficiency (Chen, 1986). The preliminary model formulations were provided by Chen and Gomez (1987). A brief discussion of the model and its formulations was presented at the IUFRO Conference on Forest Growth Modelling and Prediction (Chen et al., 1987).

The tree model was designed to simulate a single tree. Figure 1 presents the air-plant-soil system simulated by the model. The single tree is surrounded by neighboring trees. A percent of shading factor defines the fraction of incident light absorbed by the neighboring trees. The area of potential availability (APA) defines the soil region where the tree roots extract nutrients and water. The roots may spread horizontally beyond the area of potential availability, or the area may be invaded by the roots of neighboring trees. The model assumes that spreading and invading roots compensate each other in terms of nutrient and water uptake.

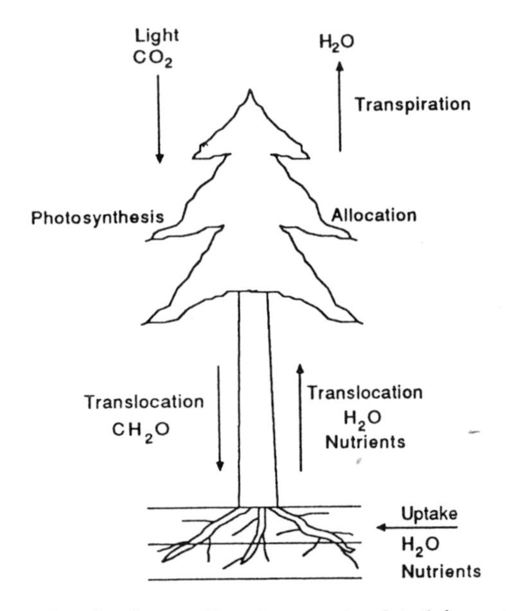

Figure 1. The air-plant-soil system as simulated for a single tree.

State variables for the tree are organized at the whole-plant level (i.e., canopy, stem, and root). The canopy has a mass, a leaf area, and an age class of needles. The stem has a mass, a height, and a diameter at breast height (dbh). The mass of the stem is divided into heartwood and sapwood. The root has two classes (coarse and fine) and is distributed among soil horizons. All the masses of canopy, stem, and root have their stoichiometric contents of nutrients (Ca, Mg, K, C, N, P, etc.). The soil is divided into horizons. Each horizon has its mineral composition (including organic matter) and soil solution chemistry.

The model simulates a tree according to its genetic blueprint. The genetic blueprint is expressed by a series of equations that define the rates of various physiological processes, i.e., light interception, photosynthesis, respiration, evapotranspiration, nutrient uptake, water uptake, and the growth and mortality of plant parts (Chen and Gomez, 1988; Chen et al., 1987). These equations are assembled in a computer model and solved simultaneously by a suitable numerical method. The results are the time series of tree diameter, tree height, and biomass of needles, roots, and stem. These time series represent the progression of tree growth.

The effects of environmental stresses are defined by equations which modify the rates of the physiological processes given above. For example, ozone and water deficit, at the sublethal level, cause the closure of stomata, resulting in the reduction of CO_2 uptake and photosynthesis rate. The cumulative dose of ozone amplifies the mortality rate of needles and, therefore, the rate of litterfall. Likewise, the acidification of the soil raises the inorganic aluminum concentration at the root zone which, at sublethal concentration, reduces the cation uptake rate and, at lethal concentration, increases the root mortality rate.

When the equations of growth physiology are solved together with the equations of stress physiology, the model generates the time series of tree responses that deviate from the normal tree growth. The magnitude of deviations represents the combined impacts of environmental stresses.

Figure 2 shows the input-output relationship of the tree model. The model provides input files to accept various data collected in a research program. These data are integrated by the model to produce the plant responses.

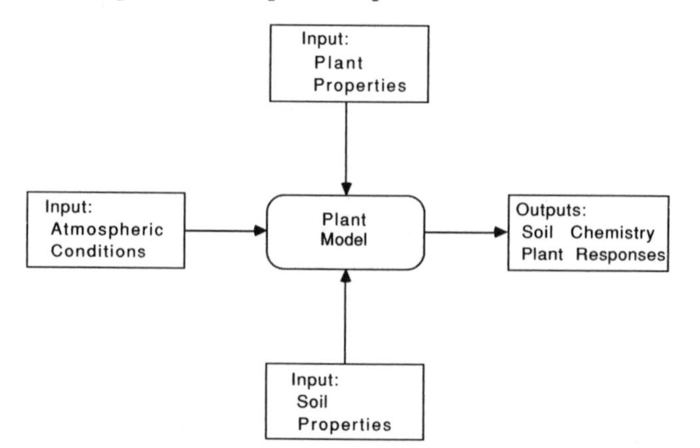

Figure 2. The input-output relationship of the tree model.

The model is driven by daily maximum and minimum air temperatures, wet bulb temperatures (or relative humidity), precipitation, precipitation chemistry, solar radiation (or percent of cloud cover), and maximum hourly ozone concentration. The model output includes the time history of soil, plant, and diagnostic parameters. The soil conditions are described by the daily values of soil temperature, soil moisture, and soil solution chemistry. The plant responses are described by the daily values of stem diameter, tree height, litterfall, canopy biomass, root biomass, root distribution among soil layers, nutrient content of the biomass, and mass of storage reserve. The diagnostic history is documented by the times of occurrences of ozone damage, water deficit, nutrient deficit, and aluminum toxicity, which are useful in diagnosing the plant-level responses. It is important to point out that all these parameters can be measured in order to verify model results.

HYPOTHESIS TESTING

In a scientific study, an investigator often formulates an hypothesis for research. The hypothesis can be expressed in a statement: A certain phenomenon should happen if such and such are true. The investigator can then design an experiment to collect data that may prove or disprove the hypothesis. Hypothesis testing with the model can be performed in an analogous manner, except that the investigator uses the model to simulate tree responses instead of using experiments to measure these responses.

Not all scientific hypotheses are proven correct by experimental data (Chen et al., 1983). Likewise, the model results may show unexpected responses. In that case, the investigators must reconcile the differences. If the errors were in the model formulation, corrections will be made. Through this procedure, the model can be improved to contain only supportable hypotheses.

A worksheet has been designed to guide the hypothesis formulations. On the worksheet, the workshop participants can write down the statement of hypothesis and expected results based on their intuitive reasoning. The modeler can provide the testing procedure and model result for comparison against the anticipated outcome.

To facilitate hypothesis testing, the model was developed with a number of factors that can be used to alter the environment from the base condition. For example, there is a scaling factor for daily precipitation. This factor can be adjusted up or down to study tree response under wet-year versus drought-year conditions. Other scaling factors can be adjusted to different levels of air pollution. In addition, a shading factor can be adjusted to study tree responses under shaded versus unshaded conditions.

RESULTS

Base Case

The base case was set up for a ponderosa pine (*Pinus ponderosa* Laws.) growing in the Sierra Nevada Mountains, U.S.A. The initial condition of the tree had a dbh of 15 cm. The area of potential availability (APA) was 10 m^2. The shading factor from neighboring trees was 40%.

The soil characteristics were estimated from a soil pedon measured in the watershed of the Eastern Brook Lake in the eastern Sierra Nevada Mountains. The meteorological data were taken from measurements performed in the Eastern Brook Lake Watershed Acidification Study (Chen and Gomez, 1988). The ozone concentrations were taken from the Rock Creek Station near the Eastern Brook Lake.

The base case is hypothetical. It was set up with data that happened to be available to us at a certain point in time. The base case does not represent any specific site. However, the deviation from the base case under hypothesized conditions can provide information about the general behavior of the tree model.

Hypothesis 1

Statement of hypothesis Trees growing in a small space, due to competition with neighboring trees, will grow to be smaller trees.

Testing procedure Run the case for the crowded tree by halving the area of potential availability (APA).

Expected results Trees compete with neighboring trees for light, water, and nutrients. A tree growing in a crowded stand will have a smaller APA. A smaller APA will cause one of the ingredients needed for growth to become limiting, which produces a smaller tree.

Model results Figure 3 shows the effect of crowding on dbh. The APA for the base case is 10 m^2 and for the crowded case is 5 m^2. Under the base condition, dbh of the tree grew to 37 cm over 10 years. When the APA was halved to 5 m^2, the tree grew to only 16 cm over the same 10 years.

Evaluation The model has simulated the growth of a smaller tree in a crowded condition as expected. The amount of growth reduction appears to be non-linear with respect to APA. Light becomes very limiting when the APA is halved from 10 to 5 m^2.

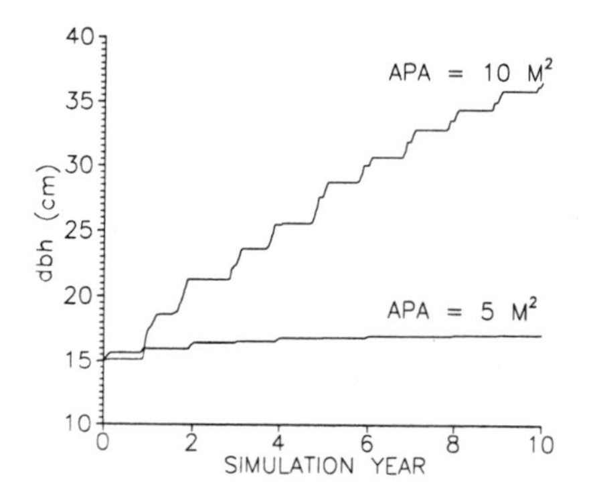

Figure 3. The effect of crowding on dbh growth.

Hypothesis 2

Statement of hypothesis Drought conditions will reduce tree growth and produce deeper roots.

Testing procedure Run the drought case by halving the precipitation of the base case.

Expected results The drought conditions will increase the amount of water stress realized by the tree. This stress will lead to lower stomatal conductance for CO_2 uptake, decrease canopy photosynthesis, and increase leaf mortality. The tree will respond to water stress by extending its roots into deeper soil layers.

Model results The results shows that dbh of the tree grew from 15 to 37 cm over 10 years. Under the drought condition (0.5x precipitation), dbh of the tree grew from 15 to 25 cm in the same period. The model did simulate a reduced diameter growth under the drought condition.

Figure 4 shows the effect of drought on the growth of canopy and roots. The biomass of both canopy and roots is reduced by the drought. Under the drought condition, however, the model did not grow more roots at the expense of the canopy as anticipated.

Figure 5 shows the effect of drought on root distribution among soil layers. Adventi-

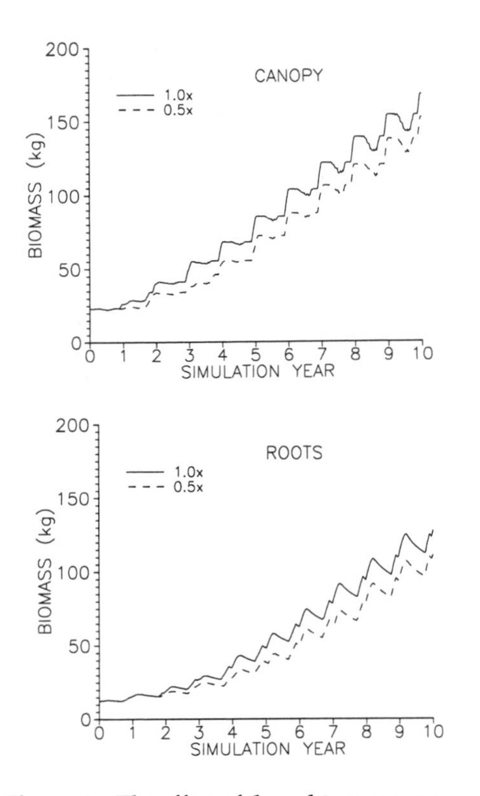

Figure 4. The effect of drought on canopy and root growth. Normal precipitation is expressed as 1.0x; the drought case is expressed as 0.5x.

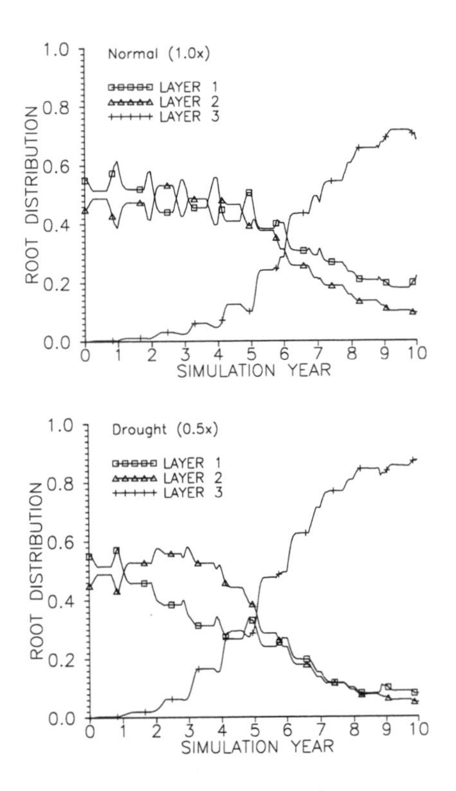

Figure 5. The effect of drought on root distribution. Normal precipitation is expressed as 1.0x; the drought case is expressed as 0.5x.

tious roots are assumed to be 5% in this simulation. Under this assumption, the tree continues to put more roots into layer 3 where moisture is available. In 6 years, there are more roots in layer 3 than in layers 1 and 2. Under the drought condition, root growth in layer 3 is accelerated even more: It takes only 5 years for the roots in layer 3 to outgrow those in layers 1 and 2.

Evaluation The model broadcasts a percent of adventitious roots to deeper soil horizons during the growing season. If the moisture condition in a soil horizon is adequate, the broadcasted roots will survive; otherwise they will die. As a result, the model simulated an increase of deep roots under a drought condition. Currently, the allocation priority for growth is canopy first, root second, and stem third. The model formulation needs to be modified to shift the priority to root first, canopy second, and stem third under conditions of water stress and nutrient deficiency. The simulation case was for drought years occurring as 10 consecutive years. The conditions were not realistic. Future tests should consider different combinations of wet years and dry years.

Hypothesis 3

Statement of hypothesis Higher ozone concentration in the air will reduce tree growth and needle retention.

Testing procedure Run the high ozone case by doubling the ozone concentration in the air.

Expected results At the sublethal level, ozone causes stomatal closure and reduces photosynthesis. At high concentration, ozone causes leaf senescence which in turn reduces the biomass of leaves for photosynthesis (lethal effect). All these conditions result in reduced stem growth and a loss of older needles.

Model results During the model simulation, it was found that the ozone concentration in the base case was relatively low. To demonstrate the ozone effect, it was decided to triple the ozone level instead of doubling it as suggested in the testing procedure. We also found that the original formulation for leaf mortality was based on the first order kinetic. Such a formulation could not eliminate older needles completely due to its mathematical property. The formulation was changed to zero order kinetic in the model.

The model results indicate that dbh of the tree grew from 15 cm to 37 cm over 10 years. When the ozone level was tripled, dbh increased from 15 cm to 21 cm over the same period. The model did simulate a growth reduction due to high ozone concentration in the air. Figure 6 shows the effect of ozone on needle retention in the tree. The upper figure shows the needle biomass under normal ozone level (1.0x) and normal precipitation (1.0x). The lower figure shows the needle biomass under high ozone (3.0x) and normal precipitation (1.0x).

Evaluation In the model simulation, the tree retained all four year classes of the needles under normal precipitation (1.0x) and ozone concentration (1.0x). When ozone concentration was tripled, year class 4 needles disappeared and only a small part of year class 3 needles remained. Thus, the model simulated a reduction of needles for all year classes and a loss of needles for older year classes.

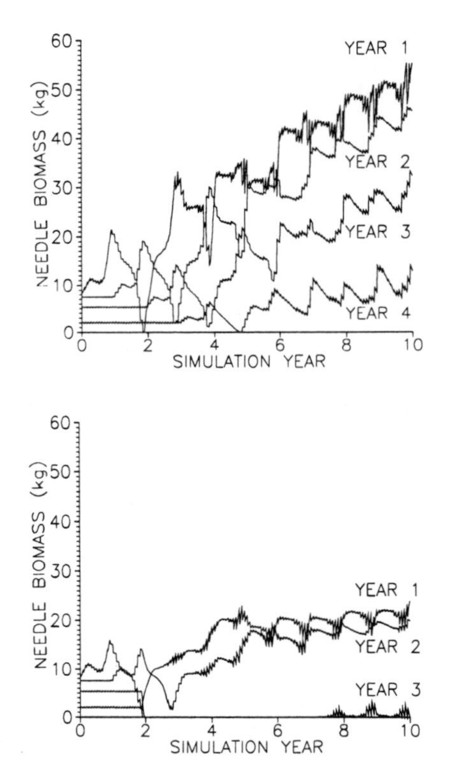

Figure 6. The effect of ozone on needle retention. Normal ozone is expressed as 1.0x (upper figure) and high ozone as 3.0x (lower figure). Both figures assume a normal precipitation of 1.0x.

Hypothesis 4

Statement of hypothesis The incremental effect of growth reduction by higher ozone concentration in the air will be less under a drought condition than under normal precipitation.

Testing procedure This hypothesis requires four simulations. To obtain the growth reduction of ozone under normal precipitation, we need to run the model under high ozone (3.0x) and normal precipitation (1.0x) for the hypothesis case and under normal ozone (1.0x) and normal precipitation (1.0x) for the base case. To obtain the growth reduction of ozone under a drought condition, we need to run the model under high ozone (3.0x) and drought (0.5x precipitation) for the hypothesis case and under normal ozone (1.0x) and drought condition (0.5x precipitation) for the base case.

Expected results Since the drought condition already causes a stomatal closure and reduced growth, it is reasonable to expect a lower incremental growth reduction by ozone under the drought condition.

Model results Figure 7 shows the effect of ozone and precipitation on dbh growth. Under normal precipitation, dbh increased to 37 cm under normal ozone concentration (1.0x). When the ozone concentration was tripled, dbh increased to only 21 cm. The growth reduction due to ozone was 16 cm. Under the drought condition, dbh grew to 25 cm under normal ozone concentration (1.0x). When the ozone concentration was tripled, dbh increased to only 18 cm. The growth reduction due to ozone was 7 cm. Thus, the model did simulate a lower growth reduction (7 cm) due to ozone under the drought condition than that (16 cm) under the normal precipitation.

Evaluation The model seems to behave as expected.

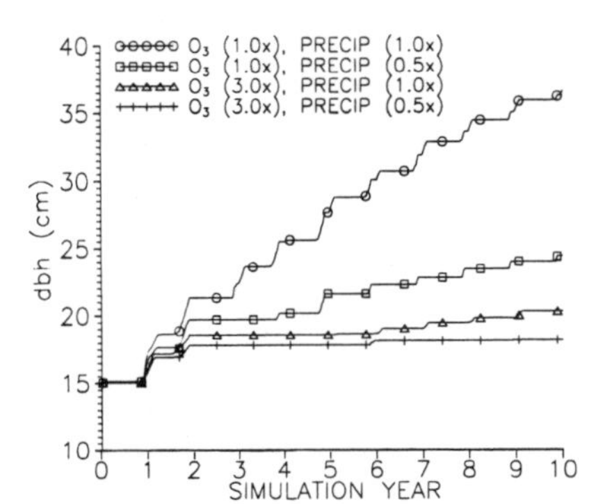

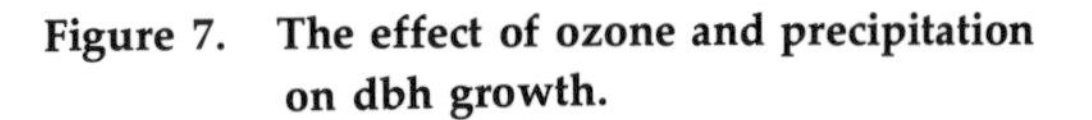
Figure 7. The effect of ozone and precipitation on dbh growth.

Hypothesis 5

Statement of hypothesis The effect of ozone on a tree under a more shaded condition will be higher than the effect of ozone on a tree under a less shaded condition.

Testing procedure This hypothesis also requires four simulation runs. To obtain the growth reduction due to ozone under the original shading, we need to run the model under original shading (40%) and normal ozone (1.0x) for the base case and under original shading (40%) and high ozone (3.0x) for the hypothesis case. To obtain the growth reduction due to ozone under the higher shading condition, we need to run the model under higher shading (60%) and normal ozone (1.0x) for the base case and under higher shading (60%) and higher ozone (3.0x) for the hypothesis case.

Expected results Since the tree under a more shaded condition is already stressed by the light limitation, it is expected that ozone effect on the tree will be exacerbated.

Model results Figure 8 shows the effect of ozone and shading on dbh growth. Under the base shading condition (40%), dbh of the tree increased to 37 cm under normal ozone level (1.0x) and to 21 cm under higher ozone concentration (3.0x). The growth reduction due to ozone was 16 cm. Under a higher shading condition (60%) dbh of the tree increased to 23 cm under normal ozone level (1.0x) and to 16 cm under higher ozone concentration (3.0x). The growth reduction due to ozone was 7 cm. The model simulated a higher growth reduction (16 cm) under a less shaded condition (40%) than the growth reduction (7 cm) under a more shaded condition (60%).

Evaluation The growth reduction due to ozone was expected to be higher under the more shaded condition than under the less shaded condition. The model results were opposite to this expectation. The model suggested that under conditions of more light the tree might have a higher photosynthesis rate allowing ozone to exert a greater influence. Under the more shaded condition, tree growth was already stunted by the light limitation and therefore the incremental growth reduction due to ozone was smaller. From this observation, the model appeared to be reasonable.

The model did show smaller diameter growth increments for the tree under the condition of more shade and higher ozone. This was consistent with observations of Dr. Paul Miller of the USDA Forest Service at Riverside, California, U.S.A.

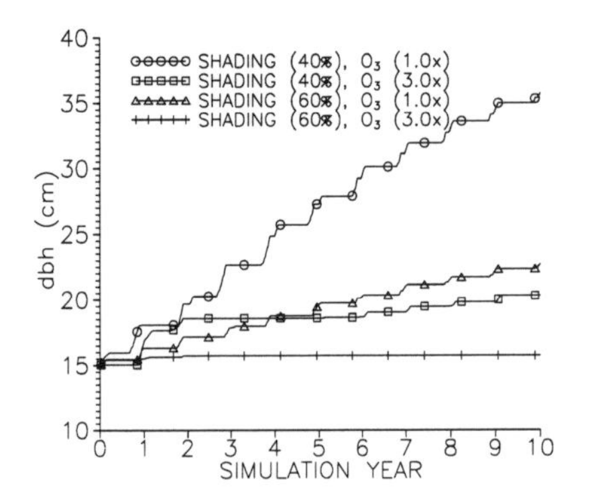

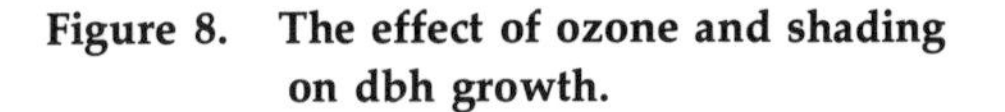
Figure 8. The effect of ozone and shading on dbh growth.

MODEL CRITIQUES

One purpose of the workshops was to receive model critiques from plant scientists. The workshop participants agreed that the present formulation provided a good starting point for further modifications. There were some disagreements as to the level of details in the formulations of the model. Individuals who were knowledgeable about the canopy processes wanted to see the canopy module expanded. Scientists with interest in roots wanted to see more emphasis on the rooting algorithm. However, all agreed that the model, at the present, had a fairly balanced level of detail among soil, air, and tree (canopy, roots, and stem). The model could track parameters that were measurable. A more detailed model would increase the data requirements. One must balance the need for more detail in the model with the ability to supply the required data.

Specific recommendations for improving the model formulations are summarized as follows.

Calculation of STOPEN

In the earlier formulation, the stomata opening (STOPEN) was made to respond instantaneously to new soil moisture and ozone concentration. The scientists pointed out that STOPEN should close slowly in response to high ozone and recover slowly in response to low ozone. The formulation was therefore modified to provide an inertia for the STOPEN to adjust to the new ozone concentration. The STOPEN calculated for the pervious 2 days and the instantaneous STOPEN calculated for the day were used in a moving average calculation of new STOPEN. Furthermore, it was recommended that STOPEN should first be calculated as a function of the relative humidity and cloud cover and then adjusted for sublethal stresses from that point.

Light Use Efficiency

At the present, the model calculates the hourly light interceptions and accumulates them for the daily light interception. The daily light interception is then used to calculate the daily photosynthesis. Such a formulation would calculate the same photosynthesis rate to the low-light-saturation trees and the high-light-saturation trees.

It was recommended to calculate the photosynthesis rates hourly instead of daily. To do that, we need to introduce a term, the relative light use efficiency, to convert the hourly light interception to the hourly photosynthesis rate. The relative light use efficiency will mimic the light saturation characteristic of a tree.

Photosynthesis of Needles

At the present, the model includes age classes of needles. However, the photosynthesis rate is calculated by the light interception regardless of the age class of needles involved.

Some workshop participants expressed the desire to use a different photosynthesis rate for each age class of needles. Such a demand, however, requires the development of a detailed canopy module. To develop such a model, we need to know how to position the needles in order to keep track of their shifting positions relative to sun angle and to calculate the average light intensity of each age class, among other considerations. At this point, we are not certain how to accommodate the request to use a different photosynthesis rate for each age class of needles. More work is needed in this area.

CO_2 Effects

At the present, the model calculates the photosynthesis of a tree as a function of light interception and STOPEN. This formulation implies that STOPEN alone regulates the CO_2 uptake. The CO_2 concentration in the ambient air is assumed constant. This will not be satisfactory, however, if we wish to use the model to evaluate the impact of elevated CO_2 concentration. Increased CO_2 concentration due to the greenhouse effect might become a very important environmental issue in the future. Furthermore, the CO_2 concentration at the canopy is not really constant. The variation of CO_2 concentration at the canopy may have an important effect on CO_2 uptake.

It is our intent to include the CO_2 effect. Although the specific formulation has not been finalized, the formulation will consider the CO_2 concentrations of air masses in the far field and at the canopy. The formulation will make CO_2 concentration very limiting under conditions of low wind and high light intensity. Under conditions of high wind and low light intensity, the CO_2 concentration at the canopy will probably be close to that in the far field.

Lethal Effects

In the earlier version of the model, an ozone concentration exceeding a threshold value would cause an amplification of the canopy mortality rate. When the ozone concentration decreased, the mortality amplification disappeared instantaneously.

It was suggested that the model should calculate the cumulative dose of ozone for yearly classes (monthly cohort) of leaves (needles). The mortality amplification factor for each year class could be made proportional to the cumulative dose of ozone.

A question was raised as to whether the canopy could recover from the damage. Some said yes and some said no. We believe that the model should allow for recovery. However, the recovery rate can be set to zero to simulate the condition of non-recovery.

Allometric Relationship

Two issues were raised about the allometric relationships. Should the allometric relationships be species-specific regardless of their site locations? Should the allometric relationships be applicable to a tree from sapling to maturity?

With respect to the first issue, we are in favor of the notion that the allometric relationships are the genetic property of the tree, that is, that they are species-specific. However, trees of the same species grown at different sites will have slightly different shapes. This is caused by the various limiting factors that may develop during the life of a tree, resulting from such environmental conditions as climate, soil characteristics, and shading by neighboring trees.

This discussion brings up the question: Which trees should be used to establish the allometric relationships for the model? In principle, the trees for this purpose should be grown in a stress-free environment. In the real world, such a tree can never be found. We may have to estimate the ideal allometric relationships based on less than ideal trees.

With respect to the second issue, it was suggested that the model use a variable allometric function of the form described by Ruark, Martin, and Brockeim (1987):

$$X = (a\ \mathrm{dbh})^{b} e^{(c\ \mathrm{dbh})} \tag{1}$$

where X = a plant part parameter; dbh = diameter of the stem at breast height; and a, b, and c are constants. In this formulation, the allometric relationship will automatically be adjusted as the tree grows bigger. Hopefully, such a formulation will allow us to run the model from sapling to mature tree.

SUMMARY AND CONCLUSIONS

A large number of hypotheses were formulated and tested with the tree model. The hypotheses started with the single-parameter variations of APA, drought, ozone, and shading. They ended with the two-parameter variations of drought-ozone and shading-ozone combinations. Table 1 presents a summary of simulations performed in the hypothesis tests. Table 2 summarizes the effects on the final stem diameter of the tree after 10 years of simulation under various combinations of precipitation and ozone conditions. Table 3 summarizes the effects on the final stem diameter of the tree after 10 years of simulation under various combination of shading and ozone conditions.

The model results show that crowding produces a smaller tree. Drought reduces tree growth. Elevated ozone reduces tree growth and needle retention. The combination of drought and high ozone and the combination of high ozone and high shading produce the smallest tree at the end of a 10-year simulation period.

It was concluded that the model results were reasonable. The model could be used as a starting point and further refined. Specific recommendations were made for the modification of some model formulations. The model, after improvement and more testing, will be calibrated to ponderosa pine (*Pinus ponderosa* Laws.) in California and loblolly pine (*Pinus taeda* L.) in North Carolina, U.S.A. The workshops, in providing for a fruitful exchange of ideas and information, serve a valuable purpose in the development of a sound and useful plant-growth–stress model.

Table 1. Hypotheses tested with the free model.

Hypotheses	APA	Precipitation	Ozone	Shading
Base case	1.0x	1.0x	1.0x	40%
Crowding	0.5x	1.0x	1.0x	40%
Drought	1.0x	0.5x	1.0x	40%
Ozone	1.0x	1.0x	3.0x	40%
Ozone-Drought	1.0x	0.5x	3.0x	40%
Shading	1.0x	1.0x	1.0x	60%
Shading-Ozone	1.0x	1.0x	3.0x	60%

Table 2. Dbh (cm) at the end of a 10-year simulation for various ozone and drought combinations.

	Precipitation		
	1.0x	0.5x	Δdbh
Ozone (1.0x)	37	25	12
Ozone (3.0x)	21	18	3
Δdbh	16	7	---

Table 3. Dbh (cm) at the end of a 10-year simulation for various ozone and shade combinations.

	Shading		
	40%	60%	Δdbh
Ozone (1.0x)	37	23	14
Ozone (3.0x)	21	16	5
Δdbh	16	7	---

LITERATURE CITED

Chen, C. W. 1986. The development of the Plant-Growth–Stress Model 1: conceptual model and developmental plan. Report to the Electric Power Research Institute. EPRI RP-2365, Systech Engineering, Inc., Lafayette, CA, U.S.A.

Chen, C. W., S. A. Gherini, R. J. M. Hudson, and J. D. Dean. 1983. *The Integrated Lake-Watershed Acidification Study,* Vol. I. *Model Principles and Application Procedures.* EA-3221, Electric Power Research Institute, Palo Alto, CA, U.S.A. 214 pp.

Chen, C. W., and R. A. Goldstein. 1985. Techniques for assessing ecosystem inpacts of air pollutants. In: Legge, A. H., and S. V. Krupa, eds. *Air Pollutants and Their Effects on the Terrestrial Ecosystem.* John Wiley and Sons, Chichester, England, U.K. Pp. 603–630.

Chen, C. W., and L. E. Gomez. 1987. Formulations of the Plant-Growth–Stress Model. Progress Report to the Southern California Edison Company and the National Council of the Paper Industry for Air and Stream Improvement. Systech Engineering, Inc., Lafayette, Ca, U.S.A. 30 pp.

Chen, C. W., and L. E. Gomez. 1988. Application of the ILWAS Model to Eastern Brook Lake Watershed in the Sierra Nevada Mountains. Report to the Southern California Edison Company, West Associates, and the Electric Power Research Institute, Palo Alto, CA, U.S.A.

Chen, C. W., and L. E. Gomez, C. A. Fox, R. A. Goldstein, and A. A. Lucier. 1987. A tree model with multiple stresses. In: Ek, A. R., S. R. Shifley, and T. E. Burk, eds. *Forest Growth Modelling and Prediction,* Vol. 1. General Technical Report NC-120, USDA Forest Service North Central Forest Experiment Station, St. Paul, MN, U.S.A. Pp. 270–277.

EPRI. 1985. Forest stress and acid rain. *EPRI Journal* (September): 16–25.

Hakkarinen, C. 1987. Forest health and ozone. EA-5135-SR, Electric Power Research Institute, Palo Alto, CA, U.S.A.

Hakkarinen, C., and M. A. Allan. 1986. Forest health and acidic deposition. EA-4813-SR, Electric Power Research Institute, Palo Alto, CA, U.S.A.

Ruark, G. A., G. L. Martin, and J. G. Brockeim. 1987. Comparison of constant and variable allometric ratios for estimating *Populus tremuloides* biomass. *Forest Science* 33(2):294–300.

CHAPTER 31

MODELING POLLUTANT GAS UPTAKE BY LEAVES: AN APPROACH BASED ON PHYSICOCHEMICAL PROPERTIES

Paul J. Hanson and George E. Taylor, Jr.

Abstract. A model was developed to characterize the location and extent of pollutant gas deposition to leaf surfaces and interiors. The model requires inputs of a gas's ambient concentration, Henry's law coefficient, and diffusivity in air and water. For a 3 ppbv (nL/L) gradient, the model predicted similar stomatal fluxes for HNO_3 (nitric acid vapor), NO (nitric oxide), O_3, and SO_2, to oak leaves. However, surface flux was predicted to be from 5 to 8 orders of magnitude higher for HNO_3 than for the other gases. For NO, O_3, and SO_2, essentially all predicted deposition was internal, while 29% of HNO_3 deposition was through the cuticle. Gradients of pollutant gas within the leaf were predicted for all gases. A theoretical basis for modeling pollutant sorption to plant cuticular surfaces is needed. With further refinement, the model or its projections could be adapted at both lower (cellular) and higher (whole plant, forest stand) scales of resolution.

INTRODUCTION

Many effects of pollutant gases on plant physiological processes and growth have been related to the gases' diffusion into the leaf interior and subsequent deposition to surfaces of the active mesophyll cells (Reich, 1987; Tingey and Taylor, 1982). Internal deposition of some gases can be measured using tracers, or the mass balance approach via subtraction of surface deposition from the total flux to leaves or whole shoots (Taylor et al., 1988). However, the uptake of other gases (e.g., nitric acid vapor) cannot be easily measured using existing methodologies because instrumentation for real-time analysis is not readily available. In the absence of accurate experimental methods, theoretical models of pollutant gas deposition represent a conceptual framework with which to evaluate hypotheses concerning gas uptake.

Physicochemical characteristics of pollutant gases have previously been used in simple resistance analog models to drive gas deposition to leaves (Bennett et al., 1973; O'Dell et al., 1977). These models divided the leaf into only a few tissue types (i.e., stomata, internal air spaces, and mesophyll cells). Parkhurst (1986) demonstrated the limitation of

---

Drs. Hanson and Taylor are in the Environmental Sciences Division, Oak Ridge National Laboratory, Oak Ridge, TN, 37831-6034 U.S.A. This research was sponsored by the Electric Power Research Institute, Integrated Forest Study under Interagency Agreement ERD-83-321 with the Office of Health and Environmental Research, U.S. Department of Energy, under Contract No. DE-ACO5-84OR21400 with Martin Marietta Energy Systems, Inc. The senior author was supported by Oak Ridge Associated Universities and Automated Sciences Group, Inc., Publication No. 3149, Environmental Sciences Division, Oak Ridge National Laboratory.

models which grouped tissues into only a few compartments by estimating gradients of CO_2 within leaf air spaces and mesophyll cells—an observation recently confirmed experimentally (Parkhurst et al., 1988). In order to characterize the location and extent of deposition of nitric acid vapor to leaf surfaces and interiors, we constructed a model based on physicochemical properties (i.e., diffusivities and Henry's law coefficients) that has the capacity to resolve spatial distributions of pollutant gas molecules within a hypothetical leaf stomatal unit, i.e., the area of a leaf influenced by a single stomate.

MODEL DESCRIPTION

The model is solved by the finite difference method and provides flux and concentration data for 50 compartments (i.e., air spaces, epidermal cells, and mesophyll cells) making up the tissue subtending one stomate of a hypostomatous leaf (Figure 1).

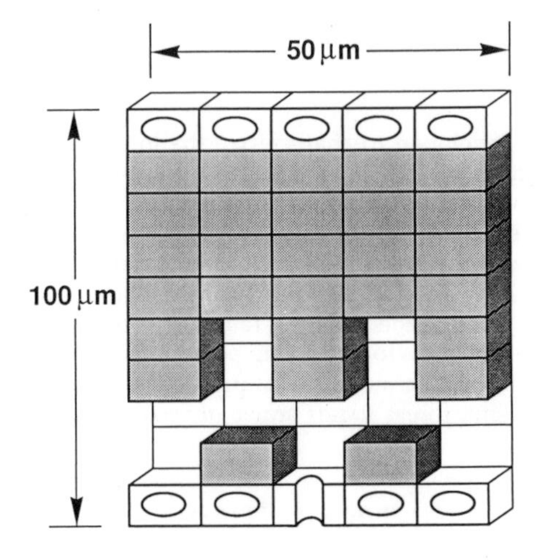

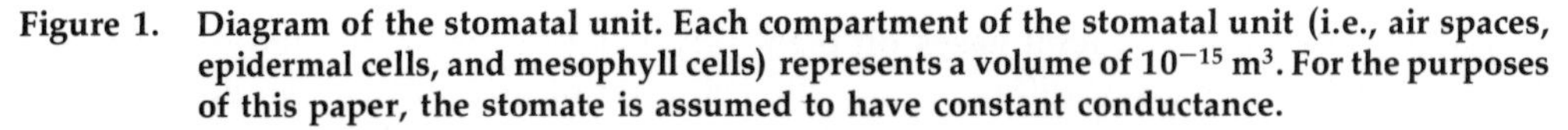
Figure 1. Diagram of the stomatal unit. Each compartment of the stomatal unit (i.e., air spaces, epidermal cells, and mesophyll cells) represents a volume of 10^{-15} m^3. For the purposes of this paper, the stomate is assumed to have constant conductance.

Based on data from the literature for a typical oak (*Quercus*) leaf, the hypothetical leaf is 10^{-4} m thick, each stomatal unit has an external surface area of 3.2×10^{-9} m^2 (450 stomata per mm^2), and all compartments within the stomatal unit are defined to have a volume of 10^{-15} m^3. The model calculates fluxes and concentrations in only two dimensions, yielding information for the central slice of the stomatal unit. For calculating horizontal fluxes at the edges of the stomatal unit (i.e., right and left edges of the diagram in Figure 1), the spatial distribution of mesophyll cells and air spaces is assumed to repeat itself. For simplicity, the aerodynamic boundary layer outside the leaf is set equal to zero, and leaf conductance to water vapor is arbitrarily set at a typical high value for oak seedlings (0.003 m/s for total leaf area; Körner et al., 1979).

The model iterates on a very short time scale (10^{-10} s) so that diffusion between air compartments does not completely deplete the pollutant gas concentration of an air space (an impossible condition). The model calculates between-compartment fluxes of pollutant gas starting at the adaxial surface, and finishes at the abaxial surface of the hypostomatous leaf. After each iteration, two outputs are recorded: (1) the concentration and/or partial pressure of the gas in each compartment of the stomatal unit, and (2) the flux (femto-moles [10^{-15}] per second; fmole/s) of the pollutant gas between every adjacent compartment of the stomatal unit, and between external leaf surfaces of the stomatal unit and ambient air.

Four physicochemical variables specific to a pollutant gas are required to run the model: ambient pollutant concentration in μatm (P_a), Henry's law coefficient in m^3 μatm/μmole (K_h), diffusivity in air in m^2/s (D_a), and diffusivity in water in m^2/s (D_w). Representative diffusivity and Henry's law coefficients for HNO_3, NO, O_3, and SO_2 are shown in Table 1. Calculations of pollutant gas flux to the stomatal unit are derived from two equations describing flux between homogeneous (i.e., air-air or solution-solution) or heterogeneous phases (i.e., air-solution).

Table 1. Physicochemical constants for CO_2 and four pollutant gases used as model inputs. A stomatal unit is the area of a leaf served by a single stomate. Constant temperature conditions are assumed.

Physicochemical property	Pollutant gas				
	CO_2	HNO_3	NO	O_3	SO_2
Diffusivity in air (m^2/s)[1]	1.65E-5	1.18E-5	1.99E-5	1.58E-5	1.22E-5
Diffusivity in water (m^2/s)[2]	1.65E-9	1.18E-9	1.99E-9	1.58E-9	1.22E-9
Henry's law coefficient (m^3/μatm/μmole)[3]	2.997E-2	7.00E-9	5.69E-1	2.29E0	7.88E-4
Maximum stomatal conductance (m/s)[4]	3.8E-2	3.2E-2	4.7E-2	3.7E-2	3.25E-2

[1] Fuller et al. (1966); Durham and Stockburger (1986); and numerical approximation techniques based on the square root of ratios of molecular weights (Lyman et al., 1982).
[2] Assumptions for CO_2 of 10^4 times lower diffusivity in water (Jarvis, 1971) applied to all gases. See also Lyman et al., 1982.
[3] CO_2 from Skirrow (1975); HNO_3 approximate from Davis and De Bruin (1964); others from Stephen and Stephen (1963).
[4] Körner et al. (1979) and numerical approximation methods (Lyman et al., 1982).

The homogeneous phase equation takes the following form for transport between air-containing compartments:

$$J_p = \left(\frac{D_a}{h}\right) \times (P_1 - P_2) \tag{1}$$

where J_p is the flux between compartments (fmole/m^2/s), h is the distance across which diffusion takes place (m), and D_a and P_x are as defined previously. For transport between solution-containing compartments, concentrations (C_x—fmole per "cell") replace partial pressures, and the diffusion coefficient (D_w) is adjusted for transport through solution (i.e., water). In Equation 1, the quotient of D_a and h is the conductance (m/s), and the difference in partial pressure (or concentration) is the gradient for flux between compartments. The heterogeneous phase equation is a simplification of the two-layer stagnant film model used to describe the flux of CO_2 across an air-water interface (Smith, 1985):

$$J_p = \left(\frac{D_w}{K_h \times h}\right) \times (P_a - P_w) \tag{2}$$

Equations 1 and 2 assume instantaneous mixing within each compartment of the stomatal unit. Low conductances associated with transport across epidermal cuticles (Lendzian, 1984) were approximated in our model by increasing the distance over which diffusion took place (i.e., we made the cuticle mathematically thick).

The computer program was written in version 3.2 BASICA for execution on a DOS-based microcomputer. Additional information about the program code is available from the senior author.

RESULTS AND DISCUSSION

To determine the utility of the model in estimating gas exchange within a leaf, physicochemical variables and empirical gas exchange equations for CO_2 were used in the model. These analyses showed the gradients of internal CO_2 concentrations predicted by the model to be similar to results published elsewhere (Parkhurst, 1986).

For a 3 ppbv (nL/L) concentration gradient, the model predicted pollutant fluxes to leaf interiors to be of similar magnitude for the four gases (Table 2), but those gases with the highest diffusivities showed the greatest stomatal flux ($NO > O_3 > SO_2 = HNO_3$). Furthermore, flux of pollutant gas to the leaf surface was predicted to be from 5 to 8 orders of magnitude higher for HNO_3 than for the other gases. With the exception of HNO_3, the model predicted decreasing concentrations of pollutant gas within the leaf with distance from the stomata. The HNO_3 results showed a deviation from this pattern near epidermal surfaces (data not shown) resulting from appreciable transport of HNO_3 gas across the epidermis to the leaf air spaces (Table 2).

Table 2. Flux of four pollutant gases (10^{-15} mol/s) to external and internal surfaces (epidermal and stomatal fluxes respectively) of the hypothetical stomatal unit after one nanosecond. All pollutant gases were defined to be present at 3 ppbv (nL/L) concentrations at all external leaf surfaces. Data in parentheses are the percent of total deposition to the stomatal unit.

Pollutant gas	Epidermal flux	Stomatal flux
	fmole/s	
HNO_3	1.6E-3	3.9E-4
	(29)	(71)
NO	3.3E-11	5.8E-4
	(0)	(100)
O_3	6.4E-12	4.5E-4
	(0)	(100)
SO_2	1.4E-8	4.0E-4
	(0.003)	(99.9)

The percent contributions of epidermal and stomatal flux to total pollutant uptake (Table 2) reflect experimentally observed trends among gases (Taylor et al., 1988), but the model's projections currently lack quantitative accuracy for gases exhibiting significant surface (cuticle) sorption. Mathematical functions derived from theories of gas sorption to solids (Adamson and Massoudi, 1984) and future experimental observations designed to characterize the reactive, sorptive, and diffusive properties of surface deposition should allow our model to be improved in this area. Application of the model is also limited by the availability of appropriate chemical constants for pollutant gases.

The current version of our model does not simulate chemical reactions of pollutant gases with plant surfaces. Therefore, the model's assumptions limit the influence of different gas concentrations to changes in the gradient which drives deposition. Realistically, pollutant gas deposition would be driven by solubility in plant cuticles and cell solutions, and further enhanced through chemical reactions with those same plant materials (Taylor et al., 1988). The conductance portions of Equations 1 and 2 can be modified to include the phenomenon of chemical reactivity by adding a coefficient (K_c) to the numerator (a modified Equation 1 is shown as an example):

$$J_p = \left(\frac{D_a \times K_c}{h}\right) \times (P_1 - P_2) \quad (3)$$

Although this would extend the current model's mechanistic accuracy, the absence of experimentally derived estimates of pollutant gas chemical reactivity with various plant materials precludes this modification at the present time.

The inclusion of a boundary layer in the model could change the results in two ways. An aerodynamic boundary layer would reduce pollutant concentrations at the surface of the leaf below concentrations present in the well-stirred atmosphere around the leaf. Furthermore, a boundary layer would cause localized reductions in pollutant concentration above and around individual stomates effectively limiting access of pollutant gas to cuticular surfaces between stomata.

The model or its projections can be adapted at both lower (cellular) and higher (whole plant, forest stand) scales of resolution. The ability to model gradients of concentration within the leaf could elucidate likely biochemical sites of action. Conversely, at higher scales of resolution, predicted relationships between surface and internal deposition could be used to determine canopy-level processes important to estimates of gas uptake (e.g., stomatal conductance). The projections of our model indicate that stand-level models of atmospheric deposition (e.g., Baldocchi et al., 1987) should consider the physicochemical characteristics of the pollutant gases. For example, because HNO_3 was projected to have the capacity to reach the leaf interior through the epidermis, a model incorporating only stomatal control would be inadequate.

LITERATURE CITED

Adamson, A. W., and R. Massoudi. 1984. Some facts and fancies about the physical adsorption of vapors. In: Myers, A. L., and G. Belfort, eds. *Fundamentals of Adsorption.* American Institute of Chemical Engineering, New York, NY, U.S.A. Pp. 23–37.

Baldocchi, D. D., B. B. Hicks, and P. Camara. 1987. A canopy stomatal resistance model for gaseous deposition to vegetated surfaces. *Atmospheric Environment* 21:91–101.

Bennett, J. H., A. C. Hill, and D. M. Gates. 1973. A model for gaseous pollutant sorption by leaves. *Journal of the Air Pollution Control Association* 23:957–962.

Davis, W., Jr., and H. J. de Bruin. 1964. New activity coefficients of 0–100 percent aqueous nitric acid. *Journal of Inorganic and Nuclear Chemistry* 26:1069–1083.

Durham, J. L., and L. Stockburger. 1986. Nitric-acid diffusion coefficient: experimental determination. *Atmospheric Environment* 20:559–563.

Fuller, E. N., P. D. Schettler, and J. C. Giddings. 1966. A new method for prediction of binary gas-phase diffusion coefficients. *Industrial and Engineering Chemistry* 58:19–27.

Jarvis, P. G., 1971. The estimation of resistances to carbon dioxide transfer. In: Sestak, Z., J. Catsky, and P. G. Jarvis, eds. *Plant Photosynthetic Production: Manual of Methods.* Dr. W. Junk, the Hague, the Netherlands. Pp. 566–631.

Körner, C., J. A. Scheel, and H. Bauer. 1979. Maximum leaf diffusive conductance in vascular plants. *Photosynthetica* 13:45–82.

Lendzian, K. J. 1984. Permeability of plant cuticles to gaseous air pollutants. In: Koziol, M. J., and F. R. Whatley, eds. *Gaseous Air Pollutants and Plant Metabolism.* Butterworth's, London, England, U.K. Pp. 77–81.

Lyman, W. J., W. F. Reehl, and D. H. Rosenblatt. 1982. *Handbook of Chemical Property Estimation Methods.* McGraw-Hill Book Company, New York, NY, U.S.A. 977 pp.

O'Dell, R. A., M. Taheri, and R. L. Kabel. 1977. A model for uptake of pollutants by vegetation. *Journal of the Air Pollution Control Association* 27:1104–1109.

Parkhurst, D. F. 1986. Internal leaf structure: a three-dimensional perspective. In: Givnish, T. J., ed. *On the Economy of Plant Form and Function.* Cambridge University Press, Cambridge, England, U.K. Pp. 215–249.

Parkhurst, D. F., S. Wong, G. D. Farquhar, and I. R. Cowan. 1988. Gradients of intercellular CO_2 levels across the leaf mesophyll. *Plant Physiology* 86:1032–1037.

Reich, P. B. 1987. Quantifying plant response to ozone: a unifying theory. *Tree Physiology* 3:63–91.
Skirrow, G. 1975. The dissolved gasses—carbon dioxide. In: Ripley, J. P., and G. Skirrow, eds. *Chemical Oceanography.* Academic Press, New York, NY, U.S.A. Ch. 9, pp. 1–181.
Smith, S. V. 1985. Physical, chemical and biological characteristics of CO_2 gas flux across the air-water interface. *Plant Cell and Environment* 8:387–398.
Stephen, H., and T. Stephen. 1963. *Solubilities of Inorganic and Organic Compounds.* Macmillan Publishing, New York, NY, U.S.A.
Taylor, G. E., Jr., P. J. Hanson, and D. D. Baldocchi, 1988. Pollutant deposition to individual leaves and plant canopies: sites of regulation and relationship to injury. In: Heck, W. W., O. C. Taylor, and D. T. Tingey, eds. *Assessment of Crop Loss from Air Pollutants.* Elsevier Science Publishers, New York, NY, U.S.A. Pp. 227–257.
Tingey, D. T., and G. E. Taylor, Jr. 1982. Variation in plant response to ozone: a conceptual model of physiological events. In: Unsworth, M. H., and D. P. Ormrod, eds. *Air Pollution in Agriculture and Horticulture.* Butterworth's, London, England, U.K. Pp. 113–138.

CHAPTER 32

A SIMULTANEOUS APPROACH TO TREE-RING ANALYSIS

Paul C. Van Deusen

Abstract. A method is described for studying the effect of climate on tree-ring width by incorporating all the traditional steps in tree-ring analysis into a single, simultaneous process. The usual standardization procedures of dendrochronology are replaced with the assumption that ring width is described by an exponential model with random parameters. Climate parameters are allowed to vary over time, making the uniformitarian supposition unnecessary. The method is applied to slash pine (*Pinus elliottii* var. *elliottii* Engelm.) tree cores from a nearly even-aged stand in Mississippi, U.S.A. The Kalman Filter is used in the implementation.

INTRODUCTION

The field of dendrochronology is currently in a period of rapid development. This is due in part to emerging concern about air pollution effects on forests and awareness that tree rings represent a long-term record that may yield information on these effects. Much of the current research will be conducted in closed-canopy forests, whereas traditionally, tree-ring research centered in the arid forests of the southwestern United States where trees are less affected by stand dynamics. Therefore, one of the mainstays of dendrochronology, the uniformitarian assumption, comes under question. Trees heavily affected by stand dynamics are likely to exhibit changing relationships with the environment over time, even though the environment itself has remained stable.

A method for dynamically modeling the relationship between climate and ring width (Van Deusen, 1987a) is expanded here to include more individual-tree information, such as diameter and age. The procedure presented herein has a number of noteworthy features:

1. The uniformitarian assumption is not required. Climate models with dynamic parameters are employed, so that changing climatic relationships can be modeled. The Kalman Filter (Kalman, 1960) becomes important in the estimation process, but it should be noted that any linear model could be formulated (as a state-space model) in a manner suitable for estimation using the Kalman Filter. Recursive estimation and limited memory are also implicit in this approach. Past data are given less weight than current data in determining the current parameter estimates. This seems biologically reasonable, because when the tree was growing, it

Dr. Van Deusen is a Mathematical Statistician at the Institute for Quantitative Studies, USDA Forest Service Southern Forest Experiment Station, 701 Loyola Avenue, New Orleans, LA, 70113 U.S.A.

was presumably influenced more by recent circumstances than by the distant past. Other applications of the Kalman Filter have been made to tree-ring analysis (e.g., Visser and Molenaar, 1987; and Van Deusen, 1987b), but they are quite different from the one discussed below.

2. All tree-ring series are maintained as individuals, and data reduction is achieved by estimating parameters. This differs from the traditional method of modeling climate with a single average ring-width or index series, although the results could be quite similar in some cases.
3. The usual steps in a tree-ring analysis (Fritts, 1976) are encompassed simultaneously in the procedure to be discussed. Specifically, the mean value function, the climate model, model validation, and standardization are all achieved simultaneously.

REVIEW OF STEPS IN TREE-RING ANALYSIS

An important part of the tree-ring analyst's work is to remove "noise" from the data. Much of the stand-level noise is, hopefully, removed by selecting dominant and codominant trees, although this is likely to be more successful in open-canopy than closed-canopy forests. A tree's underlying biological growth trend is one component of this noise that cannot be removed during sample selection. Prepocessing tree-ring data to remove components of growth, the biological component in particular is called *standardization* in the dendrochronology literature and is one of the most important aspects of tree-ring analysis.

A great deal has been written about standardization beginning with Douglass (1919), the founder of modern dendrochronology, and more recently by Fritts (1976), Graybill (1982), Wigley et al. (1984), and Cook (1985b). Sundberg (1974) presented a log-linear aggregate model, and Graybill (1982) and Cook (1985a, b) have presented a linear aggregate model (LAM) that facilitates discussion of standardization. A LAM for ring width is

$$R_{it} = B_{it} + P_t + C_t + O_t + e_{it} \tag{1}$$

where R_{it} is the ring width of tree i at time t, B_{it} is the biological component for tree i at time t, P_t is the stand-level perturbation component at time t, C_t is the stand- or regional-level climate component at time t, O_t is a regional-level component for other distrubances at time t (e.g. atmospheric deposition or sunspots), and e_{it} is a catchall for disturbances at the tree level that may be due to model errors, measurement errors, genetic factors, or local disturbances.

The usual approach to standardization is to fit a curve to the series R_{it}, $t = 1, \ldots, T$ to obtain a predicted ring width for each time. An index value is then formed by dividing the actual ring width by the predicted ring width. Monserud (1986) motivates this procedure by pointing out that any ring width can be represented as a function of its predicted value and an error term, thus

$$R_{it} = \hat{R}_{it} + e_{it} \tag{2}$$

Assuming that e_{it} has zero mean and variance $\hat{R}_{it}{}^2\sigma^2$, the standardized index could be

$$I_{it} = 1 + \frac{e_{it}}{\hat{R}_{it}} \tag{3}$$

Thus, I_{it} has an expected value of 1 with a more homogeneous variance than the original series.

Standardization is traditionally the first step in a tree-ring analysis, and usually the second step is to form a chronology by averaging over individual standardized tree-ring series. Letting Y_{it} be a standardized ring width for tree i at time t, the chronology would be

$$Y_t = \frac{\sum_i Y_{it}}{n_t} \qquad t = 1, \ldots, T \qquad (4)$$

The chronology is often used for climate modeling in a third step (Fritts, 1976, 1982) by regressing y_t on climate variables such as monthly average temperature and total rainfall. An additional step of prewhitening standardized series with Auto Regressive Moving Average (ARMA) models (Box and Jenkins, 1976) has come into use recently (e.g., Cook et al., 1987). Prewhitening is motivated by a desire to eliminate autocorrelation from the standardized data so that statistical tests in later climate modeling will be valid. The appropriate place to be concerned with autocorrelation is in the residuals of the climate model, and non-independence of the climate model residuals should be handled as part of the estimation process. Basically, prewhitening amounts to removing autocorrelation from the y-variable, rather than from the final model residuals, and can be justified only for exploratory work.

A SIMULTANEOUS APPROACH TO TREE-RING ANALYSIS

A simultaneous approach that relies on choosing biologically reasonable growth models to guide the analysis is presented in this section. The basic growth model is an exponential function, and it is shown how random parameters can be incorporated to give the model the flexibility to account for individual growth patterns. Autocorrelation is accounted for in the error terms; a random, time-varying climate component is added to the growth-curve model, and the Kalman Filter is employed to simultaneously fit the model to the ring-width data in natural log form. To assess the results, the prediction errors from the Kalman Filter are tested for serial and contemporaneous correlation. Thus, the steps described previously are reduced to a single simultaneous process. The results of the analysis can then be assessed by using goodness-of-fit statistics and valid confidence intervals and by criticizing the growth model.

Exponential Aggregate Models

An exponential aggregate model (EXAM) is appealing, because the sigmoidal nature of tree growth is automatically incorporated. If radial increment is the variable of interest, this method can lead to taking first differences of natural logarithms to remove the biological (age-related) trend, which is quite different from the standardization procedures discussed previously. Although methods based on Equation 1 are popular, EXAM-based methods deserve more consideration.

The general modified exponential function for diameter of tree i at time t takes the form

$$D_{it} = \alpha_i(1 + \beta_i e^{\gamma_i A_{it}})^{k_i} \qquad (5)$$

where A_{it} is the age of tree i at time t, $K_i = -1$ for the logistic curve, and $K_i = 1$ for the simple modified exponential. Harvey (1984) discusses the use of this model for time-series forecasting, and his development is followed here. Differentiating the model described by Equation 5, doing some algebraic manipulation, and taking logarithms leads to the following linear model for the log of ring width:

$$\log R_{it} = \pi_i \log D_{i,t-1} + \delta_i + \gamma_i A_{it} + e_{it} \qquad t = 2, \ldots, T \qquad (6)$$

where e_{it} is a random error potentially exhibiting serial correlation, $\pi_i = (k_i - 1)/k_i$, and $\delta_i = \log(k_i \beta_i \alpha_i^{1/k_i} \gamma_i)$.

This is an appealing model for ring width, because the variables in the model are the only tree-level variables usually known in a tree-ring study. Both diameter and age at time t cannot be inferred from the data in some cases, but the model described by Equation 6 can be manipulated to handle this. Set π_i to zero (assume a simple modified exponential) when diameter is not known; and when age is not known, take first differences of logarithms or set γ_i to zero.

Taking first differences results in

$$\Delta \log R_{it} = \pi_i \Delta \log D_{i,t-1} + \gamma_i + w_{it} \qquad t = 3, \ldots, T \qquad (7)$$

where w_{it} is an error term dependent on the properties of e_{it} in Equation 6. The model based on Equation 7, with π_i set to zero, was suggested by Hollstein (1980) and Van Deusen (1987)a, b) as a method to objectively standardize tree-ring series. The EXAM approach is similar to the LAM approach, but removal of the age-related component results from the initial assumption of underlying sigmoidal growth rather than from a method such as that described by Equation 3.

A Growth-Curve Formulation

Analysis of repeated-measurement or growth-curve models, in which a single characteristic has been measured at T different occasions, has been considered by many authors, including Potthoff and Roy (1964), Grizzle and Allen (1969), and Laird et al. (1987). Tree-core data fall naturally into this class of models, and a growth-curve formulation is now given as the basis of a simultaneous approach to tree-ring analysis.

The classical approach to standardization recognizes that each tree's age-related component is best described by a unique curve form, although all such curves might belong to the same family. The growth-curve approach is in the same spirit in that the individuality of each tree is recognized. Consider the following model for ring width of tree i at time t:

$$r_{it} = x_{it}(\overline{\beta} + \beta_i) + v_{it} \qquad (8)$$

where x_{it} is a lxp matrix of independent variables associated with the age-related component of tree i's growth curve, $\overline{\beta} + \beta_i$ is the pxl time invariant biological component parameter for tree i, and v_{it} is a random error term. The random components in this model are β_i distributed with zero mean and pxp variance matrix Q, and v_{it} distributed with zero mean and variance σ^2_t. Therefore, the variance of r_{it} in the model described by Equation 8 is

$$E[v_{it} + x_{it}\beta_i]\,[v_{it} + x_{it}\beta_i]' = \sigma_t^2 + x_{it} Q x_{it}' \qquad (9)$$

An overall growth model that uses the exponential model described by Equation 6 in natural log form is now written as

$$Y_t = X_t \overline{\beta} + X_t^* \beta(i_t) + v_t \qquad (10)$$

where Y_t is an $n_t$xl vector of natural logs of ring widths containing all the data for year t, X_t is an $n_t$xp matrix containing the independent variables from the model described by Equation 8, $\overline{\beta}$ is the pxl fixed component of the parameter vector, X^*_t is an n_txn_tp matrix where

$$X_t^* = \begin{bmatrix} x_{it}\ 0 \ldots 0 \\ 0\ x_{2t} \ldots 0 \\ 0\ 0 \ldots \\ 0 \ldots x_{n_t,t} \end{bmatrix} \qquad (11)$$

$\beta(i_t)$ is the random component of the parameter vector that includes a random term for all trees having a ring width at time t, and v_t is a random error vector. The random term $\beta(i_t)$ has become time-varying now only because trees tend to be of different ages and do not have ring widths for all periods. The variance of Y_t under the model described by Equation 10 is therefore

$$V_t = E[v_t + X_t^*\beta(i_t)]\ [v_t + X_t^*\beta(i_t)\ ]' = \sigma_t^2 I_t + X_t^*(I_t \otimes Q)X_t^{*\prime} \tag{12}$$

where I_t is an identity matrix of order n_t. It is being assumed that the v_t vector is independent of previous v_t's, which may be unrealistic.

Autocorrelation, Modeling Climate, and the Final Formulation

Equation 10 represents a method for simultaneously modeling the biological component of a number of tree-ring series. However, it is likely that the differences between the predicted average growth curve and the measurements for an individual tree will be serially correlated. A way of accounting for this problem is to model the error vector in Equation 10 as a vector autoregressive process.

Equation 10 is rewritten with a new error term as follows:

$$Y_t = X_t\overline{\beta} + X_t^*\beta(i_t) + v_t^* \tag{13}$$

where $v_t^* = \rho v_{t-1}^* + u_t$. Error vector v_t^* is parsimoniously assumed to be a first-order vector autoregressive process defined by a single autocorrelation coefficient, ρ. Error vector u_t is now further broken into components to allow for bringing climate and other supra–stand-level effects into the model. Assume that ith row of u_t is

$$u_{it} = m_t + v_{it} \tag{14}$$

where m_t is a common component of variance shared by all trees, and v_{it} is a random, serially uncorrelated error term as it was in Equation 10. The variance component, m_t, is where supra–stand-level effects enter into the model.

A general equation for modeling climatic effects on tree rings with a serially uncorrelated error vector v_t follows from the assumptions made on the error vector v_t^* in Equation 13:

$$\tilde{Y}_t = \tilde{X}_t\overline{\beta} + m_t + \tilde{X}_t^*\beta(i_t) + v_t \tag{15}$$

where $\tilde{Y}_t = Y_t - \rho Y_{t-1}$, $\tilde{X}_t = X_t - \rho X_{t-1}$, and $\tilde{X}_t^* = X_t^* - \rho X_{t-1}^*$. Assuming that m_t is partially explained by climate, a transition equation is proposed as follows:

$$m_t = a + c_t\lambda_t + w_{1t} \tag{16}$$

where a is an intercept that replaces the intercept in the growth model so that the column dimension of X_t has been reduced by one; c_t is lxq row vector containing climate variables; and λ_t is a qxl time-varying, climate-parameter vector assumed to follow a random walk. Thus

$$\lambda_t = \lambda_{t-1} + w_{2t} \tag{17}$$

with w_{1t} being a random error and w_{2t} a qxl random vector jointly distributed with zero mean and $(q + l) \times (q + l)$ variance matrix $\overline{W}_t$. The climate parameters, λ_t, should be allowed to vary over time to account for changes in climate response due to changing stand density or other changing regional effects such as pollution. Plots of the λ_t's versus time could reveal interpretable trends (Van Deusen, 1987a). Intercept a is not allowed to vary because it might absorb long-term trends that the growth curve should account for. However, there is no statistical restriction preventing a time-varying a-parameter, and in some cases this could be appropriate.

If Equations 16 and 17 are augmented with transition equations for a and $\overline{\beta}$,

$$a_t = a_{t-1} + 0 \tag{18}$$

and

$$\overline{\beta}_t = \overline{\beta}_{t-1} + 0 \tag{19}$$

then Equation 15 becomes the observation equation, and Equations 16–19 comprise the transition equations for a Kalman Filter (Kalman, 1960; Harrison and Stevens, 1976; Harvey, 1981). Equations 18 and 19 depict time-constant parameters because the error term identically equals 0; the t-subscripts are added to conform to Kalman Filter notation. The overall variance matrix associated with the transition equations (Equations 16–19) is

$$W_t = \begin{bmatrix} \overline{W}_t & 0 \\ 0 & 0 \end{bmatrix} \tag{20}$$

KALMAN FILTER

The Kalman Filter is briefly developed in generic form before continuing with the development begun with Equations 15–19. The relationship between the vector of observations at time t, Y_t, and the parameters, α_t, is called the *observation* or *measurement* equation:

$$Y_t = F_t \alpha_t + v_t \tag{21}$$

where Y_t is a vector of length n_t of transformed ring-widths for the current application, F_t is a fixed $n_t$xp matrix, n_t is the sample size at time t, α_t is a pxl vector of state parameters, and v_t is an $n_t$xl vector of residuals with zero expectation and variance matrix V_t.

The state parameters are variables that evolve over time according to a first-order Markov process as defined by the transition or system equation:

$$\alpha_t = G_t \alpha_{t-1} + w_t \tag{22}$$

where G_t is fixed pxp matrix, and w_t is a pxl vector of residuals with zero expectation and variance matrix W_t. These equations are initialized with α_0, which is distributed independently of the disturbances in Equations 21 and 22, which are uncorrelated with each other for all time periods.

The Kalman Filter solution requires knowledge of variance matrices W_t and V_t that is not often available in statistical applications. A number of suggested methods exist for estimating the unknown parameters in these matrices, including a maximum likelihood approach. Schweppe (1965) has shown that the log-likelihood of a sample from a population described by Equations 21 and 22 is

$$L = -\frac{1}{2} \sum_t \log |H_t| + E_t' H_t^{-1} E_t \tag{23}$$

where E_t are one-step-ahead prediction errors, and H_t is the covariance matrix of E_t. The likelihood function is specified completely in terms of the prediction errors and their covariance matrices, which are natural outputs of the Kalman Filter solution equations. The only assumptions required for this approach are that w_t and v_t be normally and independently distributed. Harvey (1981) and Engle and Watson (1981) give good developments of this method. The likelihood function (Equation 23) is evaluated iteratively by solving for the state parameters conditional on the most recent values for the unknown parameters in the model. The log-likelihood is maximized by using a grid search for the variance parameters or the method of scoring (Engle and Watson, 1981).

If the model is correct, the prediction errors form an innovations sequence (Anderson and Moore, 1979) in that E_t represents the new information in Y_t in the sense that E_t is orthogonal to all previous E_t's. This is true with or without the normality assumption. A diagnostic check for a correctly specified model involves checking to see if the prediction errors form a white noise sequence with no lagged cross-correlations. Thus, the Kalman Filter in combination with the method of maximum likelihood provides a feasible algorithm for fitting and testing the validity of Equations 15–19 with tree-ring data.

A SIMPLIFIED VERSION OF THE GENERAL FORMULATION

A special case of the general formulation is now presented and applied to a slash pine (*Pinus elliottii* var. *elliottii* Engelm.) data set. The general form of the measurement equation was presented above as Equation 21, and its components are now specified for this simplified application. The vector, Y_t, is $n_t$xl, which consists of the first differences of natural logarithms of ring widths. Thus, the growth model being used is Equation 7 with π_i set to zero; this is the simplest version of the EXAM approach. Since age and diameter are not included, the variance components due to random parameters in the growth-curve formulation (Equation 10) are not present. The matrix, F_t, is $n_t$xp, and for this formulation has l's in the first column and O's elsewhere. The state parameters, α_t, as a pxl parameter vector, and v_t is an $n_t$xl error vector with variance matrix V_t.

More specifically, the parameter vector in our suggested formulation is

$$\alpha_t = \begin{bmatrix} m_t \\ b_0 \\ b_{1t} \\ \cdot \\ \cdot \\ \cdot \\ b_{p-2,t} \end{bmatrix} \qquad (24)$$

where m_t is roughly equivalent to the average chronology in classical dendrochronology, b_O is a fixed intercept in the climate model, and the other b-parameters are multiplied by p−2 climate variables in the transition equation to partially explain the m_t parameter trend. Notice that only the m-parameter appears in the observation equation, because the F-matrix consists of O's except for the first column of 1's which zeroes out the b-parameters.

The transition equations for each parameter are

$$\begin{aligned} m_t = p_{0,t} + b_{1,t-1}c_{1,t} + \ldots + b_{b-2,t-1}c_{p-2,t} + w_{mt} \qquad (25) \\ b_{0,t} &= b_{0,t-1} + 0 \\ b_{1,t} &= b_{1,t-1} + w_{1t} \\ b_{2,t} &= b_{2,t-1} + w_{2t} \\ &\cdot \\ &\cdot \\ &\cdot \\ b_{p-2,t} &= b_{p-2,t-1} + w_{p-2,t} \end{aligned}$$

where the $c_{i,t}$, $i = 1, \ldots, p-2$ represent p−2 climate variables. Thus, Equation 25 is analogous to the climate model from classical methods, with the important difference being that the parameters are time-varying. The following G-matrix will result in the above tansition equations when substituted into Equation 22:

$$G_t = \begin{bmatrix} 0 & 1 & c_{1,t} \dots c_{p-2,t} \\ 0 & 1 & 0 \dots 0 \\ \cdot & \cdot & \\ \cdot & & \cdot \\ \cdot & & \quad \cdot \\ 0 & \dots & 1 \end{bmatrix} \tag{26}$$

Basically, the G-matrix is created by altering the first row of an identity matrix of order p.

Variance Specification

The observation variance matrix, V_t, and the transition variance matrix, W_t, must be specified before the Kalman Filter equations can be solved. We specify V_t to be $\sigma_t^2 I_t$, where I_t is an identity matrix of order n_t and

$$\hat{\sigma}_t^2 = \frac{\sum_{i=1}^{n_t} (y_{it} - \bar{y}_t)^2}{n_t - 1} \tag{27}$$

Thus, the error terms in the observation equation for year t are assumed to be independent, with variance estimated from the standardized ring widths for year t. However, the fact that m_t is random and shared by all observations means that contemporaneous correlation is accounted for in this formulation.

The transition variance matrix, W_t, is assumed to be

$$W_t = \frac{\sigma_t^2}{n_t} \text{DIAG}[q_m, O, q_1, \dots, q_{p-2}] \tag{28}$$

where DIAG[.] means to form a diagonal matrix from the arguments. Therefore, the parameters in the model are assumed *a priori* to be independent, and the intercept term in Equation 25 is fixed.

The reason for multiplying the transition variance matrix by σ_t^2/n_t is that the m_t parameter should be roughly equal to the mean value in the vector Y_t, and the proper choice for q_m allows this to happen even when the climate model is poor. In fact, it can be shown that when all the b-parameters equal zero, the mean value parameters are estimated as follows:

$$m_t = \frac{q_m}{q_m + 1} \bar{y}_t \tag{29}$$

with variance

$$V(m_t) = \frac{q_m}{q_m + 1} \frac{\sigma_t^2}{n_t} \tag{30}$$

Thus, a large value for q_m allows the system to produce reasonable estimates of the mean value function even when the climate relationships are poor. When the climate model predicts well, less weight is given to the data and more to the climate model. The parameters $q_m, \dots, q_{p-2}$ can be estimated via the method of maximum likelihood discussed above. The q_m-parameter was assumed equal to 1 in Van Deusen (1987a), but allowing more flexibility can considerably improve the model performance as shown.

RESULTS OF THE SLASH PINE APPLICATION

A tree-ring data set for slash pine from the Gulf Coast region of Mississippi, U.S.A., was used for this analysis. Two cores were taken from each of 33 trees in a nearly even-aged stand. The average age of these trees in 1985 was 41 years with a standard error of 1 year. The youngest tree was 26 years, and the oldest was 53 years. The average diameter was 27 cm. ranging from 17 cm to 48 cm. The climatic data used here are from average temperature data provided by the National Environmental Satellite, Data, and Information Service for national climatological division 10 in Mississippi, U.S.A.

These cores were originally collected to determine whether evidence of a widespread needle blight that occurred in the late fall and winter of 1970–1971 (Czabator et al, 1971) could be detected. Although the results are not conclusive, they are promising and suggest that tree rings are sensitive historical records. After trying a number of climate variables, November temperature emerged as the most important variable as judged by the likelihood criterion. Specifically, the first difference of November temperature resulted in a much larger value of the log-likelihood function given in Equation 23 than any other variable alone or in combination. This is not surprising because the Gulf Coast of the U.S.A. receives significant rainfall and experiences a relatively constant temperature throughout the summer. However, November temperature can be quite variable; this is an example of the principle of limiting factors (Fritts, 1976).

Two figures summarize the results of this analysis. Figure 1 shows the time-varying parameter on November temperature differences surrounded by approximate 95% confidence intervals. This corresponds to the b_1 parameter in Equation 25, and the first difference of November temperature corresponds to c_1 in the same equation. Around 1960, November temperature becomes significant, which is not surprising because nearly all earlier data are from juvenile wood.

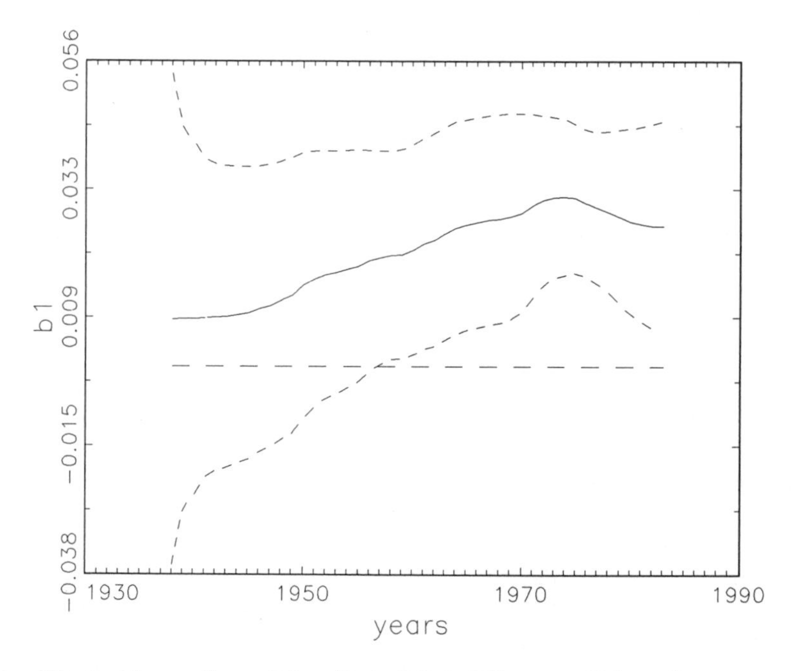

Figure 1. The trend over time of the effect of first difference of November temperature on the natural log of ring width for slash pine from the Gulf Coast of Mississippi, U.S.A. The dashed lines are approximate 95% confidence intervals.

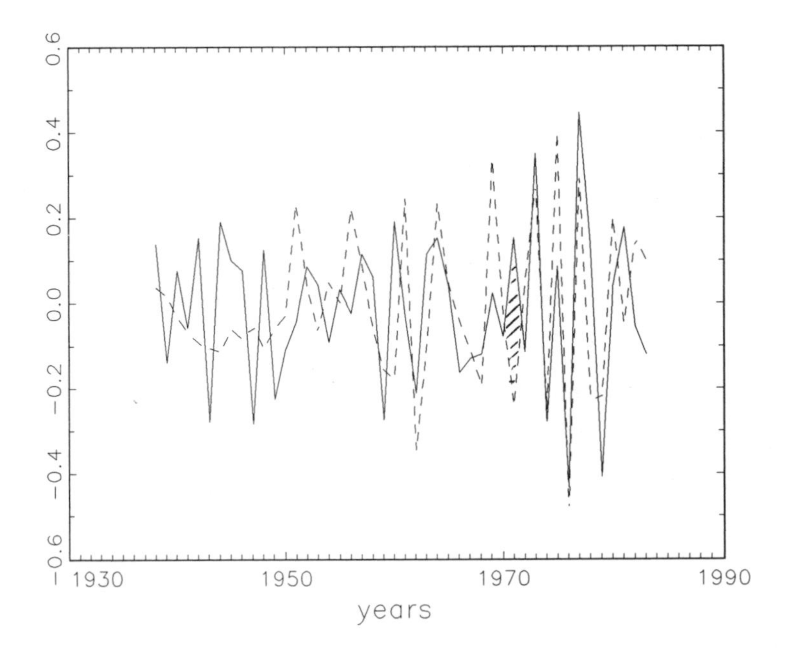

Figure 2. **First difference of November temperature (*solid line*) overlaid on the first difference of the natural log of ring width (*dashed line*) averaged over all trees. The temperature variable is rescaled to have the same mean and variance as the average tree-ring variable.**

Figure 2 shows the first difference of November temperature overlaid on the average chronology; the chronology was formed by averaging the first difference of the natural logarithm of ring width. Figure 2 verifies that the chronology and the temperature variable become synchronized around 1960. Notice that 1971 is the only year between 1960 and 1981 where the chronology turns down and the temperature variable turns up. This is purely circumstantial evidence, but it does suggest that tree rings can be rather sensitive recorders of environmental stress. Figure 2 also suggests that 1981 and 1983 are years where the tree-ring and temperature variables are not synchronized.

Regardless of the interpretation, these figures demonstrate that the analytical procedures outlined here are promising. A procedure that assumed uniform climate response through time could have failed to discover the relationship between ring width and November temperature.

CONCLUSIONS

A method for analyzing tree-ring data that handles all the traditional steps simultaneously has been described. The method also allows the tree-ring response to climate to change over time, which is important when working in closed-canopy forests with time-varying competition effects, or with data spanning both juvenile and mature stages of tree development.

An example was given where the response to climate changes as the trees matured. A deviation occurred between the climate-variable trend and the average ring-width variable after a documented incidence of severe needle blight. If this deviation is in fact related to the needle blight, tree rings are more sensitive recorders of environmental stress than is currently believed. Due to the lack of other types of historical records for forests, the methods presented herein hold promise for the study of forest response to environmental stress.

LITERATURE CITED

Anderson, B. D., and J. B. Moore. 1979. *Optimal Filtering.* Prentice Hall, Englewood Cliffs, NJ, U.S.A.

Box, G. E. .P., and G. M. Jenkins. 1976. *Time Series Analysis: Forecasting and Control.* Holden-Day, Oakland, CA, U.S.A.

Cook, E. R. 1985a. A time series analysis approach to tree-ring standardization. Ph.D. Thesis, University of Arizona, Tucson, AZ, U.S.A.

Cook, E. R. 1985b. The use and limitations of dendrochronology in studying effects of air pollution on forests. In: Hutchinson, T., ed. *NATO Advanced Research Workshop on Effects of Air Pollution on Forest, Wetland, and Agricultural Ecosystems.* Toronto, Canada.

Cook, E. R., A. H. Johnson, and T. J. Blasing. 1987. Forest decline: modeling the effect of climate in tree rings. *Tree Physiology* 3:27–40.

Czabator, F. J., J. M. Staley, and G. A. Snow. 1971. Extensive southern pine needle blight during 1970–1971, and associated fungi. *Plant Disease Reporter* 55(9):764–766.

Douglass, A. E. 1919. *Climatic cycles and tree growth,* Vol. 1. Publication 289, Carnegie Institute, Washington, DC, U.S.A.

Engle, R., and M. Watson. 1981. A one-factor multivariate time series model of metropolitan wage rates. *Journal of the American Statistical Association* 76(376):774–781.

Fritts, H. C. 1976. *Tree Rings and Climate.* Academic Press, London, England, U.K.

Fritts, H. C. 1982. The climate growth response. In: Hughes, M. K., P. M. Kelly, J. R. Pilcher, and V. C. Lamarche, Jr., eds. *Climate from Tree Rings.* Cambridge University Press, Cambridge, England, U.K. Pp. 33–37.

Graybill, D. A. 1982. Chronology development and analysis. In: Hughes, M. K., P. M. Kelly, J. R. Pilcher, and V. C. Lamarche, Jr., eds. *Climate from Tree Rings.* Cambridge University Press, Cambridge, England, U.K. Pp. 21–28.

Grizzle, J. E., and D. M. Allen. 1969. Analysis of growth and dose response curves. *Biometrics* 25:357–382.

Harrison, P. J., and C. F. Stevens. 1976. Bayesian forecasting (with discussion). *Journal of the Royal Statistical Society,* Series B, 38:205–247.

Harvey, A. C. 1981. *Time Series Models.* Philip Allan Publishers Ltd., Oxford, England, U.K.

Harvey, A. C. 1984. Time series forecasting based on the logistic curve. *Journal of the Operational Research Society* 35(7):641–646.

Hollstein, E. 1980. *Mitteleuropa † ische Eichenchronologie* (Trierer Grabungen und Forschungen XI). Philip von Zabern, Mainz, F.R.G.

Kalman, R. E. 1960. A new approach to linear filtering and prediction problems. *Journal of Basic Engineering* 82:34–45.

Laird, N., N. Lange, and D. Stram. 1987. Maximum likelihood computations with repeated measures: application of the EM algorithm. *Journal of the American Statistical Association* 82(397):97–105.

Monserud, R. A. 1986. Time series analysis of tree ring chronologies. *Forest Science* 32:349–372.

Potthoff, R. F., and S. N. Roy. 1964. A generalized multivariate analysis of variance models useful especially for growth curve problems. *Biometrika* 51:313–326.

Schweppe, F. C. 1965. Evaluation of likelihood functions for gaussian signals. *IEEE Transactions on Information Theory* 11:61–70.

Sundberg, R. H. 1974. On the estimation of pollution-caused growth reduction in forest trees. In: Pratt, J. W., ed. *Statistical and Mathematical Aspects of Pollution Problems.* Marcel Dekker, New York, NY, U.S.A. Pp. 167–175.

Van Deusen, P. C. 1987a. Some applications of the Kalman Filter to tree ring analysis. In: *Proceedings, an International Symposium on Ecological Aspects of Tree Ring Analysis,* August, 1986, Marymount College, Tarreytown, NY, U.S.A. Pp. 566–578.

Van Deusen, P. C. 1987b. Testing for stand dynamics effects on red spruce growth trends. *Canadian Journal of Forest Research* (December).

Visser, H., and J. Molenaar. 1987. Time dependent responses of trees to weather variations: an application of the Kalman Filter. In: *Proceedings, an International Symposium on Ecological Aspects of Tree Ring Analysis,* August, 1986, Marymount College, Tarreyton, NY, U.S.A. Pp. 579–590.

Wigley, T. M. L., K. R. Briffa, and P. D. Jones. 1984. On the average value of correlated time series, with applications in dendroclimatology and hydrometeorology. *Journal of Climatology and Applied Meteorology* 23:201–213.

CHAPTER 33

MODELING TREE-RING AND ENVIRONMENTAL RELATIONSHIPS FOR DENDROCHRONOLOGICAL ANALYSIS

Harold C. Fritts

Abstract. Certain unique properties of dendrochronological time series that make them highly relevant to simulation modeling are: (1) sample stratification that enhances the amount of information in the ring characteristics of the sample, (2) absolute dating accuracy that assures that all rings are correctly identified as to the year in which they grew, and (3) well-behaved time-series characteristics. Dendrochronological investigations have made contributions to stem analysis, statistical analysis of environmental relationships, and field studies of environmental, physiological, and growth processes throughout both the growing and the dormant seasons. A simulation model, ARBOR, emulates the growth of an annual ring through the proceses governing cambial activity. Current efforts are directed at modeling the environmental controls of the cambial processes to allow for simulation of ring features that ultimately can be used to diagnose past problems and to estimate future probabilities.

INTRODUCTION

Well-dated annual growth layers in trees have been used as a dendrochronological tool for many years (Fritts, 1976) and are now being applied to a wide variety of forest problems (Fritts and Swetnam, in press; Schweingruber, 1988). Dendrochronology is the science of dating the annual rings of trees and the application of the dated rings to a variety of scientific questions (Fritts, 1976). Dendroclimatology and dendroecology are subfields which apply dendrochronologically dated ring sequences to the study of past and present climates and environments (Fritts, 1976; Stockton et al., 1985; Schweingruber, 1988; Fritts and Swetnam, in press). Dating is a visual and computer-assisted procedure that uses large-scale patterns of ring-width and wood-density variations to identify the actual year in which each ring was formed (Fritts, 1976; Holmes, 1983). In addition, the ring-width and wood-density changes within the dated rings are of value because they contain information on past variations in environmental conditions that have limited past growth.

SOME UNIQUE CHARACTERISTICS OF DENDROCHRONOLOGY

A dendrochronologist often asks questions that are different from those of the forest scientist (Fritts, 1976) and uses unfamiliar sampling procedures as well. For example, a forester who is interested in the wood volume growth of a stand might core randomly

Dr. Fritts is Professor of Dendrochronology at the Laboratory of Tree-Ring Research, University of Arizona, Tucson, AZ, 85721 U.S.A.

selected trees. A dendrochronologist, on the other hand, interested in the longest possible growth record of past environmental conditions (LaMarche, 1982), might stratify the sample by coring only the oldest trees growing in stress sites, since such trees provide the longest record with the greatest ring-width response to climate. Consequently, the ring-width variations from these two data sets would contain different types of information, but both procedures are scientifically sound approaches to the problems being studied and contribute valuable information about growth and the forest environment. For example, both kinds of information would be needed to model the possible impact of future climatic change on forest productivity.

A dendrochronologist interested in the history of past droughts may sample trees growing on drought-limiting sites in order to maximize the amount of information on drought in the tree-ring chronolgy (Fritts, 1976). The growth of trees growing near the upper altitudinal limits of a species would be sampled if a temperature record was desired, because the growth of these individuals is known to be highly limited by low temperature (LaMarche, 1978, 1982). Trees that are likely to be affected by an insect infestation would be sampled to reconstruct past insect activity (Swetnam, 1987; Fritts and Swetnam, in press). Trees with the greatest number of fire scars would be selected to obtain the most complete record of the fire history (Dieterich and Swetnam, 1984). Sampling may be more or less random within the criterion of selection. In fact, dendroecological questions often require sampling dense stands where climate is not highly limiting. Such sampling would more closely match the kinds of habitats that are usually modeled.

However, there must be a discernible pattern of large-scale variability of ring widths or some other feature that can be used for the dendrochronological dating (Fritts, 1976; Brubaker and Cook, 1983; Stockton et al., 1985; Schweingruber, 1988). Dating is a requirement that assures that every ring is placed accurately in time (LaMarche and Harlan, 1973), so that the features of variability can be safely pooled and averaged for many trees and related to yearly variations in conditions on the respective sites (Fritts, 1976). If samples can be dated, that fact alone indicates that some of the variance is attributable to a large-scale factor that is likely to be climatic variation. If the climate-related variance is insufficient to allow the rings to be dated, then the procedures and practices of dendrochronology cannot be applied to the ring information from those particular trees. Mutual understanding and cooperation between our two disciplines is growing as more workers become familiar with these differences and appreciate the unique contributions that both fields can make to the study of forest influences (Fritts and Swetnam, in press; Schweingruber, 1988).

Usually, many trees are cored from each dendrochronological study site, the cores are cross-dated, and characteristics of the rings such as width and wood density are measured, standardized, and averaged to obtain a ring-width chronology (Figure 1). This chronology maximizes the year-by-year ring response that is common to all sampled trees and minimizes the between-tree responses attributable to increasing tree age, variable site quality, and other similar growth influences (Fritts, 1976). An analysis of variance can be used to examine chronological differences among trees, but it is difficult to identify the exact source of this between-tree variation. Auto Regressive Moving Average (ARMA) modeling (Cook, 1985; Guiot et al., 1982; and Guiot, 1986) and the Kalman Filter (Van Deusen, this volume) are techniques that may help one to analyze these variance sources.

From a statistical point of view, standardization is necessary to convert the non-stationary ring-width measurements with age-related trends into a stationary time series without these trends. The variances of the departures from the fitted curves are not homogeneous over time, so ratios rather than differences are calculated. The result is an index that can be multiplied by 100 to express the growth as a percentage of the fitted

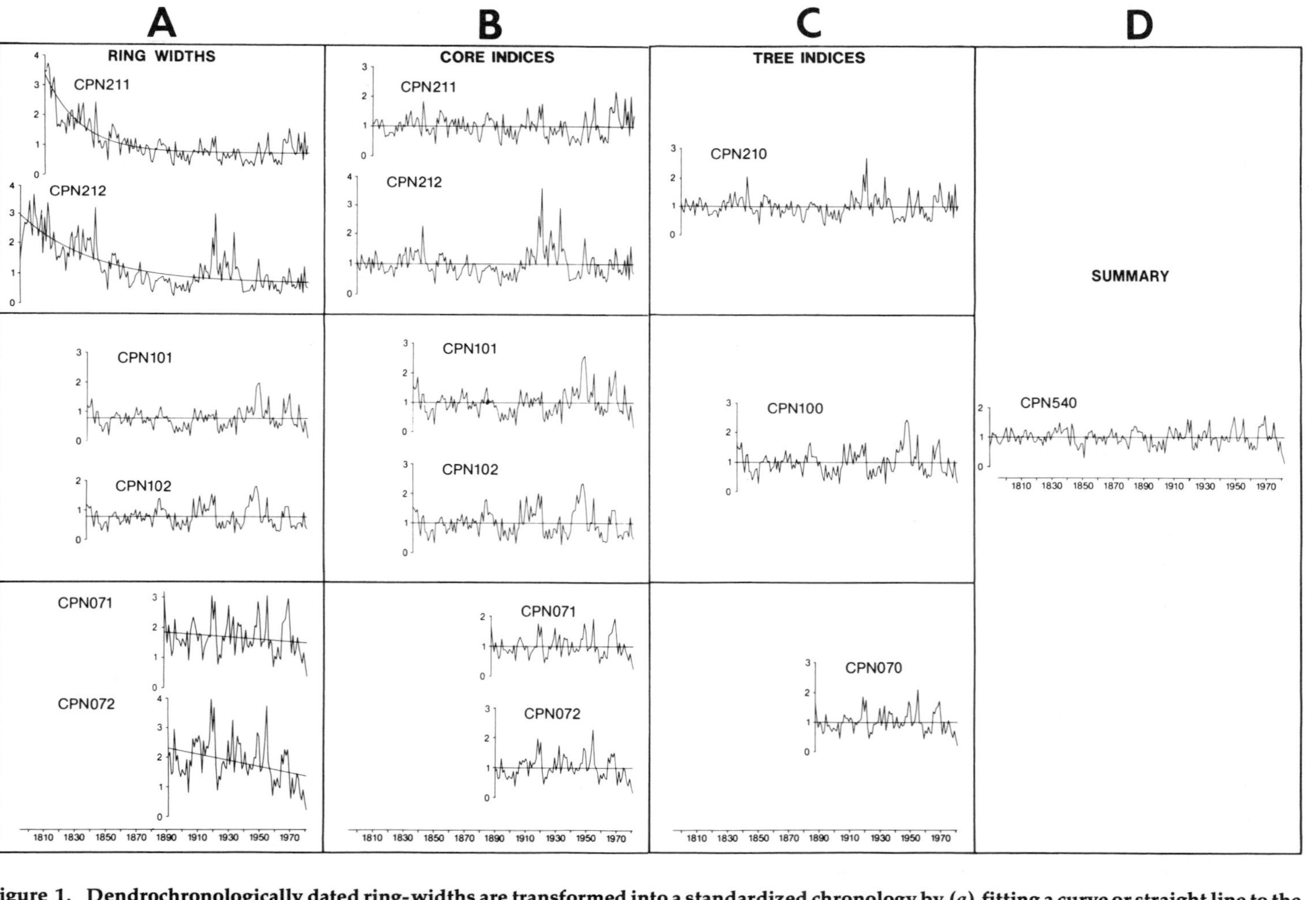

Figure 1. Dendrochronologically dated ring-widths are transformed into a standardized chronology by (*a*) fitting a curve or straight line to the ring widths from each core, (*b*) dividing by the values of the fitted curve to obtain the core indices, (*c*) averaging the core indices for each tree to obtain the tree indices, and (*d*) averaging the tree indices to obtain the summary of the site chronology (Fritts and Swetnam, in press).

curve estimate. At present, this is usually accomplished by dividing the measurement by the value of a fitted negative exponential curve with a constant (Graybill, 1979). Cook and Peters (1981) and Briffa (1984) apply cubic splines or digital filters to estimate the curve, but this approach must be used with care to avoid removing excessive amounts of low-frequency variation. ARMA techniques (Cook, 1985; Holmes et al., 1986; Guiot, 1986) and other proposed standardizing techniques offer possibilities that have not been fully implemented (Warren and MacWilliam, 1981). Cook's Auto Regressive Standardization (ARSTAN) program (Cook, 1985) uses the same negative exponential curve-fitting options, a bivariate mean, and ARMA modeling of the standardized values from each measured radius before combining the data into a site chronology. The average persistence of all series is added to the white noise mean chronology to obtain the ARSTAN chronology. Both the white noise and the ARSTAN chronologies are included in the output for use in different kinds of analysis. Holmes et al. (1986) found Cook's ARSTAN standardizing procedure to be superior. They added a double-detrending option that corrected for slightly higher errors in the outside indices caused by the least squares algorithm used to estimate the exponential curve for standardization.

STEM ANALYSIS

The evidence and rationale for many dendrochronological procedures is well documented in the dendrochronological literature (Fritts, 1969, 1976; Fritts et al., 1965a, 1965c). One example (Fritts et al., 1965b) may be of special interest to forest scientists. It applied the stem analysis techniques of Duff and Nolan (1953) to an evaluation of dendrochronological statistics of rings throughout the main stem of an old, drought-subjected ponderosa pine (*Pinus ponderosa* Laws) growing in northern Arizona, U.S.A. It differed from the usual Duff and Nolan analysis of individual annual rings in that statistics were generated from the data in successive 20-year stem segments.

The values of these statistics were aranged according to four sequences at 20-year intervals following the classification described by Duff and Nolan (1953), and the average sequence was computed simply to reduce the volume of the analysis for publication. Thus the power of the Duff and Nolan approach, which was developed for analysis of growth in young trees, was adapted to study the dendrochronological characteristics of rings throughout the main stem of a mature tree.

The ring-width curve plotted in series A (Figure 2) is similar in shape to the Duff and Nolan curves plotted at 20-year intervals and approximates the negative exponential curve associated with cambial age that is sometimes used for dendrochronological standardization (Figure 1). The low percentages of rings absent for cambial ages from 1–60 years and the low intra- and inter-correlation (correlation of widths around the circuit of that section and correlation with the rings formed on the seven sections above or below them) indicate that the very young rings in the stem are influenced largely by local conditions of the crown, while the older rings throughout the remainder of the main stem provide a more consistent and reliable record of the overall tree growth. The intra- and inter-correlation in series B and C show general agreement in ring patterns at mid-stem heights, although there are lower correlations within the crown, near the root flare, and at mid-stem locations below the junction of major branches. Series D, which is traditionally studied by dendrochronologists, exhibits very litle trend and corresponds to the relatively flat part of the exponential curve used for standardization. The significance of this and related analyses to dendrochronological methodologies is described fully by Fritts (1976). For a different example of a stem analysis of dendrochronologically dated materials, see LeBlanc et al., (1987). Similar blending of forestry and dendrochronological

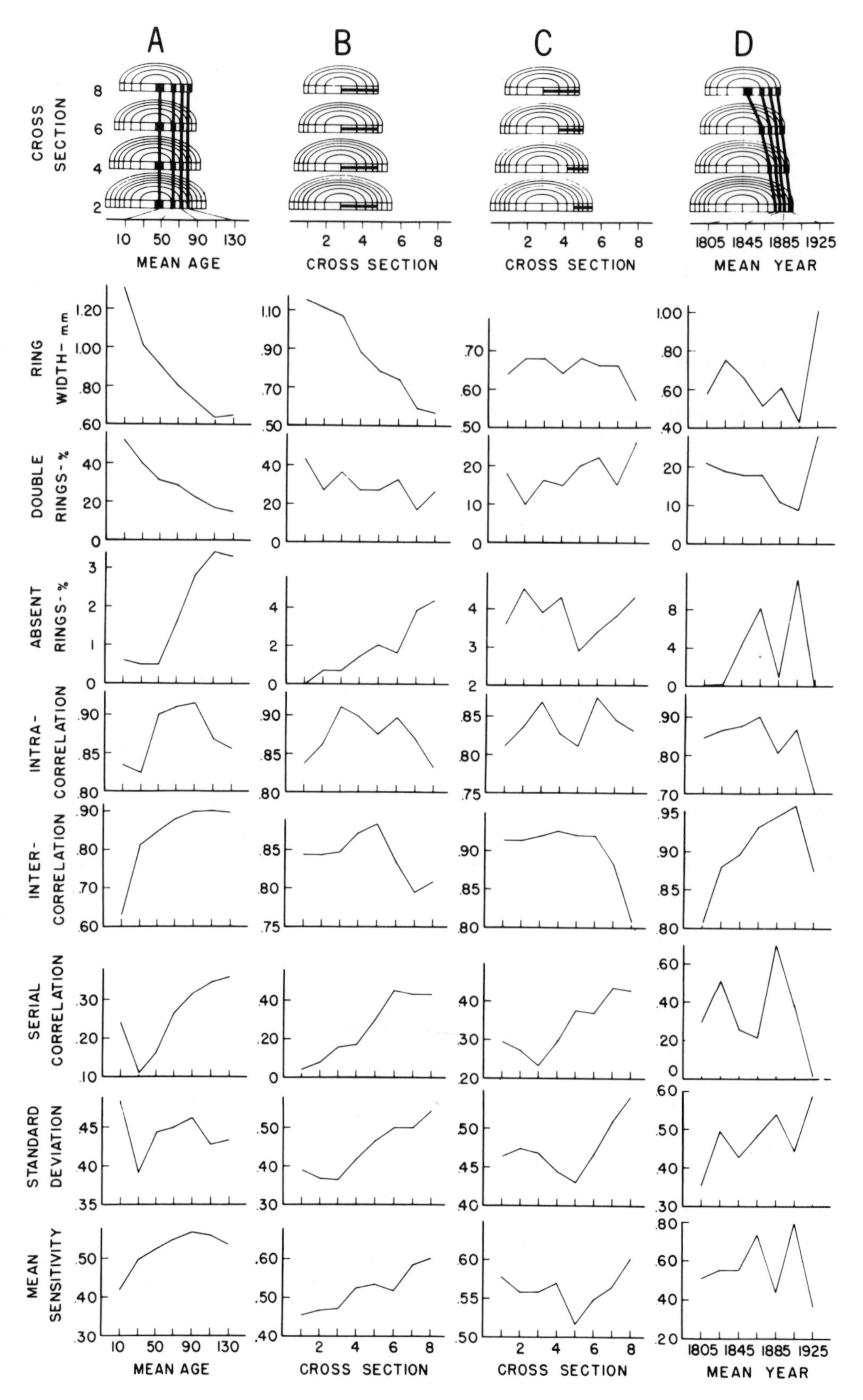
A
B
C
D
CROSS SECTION
8
6
4
2
10 50 90 130
MEAN AGE
2 4 6 8
CROSS SECTION
2 4 6 8
CROSS SECTION
1805 1845 1885 1925
MEAN YEAR
RING WIDTH- mm
DOUBLE RINGS-%
ABSENT RINGS-%
INTRA-CORRELATION
INTER-CORRELATION
SERIAL CORRELATION
STANDARD DEVIATION
MEAN SENSITIVITY
10 50 90 130
MEAN AGE
2 4 6 8
CROSS SECTION
2 4 6 8
CROSS SECTION
1805 1845 1885 1925
MEAN YEAR

approaches in the future will undoubtedly enhance our understanding of forest influences and lead to unique insights in forest modeling work. For this reason, dendrochronologists must actively participate in forest modeling efforts.

DENDROCHRONOLOGICAL STATISTICS

Since tree-growth and climatic relationships often involve many casual factors and complex relationships, linear multivariate models rather than simple correlation models have been used to statistically relate ring-width chronologies to monthly and seasonal environmental conditions thought to influence growth (Fritts, 1962; Fritts et al., 1971; Hughes et al., 1982). Response function analysis is one multivariate technique that uses a principal component regression of monthly climatic data to statistically estimate the ring-width chronology for a site (Fritts et al., 1971). The coefficients of the principal component regression are multiplied by the eigenvectors of climate to obtain the response function coefficients that are interpreted as a growth response (Figure 3). This technique appears to underestimate the size of the confidence limits (Cropper, 1985), and there are a variety of opinions as to which alternative aproach is best (Hughes et al., 1982; Guiot et al., 1982); but many of the techniques produce very similar results (Fritts and Wu, 1986). These multivariate models have been favored over bivariate models because it was thought that a wide number of candidate variables could be considered and screened in a stepwise analysis (Fritts, 1962, 1976). It is now thought that this stepwise selection process substantially biases the probability and overestimates the significance of the final estimates (Draper et al., 1971); Rencher and Pun, 1980). Nevertheless, response functions can reveal possible linear relationships between growth and environmental variables that should be investigated and modeled.

Two response functions are shown in Figure 3, "A" is a typical response function using monthly precipitation and temperature as predictors of ponderosa pine growth (Fritts, 1976). "B" is a response function analysis on the same tree-ring data but using monthly Palmer Drought Severity Indices instead of precipitation and temperature amounts.

Table 1 sumarizes response function results using monthly temperature and precipitation as predictors of 127 western North American chronologies (Fritts, 1974). The percentages in the table reflect the number of response functions with at least one significant coefficient ($p \cong 0.66$) associated with that variable, season, and sign. For example, 87 to 93% of the response function weights were significant and positive for precipitation in autumn, winter, and spring. Only 46 and 49% were significant and positive for the prior and current summer months. This emphasizes the importance of autumn, winter, and spring precipitation to ring-width growth in the arid sites of the West. A significant negative coefficient for precipitation is rare except for summer months.

Figure 2 (facing page).

Changes in ring-width statistics associated with four sequences described by Duff and Nolan (1953) using the ring measurements from eight sections (four radial measurements per section) of ponderosa pine (*Pinus ponderosa* Laws.). The statistics were calculated from each 20-year segment and then averaged according to the sequence as shown by the direction of the heavy lines in the diagrams at the top of the figure. The odd-numbered sections and every other 20-year segment are omitted from the diagram for clarity. The plots express the averages of each statistic as a function of the 20-year segments, from the inside toward the outside, or the cross-section number, ordered from the bottom to the top. Series A is Duff and Nolan's first horizontal sequence except that its units are 20-year segments rather than yearly increments. Series B is the vertical sequence, series C is the diagonal sequence, and series D is the second horizontal sequence.

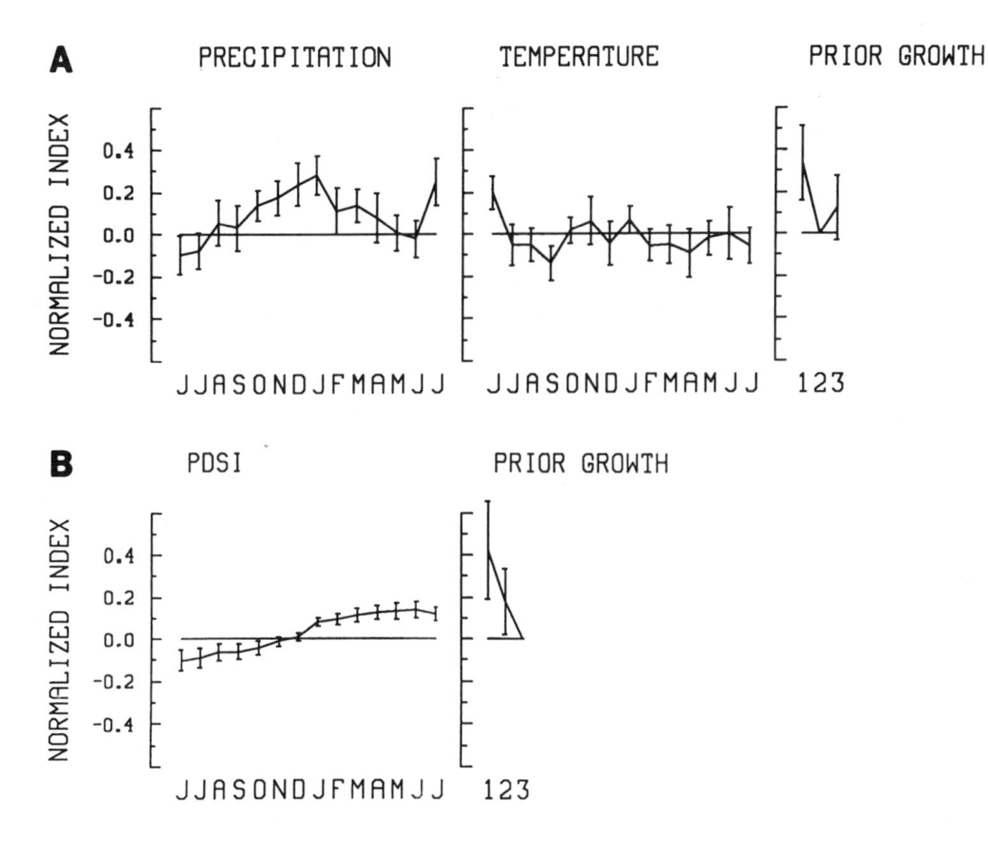

Figure 3. **The weights of two response functions calculated for the average values of the Walnut Canyon and Show Low chronologies of Arizona, U.S.A., ponderosa pine. Each weight can be interpreted as the relative effect of monthly precipitation, temperature, or Palmer Drought Severity Index (PDSI) on growth. (*a*) The response to temperature, precipitation, and prior growth for 14 months starting with June of the year prior to growth. Climate explains 66% of the growth variance, and prior growth explains an additional 17% of the variance. (*b*) The response to PDSI and prior growth for the same 14 months. PDSI explains 45% of the growth variance, and prior growth explains an additional 29% of the variance. The vertical lines are the 0.67 probability limits. Note that the weights for precipitation are significantly positive from October through March (i.e., their error bars do not intercept zero), and nine of the temperature weights are negative although only two are significant. Prior growth at a 1-year lag is most consistently significant. This suggests that high precipitation and low temperatures favor high growth through replenishment of soil moisture and reduced water stress. The PDSI response probably reflects the lag built into the index and the high colinearity in the values for different months.**

Table 1. Significant response function coefficients.

	Temperature		Precipitation	
Season	% Positive	% Negative	% Positive	% Negative
Prior summer	35	57	49	40
Autumn	31	69	87	28
Winter	33	57	90	18
Spring	22	73	93	8
June–July	36	71	46	31
Weighted average	31	65	75	25

With respect to temperature (Table 1), the negative relationships dominate and occur at a frequency of 57 to 73%. Some positive temperature relationships pass the significance test, but the frequency is half that of the inverse temperature effects. This suggests that, at least on these stress sites, temperature limits growth most often through its effects on water balance, although some direct temperature effects are not ruled out (Fritts, 1976).

A useful technique for response-function modeling is to construct a response surface, which is a three-dimensional representation of a relationship helpful in identifying nonlinearities, variable interactions, and unmodeled sources of variation affecting growth (Fritts et al., 1965a; Graumlich and Brubaker, 1986). Figure 4 includes two unsmoothed response surfaces for the same arid-site ponderosa pine considered in Figure 3. Surface A shows an inverse response to spring temperature, but the relationship with spring precipitation appears either weakly inverse or uncertain at best. There is no warping or twist in the surface, so it may be inferred that no variable interaction is present.

Surface B includes a simulation of pine (*Pinus*) growth using the Drought Index for Southern Pine (DISP) model (see Zahner and Grier, this volume) applied to Arizona Palmer Drought Severity Index (PDSI) data for the Arizona climatic division most proximate to the trees. These results are plotted against the same spring temperature and same ponderosa pine growth that was shown in the A surface. Because the model was originally developed for southern pine and for associated climatic regions, we did not know what to expect. However, it was an exercise to examine how well the model might apply to arid-site trees in the western U.S.A.

The surface associated with the simulation, which takes soil moisture into account by using the PDSI, exhibits a steeper slope than that shown for precipitation and temperature in the A surface. A regression analysis indicates that 51% of the ponderosa pine ring-width variance was reduced by the simulation, and that it was highly significant. An interaction is apparent by the twist in the surface associated with the simulation and spring temperature. This relationship was tested by adding temperature and a cross-product

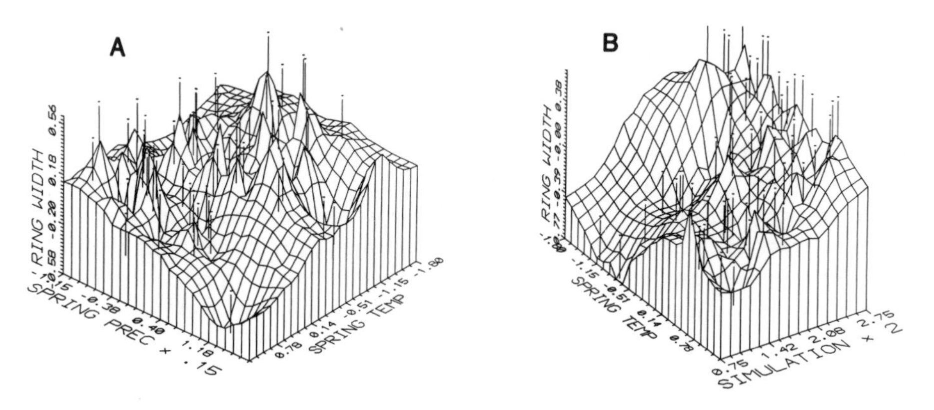

Figure 4. Two unsmoothed response surfaces for the arid-site ponderosa pine chronologies used in Figure 3. Surface A displays only a weak inverse response to spring (March–June) temperature as in the response function, but the relationship with precipitation is unclear or weakly inverse. No interaction between these variables is apparent in this plot. Surface B includes a simulation of pine growth using the DISP model (see Zahner and Grier, this volume) applied to Arizona, U.S.A., PDSI amounts and a plot of the spring temperature effects. 51% of the growth variance was reduced by the simulation, and a significant interaction with temperature was evident which accounted for an additional 5% of the variance. Spring precipitation was multiplied by 0.15 and the simulation multiplied by 2 to scale the axes for plotting purposes.

term to the regression, and all three variables had significant t values ($p > 0.95$) and reduced 55.5% of the variance.

In addition to indicating the presence of interaction, these results, along with the response functions shown in Figure 3, suggest that the DISP model may be useful outside the southern U.S.A.; but as one would expect, the model needs to be adapted to these arid sites by considering possible interactions with temperature not involving the PDSI. The surface indicates that actual growth is lower than simulated growth when spring temperatures are low. Although not particularly evident in the plots shown, this is in agreement with many response functions for ponderosa pine which suggest that a direct response of growth to temperature in April and May is sometimes present (Fritts, 1974). This may reflect the inhibition of growth in the early season, when near-freezing temperatures become more limiting to cambial activity, rather than inhibition later in the season, when warm temperatures might lead to high water stress (Fritts et al., 1965a; Fritts, 1976). These kinds of dendrochronological techniques and data offer a valuable source of information that may be used in the development, improvement, and testing of certain types of forest models.

ENVIRONMENTAL-GROWTH STUDIES

Two decades ago, response function results suggested that winter and autumn precipitation was as important to ring-width growth in arid-site trees as precipitation in spring and summer. It was hypothesized that more was involved in this relationship than a simple soil moisture storage. Photosynthesis throughout the autumn, winter, and spring seemed to be the most likely candidate (Fritts et al., 1965a; Fritts, 1976). Active photosynthesis during the winter months had not been demonstrated under field conditions.

A series of investigations were designed to test this hypothesis. Radial growth and apical growth were measured at several locations using dendographs, dendrometers, anatomical sampling, and phenological observations (Fritts, 1976). A variety of species and environmental variables were monitored (Fritts et al., 1965a; Fritts, 1969), and soil moisture changes, water balance, and net photosynthesis were studied throughout the autumn, winter, and spring as well as during the growing season (Brown, 1968; Budelsky, 1969; Drew, 1967; Drew et al., 1972; Fritts, 1976) (Figure 5).

Radial growth began from late April through late June, depending upon altitude and tree exposure, and often continued into August or sometimes later. However, the rate of growth rather than the length of the growing season appeared to be the primary determinant of variations in ring width (Fritts, 1976; Vaganov et al., 1985). Net photosynthesis was found to occur throughout the entire year in semi-arid sites as long as daytime temperatures were above freezing (Figure 5). However, at high altitudes (3,000 m or higher) or during extremely cold weather (when temperatures were consistently well below freezing), wintertime photosynthesis was not measured even when daytime temperatures were above freezing. With higher daytime temperatures, net photosynthesis increased along with greater loss of moisture from both evaporation and transpiration. In general, net photosynthesis was rapid until there was a substantial decline in available soil moisture or until high daytime temperatures became the most limiting factor.

In addition to measuring stem size changes over the growing season, cores were extracted at the end of each study from the stem radii that had been monitored, and cell size changes across the rings were carefully measured. Some of the relationships derived from these growth and environmental data were used to develop conceptual models of growth (Fritts, 1971, 1974, 1976), and the data now serve as base-line information used to develop and evaluate the following simulation models of cambial activity.

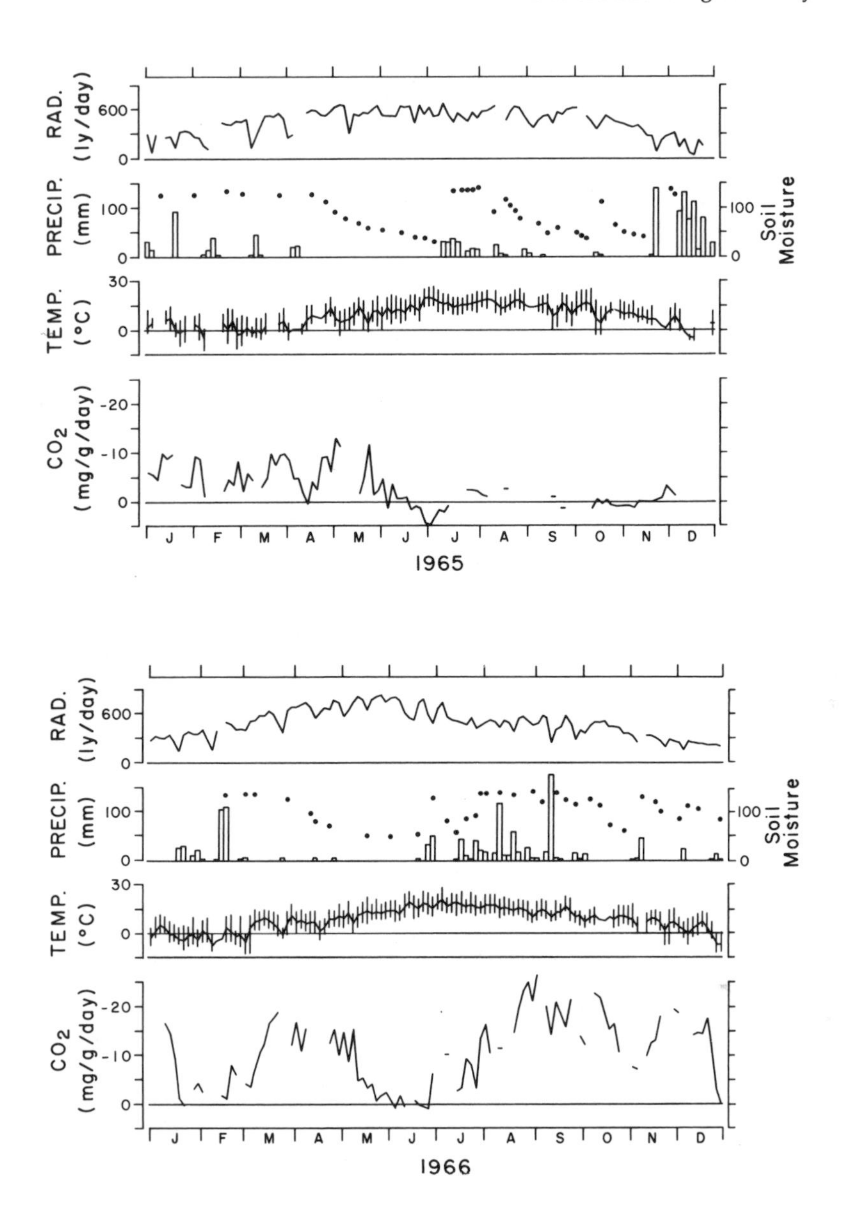

Figure 5. **An example of two years of CO_2 flux and other measurements from ponderosa pine growing on a south-facing slope at 2600 m in southern Arizona, U.S.A. 1966 was much wetter than 1965, but 1965 was not exceptionally dry; it was primarily lacking in summer and autumn moisture. Heavy precipitation late in November and December replenished soil moisture for the 1966 season. Note that net photosynthesis was rapid in winter whenever daytime temperatures were not limiting; higher rates were noted in 1966 when soil moisture levels were high. Rates of net photosynthesis decline with the approach of the warmer and dryer conditions of late May and June. By late June, water stress is so extreme that respiration exceeds photosynthesis, yet the cambium was still active although the rate appeared to be reduced by high water deficits in the growing tissues. After the summer rains, the photosynthetic activity rises but continues to depend upon the amount of soil moisture. The lower amounts of summer rainfall and associated soil moisture in 1965 greatly reduced net photosynthesis during late summer continuing well into the autumn period. In contrast, the rates of net photosynthesis attained the highest levels in late August and September of 1966, as daytime temperatures were moderated by declining solar radiation at a time when the soil still had sufficient moisture.**

SIMULATING TREE-RING DATA

I began computer modeling in 1969 when I was working on stochastic methods for generating synthetic tree-ring series from random representations of the climatic signal using time-series statistics. This work resulted in program RAND, which is capable of generating synthetic time-series of known characteristics. Such synthetic data can be used in Monte Carlo tests of statistics and calibration techniques used in dendrochronology. The program simulates the transfer of yearly limiting climatic conditions over a long period of time into a time series mimicking real-world tree-ring chronologies.

In 1975, my associate, Donald Stevens, and I developed ARBOR, which is a modification and extension of the cambial model of Wilson and Howard (1968), to simulate realistic features of the ring caused by variations in cell numbers, cell sizes, and wall thickness (Stevens, 1975). The original model has been modified for a personal computer by my associate, Martin Rose, and we hope to expand the model to simulate some of the most important effects of environmental conditions on cambial activity and ring character.

The cambium is considered to be a single layer of small dividing cells producing an adjoining layer of phloem mother cells on the outside and xylem mother cells on the inside. The thicknesses of these mother-cell layers are specified by the model, and as new cells are produced, the outermost individuals are forced outside the layer where they lose their ability to divide and begin to enlarge. When a phloem cell reaches its maximum size, it is mature; but the fully enlarged xylem cell enters one more phase of growth, wall thickening, and it matures when its wall attains the maximum thickness allowed at that time.

ARBOR simulates the following processes day by day as the season progresses:

1. Radial growth, elongation and division of the cambial initial cell.
2. Radial growth and division of the mother cells.
3. Differentiation of mother cells into enlarging and maturing phloem and xylem cells.
4. Radial enlargement of phloem and xylem cells.
5. Thickening of the xylem cell walls after enlargement.

The model uses 11 parameters and 11 time-dependent variables to control these processes (Figure 6). The program identifies each cell by its position in the ring and keeps track of its characteristics as it passes through the various stages in its development (Figure 7).

The cell sizes, wall thicknesses, and relative densities of the synthesized cells in the ring are shown in Figure 8. This is a rather ordinary ring with diminishing cell size and increasing cell-wall thickness across the ring surface. The model is more interesting and challenging when one tries to generate false rings and the varying cell charateristics of a given species growing under prescribed conditions.

Vaganov et al. (1981, 1985) have developed a biophysical model for cell number and size production by the cambium that is based upon the kinetics of xylem mother cell formation. The diameters of the mature xylem cells are modeled as a function of the specific growth rates of the xylem mother cells active in the cambium. They also relate these processes to controlling environmental factors in attempts to simulate the effects of climate on the cell size changes within the annual ring.

Our current efforts involve collaboration with Vaganov* in an attempt to use the strengths of the Vaganov and ARBOR models to improve our capacity to simulate the processes of cell division as they are influenced by the most important environmental controls that in turn influence the structure of the annual ring.

---

*Dr. Vaganov was under the auspices of the U.S.-U.S.S.R. Environmental Protection Agreement (EPA Project #02.03-21), Soviet Union Scientific Exchange.

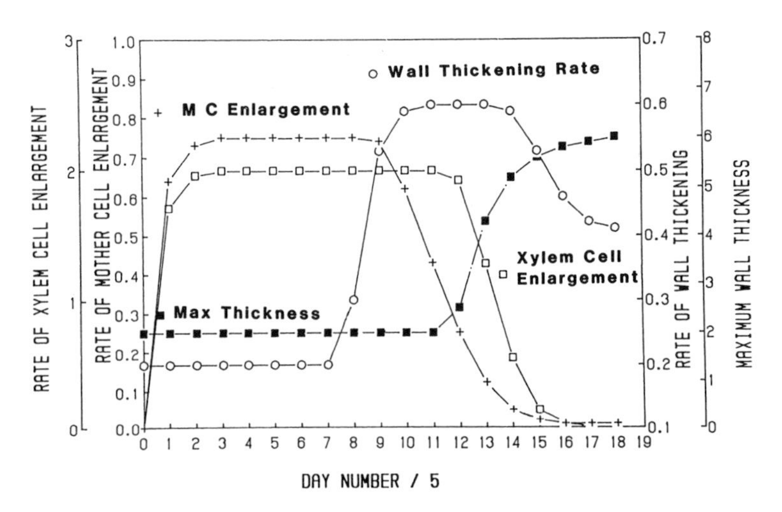

Figure 6. An example of the values assigned to four variables used in the simulation of a 90-day growing season. The rates of enlargement of both mother cells an xylem cells rise to maximum values early in the growing season, which continue until intervals 9 or 11, when the rates begin to decline, reaching zero levels before the end of the growing season. The maximum wall thickness remains low until interval 11 and then increases. The rate of wall thickening is modeled to be low at first, rises to maximum rates after interval 10, and after 14 is modeled to decline. The units are in microns or microns per day.

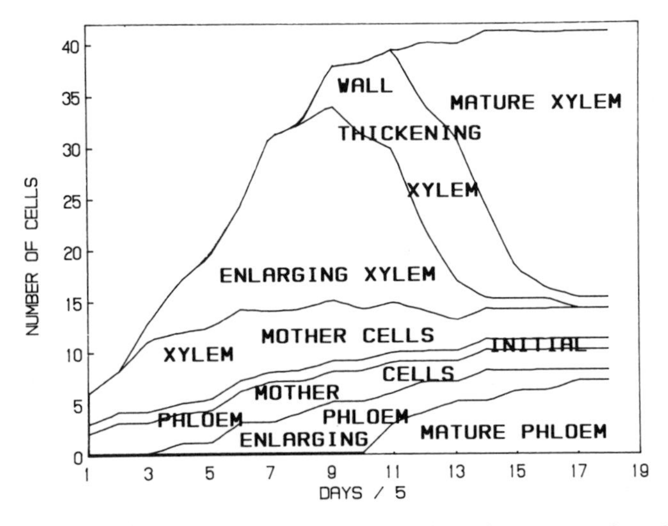

Figure 7. The accumulative number of cells modeled in each category plotted as a function of 5-day intervals over the growing season. The top line is the total number of cells produced by the model. Enlarging phloem and xylem appear after the 20th day (that is, on interval 4 in the figure). The first xylem wall-thickening begins after interval 8, and the first mature xylem was noted at interval 10 (50 days from the beginning of the growing season). By interval 15, all cell division ceases, and the enlarging and thickening xylem cells eventually mature by interval 19, which is the end of the growing season of 90 days. The fact that the wall of one xylem cell is still thickening and one phloem was enlarging at the end of the season suggests that a slightly longer growing season should have been modeled.

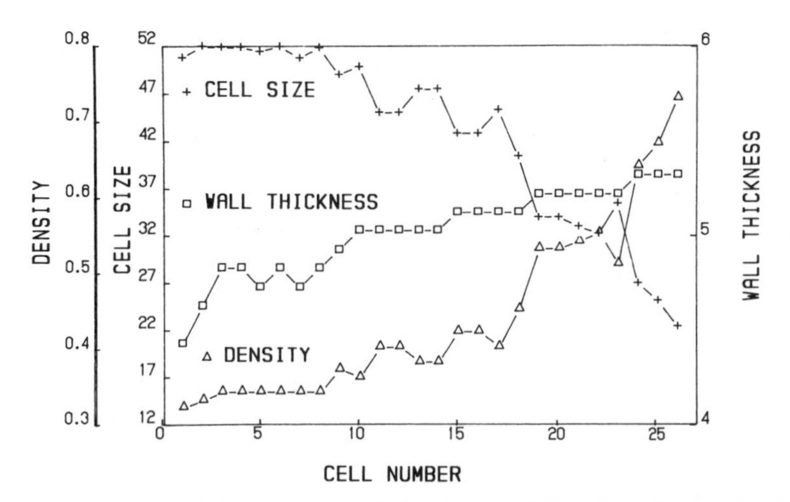

Figure 8. **The cell size, wall thickness, and relative density of the cells in the simulated ring, showing a gradual transition from the low-density earlywood to the high-density latewood. The units are in microns. Density is the fraction of cell wall compared to cell volume.**

FUTURE DIRECTIONS

It would be presumptuous to suggest that these simple cambial models are comparable in complexity to, or in competition with, any existing forest process models in use today. However, I believe it is realistic for dendrochronologists to experiment with the Wilson and Vaganov models, and any other model of cambial activity, as a means of understanding the fundamental principles governing ring width, density changes, and the quality of wood. As these models are developed more fully, perhaps they could be linked to state-of-the-art process models simulating photosynthesis, water stress, and other basic conditions affecting tree growth to see if it is possible to generate meaningful representations of tree-ring features, including the density of the wood. Such efforts could advance our understanding of the important stem-growth process and allow model outputs to be expressed in terms of realistic features of cell-size variations, density changes, and ring width.

LITERATURE CITED

Briffa, K. R. 1984. Tree-climate relationships and dendroclimatological reconstruction in the British Isles. Ph.D. Thesis, University of East Anglia, Norwich, England, U.K.

Brown, J. M. 1968. The photosynthetic regime of some southern Arizona ponderosa pine. Ph.D. Thesis, University of Arizona, Tucson, AZ, U.S.A. 99 pp.

Brubaker, L. B., and E. R. Cook. 1983. Tree-ring studies of holocene environments. In: Wright, H. E., Jr., ed. *Late Quaternary Environments of the United States,* Vol. 2. *The Holocene.* University of Minnesota Press, Minneapolis, MN, U.S.A. Pp. 222–235.

Budelsky, C. A. 1969. Variation in transpiration and its relationship with growth for *Pinus ponderosa* 'Lawson' in southern Arizona. Ph.D. Thesis, University of Arizona, Tucson, AZ, U.S.A.

Cook, E. R. 1985. A time-series analysis approach to tree-ring standardization. Ph.D. Thesis, University of Arizona, Tucson, AZ, U.S.A. 104 pp.

Cook, E. R., and K. Peters. 1981. The smoothing spline: a new approach to standardizing forest interior tree-ring width series for dendroclimatic studies. *Tree-Ring Bulletin* 41:45–53.

Cropper, J. P. 1985. Tree-ring response functions: an evaluation by means of simulations. Ph.D. Thesis, University of Arizona, Tucson, AZ, U.S.A. 132 pp.

Dieterich, J. H., and T. W. Swetnam. 1984. Dendrochronology of a fire-scarred ponderosa pine. *Forest Science* 30(1):238–247.
Draper, N. R., I. Guttman, and H. Kanemasu. 1971. The distribution of certain regression statistics. *Biometrika* 58(2):295–298.
Drew, A. P., 1967 Stomatal activity in semi-arid site ponderosa pine. M.S. Thesis, University of Arizona, Tucson, AZ, U.S.A. 121 pp.
Drew, A. P., L. G. Drew, and H. C. Fritts. 1972. Environmental control of stomatal activity in mature semiarid site ponderosa pine. *Journal of the Arizona Academy of Sciences* 7(2):85–93.
Duff, G. H., and N. J. Nolan. 1953. Growth and morphogenesis in the Canadian forest species, I. The controls of cambial and apical activity in *Pinus resinosa* Ait. *Canadian Journal of Botany* 31:471–513.
Fritts, H. C. 1962. An approach to dendroclimatology: screening by means of multiple regression techniques. *Journal of Geophysical Research* 67(4):1413–1420.
Fritts, H. C. 1969. Bristlecone pine in the White Mountains of California: growth and ring-width characteristics. Papers of the Laboratory of Tree-Ring Research, 4. University of Arizona Press, Tucson, AZ, U.S.A.
Fritts, H. C. 1971. Dendroclimatology and dendroecology. *Quaternary Research* 1(4):419–449.
Fritts, H. C. 1974. Relationships of ring widths in arid-site conifers to variations in monthly temperature and precipitation. *Ecological Monograph* 44(4):411–440.
Fritts, H. C. 1976. *Tree Rings and Climate.* Academic Press, London, England, U.K. 567 pp. Reprinted 1987 as *Tree Rings and Climate,* In: Kairiukstis, L., Z. Bednarz, and E. Feliksik, eds. *Methods of Dendrochronology,* Vols. II and III. International Institute for Applied Systems Analysis and the Polish Academy of Sciences, Warsaw, Poland.
Fritts, H. C., T. J. Blasing, B. P. Hayden, and J. E. Kutzbach. 1971. Multivariate techniques for specifying tree-growth and climate relationships and for reconstructing anomalies in paleoclimate. *Journal of Applied Meteorology* 10(5):845–864.
Fritts, H. C., D. G. Smith, C. A. Budelsky, and J. W. Cardis. 1965b. The variability of ring characteristics within trees as shown by a reanalysis of four ponderosa pine. *Tree-Ring Bulletin* 27(1–2):3–18.
Fritts, H. C., D. G. Smith, J. W. Cardis, and C. A. Budelsky. 1965c. Tree-ring characteristics along a vegetation gradient in northern Arizona. *Ecology* 46(4):393–401.
Fritts, H. C., D. G. Smith, and M. A. Stokes. 1965a. The biological model for paleoclimatic interpretation of Mesa Verde tree-ring series. *American Antiquity* 31(2, Part 2):101–121.
Fritts, H. C., and T. W. Swetnam. In press. Dendroecology: a tool for evaluating variations in past and present forest environments. *Advances in Ecological Research* 19.
Fritts, H. C., and Wu Xiangding. 1986. A comparison between response function analysis and other regression techniques. *Tree-Ring Bulletin* 46:31–46.
Graumlich, L., and L. B. Brubaker. 1986. Reconstruction of annual temperatures (1590–1979) for Longmire, Washington, derived from tree-rings. *Quaternary Research* 25:223–234.
Graybill, D. A. 1979. Revised computer programs for tree-ring research. *Tree-Ring Bulletin* 39:77–82.
Guiot, J. 1986. ARMA techniques for modeling tree-ring response to climate and for reconstructing variations of paleoclimates. *Ecological Modeling* 33:149–171.
Guiot, J., F. Serre-Bachet, and L. Tessier. 1982. Application of ARMA modeling in dendroclimatology. *Comptes Rendus de l'Académie des Sciences de Paris,* Series III, 294:133–136. (In French with English abstract.)
Holmes, R. L. 1983. Computer-assisted quality control in tree-ring dating and measurement. *Tree-Ring Bulletin* 43:69–78.
Holmes, R. L., R. K. Adams, and H. C. Fritts. 1986. Tree-ring chronologies of western North America: California, eastern Oregon and northern Great Basin. Chronology Series VI, Laboratory of Tree-Ring Research, University of Arizona, Tucson, AZ, U.S.A.
Hughes, M. K., P. M. Kelly, J. R. Pilcher, and V. C. LaMarche, Jr., eds. 1982. *Climate from Tree Rings.* Cambridge University Press, Cambridge, England, U.K. 223 pp.
LaMarche, V. C., Jr. 1978. Tree-ring evidence of past climatic variability. *Nature* 276:334–338.
LaMarche, V. C., Jr. 1982. Sampling strategies. In: Hughes, M. K., P. M. Kelly, J. R. Pilcher, and V. C. LaMarche, Jr., eds. *Climate from Tree Rings.* Cambridge University Press, Cambridge, England, U.K. Pp. 2–6.
LaMarche, V. C., Jr., and T. P. Harlan. 1973. Accuracy of tree-ring dating of bristlecone pine for calibration of the radiocarbon time scale. *Journal of Geophysical Research* 78(36):8849–8858.
LeBlanc, D. C., D. J. Raynal, E. H. White, and E. H. Ketchledge. 1987. Characterization of historical growth patterns in declining red spruce trees. In: *Proceedings of the International Symposium on Ecological Aspects of Tree-Ring Analysis,* August 17–21, 1986, Marymount College, Tarrytown, NY, U.S.A. U.S. Department of Energy CONF-8608144. Pp. 360–371.

Rencher, A. C., and F. C. Pun. 1980. Inflation of R^2 in best subset regression. *Technometrics* 22(1):49–53.

Schweingruber, F. H. 1988. *Tree Rings: Basics and Applications of Dendrochronology.* D. Reidel Publishing Co., Dordrecht, the Netherlands. 292 pp.

Stevens, D. 1975. A computer program for simulating cambium activity and ring growth. *Tree-Ring Bulletin* 35:49–56.

Stockton, C. W., W. R. Boggess, and D. M. Meko. 1985. Climate and tree rings. In: Hecht, A. D., ed. *Paleoclimate Analysis and Modeling.* John Wiley and Sons, New York, NY, U.S.A. Pp. 71–161.

Swetnam T. W. 1987. A dendrochronological assessment of Western spruce budworm, *Choristoneura occidentalis* Freeman, in the southern Rocky Mountains. Ph.D. Thesis, University of Arizona, Tucson, AZ, U.S.A. 213 pp.

Vaganov, E. A., A. V. Shashkin, I. V. Sviderskaya, and L. G. Vysotskaya. 1985. *Histometric Analysis of Woody Plant Growth.* Nauka, Novosibirsk, U.S.S.R. 102 pp.

Vaganov, E. A., L. P. Starova, and A. V. Shashkin. 1981. Seasonal tree-growth modelling using the number and size of cells in tree rings. In: Terskov, I. A., ed. *Research of the Organism Growth Dynamics.* Nauka, Novosibirsk, U.S.S.R. Pp. 67–78. (In Russian.)

Van Deusen, P. C. 1989. A simultaneous approach to tree-ring analysis. In: Dixon, R. K., R. S. Meldahl, G. A. Ruark, and W. G. Warren, eds. *Process Modeling of Forest Growth Responses to Environmental Stress.* Timber Press, Portland, OR, U.S.A.

Warren, W. G., and S. L. MacWilliam. 1981. Test of a new method for removing the growth trend from dendrochronological data. *Tree-Ring Bulletin* 41:55–66.

Wilson, B. F., and R. A. Howard. 1968. A computer model for cambial activity. *Forest Science* 14(1):77–90.

Zahner, R., and C. E. Grier. 1989. Concept for a model to assess the impact of climate on the growth of the southern pines. In: Dixon, R. K., R. S. Meldahl, G. A. Ruark, and W. G. Warren, eds. *Process Modeling of Forest Growth Responses to Environmental Stress.* Timber Press, Portland, OR, U.S.A.

CHAPTER 34

CONCEPT FOR A MODEL TO ASSESS THE IMPACT OF CLIMATE ON THE GROWTH OF THE SOUTHERN PINES

Robert Zahner and Charles E. Grier

Abstract. A new model, Drought Index for Southern Pines (DISP), makes use of both the timing and intensity of wet and dry weather during the growing season to calculate an index of the monthly and annual departures from the long-term average, or normal, pattern of radial growth. The proposed index is intended for stand-level growth responses in adequately stocked, even-aged stands of any of the four major species of U.S.A. southern pines, beyond the juvenile stage, growing on typical upland sites in their natural ranges. The method integrates the impacts of both the current moisture condition and preconditioning due to antecedent soil water deficits or excesses. The DISP model uses monthly values of the Palmer Drought Severity Index (PDSI) as the sole climate input variable because this index is an integrated measure of the daily and monthly maximum and minimum temperatures, precipitation, and soil moisture regimes impinging on forest vegetation at a given season of the year at a given geographical location. Examples are given to demonstrate the computations used in the DISP model and the relation of DISP values to annual and periodic forest growth inventories. DISP software is available for personal computers.

INTRODUCTION

Annual variation in radial growth of the southern pines in the United States is more sensitive to soil water deficits than to any other climatic stress (Zahner, 1968). The impact of water stress, or the lack of it, on monthly, annual, and periodic radial growth can be assessed indirectly from weather data that reflect the soil water and evapotranspiration regimes in a given region (Zahner and Stage, 1966; Federer, 1982). Forest inventory and growth analyses obviously are more precise when they account for this climate impact in addition to the usual stand density and site quality impacts.

This study develops the concept for an index of annual departure from long-term, average, radial growth for the southern pines of the U.S.A., based on the response of current growth to previous and current climate. The new index, called the Drought Index for Southern Pines (DISP), makes use of published monthly values of the Palmer Drought Severity Index (PDSI) adapted to impact the seasonal pattern of southern pine growth. The DISP model accounts for both the timing and intensity of wet and dry periods during the growing season, integrating these impacts into a measure of the annual departure from

---

Dr. Zahner is at the Department of Forestry, Clemson University, Clemson, SC, 29634, U.S.A. and Mr. Grier is at the Southeastern Forest Experiment Station, Athens, GA, 30602 U.S.A.
Research supported by the National Vegetation Survey, National Acid Precipitation Assessment Program, through the USDA Forest Service Southern Forest Experiment Station, New Orleans, LA, U.S.A.

the long-term average, or departure from normal growth.

The southern pines are capable of utilizing long growing seasons for radial growth. Cambial activity is initiated along with expansion of terminal buds in late winter. Cambial growth processes reach maximum rates within a few weeks, sustain relatively high rates into early summer, then decline in early fall. This annual cycle of radial growth can be greatly modified by soil water deficits (Moehring and Ralston, 1967; Zahner, 1968; Fritts, 1976; Kramer and Kozlowski, 1979; Zahner and Myers, 1986). Cambial activity accelerates during summer periods of excessive rainfall and slows during periods of moderate drought. Radial growth ceases during severe summer drought but may resume at significant rates when ample rainfall recharges the soil late in the season.

The southern pines are opportunistic species, responding quickly to favorable current conditions, both in photosynthate production and in cambial activity. If late season water deficits have resulted in low food reserves going into the dormant season, for example, active photosynthesis during warm winter months replenishes reserves (Allen, 1976; Reynolds et al., 1980) and mitigates the effects of previous drought. Earlywood growth is intrinsically uniform, while latewood growth is strongly regulated by the immediate, current moisture conditions (Zahner, 1963, 1968).

The Palmer Drought Severity Index is the U.S. National Weather Service and U.S. Department of Agriculture standard indicator of relatively wet and dry local conditions, and monthly values are readily available for all geographical locations throughout the United States (Palmer, 1965; Karl and Knight, 1985). No single value of the Palmer index, however, is indicative of climatic impacts for a complete growing season. Like other weather variables, Palmer indices must be correlated with vegetative responses before their impacts on growth can be assessed.

THE PALMER DROUGHT SEVERITY INDEX

Drought is a prolonged and abnormal period of water deficit, as compared with the long-term average condition for a given geographical location. Relative drought and relative wetness are measured quantitatively by water deficits and water excesses, a complex numerical combination of potential evapotranspiration, soil water availability, precipitation, and actual evapotranspiration. The PDSI emphasizes the current moisture condition (Palmer, 1965) relative to the normal. It is calculated from daily and monthly weather measurements of precipitation and maximum and minimum temperatures at specific geographical locations. The procedure uses both antecedent and current weather data to establish the current moisture condition. The PDSI relates the current water deficit or water excess to the long-term deficit or excess for that calendar date at each location. A PDSI current value of 0 indicates that current weather is normal at that geographical location at that point in time. Values of PDSI greater than 0 give relative wetness conditions, and values less than 0 give relative drought condition.

A single PDSI value has meaning only for the current condition. PDSI averages over time periods longer than one month often cannot be correlated with vegetative response because they can embrace contrary conditions on opposite sides of 0 that are not reflected in the average. It is assumed that monthly values of the PDSI satisfactorily reflect the true moisture condition for the given month, and that sequential monthly values reflect changing climatic conditions through a season.

The U.S. National Weather Service has identified geographic regions of uniform climate and has summarized PDSI values from all reporting stations within each such region. These regions are political units for bookkeeping purposes, based on county and state lines, but the boundaries closely follow appropriate physical and climatic features. Thus, each state is subdivided into uniform climatic divisions.

The PDSI does not permit the seperation of sites or soil types. The water balance calculations are based on the average site within a region and are computed for potential evapotranspiration on level terrain. Soil water deficits and excesses are calculated for medium-textured soils of moderate rooting depth, storing a maximum of 15 cm of available water, yielding stored water to evapotranspiration at a rate proportional to that available. The method permits all rainfall to infiltrate, enter soil storage as retention and detention water, and become part of the water balance through alleviating deficits or adding to excesses.

These soil and site limitations do not seriously reduce the validity of the PDSI as an effective indicator of month-to-month climatic impacts on forest growth averaged over a region of uniform climate. Many forest sites and soil types throughout the range of the southern pines match the conditions described for the PDSI; thus, the regional climatic imact on vegetation may be accurately quantified using the PDSI for each region. It must be stressed, however, that forest stands on specific sites have diverse water regimes and therefore experience quite variable water deficits and water excesses within the same climatic region (Zahner and Myers, 1986). Whereas growth of forest stands across a broad spectrum of sites is adversely affected by extreme drought, recovery is normally rapid when the stress is alleviated on average and better sites, but a growth decline may be triggered on sites already under predisposing stress.

The PDSI is calculated from the water balance model of Thornthwaite and Mather (1955), which assumes a complete vegetative cover over the landscape with roots fully occupying the soil. In general, forest vegetation meets this condition, and the PDSI yields a realistic estimate of relative water stress for stands of trees on average sites. Certainly, young managed stands of southern pine are stocked adequately to fully occupy the soil on an average site. The impact of climate on radial growth in such stands can confidently be measured by the water balance method.

THE MODEL FOR A DROUGHT INDEX FOR SOUTHERN PINES

The new simulator, DISP, makes use of both the timing and intensity of wet and dry weather during the growing season to calculate an index of the monthly and annual departures from the long-term average, or normal, pattern of radial growth. The proposed index is intended for stand-level growth responses in adequately stocked stands of any of the four major species of southern pines, beyond the juvenile stage, growing on typical upland sites in their natural ranges. The method integrates the impacts of both the current moisture condition and preconditioning due to antecedent soil water deficits or excesses, by incorporating monthly values of the PDSI into the radial growth pattern of southern pines. Monthly impacts on simulated current growth rates are integrated into a single annual index: the DISP.

The four major pine species, loblolly (*Pinus taeda* L.), shortleaf (*P. echinata* Mill.), longleaf (*P. palustris* Mill.), and slash (*P. elliotti* Engelm.), are combined into a single model. Where the ranges of two or more species overlap in a given climatic region, their intrinsic radial growth patterns are similar or even identical in most cases. That is, stand-level radial growth is initiated, reaches maximum rates, slows, and ceases on the same calendar dates for those species occurring together naturally. This is not to say the rates of growth are equal for all species occurring together; the growth pattern refers to the temporal phenology. The DISP model provides for separate patterns of radial growth between different climatic regions, whereas within a given region, a single pattern is used for any of the four species occurring there.

The model is not concerned with growth rates per se, but with growth at any given time relative to the long-term average growth. Therefore, it matters not that trees in one specific stand grow at different actual rates than trees in another specific stand of different stocking, at a different age, on a different site. The DISP model estimates year-to-year variance of stand-level radial growth in any stand relative to itself.

Figure 1 shows seasonal trends of normal, monthly radial increments for dominant and codominant southern pine trees in three climatic zones. The points plotted on the curve for each zone represent that proportion of total annual growth normally occurring during each month of the growing season. The curves represent long-term average trends for the three climatic zones, based on many previous studies, cited earlier. These zones are assigned to U.S. National Weather Service climatic divisions throughout the southern pine region, accounting for the irregular boundaries shown in Figure 2.

The long-term average monthly growth trends (Figure 1 and Table 1) represent values when the PDSI is normal or nearly normal for all months of the growing season. Because earlywood growth is relatively unaffected by moderate climatic variation, the average growth trends prevail during the months of March and April in any given year, even when PDSI values are moderately greater or less than normal. Large climatic variation, in particular strongly negative values of PDSI, begins to influence intrinsic growth during April, becoming more important in May. Then, in June, moderate variation to either positive or negative values of PDSI has a highly significant impact on radial growth, as the transition to latewood occurs. During July through October, all deviations from normal PDSI impact latewood growth, roughly in proportion to the PDSI variation, both positive and negative. Large variation in PDSI has its greatest impact on the proportion of monthly growth during this period from July through October.

The following assumed impacts of radial growth are based on many field studies with dendrometers (e.g., Zahner, 1963, 1968; Moehring and Ralston, 1967; Fritts, 1976; Kramer and Kozlowski, 1979): The greatest negative impact of PDSI on total annual growth is in the month of June, because the intrinsic, potential growth rate is high, and the effect of severe drought is to reduce this rate to zero. The greatest positive impact of PDSI on total annual growth is in the month of July, when the long-term average rate is normally only moderate but the potential for latewood growth is high; an unusually wet PDSI value can double the average rate of latewood growth at this time. Continued positive values of PDSI maintain accelerated latewood growth until early fall.

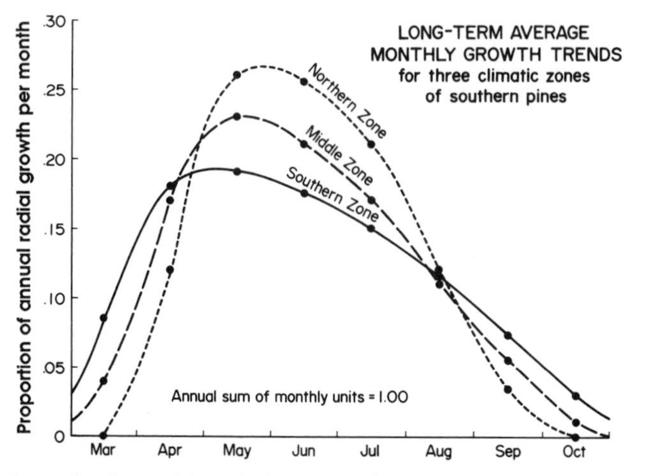

Figure 1. Seasonal trends of monthly radial growth of dominant and codominant southern pine trees in three climatic zones. The plotted points on each curve represent the normal proportion of total annual growth occurring during each month.

Table 1. Monthly proportions of annual radial increment for the southern pines for three climatic zones, given as the intrinsic long-term averages. Growth units are proportions of long-term average annual increment for the months and climatic zones plotted in Figure 1.

Month	Climatic Zone: Northern	Middle	Southern
	Growth units		
March	0.000	0.040	0.085
April	0.120	0.170	0.180
May	0.260	0.230	0.190
June	0.255	0.210	0.175
July	0.210	0.170	0.150
August	0.120	0.110	0.115
September	0.035	0.060	0.075
October	0.000	0.010	0.030
Annual	1.000	1.000	1.000

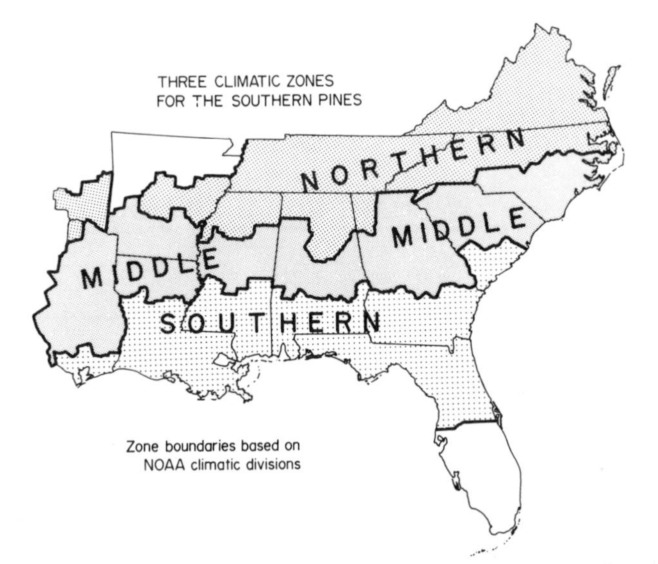

Figure 2. Map of three climatic zones for the southern pines, based on the climatic divisions of each state as delineated by the U.S. National Weather Service.

Table 2. Climate impact factors for southern pine radial growth, based on monthly values of PDSI.

A. MARCH — factor = 1.00 for all values of PDSI

B. APRIL
1. factor = 1.00 for all positive values of PDSI
2. factor = 1.00 + 0.067 × PDSI for values of PDSI between −3.00 and 0
3. factor = 0.80 for all values of PDSI less than −3.00

C. MAY
1. factor = 1.20 for all values of PDSI greater than 1.70
2. factor = 1.00 + 0.118 × PDSI for values of PDSI between −3.30 and 1.70
3. factor = 0.60 for all values of PDSI less than −3.30

D. JUNE
1. factor = 1.40 for all values of PDSI greater than 3.00
2. factor = 1.00 + 0.133 × PDSI for values of PDSI between −1.49 and 3.00
3. factor = 1.10 + 0.200 × PDSI for values of PDSI between −4.50 and −1.50
4. factor = 0.20 for all values of PDSI less than −4.50

E. JULY-AUGUST-SEPTEMBER-OCTOBER
1. factor = 1.60 for all values of PDSI greater than 3.00
2. factor = 1.00 + 0.200 × PDSI for values of PDSI between −4.00 and 3.00
3. factor = 0.20 for all values of PDSI less than −4.00

Based on these assumed responses in radial growth to mild, moderate, severe, and extreme wet and dry periods during early, middle, and late portions of the growing season, climate impacts on radial growth of the southern pines can be modeled as monthly impact factors calculated from monthly values of PDSI. Such an array of climate impact factors has been calculated for each month of the growing season (Table 2). Derivations of the climate impact values of Table 2 are the best algebraic estimates that reproduce the qualitative relationships described in the paragraph above.

In the DISP model, these factors are applied to the corresponding monthly growth proportions (units) of Table 1 to simulate the impact of measured values of monthly PDSI on growth throughout a given calendar year. These adjusted units thus convert the long-term average growth trend to a specific simulated trend that reflects the impact of climate on the monthly proportions of radial increment for a given year. The sum of the March through October adjusted growth units, then, is a measure of the overall annual impact on radial increment for a given year, the DISP (Figure 3, Table 3). The algebraic form of the DISP simulator = sum of 8 monthly adjusted growth units = sum of 8 monthly growth units, each multiplied by the corresponding climate impact factor for that month, where the monthly growth unit depends on the climatic zone (Table 1), and the monthly climate impact factor depends on monthly values of PDSI (Table 2).

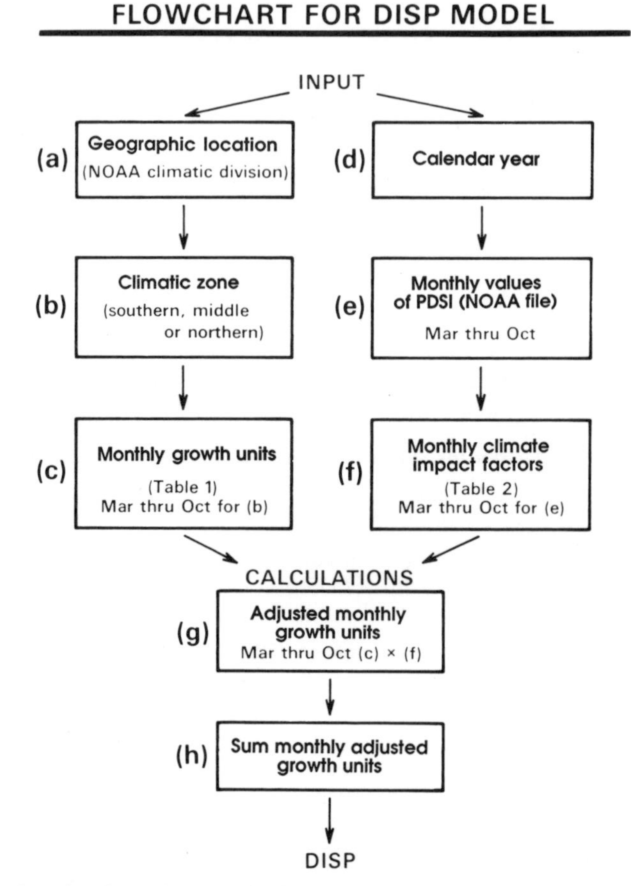

Figure 3. Flow chart for the DISP model, illustrating the algebraic form of the simulator: DISP = sum of 8 monthly growth units each multiplied by the corresponding climatic impact factor for that month.

Table 3. Annual climate impact rating based on DISP values.

DISP value	Climate-growth impact rating
> 1.300	Growth potential maximum
1.201–1.300	Growth substantially greater than average
1.101–1.200	Growth moderately greater than average
0.900–1.100	Average growth (±10%)
0.750–0.899	Growth moderately less than average
0.600–0.749	Growth substantially less than average
< 0.600	Growth severely limited

Since the PDSI is itself an index of short-term local climate relative to the long-term average climate for that region, the DISP in turn is also an index of short-term growth relative to long-term average growth for a given climatic region. Note that the DISP scale of values (Table 3) has greater potential for negative impacts on growth than for positive impacts.

Figure 4 shows the effect on growth of three examples of PDSI values in the Georgia Central climatic division of the U.S.A. Two of these are typical extremes of wet and dry years in comparison to the long-term average. The corresponding DISP values are 1.34 for 1975 (the wet year) and 0.51 for 1981 (the dry year). These DISP values are interpreted to mean that simulated annual radial increment is 34% greater than the long-term average for the wet year, and 49% less for the dry year. Calculations for these examples are given in Table 4.

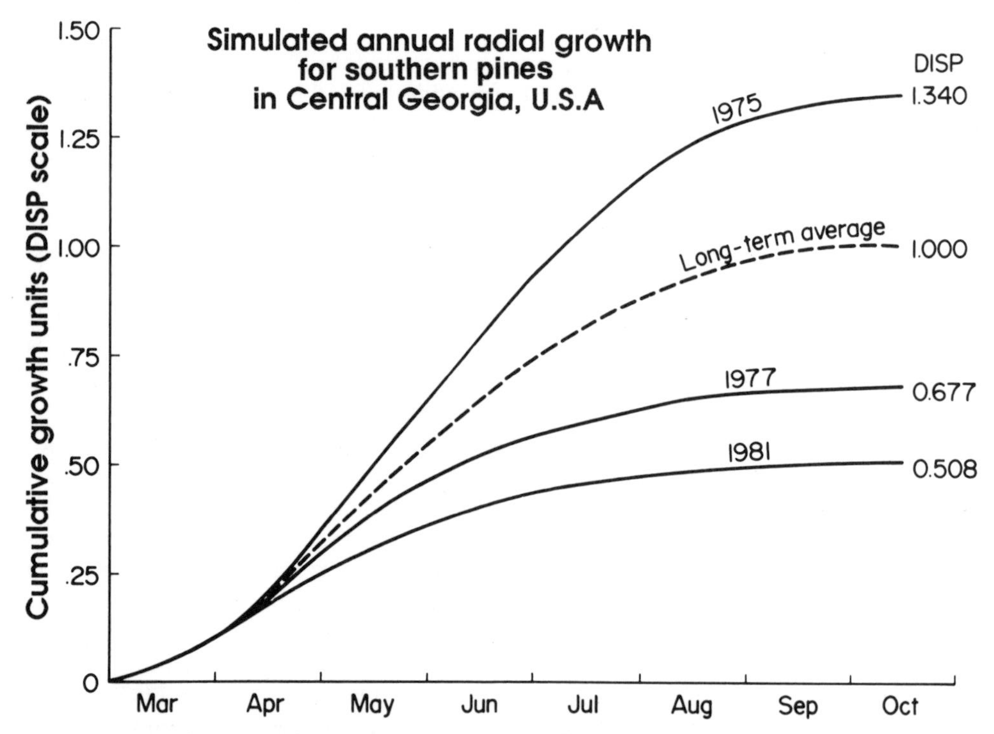

Figure 4. Simulated annual radial growth trends for three growing seasons in central Georgia, U.S.A., as compared to the long-term average trend for the middle climatic zone illustrated in Figure 1. The dry year (1981) and wet year (1975) curves are typical extremes of monthly PDSI values for this climatic division.

Table 4. Examples of the computations for DISP for calendar years 1975 and 1981 for the Central climatic division of the state of Georgia, U.S.A.

		1975			1981		
Month (1)	Average growth units (2)	PDSI (3)	Climate impact factor (4)	Adjusted growth units (5)	PDSI (3)	Climate impact factor (4)	Adjusted growth units (5)
March	0.04	2.65	1.00	0.040	−3.69	1.00	0.040
April	0.17	3.04	1.00	0.170	−3.42	0.80	0.136
May	0.23	3.35	1.20	0.276	−3.59	0.60	0.138
June	0.21	3.38	1.40	0.294	−3.25	0.45	0.094
July	0.17	3.90	1.60	0.272	−3.92	0.22	0.037
August	0.11	3.47	1.60	0.176	−3.36	0.33	0.036
September	0.06	3.64	1.60	0.096	−3.72	0.26	0.015
October	0.01	4.03	1.60	0.016	0.60	1.12	0.011
Annual	1.00			DISP = 1.340			DISP = 0.508

(2) From Table 1, Middle Climatic Zone.
(3) From U.S. National Weather Service file TD-9640.
(4) From Table 2, for corresponding values of PDSI, column (3).
(5) Column (2) × column (4). DISP is sum of monthly values.

APPLICATION OF THE MODEL

A software program (in PASCAL for personal computers) is available without charge from the authors for the computation of DISP values for any climatic division in the natural range of the southern pines. The program utilizes monthly PDSI values from the U.S. National Weather Service data file TD-9640 (provided with the DISP program) to calculate annual DISP values for any year from 1931 through 1987. The program also calculates DISP values from any March through October PDSI values entered manually for any year and weather station, so that current annual growth can be monitored as an ongoing part of continuous forest inventories. The program output includes within-season, monthly trends of radial growth for any year (as in Table 4) as well as average DISP values for multiple-year, periodic inventories.

Further refinement of the DISP model might include a special case in addition to the general case presented here. For example, on poor upland sites throughout the southern U.S.A., especially on droughty Piedmont soils, several consecutive growing seasons with favorable moisture conditions may result in the accumulation of sufficient food reserves to produce growth in excess of that simulated by the DISP model for a single year. To give another example, where the winter climate is not sufficiently favorable to ameliorate a previous year's drought, the effect of a severe drought year may carry over into the subsequent year (Kramer and Kozlowski, 1979). The DISP model should be modified to accomodate such preconditioning effects when judged appropriate due to the geographic location and sequence of climatic conditions. Negative preconditioning is important to consider in northerly climatic zones where overwinter photosynthesis is not as effective in damping out the effect of drought years as in southern zones. When adapting the DISP model to deciduous tree species, such as upland oaks (*Quercus*), a preconditioning effect should always be simulated because there is no overwinter compensation for limited food reserves from the previous year.

The normal water balance is locally upset by disturbances that permanently or even temporarily reduce stand density (Zahner, 1959, 1968). Natural causes, such as the southern pine beetle or wildfire, and man-caused removal of trees through harvesting and heavy thinning, reduce soil occupancy by roots and reduce transpirational crown cover. In all such cases, residual trees are temporarily and locally under reduced water stress due to reduced competition for soil moisture. Thus, immediately following disturbance, radial growth of individual trees is not sensitive to climatic impacts, either through water deficits or water excesses. Consequently, the PDSI water balance model does not provide a realistic basis for evaluating climatic impacts on disturbed forest stands. Time-related adjustments must be made to regional climatic models to account for the accelerated growth of residual trees in locally disturbed stands until vegetation again fully occupies the site.

The DISP model has been tested for correlations with annual and periodic increments of radial growth and tree rings as reported in many publications from throughout the southern pine region over the past 50 years (four examples are shown in Figure 5). These procedures are too lengthy to include in this paper, but the overall conclusion is that DISP simulations are well correlated with measured radial growth, especially in pine

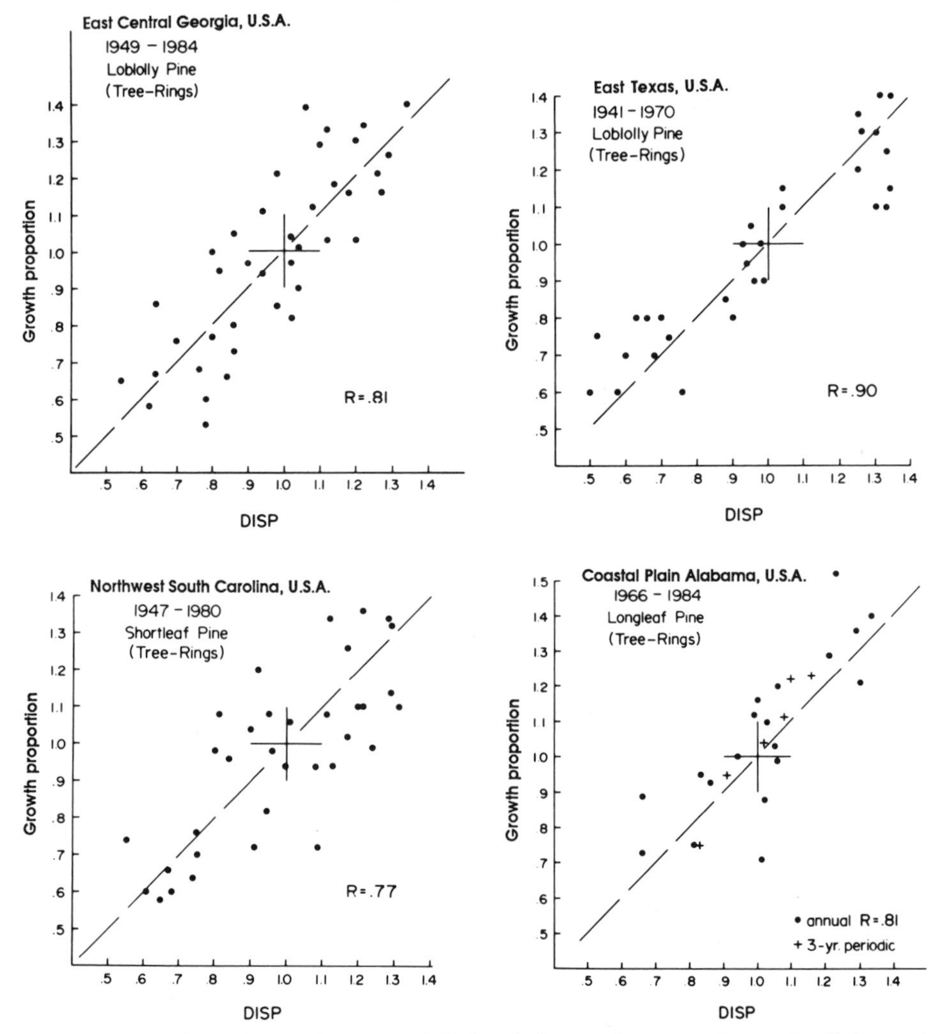

Figure 5. Examples of correlations between DISP simulations and measured annual radial growth for southern pines, in stands 20 to 50 years of age in four locations throughout the southeastern U.S.A. Growth proportions are departures from average growth trends established by the senior author from tree-ring time series in dominant and codominant trees.

stands between 20 and 50 years of age, exhibiting correlation coefficients from 0.75 to over 0.90 for annual ring widths, within the stated limitations of site, stocking levels, and stand disturbance. Poorer correlations are found in old growth stands, perhaps because biologically overmature trees respond to a different arrangement of climatic variables.

The authors believe the DISP simulation provides a satisfactory model for further research and modification. It is intended to serve as an example for the development of similar models for other forest types in other geographic locations, where it is important to assess the impact of climate on the growth of forests.

LITERATURE CITED

Allen, R. M. 1976. *Gas Exchange of Detached Needles of Southern Pines and Certain Hybrids.* Paper SO-125, USDA Forest Service Southern Forest Experiment Station, New Orleans, LA, U.S.A. 10 pp.

Federer, C. A. 1982. Frequency and intensity of drought in New Hampshire forests: evaluation by the BROOK model. In: *Applied Modeling in Catchment Hydrology.* Water Resources Publication, Littleton, CO, U.S.A. Pp. 457–470.

Fritts, H. C. 1976. *Tree Rings and Climate.* Academic Press, London, England, U.K. 567 pp.

Karl, T. R., and R. W. Knight. 1985. *Atlas of Monthly Palmer Drought Severity Indices (1931–1983) for the Contiguous United States.* U.S. Department of Commerce Weather Bureau Historical Climatological Series 3–11. 319 pp.

Kramer, P. J., and T. T. Kozlowski. 1979. *Physiology of Woody Plants.* Academic Press, New York, NY, U.S.A. 811 pp.

Moehring, D. M., and C. W. Ralston. 1967. Diameter growth of loblolly pine related to available soil moisture and rate of soil moisture loss. *Soil Science Society of America Proceedings* 31:560–562.

Palmer, W. C. 1965. *Meteorological Drought.* Research Paper No. 45, U.S. Department of Commerce Weather Bureau. 58 pp.

Reynolds, J. F., B. R. Strain, G. L. Cunningham, and K. R. Knoerr. 1980. Predicting primary productivity for forest and desert ecosystem models. In: Hesketh, J. D., and J. W. Jones, eds. *Predicting Photosynthesis for Ecosystem Models,* Vol. II. CRC Press, Boca Raton, FL, U.S.A. Pp. 169–189.

Thornthwaite, C. W., and J. R. Mather. 1955. The water balance. *Publications in Climatology* 8:1–104. Drexel Institute Laboratory of Climatology, Centerton, NJ, U.S.A.

Zahner, R. 1959. Soil moisture utilization by southern forests. In: *Symposium: Southern Forest Soils.* Louisiana State University, Baton Rouge, LA, U.S.A. Pp. 25–30.

Zahner, R. 1963. Internal moisture stress and wood formation in conifers. *Forest Products Journal* 13:240–247.

Zahner, R. 1968. Water deficits and growth of trees. In: Kozlowski, T. T., ed. *Water Deficits and Plant Growth,* Vol. II. Academic Press, New York, NY, U.S.A. Pp. 191–254.

Zahner, R., and R. K. Myers. 1986. Assessing the impact of drought on forest health. *Proceedings, Society of American Foresters Annual Meeting,* Birmingham, AL, U.S.A. Pp. 227–234.

Zahner, R., and A. R. Stage. 1966. A procedure for calculating daily moisture stress and its utility in regressions of tree growth on weather. *Ecology* 47:64–74.

CHAPTER 35

SIMULATING THE PHYSIOLOGICAL BASIS OF TREE-RING RESPONSES TO ENVIRONMENTAL CHANGES

Robert J. Luxmoore, M. Lynn Tharp, and Darrell C. West

Abstract. The detection of possible forest growth responses to changes in atmospheric CO_2 or air pollutants is very difficult by statistical analysis of tree-ring chronologies, and a complementary modeling approach has been initially tested. In this new approach, a linked set of mechanistic unified transport models (UTM) of carbon, water, and chemical dynamics in soil-plant-litter systems is used to generate a matrix of simulated annual stemwood increment and winter carbon storage values for a range of degree day, water stress, and atmospheric CO_2 concentrations. These values represent potential tree growth responses as determined by hourly time-step physiological processes. The matrix is accessed by a forest succession model (Forests of East Tennessee, FORET) according to selected degree day and water stress values or by use of actual site data. These potential growth responses are modified to realized annual increments according to the competition algorithms of the succession simulator using yearly time-steps. A 12% increase in stemwood production was predicted for an oak-hickory (*Quercus-Carya* sp.) forest in eastern Tennessee by the UTM for a change in atmospheric CO_2 both from 260 to 340 and from 340 to 600 ppmv (μL/L). A signal of $\pm \sqrt{12\%}$ was incorporated into the diameter growth algorithms for the species represented in Forests of East Tennessee (FORET), and simulations were conducted for 32 plots with slightly different initial species composition representative of the oak-hickory forest (Shugart and West, 1977). Preliminary results suggest that the spatial variation in the species complement for the 32 plots masked the detection of the CO_2 signal in 200-year simulations. In a repeat analysis eliminating spatial variability, 25 replicate simulations were conducted for a single plot, and again there were no simulation responses that could be attributed to CO_2 enrichment for the five plots evaluated in this manner. Temporal variability due to establishment and mortality algorithms in FORET probably masked the CO_2 signal from the Unified Transport Model (UTM). Spatial and/or temporal variability in forest-stand dynamics may mask the detection of tree response to atmospheric CO_2 enrichment. A backcast simulation procedure that largely eliminates spatial and temporal effects is recommended for further testing of the linked-modeling method of tree-ring chronology analysis.

INTRODUCTION

Larson (1963) recommended that measurements of annual increments in xylem tissue (tree ring) in the lower stem of a tree be used as sensitive indicators of response to environmental changes. This recommendation was based on the hierarchical pattern of

Drs. Luxmoore, Tharp, and West are Soil and Plant Scientist, Computer Programmer, and Forest Ecologist, respectively, at the Environmental Sciences Division, Oak Ridge National Laboratory, Oak Ridge, TN, 37831-6038 U.S.A. This research was sponsored by the Carbon Dioxide Research Program, U.S. Department of Energy, under Contract DE-AC05-840R21400 with Martin Marietta Energy Systems, Inc. Publication No. 3291, Environmental Sciences Division, Oak Ridge National Laboratory.

carbon allocation; sapwood having a low allocation priority, particularly at dbh 1.4 m, gives an amplified response to favorable or harsh conditions. Tree-ring chronologies are valuable records of tree growth that have been used to reconstruct historical climate regimes (Fritts, 1976) and are also being examined as a means of evaluating unexpected declines in forest growth (McLaughlin et al., 1987) and detecting possible responses to rising atmospheric CO_2 levels (Kienast and Luxmoore, 1988).

Statistically rigorous methods have been developed to identify the stand age and climate components of variation in tree-ring chronologies so that residual patterns of ring variation may be evaluated (Cook et al., 1987). Association of these residual patterns with air pollutant stress or atmospheric CO_2 enrichment may nevertheless prove to be a difficult task. For example, Kienast and Luxmoore (1988) estimated a 4% increase in annual increment as an expected response to atmospheric CO_2 enrichment (312 to 332 ppm [μL/L]) for a 20-year period from 1955. Such a response may be within the measurement error of tree-ring determinations. The equivalent response for a CO_2 increase from 260 to 340 ppm (μL/L) is a 16% increase in annual increment, which may be detectable in chronologies dating back to the early 1800s.

A complimentary method for addressing tree growth responses to environmental changes is outlined. The procedure involves the linkage between a physiologically based simulator, the UTM (Luxmoore, 1989), and FORET (Shugart and West, 1977). In this approach, knowledge of short-term physiological responses to environmental changes is integrated into a matrix of annual growth responses that become input values to the succession model. A sequence of annual growth increments simulated during successional changes over several decades represents a chronology that can be compared with an actual tree-ring chronology to test the approach. Brief descriptions of the two tree models (UTM, FORET) and their linkage are given. Preliminary forest growth simulations for changing atmospheric CO_2 and air temperature regimes illustrate some of the challenges of linking models with different scales.

UNIFIED TRANSPORT MODEL (UTM)

Five model modules describing carbon, water, and nutrient dynamics (Table 1) are based on various forms of gradient-driven flow equations, empirical relationships for soil and plant characteristics, and simplified relationships for complex processes (e.g., macropore flow, tree phenology). The codes are linked within the bookkeeping framework of a hydrologic model to form the UTM (Luxmoore, 1989). The photosynthesis equation is defined by the CO_2 gradient between the atmosphere and the chloroplast, divided by the sum of boundary layer, stomatal, and mesophyll resistances to CO_2 movement. Photosynthate allocation to stemwood production is favored under low water stress conditions. Tissue respiration is a function of tissue mass and temperature. All calculations are made on a unit-area-of-land basis which can be interpreted either as a single-tree or forest-stand simulation.

FOREST SUCCESSION MODEL (FORET)

FORET is a stochastic forest simulator designed to model stand density characteristics of deciduous and coniferous forests (Figure 1) and is based on the JAnak, BOtkin, WAllis (JABOWA) model developed by Botkin et al. (1972). Diameter of individual trees is simulated on a circular 1/12-ha plot and is incremented annually according to a function that produces ⅔ of a tree's dbh at ½ its age. Growth reduction factors are determined from

Table 1. Component model modules of the Unified Transport Model (UTM) and their attributes.

Component	Water	Soil chemicals	Carbon	Root solute uptake	Plant solutes
Name	PROSPER (AGTEHM)	CADIL	CERES	DIFMAS	DRYADS
Time-step	15 or 60 min.	15 or 60 min.	60 min.	15 or 60 min.	15 or 60 min.
Attributes	Evapotranspiration by combination equation Soil water flow by Darcy flow equation Macropore flow Uses empirical relationship between surface resistance and surface water potential Empirical data for soil hydraulic properties	Uses Freundlich isotherm First-order chemical degradation (organics) Chemical transport in matrix and macropores	CO_2 gradient equation for net photosynthesis Sucrose gradient equation for translocation Uses input values for potential growth of plant components Empirical litter decomposition relationships	Implements model of diffusion and mass flow of solutes to roots by Baldwin, Nye, and Tinker (1973)	Solute uptake by roots and leaves Gas pollutant uptake by leaves Gradient equation for phloem translocation Transpiration flux used for xylem transport Plant demand function determined by maximum solute concentration

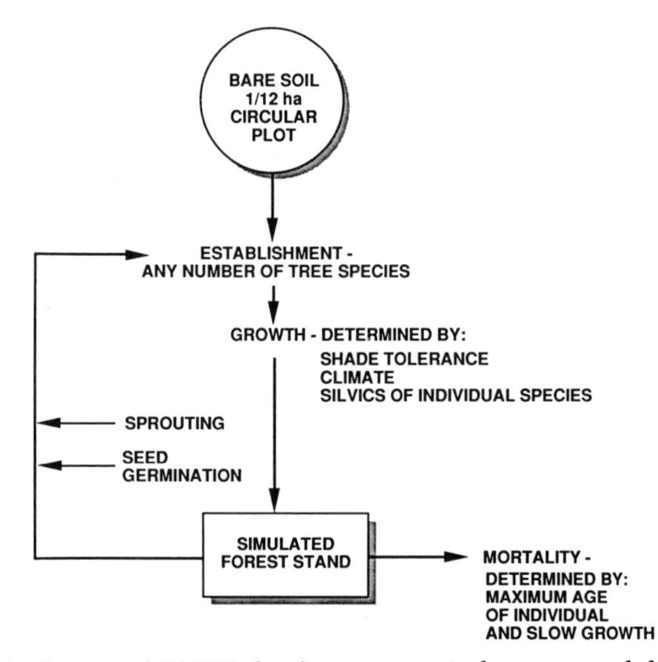

Figure 1. A schematic diagram of FORET showing component phenomena and determining factors of forest stand dynamics.

total growing degree days (5.6°C base), total leaf area of taller trees on the plot, number of trees on the plot, and available soil water. From these four reduction values, the minimum (the most limiting) is chosen as the growth reduction variable in any particular year. Recruitment, growth, and mortality of a species in a stand are determined by leaf litter, mammal browsing, sprouting characteristics, and intrinsic growth potential. Shugart and West (1977) describe FORET and provide results for a test application.

LINKAGE OF PHYSIOLOGICAL AND SUCCESSION MODELS

The UTM operates at hourly time-steps (15 minutes during rainfall) that integrate into daily, and then into annual, increments of tree growth. Inputs of soil, plant, and weather data used in the UTM determine annual increments in growth and internal carbon storage that are single-species (potential) values (Figure 2), since plant competition is not included in the UTM. The water budget and nutrient dynamics of a soil-plant system are also simulated by the UTM. Repeated simulations with the UTM for an extensive range of air temperature, precipitation, and atmospheric CO_2 levels provide output values that become matrix members associated with the growing season degree day summation, the number of dry days when water stress limits tree growth, and a particular atmospheric CO_2 concentration. As illustrated in Figure 2, each matrix member is represented by a frequency distribution of annual increment and internal (starch) storage values resulting from weather variability. For example, a single growing-degree-day value can correspond to a range of warm- and cold-period combinations coming at different times during the growing season. Since temperature has non-linear effects on physiological processes, a range of annual increment and storage values is possible for one degree-day value. Similar outcomes are possible for the timing of water stress effects on tree growth during an annual cycle.

The forest succession model (FORET) has a basic annual time-step with monthly air temperature and precipitation data being used to determine growing degree days and the number of days of soil water stress for a particular annual time-step. Actual monthly data or stochastic weather relationships may be used in long-term simulations with FORET. Potential annual increment and stroage values are selected from the matrix of UTM output and used to modify algoritms in FORET. Simulation of competition among the individuals in a forest stand modifies the potential response to a realized annual increment (Figure 2). The mean and variance of the realized response can be determined when the frequency distibution of potential responses is propagated through FORET with a proce-

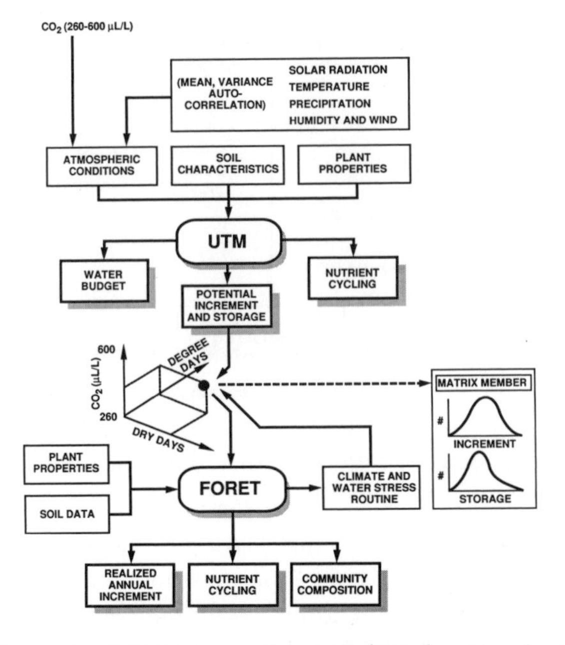

Figure 2. Linkage of a mechanistic tree growth model (UTM) with a forest succession model (FORET) through a data matrix for determination of environmental-change effects on forests.

dure such as Latin hypercube sampling (Gardner et al., 1983). The effects of environmental changes on nutrient cycling and community composition can also be determined from FORET output (Pastor and Post, 1988). Both the UTM and FORET have been under development and testing for over a decade.

EXAMPLE APPLICATION

Weather records for calendar year 1971 from Walker Branch Watershed near Oak Ridge National Laboratory in eastern Tennessee, U.S.A., were used in a UTM simulation of the water and carbon dynamics of an oak-hickory (*Quercus-Carya* sp.) stand on a ridgetop location. The precipitation for 1971 of 1372 mm/year was very close to the average for the area. Hydraulic properties for the Ultisol (podsolic) soil (Fullerton cherty silt loam) were taken from an earlier water budget study (Luxmoore, 1983).

The forest input properties needed by a component model (Table 1, CERES) were obtained from Edwards et al. (1989) for Walker Branch Watershed and from earlier simulations of a deciduous forest (Dixon et al., 1978). The nutrient dynamics models of the UTM were not used, and nutrients were considered to be non-limiting in this application. The aboveground phytomass was initiated at 157 Mg/ha. Maximum leaf area index was close to 5, and the annual aboveground increment of oak-hickory forest was estimated at 3 Mg/ha for the early 1970s (Edwards et al., 1989).

Factorial combinations of three atmospheric CO_2 concentrations (260, 340, 600 ppmv) (μL/L) and two air temperature regimes (1971 ambient, ambient + 3°C) were examined. The minimum surface (stomatal) resistance values for water vapor were set to 0.7, 1.0, and 2.0 s/cm for the 260-, 340-, and 600-ppmv (μL/L) CO_2 concentrations, respectively, based on data summarized by Cure (1985) for species with C_3 photosynthesis.

Species data for the 21 most important tree species in 32 permanent plots on Walker Branch Watershed in the oak-hickory forest community for 1967 were used as input data for FORET. The initial phytomass above stump was 100 Mg/ha in each case. The weather data for 1971 were summarized into monthly mean values and repeated annually in 200-year simulations. Two types of FORET simulations were conducted. In the first case, simulations were made for the 32 permanent plots, and the results were averaged to provide the mean forest response to the atmospheric CO_2 scenarios. In this case, the natural spatial variability in species composition on Walker Branch Watershed is included in the response in addition to temporal variability generated by the stochastic establishment and mortality algorithms. In an alternative case, the spatial variability was excluded by summing 25 repeated runs for one plot. In this case, only temporal variability is generated. Test simulations showed that 25 simulations were sufficient to obtain the mean response by averaging. This second type of analysis was conducted for five plots: one each dominated by hickory, oak, or yellow poplar, and two with mixed oak-hickory species. The UTM and FORET used the same weather data in all the computer runs summarized in this report.

PRELIMINARY RESULTS

UTM, FORET Forecasting

The UTM forecasts showed stemwood increasing with CO_2 enrichment under both temperature regimes (Figure 3a). There was a 12% increase and a 12% decrease in stemwood increment for the 600- and 260-ppmv (μL/L) CO_2 cases, respectively, relative to the 340-ppmv (μL/L) case under the 1971 weather conditions. At each CO_2 level, higher air temperature produced small effects on stemwood increment (Figure 3a). CO_2 enrichment increased stem storage, and at each CO_2 level, higher temperature greatly decreased storage (Figure 3b). Leaf water potentials showed fewer days below -1.5 MPa with increasing CO_2, being 20, 6, and 0 days at ambient air temperature, and 79, 64, and 15 days in the high-temperature case for the 260-, 340-, and 600-ppmv (μL/L) atmospheric CO_2 levels, respectively. Evapotranspiration showed a greater response to higher temperature than to the CO_2 enrichment, as was observed for the number of days of moderate leaf water stress. Seed production (acorn litter) increased dramatically with increasing CO_2, with higher temperature decreasing seed production at a particular CO_2 level (Figure 3c).

The $\pm 12\%$ change in the stemwood increment from the UTM was converted to square root values of $\pm 3.464\%$ for use in the diameter growth rate algorithm of FORET. At a fixed wood density, change in mass is equivalent to a change in wood volume, and this is depen-

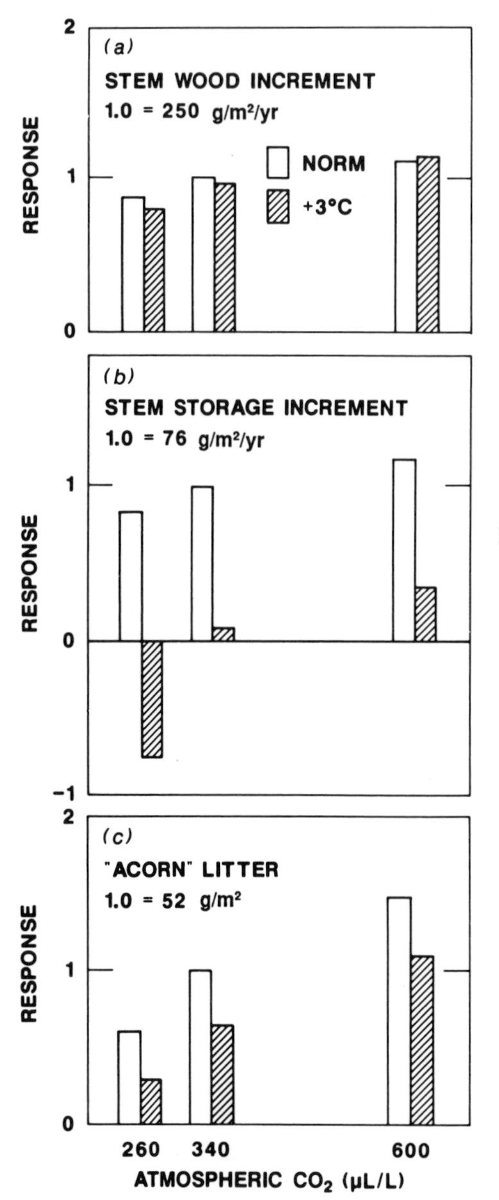

Figure 3. Relative responses in stemwood increment, stem-storage increment, and "acorn" litter predicted by the UTM for an annual simulation of an oak-hickory (*Quercus-Carya*) forest under combinations of three atmospheric CO_2 concentrations and two air temperature regimes.

dent on the square of the diameter. The diameter growth equation for each species in the stand depends on species-specific parameters; and for the present simulations, all these equations were multiplied by 0.96536, 1.00000, or 1.03464 to propagate the 260-, 340-, or 600-ppmv (μL/L) CO_2 response signal, respectively, from the UTM into FORET. The diameter growth rates were thus modified for each of the 21 species represented in the oak-hickory plots for the 260- and 600-ppmv (μL/L) atmospheric CO_2 levels. No changes were made for the 340-ppmv (μL/L) case.

Small differences were noted in the stem and branch phytomass during a 200-year simulation for the three CO_2 levels (Figure 4) and for stem number per unit area (not shown). These FORET results, based on the average response of 32 plots, showed changes in the aboveground phytomass per tree (Table 2), but there was no consistent pattern that could be attributed to a CO_2 signal, except perhaps at 200 years. At the end of the simulation, there was a slight increase in phytomass per ha and a slight decrease in tree density with CO_2 enrichment. This amounted to a $\pm$6% change in phytomass per tree (Table 2) that corresponded with the expected effects of atmospheric CO_2 change. In this case, the realized signal from FORET was less than the potential signal supplied from the UTM.

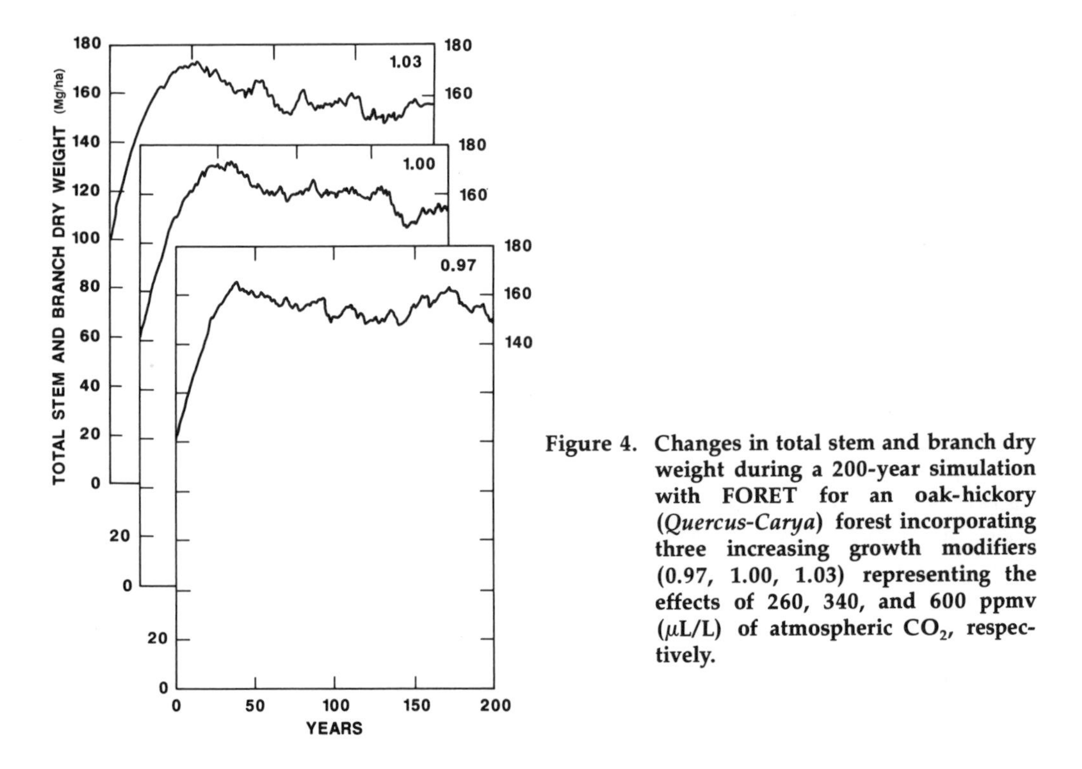

Figure 4. Changes in total stem and branch dry weight during a 200-year simulation with FORET for an oak-hickory (*Quercus-Carya*) forest incorporating three increasing growth modifiers (0.97, 1.00, 1.03) representing the effects of 260, 340, and 600 ppmv (μL/L) of atmospheric CO_2, respectively.

Table 2. Mean stem and branch dry weight determined at five time intervals for FORET simulations for three increasing growth modifiers (0.965, 1.000, 1.035) representing the effects of 260-, 340-, and 600-ppmv (μL/L) atmospheric CO_2 levels, respectively.

Time interval (y)	Mean stem and branch weight (kg/tree) Growth modifier		
	0.965	1.000	1.035
25	78	75	78
50	115	119	113
100	96	103	80
150	79	87	75
200	72	77	82

These preliminary results suggested that initial spatial and or temporal variability among 32 different plots used in the FORET simulations may have masked the physiologically predicted CO_2 responses. In the alternative FORET simulations, where spatial variability was eliminated by repeat simulation for selected plots, there were no consistent growth responses (data not presented) that could be attributed to the CO_2 signal in the five plot types evaluated. This result indicated that temporal variability introduced by establishment and mortality algorithms was also sufficient to mask the CO_2 response. These simulations suggest that a doubling of atmospheric CO_2 may produce significant physiological responses that become masked at the level of stand dynamics.

Backcasting

A third method of modeling involving backcasting rather than forecasting could be a useful approach for detection of a CO_2 response in tree-ring chronologies. Backcasting would largely eliminate the temporal variability effects of establishment and mortality in FORET by starting the simulation with known stand composition and subtracting annual increments from the component species. In this approach, the current period with the most detailed knowledge of site characteristics is used for defining the intial conditions in the model.

CONCLUSION

The concept of linking two models for the interpretation of tree-ring chronologies has been only partly explored in this initial analysis. The approach requires the scaling up of physiological information (currently the only available response data) for evaluation of forest-stand response to CO_2 enrichment over several decades. Preliminary results suggest that individual-tree responses to CO_2 enrichment could become masked at the forest-stand level. A backcasting procedure should provide more sensitive signal detection than the forecasting method used here, and a forest-stand response to CO_2 enrichment may eventually be determined from further analysis. An important feature of the linked modeling approach to the analysis of tree-ring chronologies is that it requires comparison of the simulations with actual chronologies, preventing uncontained speculation about the factors influencing forest growth.

LITERATURE CITED

Baldwin, J. P., P. B. Nye, and P. B. Tinker. 1973. Uptake of solutes by multiple root systems from soil. III. A model for calculating the solute uptake by a randomly dispersed root system developing in a finite volume of soil. *Plant Soil* 38:621–635.

Botkin, D. B., J. F. Janak, and J. R. Wallis. 1972. Some ecological consequences of a computer model of forest growth. *Journal of Ecology* 60:849–873.

Cook, E. R., A. H. Johnson, and T. J. Blasing. 1987. Forest decline: modeling the effect of climate in tree rings. *Tree Physiology* 3:27–40.

Cure, J. D. 1985. Carbon dioxide doubling responses: a crop survey. In: Strain, B. R., and J. D. Cure, eds. *Direct Effects of Increasing Carbon Dioxide on Vegetation.* DOE/ER-0238, U.S. Department of Energy, Washington, DC, U.S.A. Pp. 99–116.

Dixon, K. R., R. J. Luxmoore, and C. L. Begovich. 1978. CERES—a model of forest stand biomass dynamics for predicting trace contaminant, nutrient, and water effects. I. Model description. II. Model application. *Ecological Modelling* 5:17–38, 93–114.

Edwards, N. T., D. W. Johnson, S. B. McLaughlin, and W. F. Harris. 1989. Carbon dynamics and productivity. In: Johnson, D. W., and R. I. Van Hook, eds. *Analysis of Biogeochemical Cycling Processes in Walker Branch Watershed.* Springer-Verlag, New York, NY, U.S.A. Pp. 197–235.

Fritts, H. C. 1976. *Tree Rings and Climate.* Academic Press, New York, NY, U.S.A.
Gardner, R. H., B. Röjder, and U. Berström. 1983. PRISM—a systematic method for determining the effect of parameter uncertainties on model predictions. Studsvik Energiteknik AB Report NW-83/555. Mykoping, Sweden.
Kienast, F., and R. J. Luxmoore. 1988. Tree-ring analysis and conifer growth responses to increased atmospheric CO_2 levels. *Oecologia* 76:487–495.
Larson, P. R. 1963. Stem form development in forest trees. *Forest Science Monographs* 5:1–42.
Luxmoore, R. J. 1983. Water budget of an eastern deciduous forest stand. *Soil Science Society of America Journal* 47:785–791.
Luxmoore, R. J. 1989. Modeling chemical transport, uptake, and effects in the soil-plant-litter system. In: Johnson, D. W., and R. I. Van Hook, eds. *Analysis of Biogeochemical Cycling Processes in Walker Branch Watershed.* Springer-Verlag, New York, NY, U.S.A. Pp. 351–384.
McLaughlin, S. B., D. J. Downing, T. J. Blasing, E. R. Cook, and H. S. Adams. 1987. An analysis of climate and competition as contributors to the decline of red spruce in high elevation Appalachian forests of the eastern United States. *Oecologia* 72:487–501.
Pastor, J., and W. M. Post. 1988. Response of northern forests to CO_2 induced climate change. *Nature* 334:55–58.
Shugart, H. H., and D. C. West. 1977. Development of an Appalachian deciduous forest succession model and its application to assessment of the impact of chestnut blight. *Journal of Environmental Management* 5:161–179.

CHAPTER 36

MODELING PINE RESISTANCE TO BARK BEETLES BASED ON GROWTH AND DIFFERENTIATION BALANCE PRINCIPLES

Peter L. Lorio, Jr., Robert A. Sommers, Catalino A. Blanche, John D. Hodges, and T. Evan Nebeker

Abstract. A preliminary conceptual model of the seasonal change in pine (*Pinus*) resistance to bark beetle (*Dendroctonus frontalis* Zimm.) attack is proposed. The model characterizes the role of pine oleoresin in tree resistance to beetle attack and relationships between tree growth and differentiation processes in response to environmental conditions through a growing season. Analysis of soil water balance, tree cambial growth, vertical resin duct density in the current annual ring, and oleoresin yield from small wounds over a 2-year period support the model's applicability. However, major shifts in timing, degree of intensity, and duration of water deficits can greatly alter the seasonal pattern of changes in resistance to beetle attack. Tree responses are consistent with those predicted by plant growth and differentiation balance principles. We suggest that these principles can assist in process-based modeling of tree and stand growth and development, as well as in the study of interactions between bark beetles and pines.

INTRODUCTION

Tree susceptibility to insect attack is commonly hypothesized to be based on stress effects, such as water deficits, interfering with normal physiological processes (Berryman, 1972; Mattson and Haack, 1987). Another hypothesis is suggested by W. E. Loomis's (1932, 1953) concept of plant growth and differentiation balance. His concept indicates that plants expend energy predominantly in growth processes when environmental conditions are favorable, but expend proportionately more energy in differentiation processes, such as oleoresin synthesis in pines (*Pinus*), when environmental conditions limit cell division and enlargement without adversely affecting photosynthesis and translocation. Several publications in recent years have suggested the potential utility of examining bark beetle (*Coleoptera: Scolytidae*) and pine (*Pinus*) tree interactions in this broader context, and have encouraged research to test hypotheses based on this concept (Lorio and Hodges, 1985; Lorio, 1986; Lorio and Sommers, 1986; Lorio, 1988). In this paper, we report test results of a preliminary conceptual model of loblolly pine (*P. taeda* L.) resistance to southern pine beetle (*Dendroctonus frontalis* Zimm.) based on Loomis's concept, host-tree seasonal ontogeny, and southern pine beetle behavior.

Dr. Lorio and Mr. Sommers are with the USDA Forest Service, Pineville, LA, 71360 U.S.A. Dr. Blanche, Dr. Hodges, and Dr. Nebeker are at Mississippi State University, Mississippi State, MS, 39762 U.S.A.

METHODS AND MATERIALS

Model Development

We consider that resistance to bark beetle attack in southern pines depends largely on oleoresin flow rate and total flow (Hodges et al., 1979) and that the seasonal changes in resin flow from small wounds indicate changes in resistance to beetle attack. Here, seasons of the year refer to the vernal (March 21) and autumnal (September 21) equinoxes, and the summer (June 21) and winter (December 21) solstices for the northern hemisphere. However, the weather associated with seasons varies considerably across regions and with latitude. In model development, we focus on seasonal changes characteristic of the West Gulf Coastal Plain, approximately latitude 31°N and longitude 92°W.

Resin flow is highly sensitive to temperature and moisture, and from winter to early spring, flow from small wounds appears to be a primary function of temperature. Once ambient temperature is not limiting, other factors (e.g., water status, substrate availability) may increasingly influence the potential yield of resin from small wounds. Toward the end of the growing season, resin flow once again becomes highly temperature dependent. Conceptually, therefore, host resistance may be represented as diagramed in Figure 1, based on expected responses to long-term average temperature, soil water regime (Buol et al., 1973), and plant growth and differentiation balance principles (Loomis, 1932, 1953). The basis of the model, and the model itself, relate closely to other theories of plant-herbivore interactions (Bryant et al., 1983; Mooney et al., 1983; Coley et al., 1985).

The conceptual model suggests relatively low tree resistance to beetle attack in the spring, followed by increasing resistance associated with transition to latewood formation and production of the current season's vertical resin duct system (Figure 1). Increasing resistance to beetle attack in late winter and early spring, and decreasing resistance in autumn, are in response to increasing temperature in the former case and

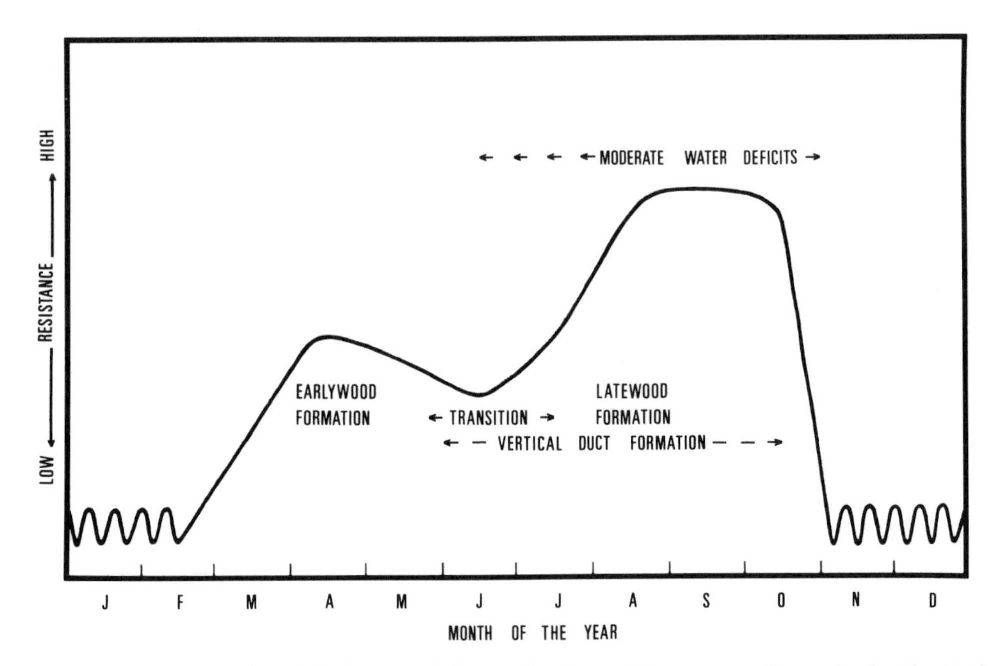

Figure 1. A conceptual model of seasonal changes in pine resistance to southern pine beetle attack for years which have soil water balance patterns similar to that of the long-term average. Resistance to the earliest attacking beetles is considered to be highly dependent on the potential flow of oleoresin at the wound sites.

decreasing temperature in the latter. During earlywood formation, tree energy allocation to oleoresin synthesis is limited by strong demands for reproductive and vegetative growth processes, and, as rows of earlywood are produced, the beetle attack site is progressively displaced from the principal oleoresin reservoir (the vertical resin ducts of the preceding and prior years). With the transition to latewood and the development of the current year's vertical resin ducts, the potential flow of resin from beetle attack sites in the cambial region increases, as does the inferred resistance to beetle attack.

The southern pine beetle produces multiple generations, but the major brood emergence, dispersal, and initiation of new infestations typically occur in the spring when tree energy demands by reproductive and vegetative growth processes are normally high. Enlargement of existing infestations, in which beetles rapidly colonize nearby host trees, predominates from the mid- to late-summer period when moderate water deficits normally have slowed growth processes and resin synthesis is favored (Lorio, 1986).

Data Collection and Model Testing

Field tests of the model were carried out near Alexandria, Louisiana, U.S.A. (latitude 31°25'N, longitude 92°17'W). Trees were chosen from a naturally established loblolly pine stand approximately 35 years old, growing on Malbis fine sandy loam, a plinthic paleudult, and Ruston fine sandy loam, a typic paleudult, with slopes of 1–5% and a site index for loblolly pine of 27 m (height at base age 50 years). Average soil texture is loam in the top meter of soil, and available soil water storage is estimated to be 20 cm. In the 1984 and 1985 tests, 10 and 16 trees were used respectively. Trees chosen were all similar in size and phenotypic characteristics (Table 1).

Table 1. Characteristics of trees studied for bole cambial growth, vertical resin duct formation and distribution, and 24-hour oleoresin yield of loblolly pine (*Pinus taeda* L.) in 1984 and 1985. n = 10 for 1984; n = 16 for 1985.

	dbh (cm)		Total tree height (m)		Rings at bh (number)		Live crown (%)	
	1984	1985	1984	1985	1984	1985	1984	1985
$\overline{X}$[1]	40.6	37.1	24.1	24.2	31.3	35.9	34.4	37.3
SE[2]	0.3	0.3	0.4	0.3	0.4	0.5	1.3	0.9

[1] $\overline{X}$ = Sample mean
[2] SE = Standard error of the mean

Tree oleoresin sampling procedures were as described by Lorio and Sommers (1986) except as indicated below. Starting in early March and continuing through December of both years, 2.5-cm–wide wounds were made in trees by removing small patches of bark with a knife in 1984, and with an arch punch in 1985, avoiding injury to the face of the xylem. In 1984, wounding was repeated biweekly through September 5; thereafter, wounds were made weekly. In 1985, wounds were made weekly on opposite sides of each tree. To minimize the potential influence of traumatic responses, wounds were offset horizontally and vertically (5 cm to the right and 2.5 cm upward) each time a new sample was taken. Aluminum troughs were pinned to the bark below wounds to channel resin flow into tapered graduated centrifuge tubes placed below the troughs and held in place with wire staples. Oleoresin was collected over a 24-hour period and the total volume determined ocularly to the nearest 0.1 ml. Resin yields for 1984 were averages of single measurements from each tree. Yields for 1985 were averages of the two measurements from each tree averaged for all 16 study trees.

After removal of resin samples from trees, wood samples were taken immediately above the wound site by extracting cores with a 1.3-cm arch punch, placed in vials containing a formalin:acetic acid:ethanol mixture, put under a partial vacuum temporarily, then capped and stored under refrigeration. Subsequently, these samples were placed in a softening solution of three parts 50% ethanol to one part glycerin and placed in an oven at 60°C for at least 3 days. Cores were sectioned with a razor blade or a sliding microtome and examined under a compound microscope. Tracheid counts were made in three files of an 8-mm–wide band of tracheids parallel to the cambium and of variable length depending on the amount of current radial growth. Earlywood and latewood were differentiated by Mork's criteria as described by Smith and Wilsie (1961). Procedures followed Whitmore and Zahner (1966) and Moehring et al. (1975); however, the various maturing stages of xylem were not studied. The approximate time of earliest latewood initiation was estimated by interpolating between the date on which no latewood tracheids were observed and the date on which one or more latewood tracheids were found.

Vertical resin ducts were counted in the same 8-mm–wide bands of the transverse sections used for the assessment of xylem formation, and their number per unit area (density) calculated for each sample date. Time of earliest initiation of vertical ducts was estimated as for the latewood tracheids.

Weather data came from a U.S. National Oceanic and Atmospheric Administration station about 11 km from the study site. These data were used to calculate and plot Thornthwaite monthly soil water balance (water regime) for long-term (1951–80) average conditions, and for 1984 and 1985, assuming 20 cm of available soil water (Thornthwaite and Mather, 1957).

Zahner and Stage's (1966) two-level soil water balance program was used to calculate daily soil water storage, daily deficits, and cumulative water deficits for 1984 and 1985, again assuming 20 cm of available soil water. Soil water depletion was assumed to equal potential evapotranspiration (PE) minus precipitation (P) until soil water was reduced to 67% of storage capacity. Subsequently, depletion was determined by multiplying the ratio of current storage to storage capacity times the excess of PE over P. The procedure allows for linear depletion over the first 33% of storage capacity, followed by a curvilinear depletion over the remainder of storage capacity, a reasonable model of the depletion pattern for loam soils. Daily water balance data were plotted to permit detailed graphical analysis of within-year changes in soil water status in relation to tree growth and development variables.

Mean values and their standard errors for cambial growth, resin duct density, and resin yield for each sample date were plotted individually over time to illustrate their seasonal trends within years.

RESULTS AND DISCUSSION

Monthly water-balance data are plotted to illustrate the great disparity in soil water regimes between the two study years and between either year and the long-term average regime (Figure 2). Neither 1984 nor 1985 approximated the long-term average patterns of P and PE relationships. The average 1951–80 annual P was 142 cm, with PE exceeding P from June through September, and with a total calculated deficit of 8.7 cm. Annual P for 1984 was 198 cm, with P exceeding PE all months except July and September, when PE slightly exceeded P. There was no calculated water deficit for any month. Annual P for 1985 was 168 cm with PE exceeding P from April through July and a total calculated deficit of 12.2 cm. Daily soil water storage and cumulative deficits, cambial growth, vertical resin duct densities, and oleoresin yields for 1984 and 1985 are shown in Figure 3.

Monthly water balance analysis indicated no soil water deficit and very little soil water depletion in 1984. However the daily water balance indicated that water deficits accumulated slowly during late June and early July, and again in September at a faster rate, to a

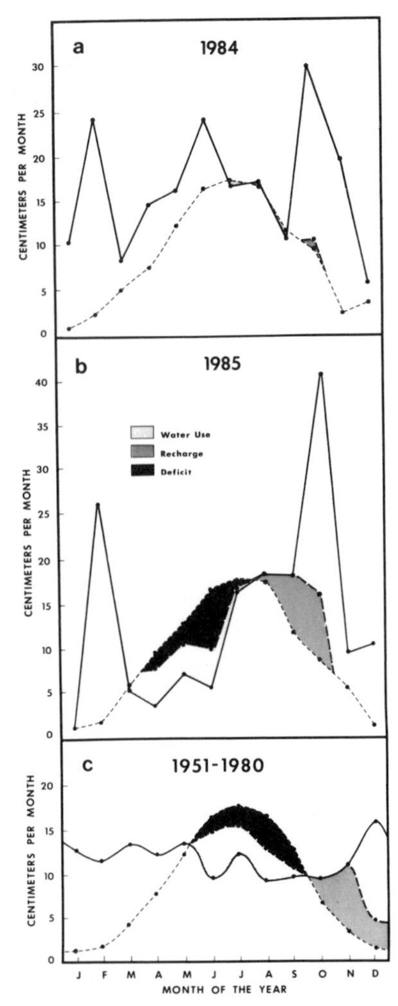

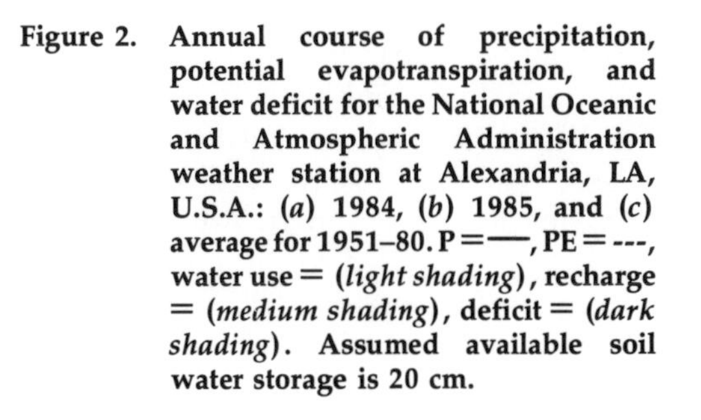
Figure 2. Annual course of precipitation, potential evapotranspiration, and water deficit for the National Oceanic and Atmospheric Administration weather station at Alexandria, LA, U.S.A.: (*a*) 1984, (*b*) 1985, and (*c*) average for 1951–80. P = —, PE = ---, water use = (*light shading*), recharge = (*medium shading*), deficit = (*dark shading*). Assumed available soil water storage is 20 cm.

total of 7.6 cm, approximating the long-term average of 8.7 cm. This regime more closely resembled the long-term average conditions than that of 1985, when available water was reduced to 25% of the total by late June and cumulative water deficits increased rapidly from late May to early August, totaling 16.3 cm. The large disparity in water regimes between years is illustrated in Figure 3 (*a, e*) in terms of soil water depletion and accretion, and the timing, degree of intensity, and duration of calculated soil water deficits.

Spring and early summer of 1984 were characterized by well-watered soil conditions, rapid cambial growth, a transition from earlywood to latewood formation between June 20 and July 4, vertical resin ducts first appearing in the May 30 sample and then slowly increasing in density until July 25, and low to moderate oleoresin yield (Figure 3, *a, b, c, d*). Minimum soil water storage for the year (45% of total available) occurred in September in correlation with maximum oleoresin yield for the year. In 1985, the spring and early summer were characterized by relatively dry conditions resulting in very slow cambial growth, transition from earlywood to latewood between June 12 and 19, vertical resin ducts appearing about 3 weeks earlier than in 1984 and rapidly increasing in density, and comparatively high oleoresin yields (Figure 3, *e, f, g, h*). Although cumulative water deficits in 1985 increased rapidly from late May to early August, oleoresin yields remained at about the same level. Recharge of soil water in August and September greatly stimulated cambial growth but had little apparent effect on oleoresin yields.

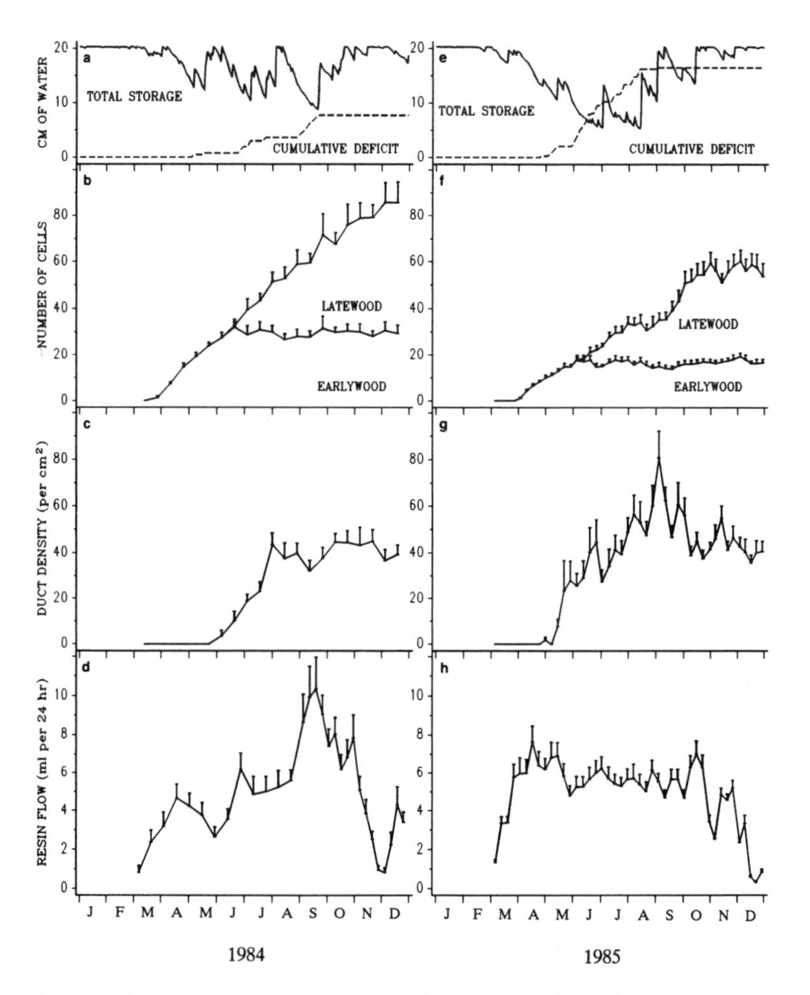

Figure 3. Graphs of water regimes, the course of xylem growth and development, vertical resin duct formation, and resin yield in 1984 (*a, b, c, d*) and 1985 (*e, f, g, h*). Daily soil water storage and cumulative water deficits (*a, e*). Tree bole cambial growth with total tracheid counts, time of transition to latewood, and amount of earlywood and latewood formed (*b, f*). Vertical resin duct densities in the current year's growth ring (*c, g*). Oleoresin yield over 24-hour periods (*d, h*). Vertical bars are standard errors; n = 10 for 1984, n = 16 for 1985.

INTERPRETATIONS

This study's results tend to support the preliminary model's applicability to long-term average soil water regimes. However, they also illustrate the potential danger of using average climatic variables to predict conditions, events, or relationships for any given year. Buol et al. (1973) discuss this problem with regard to attempts to relate climatic variables to soil development and soil characteristics.

Although 1984 was a very wet year compared to the long-term average, the results for resin yield approximate the pattern of the preliminary model. Resin yield increased through March to early April, then decreased as earlywood was being formed. Yield of resin increased abruptly and almost simultaneously with the onset of latewood formation and the first appearance of vertical resin ducts in the current year's growth. Rather than a broad plateau of high resin yield in mid- to late summer, as suggested by the model, resin yield peaked sharply in correlation with a period of increasing soil water deficit and minimum soil water storage for the year.

Results for 1985 more clearly illustrate the difficulties that may be encountered when applying relationships that are applicable to long-term average weather or climatic conditions in a year which greatly deviates from the average. Resin yield increased rapidly through March and maximized in early April, before the first appearance of vertical resin ducts in the current year's growth, or the transition to latewood formation. These results are contrary to the model's prediction and are due in part to the proximity of the resin supply in the latewood of 1984. This phenomenon is due to slow cambial growth caused by early and rapid depletion of soil water, even though soil water storage was still relatively high (Moehring and Ralston, 1967; Lorio and Hodges, 1971).

In agreement with the model, resin yield in 1985 declined with continued earlywood production and then increased for several weeks following transition to latewood formation and an increase in vertical resin duct density. However, there was no mid- to late-summer broad plateau of high resin flow. We believe these results are primarily due to two factors: (1) Early-season soil water deficits were severe enough to limit not only cambial growth, but also oleoresin synthesis, and (2) soil water recharge in August and September stimulated vegetative growth to such an extent that carbon allocation to resin synthesis was again limited. Similarly, studies of southern pines by Dewers and Moehring (1970), Shoulders (1973), and Grano (1973) showed that flower differentiation was reduced when water availability in the summer was high and vegetative growth was favored, suggesting that carbohydrates were preferentially used in growth.

Landsberg (1986) describes the potential utility of physiologically based models of forest growth and development. He advocates that research scientists in forestry look seriously at even relatively crude versions, such as our conceptual model, because they permit exploration of tree responses to changes in environmental conditions.

Crop scientists have a broad understanding of crop responses to environmental conditions. This understanding is the basis for their process-based models of crop growth and development, and their integrative analyses of host-pathogen relations (Loomis et al., 1979; Loomis and Adams, 1983). Plant growth and differentiation balance principles facilitate understanding the partitioning of new assimilates in crop plants (Loomis, 1983, pp. 350–352). They can likewise assist in understanding partitioning in forest trees, thus aiding in process-based modeling of tree and stand growth and development, and in broadening and improving the physiological basis for integrative analyses of bark beetle–host pine relations.

LITERATURE CITED

Berryman, A. A. 1972. Resistance of conifers to invasion by bark beetle–fungus associations. *BioScience* 22(10):598–602.

Bryant, J. P., F. S. Chapin, III, and D. R. Klein. 1983. Carbon/nutrient balance of boreal plants in relation to vertebrate herbivory. *Oikos* 40:357–368.

Buol, W. W., F. D. Hole, and R. J. McCracken. 1973. *Soil Genesis and Classification.* Iowa State University Press, Ames, IA, U.S.A.

Coley, P. D., J. P. Bryant, and F. S. Chapin, III. 1985. Resource availability and plant antiherbivore defense. *Science* 230:895–899.

Dewers, R. S., and D. M. Moehring. 1970. Effect of soil water stress on initiation of ovulate primordia in loblolly pine. *Forest Science* 16:219–221.

Grano, C. X. 1973. Loblolly pine fecundity in south Arkansas. Research Note SO-159, USDA Forest Service Southern Forest Experiment Station, New Orleans, LA, U.S.A. 7 pp.

Hodges, J. D., W. W. Elam, W. F. Watson, and T. E. Nebeker. 1979. Oleoresin characteristics and susceptibility of four southern pines to southern pine beetle (*Coleoptera: Scolytidae*) attacks. *Canadian Entomologist* 111:889–896.

Landsberg, J. J. 1986. *Physiological Ecology of Forest Production.* Academic Press, New York, NY, U.S.A. 198 pp.

Loomis, R. S. 1983. Crop manipulations for efficient use of water: an overview. In: Taylor, H. M., W. R. Jordan, and T. R. Sinclair, eds. *Limitations to Efficient Water Use in Crop Production.* ASA-CSSA-SSSA, Madison, WI, U.S.A. Pp. 345–374.

Loomis, R. S., and S. S. Adams. 1983. Integrative analyses of host-pathogen relations. *Annual Review of Phytopathology* 21:341–362.

Loomis, R. S., R. Rabbinge, and E. Ng. 1979. Explanatory models in crop physiology. *Annual Review of Plant Physiology* 30:339–367.

Loomis, W. E. 1932. Growth-differentiation balance vs. carbohydrate-nitrogen ratio. *Proceedings of the American Society for Horticultural Science* 29:240–245.

Loomis, W. E. 1953. Growth correlation. In: Loomis, W. E., ed. *Growth and Differentiation in Plants.* Iowa State College Press, Ames, IA, U.S.A. Pp. 197–217.

Lorio, P. L., Jr. 1986. Growth-differentiation balance: a basis for understanding southern pine beetle–tree interactions. *Forest Ecology and Management* 14:259–273.

Lorio, P. L., Jr. 1988. Growth and differentiation balance relationships in pines affect their resistance to bark beetles (*Coleoptera: Scolytidae*). In: Mattson, W. J., J. Levieux, and C. Bernard-Dagan, eds. *Mechanisms of Woody Plant Defenses Against Insects.* Springer-Verlag, New York, NY, U.S.A. Pp. 73–92.

Lorio, P. L., Jr., and J. D. Hodges. 1971. Microrelief, soil water regime, and loblolly pine growth on a wet, mounded site. *Soil Science Society of America Proceedings* 35:795–800.

Lorio, P. L., Jr., and J. D. Hodges. 1985. Theories of interactions among bark beetles, associated microorganisms, and host trees. In: Shoulders, E., ed. *Proceedings of the 3rd Biennial Southern Silvicultural Research Conference,* 7–8 November, 1984, Atlanta, GA, U.S.A. General Technical Report SO-54, USDA Forest Service Southern Forest Experiment Station, New Orleans, LA, U.S.A. Pp. 485–492.

Lorio, P. L., Jr., and R. A. Sommers. 1986. Evidence of competition for photosynthates between growth processes and oleoresin synthesis in *Pinus taeda* L. *Tree Physiology* 2:301–306.

Mattson, W. J., and R. A. Haack. 1987. The role of drought in outbreaks of plant-eating insects. *BioScience* 37(2):110–118.

Moehring, D. M., C. X. Grano, and J. R. Bassett. 1975. Xylem development of loblolly pine during irrigation and simulated drought. Research Paper SO-110, USDA Forest Service Southern Forest Experiment Station, New Orleans, LA, U.S.A. 8 pp.

Moehring, D. M., and C. W. Ralston. 1967. Diameter growth of loblolly pine related to available soil moisture and rate of soil moisture loss. *Soil Science Society of America Proceedings* 31:560–562.

Mooney, H. A., S. L. Gulmon, and N. D. Johnson. 1983. Physiological constraints on plant chemical defenses. In: Hedlin, P. A., ed. *Plant Resistance to Insects.* American Chemical Society, Washington, DC, U.S.A. Pp. 21–36.

Shoulders, E. 1973. Rainfall influences female flowering of slash pine. Research Note SO-150, USDA Forest Service Southern Forest Experiment Station, New Orleans, LA, U.S.A. 7 pp.

Smith, D. M., and M. C. Wilsie. 1961. Some anatomical responses of loblolly pine to soil-water deficiencies. *Tappi* 44:179–185.

Thornthwaite, C. W., and J. R. Mather. 1957. Instructions and tables for computing potential evapotranspiration and the water balance. Drexel Institute of Technology, Laboratory of Climatology, *Publications in Climatology,* Vol. X, No. 3. Centerton, NJ, U.S.A. 311 pp.

Whitmore, F. W., and R. Zahner. 1966. Development of the xylem ring in stems of young red pine trees. *Forest Science* 12:198–210.

Zahner, R., and A. R. Stage. 1966. A procedure for calculating daily moisture stress and its utility in regressions of tree growth on weather. *Ecology* 47:64–74.

CHAPTER 37

BALSAM FIR BOLE GROWTH AND FOLIAGE PRODUCTION IN RESPONSE TO DEFOLIATION

Dale S. Solomon and Terry D. Droessler

Abstract. Whole-tree maturation is an interdependent, dynamic system of photosynthesis source, flux, and sink relationships. A modeling link is needed to express this relationship between foliage production and bole growth. Radial growth within the bole of a forest tree increases from the tip of the crown to the lower crown and then decreases to the base of the tree. Thus, a proposed framework is presented relating the bole growth of balsam fir (*Abies balsamea* L. Mill.) trees to foliage production under pre- and post-catastrophic stress. Using a foliage production model, dynamic processes can be included to account for changes in photosynthetic efficiencies for age classes of needles and to establish changes in the transition probabilities for annual needle retention. Using the output from the foliage production model together with tree and stand characteristics, parameters can be developed to predict the radial increment along the bole of a tree as a function of the relative height above ground.

INTRODUCTION

The carbon-allocation process in forest trees is complex when structured for a modeling procedure. Modeling the allocation of photosynthate to different plant parts and the maturation process of a whole plant requires an interdependent, dynamic system of source, flux, and sink relationships (Thornley, 1976). Most conceptual tree models maintain a balance among the various compartmental processes. The detail with which these processes occur can be extended to a carbon-balance model for forest stands (McMurtrie and Wolf, 1983; Valentine, 1986). The usual priority within carbon-allocation models is to satisfy respiration and to assume a balance between root and shoot, and feeder-root and foliage production. The remaining photosynthate is allocated to the production of woody tissue (Promnitz, 1975).

Carbon-balance models have provided a theoretical framework for modeling the growth of individual trees. However, quantitative measures need to be incorporated to describe bole geometry dynamics as photosynthetic rates and foliage area change (McMurtrie and Wolf, 1983). By contrast, empirical tree growth models do not explicitly incorporate variables such as roots, soil, water, or nutrients in the growth process. Although relationships within empirical models assign carbon to the bole implicitly (Mitchell, 1975), additional components are needed to follow the process quantitatively.

Dr. Solomon is a Research Forester, USDA Forest Service Northeastern Forest Experiment Station, Durham, NH, 03824 U.S.A. Dr. Droessler is a Forest Biometrician with NSI Technology Services Corporation, U.S. EPA Environmental Research Lab, 200 SW 35th Street, Corvallis, OR, 97333 U.S.A.

Thus, one way to incorporate carbon-balance and empirical models is to include quantitative expressions of foliage photosynthate production capabilities and subsequent carbon allocation to growth regions of the tree.

Most empirical individual-tree and forest-stand models cannot modify growth response as a result of foliage stress. As physical aspects of foliage are altered and photosynthetic processes are impacted, radial-increment patterns along the bole fluctuate due to the dynamics of growth and the need to satisfy basic physiological processes.

This paper presents a model that links foliage characteristics and photosynthetic efficiencies to future foliage production and diameter increment along the bole. Our purpose is not to include all aspects of carbon allocation within a tree, but to quantify physical alterations of foliage and subsequent variation in radial increment along the bole. A conceptual-model framework provides a method of predicting changes in foliage production and how those changes can eventually be expressed in the amount of carbon allocated to the bole (Figure 1). Water, nutrients, respiration, and turnover of feeder roots are assumed sufficient over time.

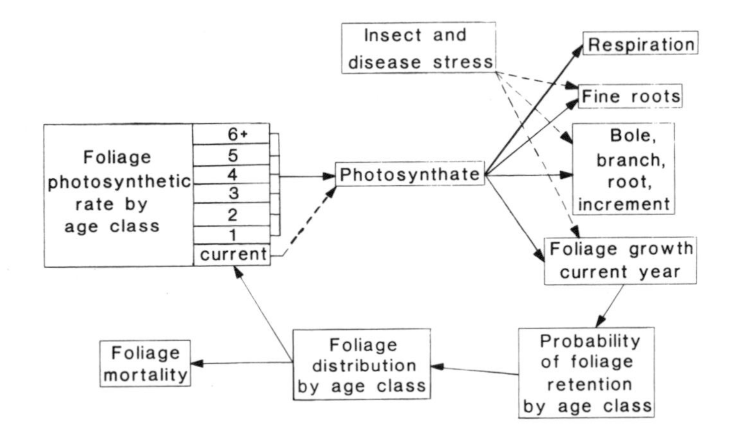

Figure 1. A carbon-allocation system indicating the changes in foliage production, bole increment, and stress by insects and disease.

BOLE INCREMENT

The configuration of growth along the bole of a tree has been documented by Onaka (1950) and Duff and Nolan (1953) and used by many. For trees in forest stands, the bole increment within the crown increases from the top of the tree to approximately the lower crown (the point of maximum branch development) and then decreases to the base of the crown. Below the crown, dominant trees with large crowns had consistent radial increments, and intermediate or suppressed trees with small to medium crowns showed decreasing radial increments to the base of the tree (Arney, 1974; Larson, 1963; Onaka, 1950). As the crown of balsam fir (*Abies balsamea* L. Mill.) is impacted by external stress such as the spruce budworm (*Choristoneura fumiferana* Clem.), reduced radial increment occurs first within the foliated length of the crown but may not be evident at breast height until 2–3 years after defoliation begins (Blais, 1958). This is caused primarily by the spruce budworm eating new foliage which is not a major contributor to branch, root, and bole growth (Clark, 1961). As the reduced amount of remaining foliage becomes older, a reduction in radial increment is evident along the bole (Mott et al., 1957). The relationship between foliage loss, defoliation history, and changes in radial increment along the bole has been shown by Solomon (1983, 1985) (Figure 2; Table 1). Radial increment is first reduced on the bole within the foliated length of the crown.

The radial increment (RI) along the bole can be considered as some exponential function of proportionate height:

$$RI = b_1(h/H)^{b_2}\, e^{b_3 (h/H)^{b_4}} \tag{1}$$

where h is a specified height above ground, H is the total tree height, e is the base of natural logarithms, and b_i (i = 1–4) are regression coefficients. These regression coefficients can then be expressed as a function of tree and foliage variables that allows adjustment when the foliage amount is altered by the physical impacts of insects, disease, frost, and so on.

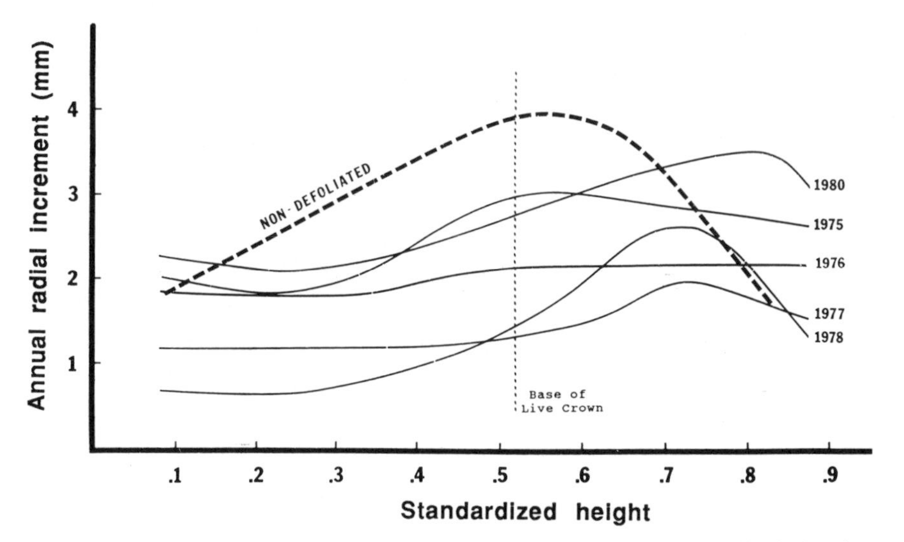

Figure 2. Changes in the amount of average radial increment along the bole of two balsam fir (*Abies balsamea* L. Mill.) trees as foliage production is impacted by defoliation (Solomon, 1985).

Table 1. Predicted average annual total foliage weight (kg) by age class for the average of two dominant defoliated balsam fir (*Abies balsamea* L. Mill.) trees with intensive protection (Solomon, 1985).

	Foliage								
Year	Current defoliation (%)	Current	1	2	Age classes 3	4	5	6+	Total[a]
1972	—	1.12	0.95	0.79	0.63	0.46	0.33	0.32	3.48
1973	15	1.07	1.08	0.91	0.72	0.52	0.34	0.39	3.96
1974	70	0.41	1.03	1.03	0.82	0.60	0.38	0.41	4.27
1975	90[b]	0.14	0.41	0.99	0.94	0.68	0.43	0.45	3.89
1976	70	0.37	0.13	0.38	0.90	0.78	0.50	0.51	3.20
1977	86[b]	0.14	0.36	0.13	0.34	0.74	0.56	0.59	2.72
1978	71[b]	0.32	0.14	0.35	0.12	0.28	0.54	0.67	2.10
1979	77[b]	0.24	0.31	0.13	0.31	0.10	0.21	0.66	1.72
1980	45[b]	0.60	0.23	0.30	0.12	0.26	0.07	0.34	1.32
1981	27[b]	0.68	0.58	0.22	0.27	0.10	0.19	0.14	1.50
1982 (Predicted)	—	—	0.66	0.56	0.20	0.22	0.07	0.07	1.78
1982 (Actual)[c]	—	—	2.61	1.53	0.56	0.37	0.12	0.03	5.22

[a]Total includes 1 to 6+ age classes.
[b]Protected.
[c]Estimated actual foliage weight by age class using branch samples.

MODELING FOLIAGE PRODUCTION

Balsam fir and other coniferous tree species can retain foliage 8–12 years, though the quantity and photosynthetic efficiency of that foliage decrease as the foliage ages (Clark, 1961). The amount of foliage dry matter by age class represents a partitioned index of the amount of photosynthate that a tree can produce in a year for use in growth, storage, and respiration. Therefore, a foliage-weight prediction system is advantageous in a tree-growth modeling system.

Estimates of foliage dry weight by age class on an unstressed balsam fir tree provide base-line information (Kleinschmidt et al., 1980). Changes in foliage amount by defoliation, for example, cause reduced photosynthetic potential in a tree. Spruce (*Picea*) and fir forests of Maine and eastern Canada are defoliated periodically by spruce budworm larvae that feed primarily on the current year's shoots, buds, needles, and staminate cones of balsam fir and, to a lesser degree, the spruces. Continued defoliation by the budworm over several years results in reduced tree growth (Piene, 1980; Solomon 1983, 1985; Mott et al., 1957). Thus, to model the relationship between foliage reduction from defoliation and subsequent tree growth, a matrix model was developed to predict production, aging, and senescence of foliage for balsam fir (Hayslett and Solomon, 1983):

$$A\tilde{w}_{\alpha} = \tilde{w}_{\alpha+1} \quad (2)$$

$$\begin{bmatrix} k\varepsilon_0 & k\varepsilon_1 & k\varepsilon_2 & k\varepsilon_3 & k\varepsilon_4 & k\varepsilon_5 & k\varepsilon_6 \\ P_0 & 0 & 0 & 0 & 0 & 0 & 0 \\ 0 & P_1 & 0 & 0 & 0 & 0 & 0 \\ 0 & 0 & P_2 & 0 & 0 & 0 & 0 \\ 0 & 0 & 0 & P_3 & 0 & 0 & 0 \\ 0 & 0 & 0 & 0 & P_4 & 0 & 0 \\ 0 & 0 & 0 & 0 & 0 & P_5 & P_{6+} \end{bmatrix} \begin{bmatrix} w_{0,\alpha} \\ w_{1,\alpha} \\ w_{2,\alpha} \\ w_{3,\alpha} \\ w_{4,\alpha} \\ w_{5,\alpha} \\ w_{6,\alpha} \end{bmatrix} = \tilde{w}_{\alpha+1}$$

where P_i is the proportion of i-year-old foliage that survives to become (i + 1)-year-old foliage, w_i is foliage weight for the ith age class, α is the age of the tree, $k\varepsilon_i$ is the relative contribution of foliage of age i to the production of new foliage, and ε_i is the relative photosynthetic efficiency of foliage at age i; k is expected to vary over time but can be assumed constant.

Solomon and Hayslett (1986) extended this matrix model to predict foliage-weight production for defoliated balsam fir trees. An application of the matrix model to two dominant defoliated balsam fir trees which received intensive protection is presented in Table 1. Note that current foliage weight increased the year following the start of protection, which resulted in a retention of foliage in the older age classes in later years. Defoliation continued to decrease the amount of current foliage produced regardless of protection. This work can be extended to a two-stage matrix model, patterned after Solomon et al. (1986), which will allow the photosynthetic efficiencies and the percent of annual needle retention to the next older age class to become dynamic. The photosynthetic efficiencies and percent of annual needle retentions can then be related to internal and external changes in foliage and to the photosynthetic process.

The process for quantifying the foliage recovery phase following years of defoliation is in need of refinement, as can be seen in Table 1. The current matrix model clearly underestimated the weight of foliage produced during recovery. Presumably, the development, maintenance, and repair of foliage during recovery result in a carbon sink. The radial increment within the crown typically shows an increase, while the growth along the lower nonfoliated bole continues to decline during the years of increased foliage production

(Figure 2). Clearly, carbon is not being allocated to the lower bole at the same rate as to the bole within the crown. The radial increment at breast height begins to increase after several years of recovery (Solomon, 1983). One possible paradigm within the two-stage matrix model would be to allow photosynthetic efficiencies to increase within the remaining foliage age classes, resulting in an increase in the photosynthate allocated to foliage production, maintenance, and repair.

As suggested by McMurtrie and Wolf (1983), it would be straightforward to extend a carbon-budget model by considering additional biomass components, foliage age classes, and dynamic photosynthetic efficiencies. A conceptual model is being constructed which will relate the changes in the radial increment along the bole to tree and stand characteristics including the foliage weight of the tree's crown by age class. As the foliage is stressed from environmental factors such as defoliation, light, atmospheric pollutants, etc., the photosynthetic capacity and the needle retention from one year to the next can be made dynamic. Then the parameters of the radial increment model can be predicted from these variables to show the changes in the radial increment along the bole of the stressed tree.

LITERATURE CITED

Arney, J. D. 1974. An individual tree model for stand simulation in Douglas-fir. In: Fries, J., ed. Growth models for tree and stand simulation. Research Note 30, Department of Forest Yield Research, Royal College of Forestry, Stockholm, Sweden. Pp. 38–46.

Blais, J. R. 1958. Effects of defoliation by spruce budworm in radial growth at breast height of balsam fir and white spruce. *Forestry Chronicle* 34:39–47.

Clark, J. 1961. *Photosynthesis and Respiration in White Spruce and Balsam Fir.* Technical Publication 85, College of Forestry, State University of New York, Syracuse, NY, U.S.A. 72 pp.

Duff, G. H., and N. J. Nolan. 1953. Growth and morphogenesis in the Canadian forest species. I. The controls of cambial and apical activity in *Pinus resinosa* Ait. *Canadian Journal of Botany* 31:471–513.

Hayslett, H. T., Jr., and D. S. Solomon. 1983. A matrix model for predicting foliage of trees by age classes. *Mathematical Bioscience* 67:113–122.

Kleinschmidt, S., G. L. Baskerville, and D. S. Solomon. 1980. Foliage weight distribution in the upper crown of balsam fir. Research Paper NE-455, USDA Forest Service Northeastern Forest Experiment Station, Broomall, PA, U.S.A. 8 pp.

Larson, P. R. 1963. Stem form development of forest trees. *Forest Science Monograph* 5. 42 pp.

McMurtrie, R., and L. Wolf. 1983. Above- and below-ground growth of forest stands: a carbon budget model. *Annals of Botany* 52:437–448.

Mitchell, K. J. 1975. Dynamics and simulated yield of Douglas-fir. *Forest Science Monograph* 17. 39 pp.

Mott, D. G., L. D. Nairn, and J. A. Cook. 1957. Radial growth in forest trees and effects of insect defoliation. *Forest Science* 3:286–304.

Onaka, F. 1950. *The Longitudinal Distribution of Radial Increments in Trees.* Forestry Bulletin 18, Kyoto University, Kyoto, Japan. (Japanese-English summary.)

Piene, H. 1980. Effects of insect defoliation on growth and foliar nutrients of young balsam fir. *Forest Science* 26:665–673.

Promnitz, L. C. 1975. A photosynthate allocation model for tree growth. *Photosynthetica* 9:1–15.

Solomon, D. S. 1983. Changes in growth of spruce-fir stands in the Northeast under varying levels of attack by the spruce budworm. In: Bell, J. F., and T. Atterbury, eds. *Conference on Renewable Resource Inventories for Monitoring Changes and Trends,* August 15–19, 1983, College of Forestry, Oregon State University, Corvallis, OR, U.S.A. Pp. 93–96.

Solomon, D. S. 1985. Growth responses of balsam fir defoliated by spruce budworm. In: *Spruce-Fir Management and Spruce Budworm.* General Technical Report NE-99, USDA Forest Service Northeastern Forest Experiment Station, Broomall, PA, U.S.A. Pp. 105–111.

Solomon, D. S., and H. T. Hayslett, Jr. 1986. Predicted foliage production for defoliated balsam fir trees using a matrix model. In: Solomon, D. S., and T. B. Brann, eds. *Conference of Environmental Influences on Tree and Stand Increments,* September 23–27, 1985, Durham, NH, U.S.A. Miscellaneous Publication 691, Maine Agricultural Experiment Station, University of Maine, Orono, ME, U.S.A. Pp. 138–145.

Solomon, D. S., R. A. Hosmer, and H. T. Hayslett, Jr. 1986. A two-stage matrix model for predicting growth of forest stands in the Northeast. *Canadian Journal of Forest Research* 16:521–528.
Thornley, J. H. M. 1976. *Mathematical Models in Plant Physiology.* Academic Press, London, England, U.K. 318 pp.
Valentine, H. T. 1986. *A Carbon-Balance Model of Stand Growth: A Derivation Employing the Pipe-Model Theory and the Self-Thinning Rule.* WP-87-56, International Institute for Applied Systems Analysis. 17 pp.

CHAPTER 38

PROCESS MODELING OF TREE AND FOREST GROWTH: CURRENT PERSPECTIVES AND FUTURE NEEDS

A. Ross Kiester

Abstract. This volume demonstrates that the modeling of tree and forest growth based on physiological processes is emerging as an important discipline in forestry research. Some possible directions for future research include the development of overall carbon budgets with maintenance and growth respiration considered separately, the development of a comparative (across species) approach in modeling, and the further analysis of the relationship between the processes of growth and those of mortality and reproduction. The scientific community must also further develop mechanisms for code and data sharing and the management of interdisciplinary teams for modeling efforts. Long-term goals of the modeling effort should be to reduce the number of assumptions used in the models and increase the number of calculated or predicted results.

INTRODUCTION

An Emerging Discipline

At a recent IUFRO conference, Rolf Leary analyzed the breakdown of traditional growth and yield modeling as a scientific discipline (Leary, 1988). He argued that as dissatisfaction with traditional growth and yield models increased, scientists would turn to, among other things, physiological process-based growth modeling. This volume amply confirms Leary's prediction. In it, we see the emergence of physiologically based process growth modeling as a major trend in the scientific study of trees and forests.

Here I will summarize some aspects of the other chapters in the book and sketch some of the issues surrounding this emerging discipline. This is a purely personal view and must also be temporary since the field continues to change rapidly. The other chapters represent many different scientific disciplines, and it is the new combination of disciplines that forms the foundation for physiologically based process growth models. This field is not simply tree physiology, but tree physiology bent to the needs of those who would understand and predict growth in trees and forests. In addition, the field merges into ecology insofar as it includes competition and the response to pests and pathogens. It also covers what has often been termed "stress physiology." This paper will first attempt to cover the biological background for understanding the growth of trees and forests. It will then discuss the modeling approaches that can be used to quantify our understanding of the

Dr. Kiester is Project Leader of the Synthesis and Integration Project, Forest Response Program, Environmental Protection Agency, Corvallis, OR, 97333 U.S.A. The author would like to thank E. D. Ford and N. V. L. Brokaw for discussions related to this chapter.

biology, followed by a discussion of some of the management aspects associated with research in this field. The paper will conclude with a concise and personal view of the direction that I believe modeling should take.

The Concept of Completeness

Perhaps the central concept that emerges in this volume is that we are now at a stage in the development of the field where we can begin to see our way clear to what would constitute a complete understanding of the processes that control growth in a single tree or group of adjacent trees. While a complete understanding is not now at hand, many of the models of whole trees presented in this volume contain an implicit concept of what a complete theory of tree growth would need to contain. To be sure, other entities besides whole trees are discussed and modeled. Canopies, for example, are frequently considered. But, increasingly, such entities are considered in relation to some concept of a whole tree and its entire physiology and environmental relations as they relate to growth.

BIOLOGY OF TREE AND FOREST GROWTH

The biology of forest growth can be organized by considering the growth of physiologically structured populations of trees. This approach (Oster, 1977; Metz and Diekmann, 1986) crosses the boundary between physiology and ecology and is to be recommended for tree physiology modeling to describe the demography of individuals characterized by their physiological traits. For our purposes, we consider that the three basic processes are reproduction, mortality, and growth. The interrelations between these three processes consequently need to be modeled.

Reproduction

Interest in reproduction has long been the province of evolutionary ecologists who seek to understand the processes controlling reproductive fitness (Harper, 1977). It has been of interest to traditional forestry growth and yield modelers only insofar as it affects growth. Generally, it appears that models of tree physiology have not focused on the relationship between growth and reproduction.

The topic of reproduction is frequently extended beyond the production of pollen and seeds to cover dispersal, germination, and seedling establishment. Again, models of dispersal are most often produced by evolutionary biologists (see Harper, 1977). The process of germination has been studied for variation in seed source and lifetimes in the seed bank, but the processes have rarely been modeled. Seedling establishment has been modeled for loblolly pine (*Pinus taeda* L.) (Blake and Hoogenboom, 1988), an approach which crosses the fuzzy boundary between reproduction and growth.

Mortality

Mortality seems to have received much more attention than reproduction from tree physiological ecologists. A recent issue of *Bioscience* (1987, Vol. 37, No. 8) reviews the causes and consequences of tree mortality. Some chapters in this volume address mortality (e.g., Gertner). Many tree ecological physiologists often consider mortality as dysfunctional growth or the consequence of premature stoppage of growth (Waring, 1987). At the stand level, growth of survivors is related to those trees that die.

Growth

Growth is the major focus of this volume, and most of the chapters address it in one way or another. Rather than attempt a comprehensive summary, I will touch on a few areas that seem to me to contain important outstanding problems.

Isebrands et al. (this volume) treat photosynthesis as the central physiological process related to growth, and most tree physiologists, I feel, would agree. Photosynthesis characterizes plants, and without it they die. Further, the general outlines of the biochemistry of this complex process are well understood, which gives workers the sense that they have a strong base on which to build. However, in contrast to photosynthesis, respiration is still poorly understood. Yet for any analysis of growth, respiration is just as important as photosynthesis. Respiration is clearly more difficult to measure. But the ability to measure maintenance and growth metabolism and differentiate them from the respiration associated with photosynthesis is one of the more difficult outstanding challenges facing experimental tree physiologists. Until the costs of growing and maintaining the functioning of various components of trees can be measured, life will continue to be difficult for even the most adventurous modeler. New field techniques (e.g., doubly labeled water; see Nagy, 1987) may offer methods that will greatly aid the modeler in understanding energy budgets.

Although carbon allocation and energy budgets are closely related, they appear to be sometimes confused. Carbon allocation is the pattern of the distribution of photosynthate to the different morphological compartments of a tree. Energy budgets, which are frequently studied by animal physiological ecologists, are the pattern of the utilization of energy by different physiological functions of a tree. Since photosynthate is often taken as the currency of energy available to a tree, the two concepts can overlap in a confusing way. The broader view of energy budgets may be more valuable as more detailed work on tree physiology proceeds (Mooney, 1975).

For convenience in modeling, carbon allocation is often taken to be a process rather than the result of processes (Bassow et al., this volume). McMurtrie and Wolf (1983) typify this approach, but it has been used by many others. Often, the process is expressed through allocation coefficients, which may be made variables in more sophisticated approaches (Weinstein and Beloin, this volume). But, in fact, there is no single physiological process that *is* carbon allocation; there is no "Maxwell's Demon" that does carbon allocation for a plant, shunting photosynthate into different compartments. Rather, there are a host of as yet poorly understood processes that result in patterns of carbon allocation.

While this volume does emphasize the physiology of growth, it also contains several chapters that discuss morphology and phenology as they relate to growth. One of the major messages that I take from the volume is that models of physiology must also be models of morphology and phenology. An analogy I like is that morphology and phenology are the behavior of plants, and, as is the case for those who model the physiological ecology of animals, behavior cannot be ignored (Bennett, 1980). In a similar vein, models of tree growth will eventually have to contend with the role of hormones and growth substances in the physiology of trees (see, e.g., Kramer and Kozlowski, 1979, Ch. 15).

An important unresolved question that is alluded to, but not directly discussed, in this volume, is that of the applicability of Liebig's "Law of the Minimum" (Liebig, 1843) or something like it to modeling growth processes. Is plant growth always regulated by the most limiting environmental factor, or have trees so evolved their adaptations as to be well buffered under all but the most extreme circumstances? In a recent paper, Tyree and Sperry (1988) ask, "Do woody plants operate near the point of catastrophic xylem dysfunction caused by water stress?" They argue that the answer is "yes"; but Zahner and

Grier (this volume) provide evidence that trees can be buffered, at least for awhile, against water stress. There is no general answer to this question at present, but somehow future modeling efforts will have to address it.

Another important unresolved issue that arises in this emerging discipline is that of how to create a concept of site that is more physiologically based. The concept of site index is relatively circular in that it is measured only by the final growth response of the tree. But the concept is undergoing a transformation from one which allows an easy, if oversimplified, way to account for environmental factors in growth and yield models to a complex ecological notion of the growth potential values of aggregated environmental factors affecting tree growth. An improved concept of site is a pivot about which the relation between growth and environmental factors can be developed. The goal is to have a model of environmental factors and physiology that predicts the entire temporal pattern of growth, without having to assume a maximum to growth dependent on a particular site.

Comparative Problems

In ecology, models that are applied to more than one species to gain insight into the processes themselves are often extremely powerful. I believe it is vital that forest process modeling develop more of a comparative approach. In tree physiology, comparative arguments are all too rare. Three axes of comparison are suggested here, but many others could be fruitfully explored. The first axis is that of evergreen versus deciduous trees. I take as an outstanding challenge to this field the fact that individuals of both types can be found in the same environment. Schulze et al. (1977a, b) have provided a possible beginning to the solution of this problem. Our models ought to be able to account for these two habits just as animal physiological ecologists can now account for the presence of both endotherms ("warm-blooded" animals) and ectotherms ("cold-blooded" animals) (Pough, 1980). The second axis is that of pioneer species versus persistent (or shade-tolerant) species with regard to response to the formation of gaps in the forest (Brokaw, 1985; Canham, 1988). Models that could account for these two syndromes and their variations would have to be considered quite powerful. Finally, the third axis is that of species that live in seasonal environments versus those that live in non-seasonal environments. Here we need to account for many of the differences between temperate and tropical forests, where the patterns of seasonality and hence of growth can be quite different (Bormann and Berlyn, 1981). (Of course, many tropical forests are seasonal, but based on precipitation more than temperature.)

MODELING OF TREE AND FOREST GROWTH

Prediction Versus Understanding

Many papers in this volume discuss the goals of modeling. One common theme is that in forestry we are generally moving from prediction to understanding as the primary goal. Of course, we do believe that better understanding will eventually lead to better predictions.

Analytical Models

Analytical models, that is, those which do the mathematics directly rather than via computer simulation, are represented in this volume (e.g., Valentine). In general, I feel that more effort needs to be devoted to using other modeling tools in addition to computer simulation. Certainly, most differential equations that arise in tree process modeling are not analytically solvable. But approximation methods and qualitative analysis methods have been used with much success in ecology (Levins, 1974). Other tools such as graphical models (Levins, 1966) or loop analysis (Puccia and Levins, 1985) need to be added to our toolboxes. To be blunt, it sometimes appears that we forest modelers rush off to create computer code without thinking about the processes in a way that working on analytical models may force us to do. The ready availability of computers tempts us all to use them without sufficiently contemplating the modeling, as apparently often happens with forestry statisticians and statistical packages (Warren, 1986).

Computer-Based Models and Languages

Since most of the models produced by tree and forest modelers are computer simulations, it is important to take advantage of recent advances in software engineering. The most important issue is expressiveness of the language chosen. With sufficient time and effort, any programming problem can probably be solved in any programming language. But a major criterion in the choice of a language should be readability. A useful ultimate goal is to have the scientific paper that reports the model and the computer code that implements it converge. This does not mean that we should be programming in English (it is not sufficiently disciplined). Rather, we should choose language with the capability of creating derived types (Ford and Kiester, this volume) and develop code in such a way that objects such as trees, compartments, and physiological processes become the derived types (not variables) of code. The intellectual process of developing the abstract entities that the models are about and the development of the types available in the computer code should become identical.

Uncertainty

Uncertainty is a major issue covered by Gardner et al., Gertner, and Luxmoore et al. (this volume), among others. Many levels of uncertainty exist. At the crudest level, there is uncertainty associated with a given physiological mechanism or theory of tree growth (Ford and Kiester, this volume). Then there are uncertainties of functional form, whether estimated by regression or derived from biological principles. Uncertainties in the data used to estimate parameters leads to parametric uncertainty, which can be propagated through the model. Finally, for computer models, there are uncertainties in the code itself. Although no chapters in the present volume address this issue, at the conference which led to this volume, tales were told of errors in coding persisting in models used for many years.

Models and Data Base Management

Many models in forestry are based on substantial empirical data sets. Often these are reduced via statistical modeling to particular parameter values. Obviously, if the data set changes, the parameters and the model will change. An idea that originated in artificial intelligence is that we should not distinguish so much between code and data. In modeling, this means that the data on which a model is based should be managed as part of the code of the model itself, or at least should be associated with it. The data, the statistical procedures, and the simulation code are then managed together as parts of a conceptual whole.

MANAGEMENT OF FOREST GROWTH MODELING RESEARCH

Management of efforts at tree and forest growth modeling is an area of concern to most workers in the field of tree growth modeling. Acock and Reynolds (this volume) raise some of these issues, but, for the most part, they have not been addressed. In discussion with many of the authors, I have noted three areas of concern, which I discuss below. This section is more subjective than the rest of the paper.

Team Approach

I believe that many of the people contributing to this volume feel a balanced, interdisciplinary team approach is the best way to manage a successful large-scale modeling effort. I feel that the team ideally should have, in addition to one or more modelers, an experimental physiologist with a short-term physiological view, a field ecologist with a long-term view, and a computer scientist (not just a programmer) to deal with issues of language, expressiveness, and communication. Thus four might be the minimum team. By Brooks's (Brooks, 1975) arguments, we can take the maximum size of a team to be eight. Management ought to strive to create such teams. Academia tends to frown on teams, and, in any event, there is usually no management entity that could encourage them. In academia, there is also a distinct tendency for likes to be grouped with likes, resulting in unbalanced teams. This is also true of some research work units in the USDA Forest Service. Fortunately, interdisciplinary academic departments and research work units are now beginning to appear.

Data Sharing

Data sharing has become an important general concern of the scientific public and has been the topic of much recent discussion (Fienberg et al., 1985). On one hand, there is the concept that science is public knowledge by definition. On the other hand, there are the career concerns of individuals who have spent time and effort in collecting data and want to benefit from that effort. I believe that the days in which an individual could hold on to a data set through an entire career are past. The goal of a working scientist should be to create commercial-quality data sets. By this, I mean that the data sets should be thoroughly cleaned, structured, and documented, so that another scientist could use them as though he or she had purchased a piece of commercial software. This goal is, at present, an ideal, but it is an ideal that should be pursued. In particular, managers in charge of those who produce data should understand and encourage this ideal.

Code Sharing

What is true for data is equally true for the computer code that makes up most of our models. Acock and Reynolds (this volume) emphasize new forms of communication such as electronic journals and computer conferences. These should be encouraged. In addition to the code being physically available, it is important that the source code be readable as discussed above. Managers need to emphasize both availability and readability.

CONCLUSIONS

I believe that a long-term goal of the discipline of modeling of tree and forest growth on the basis of physiological processes should be to reduce the number of assumptions required in the models and increase the number of aspects of tree biology that are calculated or predicted. Models should be able to predict overall growth, morphology, comparative patterns, ecological distribution, and many other facets of tree biology. The richer the set of predictions a model can make, the more useful it will be.

While this goal may lie in the future, the use of models to analyze multiple working hypotheses (Chamberlin, 1890) remains their immediate, and very real, value.

LITERATURE CITED

Bennett, A. F. 1980. The metabolic foundations of vertebrate behavior. *Bioscience* 30:452–456.
Blake, J. I., and G. Hoogenboom. 1988. A dynamic simulation of loblolly pine (*Pinus taeda* L.) seedling establishment based upon carbon and water balances. *Canadian Journal of Forest Research* 18:833–850.
Bormann, F. H., and G. Berlyn, eds. 1981. Age and Growth Rate of Tropical Trees: New Directions for Research. Bulletin 94, Yale University School of Forestry and Environmental Studies, New Haven, CT, U.S.A.
Brokaw, N. V. L. 1985. Treefalls, regrowth, and community structure in tropical forests. In: Pickett, S. T. A., and P. S. White, eds. *The Ecology of Natural Disturbance and Patch Dynamics.* Academic Press, New York, NY, U.S.A. Pp. 53–71.
Brooks, F. P. 1975. *The Mythical Man Month.* Addison-Wesley Publishing Company, Reading, MA, U.S.A.
Canham, C. D. 1988. Growth and canopy architecture of shade-tolerant trees: response to canopy gaps. *Ecology* 69:786–795.
Chamberlin, T. C. 1890. The method of multiple working hypotheses. *Science* 15:92–96.
Fienberg, S. E., M. E. Martin, and M. L. Straf, eds. 1985. *Sharing Research Data.* National Academy Press, Washington, DC, U.S.A.
Harper, J. L. 1977. *Population Biology of Plants.* Academic Press, London, England, U.K.
Kramer, P. J., and T. T. Kozlowski. 1979. *Physiology of Woody Plants.* Academic Press, New York, NY, U.S.A.
Leary, R. A. 1988. Some factors that will affect the next generation of forest growth models. In: Ek, A. R., S. R. Shifley, and T. E. Burk, eds. *Forest Growth Modelling and Prediction.* General Technical Report NC-120, USDA Forest Service North Central Forest Experiment Station, St. Paul, MN, U.S.A.
Levins, R. 1966. The strategy of model building in population biology. *American Scientist* 54:421–431.
Levins, R. 1974. The qualitative analysis of partially specified systems. *Annals of the New York Academy of Sciences* 231:123–138.
Liebig, J. 1843. *Chemistry and Its Application to Agriculture and Physiology,* 3rd Ed. Peterson, Philadelphia, PA, U.S.A.
McMurtrie, R., and L. Wolf. 1983. Above- and below-ground growth of forest stands: a carbon budget model. *Annals of Botany* 52:437–448.
Metz, J. A. J., and O. Diekmann. 1986. *The Dynamics of Physiologically Structured Populations.* Springer-Verlag, Berlin, F.R.G.
Mooney, H. A. 1975. Plant physiological ecology—a synthetic view. In: Vernberg, F. J., ed. *Physiological Adaptation to the Environment.* Intext Educational Publishers, New York, NY, U.S.A. Pp. 19–36.
Nagy, K. A. 1987. Field metabolic rate and food requirement scaling in birds and mammals. *Ecology* 57:111–128.
Oster, G. 1977. Lectures in population dynamics. In: DiPrima, R. C., ed. *Modern Modeling of Continuum Phenomena.* American Mathematical Society, Providence, RI, U.S.A. Pp. 149–190.
Pough, H. 1980. The advantages of ectothermy for tetrapods. *American Naturalist* 115:92–112.
Puccia, C. J., and R. Levins. 1985. *Qualitative Modeling of Complex Systems.* Harvard University Press, Cambridge, MA, U.S.A.
Schulze, E.-D., M. I. Fuchs, and M. Fuchs. 1977a. Spatial distribution of photosynthetic capacity and performance in a mountain red spruce forest in northern Germany. I. Biomass distribution and daily uptake in different crown layers. *Oecologia* 29:43–61.
Schulze, E.-D., M. I. Fuchs, and M. Fuchs. 1977b. Spatial distribution of photosynthetic capacity and performance in a mountain red spruce forest in northern Germany. II. The significance of evergreen habit. *Oecologia* 30:239–248.
Tyree, M. T., and J. S. Sperry. 1988. Do woody plants operate near the point of catastrophic xylem dysfunction caused by dynamic water stress? *Plant Physiology* 88:574–580.
Waring, R. H. 1987. Characteristics of trees predisposed to die. *Bioscience* 37:569–574.
Warren, W. G. 1986. On the presentation of statistical analysis: reason or ritual. *Canadian Journal of Forest Research* 16:1185–1191.

AUTHORS, REVIEWERS, AND CONFERENCE PARTICIPANTS

Many people contributed to the conference "Forest Growth: Process Modeling of Responses to Environmental Stress" and the subsequent book. The following alphabetical list of contributors recognizes the efforts of each individual. Addresses and telephone numbers were provided by each contributor and are given as a reference for others working in this field.

Dr. Basil Acock[1,3]
USDA/ARS/NRI
Systems Research Lab.
Bldg. 011A, Room 165B
BARC-W
Beltsville, MD, 20705 U.S.A.

Dr. Dave Alban[2]
USDA Forest Service
Forestry Sciences Lab.
1831 Hwy. 169 E.
Grand Rapids, MN, 55744 U.S.A.

Dr. Lee Allen[3]
Dept. of Forestry
N.C. State Univ.
Raleigh, NC, 27965 U.S.A.
919-737-3500

Dr. Mike Apps[3]
Canadian Forestry Service
Northern Forestry Center
5320-122 Street
Edmonton, Alberta, T6H 3S5
Canada
403-435-7292

Dr. Loukas Arvanitis[3]
University of Florida
School of Forest Resources
18 Newins-Ziegler Hall
Gainesville, FL, 32611 U.S.A.
904-392-1850

Dr. D. D. Baldocchi[2]
Atmospheric Turbulence & Diff. Div.
NOAA/ARL
P.O. Box 2456
Oak Ridge, TN, 37831 U.S.A.

Dr. Ann Bartuska[2]
USDA Forest Service
Southern Comm.
For. Res. Coop.
1509 Varsity Drive
Raleigh, NC, 27607 U.S.A.

Ms. Susan Bassow[1,3]
Ctr. for Quantit. Sci.
CQS HR-20
University of Washington
Seattle, WA, 98195 U.S.A.

Mr. R. Beloin[1]
Ecosystems Res. Ctr.
Cornell University
Ithaca, NY, 14853 U.S.A.

Dr. Greg Biging[2]
Dept. of Forestry & Resource Mgt.
213 Mulford Hall
University of California
Berkeley, CA, 94720 U.S.A.

Dr. Victor Bilan[3]
Stephen F. Austin State University
SFASU Sta. Box 6109
Nacogdoches, TX, 75952 U.S.A.
409-568-3301

Dr. John Blake[1,3]
School of Forestry
108 M. White Smith Hall
Auburn Univ., AL, 36849 U.S.A.
205-844-1014

Dr. Catalino A. Blanche[1,3]
Dept. of Forestry
Mississippi State University
P.O. Drawer FR
Mississippi State, MS, 39762 U.S.A.

Mr. Roger Bolton[3]
School of Forestry
108 M. White Smith Hall
Auburn Univ., AL, 36849 U.S.A.
205-844-1060

Dr. Marc Boucher[3]
Univ. of Moncton (CUSIM)
165 Blvd. Hebert
Edmundston, New Brunswick, E3V 2S8
Canada

[1]Authors [2]Reviewers [3]Conference participants

Mr. Doug Bowling[3]
International Paper Co.
Southlands Exp. Forest
Route 1, Box 421
Bainbridge, GA, 31717 U.S.A.
912-246-3642

Mr. D. G. Brand[3]
Canadian Forestry Service
Petawawa Nat. For. Inst.
Chalk River, Ontario
KOJ 1JO, Canada
613-687-8184

Mr. Gary Breece[3]
Georgia Power Co.
P.O. Box 4545
Atlanta, GA, 30302 U.S.A.
404-526-7077

Ms. Patricia Brewer[3]
Tennessee Valley Auth.
Missionary Ridge 25103B
Chattanooga, TN, 37401 U.S.A.
615-751-5680

Dr. Linda B. Brubaker[3]
College of Forest Resources AR-10
University of Washington
Seattle, WA, 98195 U.S.A.

Mr. David Bruce[1,3]
USDA Forest Service
Pacific NW Res. Sta.
P.O. Box 3890
Portland, OR, 97208 U.S.A.
503-231-2097

Dr. Thomas Burk[3]
Dept. of For. Res.
University of Minnesota
1530 N. Cleveland Ave.
St. Paul, MN, 55108 U.S.A.
612-624-6741

Mr. Jim Candler[3]
Environmental Center
Georgia Power Co.
791 Dekalb Industrial Way
Decatur, GA, 30245 U.S.A.
404-526-2956

Dr. M. Cannell[2]
Institute of Terrestrial Ecology
Bush Estate
Penicilik, Medelonian
Scotland, EH260QB, U.K.

Dr. Quang V. Cao[2]
School of Forestry
Louisiana State Univ.
Baton Rouge, LA, 70803 U.S.A.

Dr. Jon Caulfield[3]
School of Forestry
108 M. White Smith Hall
Auburn Univ., AL, 36849 U.S.A.
205-844-1067

Mr. Huang Ce[3]
School of Forestry & Environmental Studies
Duke University
Durham, NC, 27706 U.S.A.

Dr. Carl W. Chen[1,3]
Systech Engineering, Inc.
3744 Mt. Diablo Blvd.
Lafayette, CA, 94549 U.S.A.

Mr. David Chojnacky[2,3]
USDA Forest Service
Intermountain Res. Sta.
507 25th Street
Ogden, UT, 84401 U.S.A.
801-625-5388

Dr. E. Cook[2]
Lamont-Doherty Geological Observatory
Palisades, NY, 10964 U.S.A.

Dr. Richard Corey[2]
Soil Science Dept.
University of Wisconsin
1525 Observatory Dr.
Madison, WI, 53706 U.S.A.

Dr. T. R. Crow[1]
USDA Forest Service
Forestry Sciences Lab.
P.O. Box 898
Rhinelander, WI, 54501 U.S.A.

Dr. Virginia H. Dale[1]
Environ. Sciences Div.
Oak Ridge Nat. Lab.
P.O. Box 2008
Oak Ridge, TN, 37831 U.S.A.

Dr. Richard F. Daniels[2]
Westvaco
P.O. Box 1950
Summerville, SC, 29483 U.S.A.

Dr. Dean DeBell[3]
USDA Forest Service
Forestry Sciences Lab.
3625-93rd Ave. SW
Olympia, WA, 98502 U.S.A.
205-753-9470

Dr. D. I. Dickmann[1]
Dept. of Forestry
Michigan State Univ.
East Lansing, MI, 48824 U.S.A.

Dr. Robert K. Dixon[1,2,3]
Environmental Research Laboratory
Environmental Protection Agency
200 SW 35th St.
Corvallis, OR 97333
503-757-4600

Dr. John Donnelly[3]
School of Nat. Resources
University of Vermont
Burlington, VT, 05405 U.S.A.
802-656-2620

Dr. Phil Dougherty[3]
School of For. Resources
University of Georgia
Athens, GA, 30602 U.S.A.
404-542-6556

Dr. Thomas Doyle[1,3]
MAXIMA Corp.
Oak Ridge Nat. Lab.
Oak Ridge, TN, 37831 U.S.A.
615-576-8520

Dr. T. R. Droessler[1,2]
Environmental Sciences
NSI Technology Ser. Corp.
200 SW 35th St.
Corvallis, OR, 97333 U.S.A.

---

[1]Authors [2] Reviewers [3]Conference participants

Dr. Howard Duzan, Jr.[3]
Weyerhaeuser Co.
P.O. Box 2288
Columbus, MS, 39701 U.S.A.
601-245-5225

Dr. Alan Ek[2,3]
Dept. of For. Resources
University of Minnesota
1530 N. Cleveland Ave.
St. Paul, MN, 55108 U.S.A.
612-624-3400

Dr. William R. Emanuel[1]
Environ. Sciences Div.
Oak Ridge Nat. Lab.
P.O. Box 2008
Oak Ridge, TN, 37831-6038
U.S.A.

Dr. Stephen Fairweather[2,3]
Penn. State University
208 Ferguson Hall
University Park, PA, 16802
U.S.A.
814-865-1602

Dr. Peter Farnum[3]
Weyerhaeuser Co.
WTC 2H5
Tacoma, WA, 98477 U.S.A.
206-924-6318

Dr. Wilbur Farr[3]
USDA Forest Service
Forestry Sciences Lab.
P.O. Box 020909
Juneau, AK, 99802 U.S.A.

Dr. Richard Flagler[3]
Dept. of Forest Science
Texas A&M University
College Station, TX, 77843
U.S.A.
409-845-8803

Dr. David Ford[1,3]
Ctr. for Quantitat. Sci.
Univ. of Washington HR-20
Seattle, WA, 98195 U.S.A.
206-543-1191

Dr. Michael Fosberg[1,3]
USDA Forest Service
P.O. Box 96090-610 RPE
Washington, DC, 20090-6090
U.S.A.
703-235-8195

Dr. Jeffrey Foster[3]
Holcomb Res. Institute
Butler University
4600 Sunset Ave.
Indianapolis, IN, 46208
U.S.A.

Dr. Harold Fritts[1,3]
Lab. of Tree Ring Res.
University of Arizona
Tucson, AZ, 85721 U.S.A.
602-621-2223

Dr. Margaret Gale[1,2,3]
School of Forestry and Wood
Products
Michigan Tech. University
Houghton, MI, 49931 U.S.A.
906-487-2352

Dr. Robert Gardner[1,2,3]
Environ. Sciences Div.
Oak Ridge National Lab.
P.O. Box 2008
Oak Ridge, TN, 37831-6036
U.S.A.
615-574-7369

Ms. Terri Garner[3]
School of Forestry
108 M. White Smith Hall
Auburn Univ., AL, 36849
U.S.A.
205-844-1067

Dr. David Gates[2]
Biological Station
University of Michigan
Ann Arbor, MI, 48104 U.S.A.

Dr. Cheryl S. Gay[2]
NSI Tech. Serv. Corp.
200 SW 35th St.
Corvallis, OR, 97333 U.S.A.
503-757-4748

Mr. James Geisler[3]
Univ. of Arkansas Ext. Ser.
P.O. Box 391
Little Rock, AR, 72203 U.S.A.
501-373-2638

Mr. Chris Geron[3]
USDA Forest Service
P.O. Box 12254
Research Triangle Park
NC, 27709 U.S.A.
919-549-4014

Dr. George Gertner[1,3]
Dept. of Forestry
Univ. of Illinois
110 Mumford Hall
Urbana, IL, 61801 U.S.A.
217-333-9346

Dr. A. Gnanam[3]
Vice-Chancellor
University of Madras
Madras, TN 600 005
India

Dr. Robert Goldstein[3]
Electric Power Res. Inst.
P.O. Box 10412
Palo Alto, CA, 94303 U.S.A.
415-855-2593

Mr. Luis E. Gomez[1]
Systech Engineering, Inc.
3744 Mt. Diablo Blvd.
Lafayette, CA, 94549 U.S.A.

Dr. Kurt Gottschalk[3]
USDA Forest Service
NE For. Exp. Sta.
P.O. Box 4360
Morgantown, WV, 26505
U.S.A.
304-291-4613

Dr. Jennifer C. Grace[1,2,3]
Forest Research Institute
Private Bag 3020
Rotorua, New Zealand
073-475-899

Mr. Charles E. Grier[1]
USDA Forest Service
SE Forest Exp. Sta.
Athens, GA, 30602 U.S.A.

[1]Authors [2] Reviewers [3]Conference participants

Dr. David Grigal[1,3]
Dept. of Soil Science
University of Minnesota
439 Borlaug Hall
St. Paul, MN, 55108 U.S.A.
612-625-4232

Dr. Louis J. Gross[2]
Department of Mathematics
University of Tennessee
Knoxville, TN, 37916 U.S.A.

Ms. Elizabeth Groton[3]
Forestry Building
Tennessee Valley Auth.
Norris, TN, 37828 U.S.A.
615-632-1509

Mr. Heikki Hänninen[1,3]
Faculty of Forestry
University of Joensuu
P.O. Box 111
SF-80101 Joensuu
Finland

Mr. Richard P. Hans[1]
Dept. of Forestry
VPI & SU
319 Cheatham Hall
Blacksburg, VA, 24061 U.S.A.

Dr. Paul J. Hanson[1,2,3]
Environ. Sciences Div.
Oak Ridge Nat. Lab.
P.O. Box 2008, Bldg. 1506
Oak Ridge, TN, 37831-6038
U.S.A.
615-574-5361

Dr. Pertti Hari[1,3]
Forest Research Institute
Unioninkatu 40 A
SI-00170 Helsinki
Finland

Dr. William Harms[2,3]
USDA Forest Service
SE For. Exp. Sta.
Forestry Sciences Lab.
2730 Savannah Hwy.
Charleston, SC, 29407 U.S.A.
803-556-4860

Dr. Robin Harrington[3]
F/FRED Project
P.O. Box 186
Paia, HI, 96779 U.S.A.

Ms. Jacqueline L. Haskins[1,2,3]
Ctr. for Quantitat. Sci.
CQS HR-20
University of Washington
Seattle, WA, 98195 U.S.A.
206-543-5473

Dr. Tom Hennessey[2,3]
Department of Forestry
Oklahoma State University
Stillwater, OK, 74074 U.S.A.

Mr. Danny Herrin[3]
Alabama Power Co.
P.O. Box 2641
Birmingham, AL, 35291
U.S.A.
205-250-4124

Dr. Donald E. Hilt[2]
USDA Forest Service
NE For. Exp. Sta.
359 Main Road
Delaware, OH, 43015 U.S.A.

Dr. Tom Hinckley[2,3]
College of Forest Resources
AR-10
University of Washington
Seattle, WA, 98195 U.S.A.

Dr. John D. Hodges[1]
Dept. of Forestry
P.O. Drawer FR
Mississippi State University
Mississippi State, MS, 39762
U.S.A.

Dr. Margaret Holdaway[3]
USDA Forest Service
N. Cen. For. Exp. Sta.
1992 Folwell Ave.
St. Paul, MN, 55108 U.S.A.
612-649-5178

Ms. Maria Holmberg
Dept. of Silviculture
University of Helsinki
Unioninkatu 40B
SF=00170 Helsinki
Finland
0-1924344

Mr. Michael Holmes[1,3]
School of For. & Wood Pro.
Michigan Tech. Univ.
Houghton, MI, 49931 U.S.A.
906-487-2887

Dr. Peter Huang[3]
700 University Ave.
Toronto, Ontario, M5G 1X6
Canada

Dr. David Hyink[3]
Weyerhaeuser Company
WTC 2H5
Tacoma, WA, 98477 U.S.A.
206-924-6315

Dr. J. G. Isebrands[1,2,3]
USDA Forest Service
Forestry Sciences Lab.
Box 898
Rhinelander, WI, 54501
U.S.A.
715-362-7474

Mr. John Jansen[3]
Southern Co. Services
P.O. Box 2625
Birmingham, AL, 35202
U.S.A.
205-877-7698

Dr. Paul Jarvis[2]
Dept. of For. & Nat. Res.
University of Edinburgh
Edinburgh
Scotland, U.K.

Dr. Dale Johnson[3]
Oak Ridge Nat. Lab.
P.O. Box 2008, Bldg. 1505
Oak Ridge, TN, 37831-6038
U.S.A.
615-574-7362

[1]Authors [2] Reviewers [3]Conference participants

Ms. Elizabeth A. Jones[1]
Dept. of Math. Sci.
Michigan Tech. Univ.
Houghton, MI, 49931 U.S.A.
906-487-2793

Dr. James Jones[2]
Dept. of Agric. Engin.
University of Florida
Gainesville, FL, 32611 U.S.A.

Dr. Merrill R. Kaufmann[1,2,3]
USDA Forest Service
Rocky Mt. For. & Range Exp. Sta.
240 W. Prospect Rd.
Fort Collins, CO, 80526 U.S.A.
303-498-1256

Dr. Ross Kiester[1,2,3]
USDA Forest Service
c/o EPA
200 SW 35th St.
Corvallis, OR, 97333 U.S.A.

Dr. Tom Kimmerer[3]
Dept. of Forestry
University of Kentucky
Lexington, KY, 40546 U.S.A.
606-257-1824

Dr. Anthony King[1,3]
Environ. Sciences Div.
Oak Ridge Nat. Lab.
P.O. Box 2008, Bldg. 1505
Oak Ridge, TN, 37831-6038 U.S.A.
615-574-5397

Mr. Robert Kohut[3]
Boyce Thompson Institute
Cornell University
Ithaca, NY, 14853-1801 U.S.A.
607-257-2030

Dr. Werner Kurz[2]
Univ. of British Columbia
Forest. 270-2357 Main Mall
Vancouver, B.C., V6T 1W5
Canada

Mr. John Laurence[3]
Boyce Thompson Institute
Cornell University
Ithaca, NY, 14853-1801 U.S.A.
607-257-2030

Dr. Michael Lavigne[3]
Canadian Forestry Service
Newfoundland For. Ctr.
P.O. Box 6028
St. Johns, Newfoundland
A1C 5X8, Canada

Dr. David C. Leblanc[2]
Holcomb Res. Institute
Butler University
4600 Sunset Ave.
Indianapolis, IN, 46208 U.S.A.

Mr. Hal O. Liechty[1]
School of For. & Wood Pro.
Michigan Tech. Univ.
Houghton, MI, 49931 U.S.A.
906-487-2887

Dr. Nelson Loftus[2]
USDA Forest Service
P.O. Box 96090
Washington, DC, 20013-6090 U.S.A.

Dr. Peter L. Lorio, Jr.[1,3]
USDA Forest Service
Southern For. Exp. Sta.
2500 Shreveport Hwy.
Pineville, LA, 71360 U.S.A.

Dr. Gary Lovett[2,3]
Inst. of Ecosys. Studies
Cary Arboretum
P.O. Box AB
Millbrook, NY, 12545 U.S.A.
914-677-5343

Dr. Kim Lowell[2]
School of Forestry
University of Missouri
Columbia, MO, 65211 U.S.A.

Dr. Alan Lucier[2]
NCASI
260 Madison Ave.
New York, NY, 10016 U.S.A.

Dr. Anthony Ludlow[1,3]
Forestry Commission
Alice Holt Lodge
Farnham, Surrey
GU10 4LH
England, U.K.
0420-22255

Dr. Robert Luxmoore[1,2,3]
Environ. Sciences Div.
Oak Ridge National Lab.
P.O. Box 2008
Oak Ridge, TN, 37831-6038 U.S.A.
615-574-7357

Mr. Jory Lyons[3]
School of Forestry
108 M. White Smith Hall
Auburn Univ., AL, 36849 U.S.A.
205-826-4050

Dr. Doug MacGuire[2]
College of Forest Resources AR-10
University of Washington
Seattle, WA, 98195 U.S.A.

Dr. Annikki Mäkelä[1,3]
Dept. of Silviculture
University of Helsinki
Unioninkatu 40B
SF00170 Helsinki
Finland

Dr. T. Max[2]
USDA Forest Service
Pacific NW For. Exp. Sta.
P.O. Box 3890
Portland, OR, 97208 U.S.A.

Mr. Kevin McCauley[3]
School of Forestry
108 M. White Smith Hall
Auburn Univ., AL, 36849 U.S.A.
205-844-1067

Dr. Francis McCracken[3]
USDA Forest Service
P.O. Box 227
Stoneville, MS, 38776 U.S.A.

[1]Authors [2] Reviewers [3]Conference participants

Mr. Ronald E. McRoberts[2]
USDA Forest Service
N. Cen. For. Exp. Sta.
1992 Folwell Ave.
St. Paul, MN, 55108 U.S.A.

Mr. Karl Meilikainen[3]
Finnish For. Res. Inst.
Valimotie 13
00380 Helsinki
Finland
90-556276

Dr. Ralph Meldahl[2,3]
School of Forestry
108 M. White Smith Hall
Auburn Univ., AL, 36849
U.S.A.
205-844-1060

Dr. Paul Miller[2]
USDA Forest Service
Pacific SW For. Exp. Sta.
4955 Canyon Crest Dr.
Riverside, CA, 92507 U.S.A.

Dr. Kelsey Milner[3]
Champion International
P.O. Box 8
Milltown, VT, 59851 U.S.A.
406-258-5511

Dr. K. Mitchell[2]
Research Division
Ministry of Lands & For.
1450 Government St.
Victoria, B.C., V8W 3EX
Canada

Dr. G. M. J. Mohren[1,2,3]
Research Institute for
Forestry and Landscape
Planning
'De Dorschkamp'
Bosrandweg 20, Box 23
6700 AA, Wageningen
The Netherlands
31-8370-95111

Dr. Robert Monserud[2]
USDA Forest Service
Forest Sciences Lab.
1221 S. Main St.
Moscow, ID, 83843 U.S.A.

Dr. H. Todd Mowrer[2]
USDA Forest Service
Rocky Mt. For. & Range Exp.
Sta.
240 W. Prospect St.
Fort Collins, CO, 89526-2098
U.S.A.

Dr. Glenn D. Mroz[1]
School of For. & Wood Pro.
Michigan Tech. Univ.
Houghton, MI, 49931 U.S.A.
906-487-2496

Dr. M. S. R. Murthy[3]
Dept. of Botany
Osmania University
Hyderabad 500 007
India

Dr. Charles Myers[3]
Dept. of Forestry
Southern Illinois Univ.
Carbondale, IL, 62901 U.S.A.
618-453-3341

Dr. Warren L. Nance[2]
Southern For. Exp. Sta.
Forest Sciences Lab.
Box 2008 GMF
Gulfport, MS, 39503 U.S.A.

Dr. T. E. Nebeker[1]
Dept. of Entomology
Mississippi State University
Mississippi State, MS, 39762
U.S.A.

Mr. Eero Nikinmaa[1,3]
University of Helsinki
Dept. of Silviculture
Unioninkatu 40B
SF-00170 Helsinki
Finland
0-1924344

Dr. S. M. Northway[2]
Woodlands Services
MacMillian Bloedel LTD.
65 Front St.
Nanaimo, B.C., V9R 5H9
Canada

Dr. Richard Oderwald[1,3]
Dept. of Forestry
VPI & SU
319 Cheatham Hall
Blacksburg, VA, 24061 U.S.A.
703-231-5297

Dr. Robert V. O'Neill[1]
Environ. Sciences Div.
Oak Ridge Nat. Lab.
P.O. Box 2008
Oak Ridge, TN, 37831-6038
U.S.A.

Dr. Stephen G. Pallardy[2]
School of Forestry
University of Missouri
Columbia, MO, 65211 U.S.A.

Mr. Bernie Parasol[3]
USDA Forest Service
Southern For. Exp. Sta.
701 Loyola Ave.
New Orleans, LA, 70113
U.S.A.

Dr. Keith Paustian[3]
Dept. of Ecology & Environmental Res.
Swedish Univ. of Agr. Sci.
S-75007 Uppsala
Sweden

Mr. Bijan Payandeh[3]
Canadian Forestry Service
P.O. Box 490
Sault St. Marie, Ontario
P6A 5M7 Canada
705-949-9461

Dr. D. A. Perry[2]
Dept. of For. Sci.
Oregon State University
Corvallis, OR, 97331 U.S.A.

Dr. David Peterson[2]
USDA Forest Service
4955 Canyon Crest Dr.
Riverside, CA, 92507 U.S.A.

Dr. Harold Piene[2,3]
Canadian Forest Service
P.O. Box 4000
Fredericton, New Brunswick
E3B 5P7 Canada

[1]Authors [2] Reviewers [3]Conference participants

Mr. Harold Quicke[3]
School of Forestry
108 M. White Smith Hall
Auburn Univ., AL, 36849
U.S.A.
205-844-1067

Dr. R. Rabbinge[1]
Dept. of Theoretical Prod. Ecol.
Wageningen Agric. Univ.
Bornsesteeg 65
P.O. Box 430
6700 AK, Wageningen
The Netherlands

Dr. Harvey L. Ragsdale[2]
1555 Pierce Dr., Rm. 203
Emory University
Atlanta, GA, 30322 U.S.A.

Timothy J. Randle[1]
Forestry Commission
Alice Holt Lodge
Farnham, Surrey, GU10 4LH
England, U.K.

Dr. H. Michael Rauscher[1,2]
USDA Forest Service
Forestry Sciences Lab.
1831 Hwy. 169 E.
Grand Rapids, MN, 55744
U.S.A.

Mr. S. H. Raza[3]
Dept. of Botany
Osmania University
Hyderabad 500 007
India

Dr. G. Reams[2]
Oregon State University
c/o EPA
200 SW 35th St.
Corvallis, OR, 97333 U.S.A.

Dr. David Reed[1,3]
School of For. & Wood Pro.
Michigan Tech. Univ.
Houghton, MI, 49931 U.S.A.
906-487-2886

Dr. Peter Reich[2,3]
Department of Forestry
University of Wisconsin
1630 Linden Dr.
Madison, WI, 53706 U.S.A.
608-262-4754

Dr. James F. Reynolds[1]
Systems Ecology Res. Group
San Diego State Univ.
San Diego, CA, 92182 U.S.A.

Dr. Marion R. Reynolds[2]
Department of Statistics
VPI & SU
Blacksburg, VA, 24061 U.S.A.

Dr. James Richardson[3]
Canadian Forestry Service
351 Bov. St.
Joseph Hull, Quebec
K1A 1G5 Canada
819-997-1107

Dr. Susan Riha[3]
Agronomy Dept.
Cornell Univ.
Ithaca, NY, 14853 U.S.A.
607-255-1729

Dr. Edgar Robichaud[3]
Dept. of For. Res.
Univ. of New Brunswick
Fredericton, New Brunswick
E3B 6C2 Canada

Dr. Hugo Rogers[2]
USDA/ARS
Nat. Soil Dynamics Lab
P.O. Box 792
Auburn, AL, 36830 U.S.A.

Dr. Gregory Ruark[1,2,3]
USDA Forest Service
P.O. Box 12254
Research Triangle Park, NC,
27709 U.S.A.
919-549-4061

Dr. S. Running[2]
School of Forestry
University of Montana
Missouri, MT, 59812 U.S.A.

Mr. Mark Rutter[3]
Westvaco
P.O. Box 1950
Summerville, SC, 29484
U.S.A.
803-871-5000

Mr. Kevin Ryan[3]
IFSL
P.O. Box 8089
Missoula, MT, 59807 U.S.A.

Dr. Michael Ryan[3]
USDA Forest Service
Rocky Mt. Exp. Sta.
240 W. Prospect St.
Fort Collins, CO
80526-2098 U.S.A.
303-224-1237

Dr. P. V. Sane[3]
Nat. Bot. Res. Inst.
Ashok Marg
Lucknow, UP 226 001
India

Dr. Dan Santantonio[1,2,3]
Dept. of Ecology & Environmental Res.
Swedish Univ. of Ag. Sci.
Box 7072
S-75007 Uppsala
Sweden

Dr. George Schier[2,3]
USDA Forest Service
NE For. Exp. Sta.
359 Main Rd.
Delaware, OH, 43015 U.S.A.
614-369-4475

Prof. E.-D. Schulze[1,2,3]
Lehrstuhl fur
Pflanzanokologie der
Universitat Bayreuth
Universtrasse 30
D-8580 Bayreuth
F.R.G.

Dr. John Seiler[2]
Department of Forestry
VPI & SU
Blacksburg, VA, 24016 U.S.A.

[1]Authors [2] Reviewers [3]Conference participants

Dr. Peter Sharpe[1,3]
Dept. of Industrial Eng.
Biosystems Group
Texas A&M University
College Station, TX, 77843
U.S.A.
409-845-5559

Ms. Gerry Simmons[3]
Tennessee Valley Auth.
Oak Ridge Nat. Lab.
P.O. Box X, Bldg. 1506
Oak Ridge, TN, 37831-6034
U.S.A.
615-574-7968

Dr. Teja Singh[3]
Canadian For. Ser.
Northern Forestry Center
5320-122 St.
Edmonton, Alberta
T6H 3S5 Canada
403-435-7292

Mr. Rick Smith[1,2,3]
USDA Forest Service
Southern For. Exp. Sta.
701 Loyola Ave.
New Orleans, LA, 70113
U.S.A.

Dr. Dale Solomon[1,3]
NE For. Exp. Sta.
P.O. Box 640
Forest Sciences Laboratory
Durham, NH, 03824 U.S.A.
603-868-5710

Dr. Greg Somers[1,3]
School of Forestry
108 M. White Smith Hall
Auburn Univ., AL, 36849
U.S.A.
205-844-1021

Mr. Robert A. Sommers[1]
USDA Forest Service
Southern For. Exp. Sta.
2500 Shreveport Hwy.
Pineville, LA, 71360 U.S.A.

Dr. A. R. Stage[1,2]
USDA Forest Service
Forest Sciences Lab
1221 S. Main St.
Moscow, ID, 83843 U.S.A.

Dr. D. Steingraeber[3]
Dept. of Botany
Colorado State Univ.
Fort Collins, CO, 80521
U.S.A.
303-491-1923

Dr. Karl Stoszek[2,3]
Dept. of For. Res.
University of Idaho
Moscow, ID, 83843 U.S.A.

Dr. Edward Sucoff[2]
Department of Forest
Resources
Univ. of Minnesota
1530 N. Cleveland Ave.
St. Paul, MN, 55108 U.S.A.

Dr. Wayne T. Swank[1]
Coweeta Hydrologic Lab.
USDA Forest Service
999 Coweeta Lab. Rd.
Otto, NC, 28763 U.S.A.

Dr. George Taylor, Jr.[1]
Environ. Sciences Div.
Oak Ridge Nat. Lab.
P.O. Box 2008, Bldg. 1506
Oak Ridge, TN, 37831-6034
U.S.A.
615-574-7353

Dr. Robert Teskey[3]
School of For. Resources
University of Georgia
Athens, GA, 30602 U.S.A.
404-542-5055

Ms. M. L. Tharp[1]
Environ. Sciences Div.
Oak Ridge Nat. Lab.
P.O. Box 2008
Oak Ridge, TN, 37831-6038
U.S.A.

Dr. Kevin Topolniski[3]
Ecole de Sciences
Forestie⅝res
165 Blvd. Hebert
Edmundston, New
Brunswick
E3V 2S8 Canada
506-735-8804

Mr. Bjorn Tveite[3]
Norwegian For. Res. Inst.
P.O. Box 61
N-1432 AAS-NLH
Norway
001-47-2-949660

Dr. Harry Valentine[1,3]
USDA Forest Service
51 Mill Pond Road
Hamden, CT, 06514 U.S.A.
203-773-2012

Dr. Paul Van Deusen[1,2,3]
USDA Forest Service
Southern For. Exp. Sta.
701 Loyola Ave.
New Orleans, LA, 70113
U.S.A.
504-589-4546

Dr. Silvia Vega-Gonzales[2]
Ctr. for Quantita. Sci.
Univ. of Washington MR-20
Seattle, WA, 98195 U.S.A.

Dr. Kristina Vogt[2]
370 Prospect St.
Yale University
New Haven, CT, 06511
U.S.A.

Dr. James Vose[1,2,3]
Coweeta Hydrologic Lab.
USDA Forest Service
999 Coweeta Lab. Rd.
Otto, NC, 28763 U.S.A.
704-524-2128

Dr. Richard H. Waring[2]
Dept. of Forest Science
Oregon State University
Corvallis, OR, 97311 U.S.A.

[1]Authors [2] Reviewers [3]Conference participants

Dr. William Warren[2,3]
Science Branch/CODE
Fisheries and Oceans Canada
P.O. Box 5667
St. Johns
Newfoundland
A1C 5X1 Canada

Dr. Charles Webb[3]
NCASI
P.O. Box 12254
Research Triangle Park
NC, 27709 U.S.A.
919-541-9217

Dr. David Weinstein[1,3]
Ecosystems Res. Ctr.
Cornell University
Ithaca, NY, 14853 U.S.A.
607-255-3435

Dr. D. C. West[1,2]
Environ. Sciences Div.
Oak Ridge Nat. Lab.
P.O. Box 2008
Oak Ridge, TN, 37831-6038
U.S.A.

Dr. Robert Zahner[1,2,3]
Dept. of Forestry
Clemson University
Clemson, SC, 29634 U.S.A.
803-656-3302

Dr. John Zasada[3]
USDA Forest Service
Forestry Sciences Lab.
3200 Jefferson Way
Corvallis, OR, 97330 U.S.A.
503-757-4377

Dr. Boris Zeide[3]
Dept. of Forest Resources
Univ. of Ark.-Monticello
P.O. Box 3468
Monticello, AR, 71655 U.S.A.
501-460-1052

Mr. Bruce Zutter[3]
School of Forestry
108 M. White Smith Hall
Auburn Univ., AL, 36849
U.S.A.
205-844-1066

[1]Authors [2] Reviewers [3]Conference participants

CONVERSION TABLES

Length

centimeter	= 0.3937 inch	inch	= 2.5400 centimeters
meter	= 3.2808 feet	foot	= 0.3048 meter
meter	= 1.0936 yards	yard	= 0.9144 meter
kilometer	= 0.6214 mile	mile	= 1.6093 kilometers

Area

centimeter^2	= 0.1550 inch^2	inch^2	= 6.4516 centimeters^2
meter^2	= 10.7639 feet^2	foot^2	= 0.0929 meter^2
meter^2	= 1.1960 yards^2	yard^2	= 0.8361 meter^2
hectare	= 2.4710 acres	acre	= 0.4047 hectare
kilometer^2	= 0.3861 mile^2	mile^2	= 2.5900 kilometers^2
meter^2/hectare	= 4.356 feet^2/acre	foot^2/acre	= 0.2296 meter^2/hectare

Volume

centimeter^3	= 0.0610 inch^3	inches^3	= 16.3872 centimeters^3
meter^3	= 35.3145 feet^3	foot^3	= 0.0283 meter^3
meter^3	= 1.3079 yard^3	yard^3	= 0.7646 meter^3
meter^3/hectare	= 14.291 feet^3/acre	foot^3/acre	= 0.06997 meter^3/hectare

Capacity

liter	= 61.0250 inches^3	inch^3	= 0.0164 liter
liter	= 0.0353 foot^3	foot^3	= 28.3162 liters
liter	= 0.2642 gallon (U.S.)	gallon (U.S.)	= 3.7853 liters

liter = [1000.027 centimeters
[1.0567 quarts (liquid) or 0.9081 quart (dry)
[2.2046 pounds of pure water at 4°C = 1 kilogram

Weight

gram	= 0.0353 ounce	ounce	= 28.3495 grams
kilogram	= 2.2046 pounds	pound	= 0.4536 kilogram
kilogram	= 0.0011 ton (sht)	ton (sht)	= 907.1848 kilograms
ton (metric)	= 1.1023 tons (sht)	ton (sht)	= 0.9072 ton (metric)
ton (metric)	= 0.9842 ton (lg)	ton (lg)	= 1.0160 tons (metric)
metric ton	= 2204.6 pounds		

Pressure

0.1 MPa = 1.0 bar = 0.9869 atmosphere

Temperature

$$°F = °C \times 9/5 + 32$$

$$°C = [°F - 32] \times 5/9$$

ABBREVIATION AND ACRONYM GLOSSARY

AGTEHM AGricultural version of Terrestrial Ecosystem Hydrology Model

ADG Annual Diameter Growth

AI Artificial Intelligence

ANOVA ANalysis Of VAriance

APA Area Potentially Available (Doyle, Smith)

APA Area of Potential Availability (Chen and Gomez)

ARBOR For the tree-ring growth simulation program

ARMA Auto Regressive Moving Average

ARS Agricultural Research Service

ARSTAN Auto Regressive STANdardization

ATP Adenosine TriPhosphate

AWP Annual Wood Production

BLF Branch Length Function

BRANCH Model of BRANCH growth

CA Crown Area

CADIL Chemical Absorption and Degradation In Land

CAM Crassulacean Acid Metabolism

CARBON Carbon-based growth model

CD-ROM Compact Disc–Read Only Memory

CERES First of all, to nourish natural things, she the creator of all natural law (Ovid, *The Metamorphoses*)

CIO Competition Influence-zone Overlap

CLIMACS Computer Linked Integrated Model for Assessing Community Structure

COMAX COtton MAnagement eXpert

CR Crown Ratio class

CU Chilling Unit

CUcrit Critical chilling Unit sum

CUsum Chilling Unit sum

CV Crown Volume

DBH Diameter at Breast Height

DFIT Douglas Fir Interim Tables

DIFMAS DIFfusion and MAss flow of Solutes

DISP Drought Index for Southern Pines

DRYADS For wood nymphs

ECOPHYS ECOPHYSiological whole-tree growth process model

EPRI Electric Power Research Institute

ESA Exposed Surface Area

ESS Evolutionarily Stable Strategy

FAST Fourier Amplitude Sensitivity Test

FORET FORests of East Tennessee

FRP Forest Response Program

FU Forcing Unit

FUcrit Critical Forcing Unit sum

FUsum Forcing Unit sum

FV Foliar Volume

GCM General Circulation Model

GOSSYM For *Gossypium* simulation

HBLC Height to Base of Live Crown

HYDRA Model of water transport

IBSNAT International Benchmark Sites Network for Agroforestry Transfer

IFPM Integrated Forest Process Model

INTERV Time INTERVal in years

IUFRO International Union of Forest Research Organizations

JABOWA For the JAnak, BOtkin, WAllis model

LAI Leaf Area Index

Mg Magnesium or Megagrams

MOD MODifier function

MPA Megapascals

NADP Nicotinamide Adenosine Dinucleotide Phosphate

NAPAP National Acid Precipitation Assessment Program

NCASI National Council of the paper industry for Air and Stream Improvement

NIR Near Infra-red Radiant energy

NOAA National Oceanic and Atmospheric Administration

NT Number of live Trees

P Precipitation

P or p Probability

PAR Photosynthetically Active Radiant energy

PC Projected Crown

PDSI Palmer Drought Severity Index

PE Potential Evaporation

PI Productivity Index

POT POTential function

PPFD Photosynthetic Photon Flux Density

PS Probability of tree Survival

R or ρ or r Linear correlation coefficient

r^2 Multiple correlation coefficient (coefficient of multiple determination)

R^2 (partial) Additional explained variation due to the addition of one or more independent variables to the model (R^2 (full) $-$ R^2 (restricted) $=$ partial R^2)

RAND For the synthetic tree-ring series generator

RI Radial Increment

ROOTSIMU ROOT SIMUlation model

ROPIS Response Of Plants to Interacting Stresses

SCE Southern California Edison company

SDI Stand Density Index

SHAWN SHAWNigan Lake model

SI Site Index

SRIC Short Rotation Intensive Culture

STEMS Stand and Tree Evaluation and Modeling Systems

STOPEN STOmata oPEN

SWT Simple Whole Tree model

$S_{y \cdot xx}$ Standard error of the regression (square root of the Mean Square Error from the regression)

TASS Tree And Stand Simulator

TNPP Total Net Primary Production

TRANS Non-steady state models of Münch pressure flow

USDA United States Department of Agriculture

UTM Unified Transport Model

VT Cubic Volume Total

WF Weighting Factor

INDEX